The Oxford Anthology of Statistics in Sports

THE OXFORD ANTHOLOGY OF STATISTICS IN SPORTS

Volume I: 2000–2004

Edited by

JAMES J. COCHRAN

Professor of Statistics, University of Alabama

JAY BENNETT

JIM ALBERT

OXFORD
UNIVERSITY PRESS

OXFORD
UNIVERSITY PRESS

Great Clarendon Street, Oxford, OX2 6DP,
United Kingdom

Oxford University Press is a department of the University of Oxford.
It furthers the University's objective of excellence in research, scholarship,
and education by publishing worldwide. Oxford is a registered trade mark of
Oxford University Press in the UK and in certain other countries

First Edition published in 2017
Impression: 1

Published in the United States of America by Oxford University Press
198 Madison Avenue, New York, NY 10016, United States of America

British Library Cataloguing in Publication Data
Data available

Library of Congress Control Number: 2017932408

ISBN 978–0–19–872492–6

Printed and bound by
CPI Group (UK) Ltd, Croydon, CR0 4YY

CONTENTS

ACKNOWLEDGEMENTS

Only the Introduction and the Introductions to Parts I – VIII appear here for the first time. Permission to reprint original or revised versions of the following items is gratefully acknowledged:

Gary A. Simon and Jeffrey S. Simonoff, 'Were the 1996–2000 Yankees the Best Baseball Team Ever?' Chance (2002), 15(1), 23–29.

Gary Koop, 'Comparing the Performance of Baseball Players: A Multiple-Output Approach', *Journal of the American Statistical Association* (2002), 97(459), 710–720.

James J. Cochran, 'Career Records for All Modern Position Players Eligible for the Major League Baseball Hall of Fame', *Journal of Statistics Education* (2000), 8(2).

Jim Albert, 'Streakiness in Team Performance', *Chance* (2004), 17(3), 37–43.

Teddy Schall and Gary Smith, 'Do Baseball Players Regress Toward the Mean?', *The American Statistician* (2000), 54(4), 231–235.

Scott M. Berry, 'Gold Medals, Protests, and Trading Ratings: The 2002 Winter Olympics Figure Skating,' Chance (2002), 15(2), 14–18.

Scott M. Berry, 'A Statistician Reads the Sports Pages: One Modern Man or 15 Tarzans?', *Chance* (2002), 15(2), 49–53.

S.S. Wu and M.C.K. Yang, 'Evaluation of the Current Decision Rule in Figure Skating and Possible Improvements', *The American Statistician* (2004), 58(1), 46–51.

Trevor F. Cox and Ryan T. Dunn, 'An analysis of Decathlon Data', *Journal of the Royal Statistical Society: Series D* (2002), 51(2), 179–187.

Scott M. Berry, 'A Statistician Reads the Sports Pages: College Football Rankings: The BCS and the CLT', *Chance* (2003), 16(2), 46–49.

Chris S. White and Scott M. Berry, 'Tiered Polychotomous Regression: Ranking NFL Quarterbacks', *The American Statistician* (2002), 56(1), 10–21.

Scott M. Berry and Craig Wood, 'A Statistician Reads the Sports Pages: The Cold-Foot Effect', *Chance* (2004), 17(4), 47–51.

David Mease, 'A Penalized Maximum Likelihood Approach for the Ranking of College Football Teams Independent of Victory Margins', *The American Statistician* (2003), 57(4), 241–248.

Keith A. Willoughby, 'The Return of a Missed Field Goal in Canadian Football', *Chance* (2001), 14(3), 29–33.

Hal S Stern, 'Statistics and the College Football Championship', *The American Statistician* (2004), 58(3), 179–185.

Frank Duckworth, 'A Fair Result in Foul Weather', *Significance* (2004), 1(2), 94–96.

P. E. Allsopp and Stephen R. Clarke, 'Rating Teams and Analysing Outcomes in One-Day and Test Cricket', *Journal of the Royal Statistical Society: Series A* (2004), 167(4), 657–667.

Ian Preston and Jonathan Thomas, 'Rain Rules for Limited Overs Cricket and Probabilities of Victory', *Journal of the Royal Statistical Society: Series D* (2002), 51(2), 189–202.

Ian Preston and Jonathan Thomas, 'Batting Strategy in Limited Overs Cricket', *Journal of the Royal Statistical Society: Series D* (2000), 49(1), 95–106.

Lawrence L. Kupper, Leonard B. Hearne, Sandra L. Martin, and Jeffrey M. Griffin, 'Is the USGA Golf Handicap System Equitable?' *Chance* (2001), 14(1), 30–35.

Scott M. Berry, 'A Statistician Reads the Sports Pages: A Game of Which I am Not Familiar', *Chance* (2003), 16(3), 45–48.

Derek R. Bingham and Tim B. Swartz, 'Equitable Handicapping in Golf', *The American Statistician* (2000), 54(3), 170–177.

Robert Ketzscher and Trevor J. Ringrose, 'Exploratory Analysis of European Professional Golf Association Statistics', *Journal of the Royal Statistical Society: Series D* (2002), 51(2), 215–228.

Eduardo Fernandez-Cantelli and Glen Meeden, 'An Improved Award System for Soccer', *Chance* (2003), 16(1), 23–29.

Ruud H. Koning, 'Balance in Competition in Dutch Soccer', *Journal of the Royal Statistical Society. Series D* (2000), 49(3), 419–431.

Martin Crowder, Mark Dixon, Anthony Ledford, and Mike Robinson, 'Dynamic Modelling and Prediction of English Football League Matches for Betting', *Journal of the Royal Statistical Society: Series D* (2000), 51(2), 157–168.

Nobuyoshi Hirotsu and Mike Wright, 'An Evaluation of Characteristics of Teams in Association Football by Using a Markov Process Model', *Journal of the Royal Statistical Society: Series D* (2003), 52(4), 591–602.

David Forrest and Robert Simmons, 'Focus on Sport: Making up the Results: the Work of the Football Pools Panel, 1963–1997', *Journal of the Royal Statistical Society: Series D* (2000), 49(2), 253–260.

Zia Zaman, 'Coach Markov Pulls Goalie Poisson', *Chance* (2001), 14(2), 31–35.

Don Dawson and Lonnie Magee, 'The National Hockey League Entry Draft, 1969–1995: An Application of a Weighted Pool-Adjacent-Violators Algorithm', *The American Statistician*, (2001), 55(3), 194–199.

Reid Dorsey-Palmateer and Gary Smith, 'Bowlers' Hot Hands', *The American Statistician* (2004), 58(1), 38–45.

Herbert K. H. Lee, Daniel L. Cork, and David J. Algranati, 'Did Lennox Lewis Beat Evander Holyfield?: Methods for Analysing Small Sample Interrater Agreement Problems', *Journal of the Royal Statistical Society: Series D* (2002), 51(2), 129–146.

Dimitris Karlis and Ioannis Ntzoufras, 'Analysis of Sports Data by Using Bivariate Poisson Models', *Journal of the Royal Statistical Society: Series D* (2003), 52(3), 381–393.

Leonhard Knorr-Held, 'Dynamic Rating of Sports Teams', *Journal of the Royal Statistical Society: Series D* (2000), 49(2), 261–276.

Donald A. Berry and Leeann Chastain, 'Inferences about Testosterone Abuse among Athletes', *Chance* (2004), 2(17), 5–8.

Lisa Farrell and Michael A. Shields, 'Investigating the Economic and Demographic Determinants of Sporting Participation in England', *Journal of the Royal Statistical Society. Series A* (2002), 165(2), 335–348.

Jim Albert and Patricia Williamson, 'Using Model/Data Simulations to Detect Streakiness', *The American Statistician* (2001), 55(1), 41–50.

LIST OF CONTRIBUTORS

Jim Albert
Department of Mathematics and Statistics,
Bowling Green State University,
Bowling Green, OH, U.S.A.

David J. Algranati
Carnegie Mellon University, Pittsburgh, U.S.A.

P. E. Allsopp
Swinburne University of Technology,
Melbourne, Australia

Scott M. Berry
Berry Consultants, Sycamore, IL, U.S.A.

Donald A. Berry

Derek R. Bingham
Department of Statistics,
University of Michigan, Ann Arbor, MI, U.S.A.

Leeann Chastain

Stephen R. Clarke
Swinburne University of Technology,
Melbourne, Australia

James J. Cochran
Louisiana Tech University, Ruston, LA, U.S.A.

Daniel L. Cork
National Research Council,
Washington DC, U.S.A.

Trevor F. Cox
University of Newcastle, Newcastle, U.K.

Don Dawson
Department of Economics, McMaster
University, Hamilton, Ontario, Canada

Reid Dorsey-Palmateer
Fletcher Jones Professor, Department
of Economics, Pomona College,
Claremont, CA, U.S.A.

Frank Duckworth

Ryan T. Dunn
University of Newcastle, Newcastle, U.K.

Martin Crowder
Imperial College of Science, Technology
and Medicine, London, U.K.

Mark Dixon
City University, London, U.K.

Lisa Farrell
University of Melbourne,
Melbourne, Australia

David Forrest
University of Salford, Manchester, U.K.

Jeffrey M. Griffin

Leonard B. Hearne

Nobuyoshi Hirotsu
Lancaster University, Lancaster, U.K.

Eduardo Fernandez-Cantelli

Dimitris Karlis
University of the Aegean, Chios, Greece

Robert Ketzscher
Cranfield University, Shrivenham, U.K.

Leonhard Knorr-Held
Ludwig–Maximilians University,
Munich, Germany

Ruud H. Koning
University of Groningen, The Netherlands

Gary Koop
Department of Economics, University
of Glasgow, Glasgow, U.K.

Lawrence L. Kupper

Anthony Ledford
Man Investment Products Ltd, London, U.K.

Herbert K. H. Lee
Duke University, Durham, U.S.A.

Lonnie Magee
Department of Economics,
McMaster University,
Hamilton, Ontario, Canada

Sandra L. Martin

David Mease

Glen Meeden
Department of Statistics, The Wharton
School, University of Pennsylvania,
Philadelphia, PA, U.S.A.

Loannis Ntzoufras
Athens University of Economics
and Business, Greece

Ian Preston
University College London, U.K.

Trevor J. Ringrose
Cranfield University, Shrivenham, U.K.

Mike Robinson
Man Investment Products Ltd, London, U.K.

Teddy Schall
Andersen Consulting (known as Accenture
from January 1, 2001), San
Francisco, CA, U.S.A.

Michael A. Shields
University of Melbourne, Australia

Robert Simmons
University of Salford, Manchester, U.K.

Gary A. Simon

Jeffrey S. Simonoff

Gary Smith
Department of Economics, Pomona College,
Claremont, CA, U.S.A.

Hal S. Stern
Department of Statistics, Donald Bren
School of Information and Computer
Sciences, University of California,
Irvine, CA, U.S.A.

Tim B. Swartz
Department of Mathematics and Statistics,
Simon Fraser University
Burnaby BC, Canada

Jonathan Thomas
Bank of England, London, U.K.

Chris White
M. B. Flippen & Associates,
College Station, TX, U.S.A.

Patricia Williamson
Department of Mathematical Sciences,
Virginia Commonwealth University,
Richmond, VA, U.S.A.

Keith A. Willoughby

Craig Wood

Mike Wright
Lancaster University, Lancaster, U.K.

Samuel S. Wu
Department of Statistics,
University of Florida, Gainesville,
FL, U.S.A. and
Research Health Scientist at the VA Rehabilitation
Outcome Research Center of Excellence,
Gainesville, FL, U.S.A.

Mark C. K. Yang
Department of Statistics,
University of Florida,
Gainesville, FL, U.S.A.

Zia Zaman

Introduction

At the 1992 Joint Statistical Meetings (JSM), a small group of like-minded statisticians established a new section of the American Statistical Association (ASA). Joining sections that represent the interests of long-standing traditional areas of statistics such as Business and Economic Statistics, Biometrics, Survey Research Methods, Statistical Education, Physical and Engineering Sciences, and Social Statistics was the Section on Statistics in Sports (SIS). This mission of this new section included encouraging the application of statistics to sports, promoting high professional standards in the application of statistics to sports, and supporting statistical education efforts in sports both within and outside the ASA.

The establishment of the SiS did not serve as a bellwether for the application of statistics to sports; statisticians had worked on sports problems long before 1992. From Harry Chadwick's pre-Civil War efforts to develop a box score system to document baseball games to Allan Roth's work for Brach Rickey's Brooklyn Dodgers in the late 1940s to Frederick Mosteller's 1952 *Journal of the American Statistical Association (JASA)* analysis of the World Series (*JASA*, 47, 355–380) to the establishment of the Society of American Baseball Research (SABR) in 1971, statisticians have been making their mark in sports for over 150 years.

The rapid interest in the application of statistics to sports has closely followed the broadening of the world's interest in sporting interests throughout the past 50 years. Whereas early research efforts tended to focus on baseball and its wide array of readily available data, more recent efforts have been devoted to soccer, football, basketball, golf, tennis, cricket, ice hockey, rugby, and track and field. One only needs to review the table of contents of any issue of the *Journal of Quantitative Analysis of Sports* (*JQAS*), first published in 2005, to find evidence of this trend.

Prior to the establishment of the SIS, research on sports statistics was generally presented and published in areas to which it was only tangentially related. Those of us who worked on research in this area before 1992 can recall several instances in which a conference presentation on an application of statistics to sports was placed in a session in which it did not fit because it had no natural home of similarly themed presentations. Instances such as this became more frequent as interest in the application of statistics to sports grew, and this led to the creation of the SiS in 1992. This established a natural home for work of this sort and in turn encouraged further growth in this area of research, which has in turn resulted in an explosion of sports statistics papers submitted to ASA and Royal Statistical Society (RSS) publications as well as an annual volume of proceedings from the JSM (published by the SiS). The September 1994 issue of *JASA* had a section devoted exclusively to sports statistics, *The American Statistician* typically features a sports statistics paper in each issue, *Chance* often has more than one article plus a regular column "A Statistician Reads the Sports Page," the *Journal of the Royal Statistical Society: Series C* regularly features articles on statistics applied to sports, and *JQAS* now publishes quarterly volumes. Given the continued growth in interest in the application of statistics to sports, the future for this area looks very bright.

One goal of SIS since its inception has been to produce a series of collections of papers that best represents the rich history of research performed by statisticians in this field. In 2000, SIS formed a committee that was charged with reviewing sports statistics papers that had been published by ASA journals with the objective of producing a volume of selected papers that would provide a broad representation of sports, statistical concepts, and authorship. The first volume, *Anthology of Statistics in Sports* (2005, ASA-SIAM), consisted of selected papers published through 1999.

This volume takes us from 2000 through 2004; the shorter window of coverage reflects the rapid growth of research publications in this area (future volumes are planned for 2005–2009 and 2010–2014). This volume also includes both articles published by ASA journals and articles published by RSS journals, thus reflecting both the rapid worldwide spread of interest in this area of research and a broader range of sports. The volume is organized by sport. A separate section is devoted to each of several popular spectator sports (baseball, cricket, American football, golf, basketball, Olympics/track and field, and association football/soccer). Sports with less representation in the ASA and RSS literature (such as hockey, bowling, and boxing) have been collected into a separate section. Another section features articles on research that has greater breadth and generality, with each paper in this section addressing an issue that arises across several sports. Each section has been organized (with an introduction) by a notable contributor to that research in that area.

The research articles included in this anthology have been reproduced as faithfully as possible. However, we have made a few minor changes (such as renumbering figures and tables) to provide consistency throughout the anthology and enhance its readability.

It is important to note that research on applications of statistics to sports has gradually gained credibility over the past several decades as a greater appreciation for the contributions made in this area have grown. However, one only has to read the papers included in this volume to conclude that just as early researchers in applications of statistics to sports were motivated by their intellectual curiosity and the joy of working in this area, new researchers are being attracted to this area not by its increased credibility, but rather by a desire to answer fundamental questions in sports. Researchers in statistical applications to sports are drawn to work in this area because they enjoy it (the editors speak from personal experience on this subject). We hope that you enjoy reading their work as much as they enjoyed producing it.

PART I

Baseball

Introduction

HIROTSU NOBUYOSHI

In this introduction, we provide a brief overview of baseball. We then summarize statistical research in baseball and show how the articles included in this chapter fit into the general area of baseball research.

Baseball has long been one of the most popular sports in North America and Far East Asia, and the popularity of the sport is spreading across the globe. In addition to teams from these regions, the World Baseball Classic international baseball competition, established in 2006, included teams from the Netherlands, Italy, and Spain in 2015. Through its efforts to monitor the popularity of the sport in Europe, the International Baseball Federation has found that the sport is growing at its fastest pace ever in Europe, and a team of European all-stars split a series with the Japanese national team in March 2015 at the Tokyo Dome.

In the United States, baseball has been played in some form for almost two hundred years. The first professional team, the Cincinnati Redstockings, was formed in 1869. Although several professional leagues of baseball teams have been established in the U.S., two of these leagues survive and form the basis of modern Major League Baseball (MLB). These two leagues, the National and American Leagues, were established in 1869 and 1900, respectively. Each of these leagues now includes 15 teams in three divisions: East, West, and Central. In addition, an extensive network of professional minor leagues comprising teams that develop young players for MLB exists throughout the U.S. Several other professional baseball leagues exist throughout the world, most notably in Japan, Cuba, Mexico, Venezuela, Colombia, the Dominican Republic, Nicaragua, Panama, South Korea, Taiwan, China, Philippines, Israel, Austria, Croatia, France, Germany, Greece, Ireland, Italy, Netherlands, Spain, Sweden, United Kingdom, Finland, Latvia, Estonia, Lithuania, Belarus, Switzerland, Australia, and New Zealand.

Although the rules have been remarkably stable for over 100 years, the way that baseball has been played has evolved. The period prior to 1920 is known as the "deadball era" and was marked by a paucity of home runs dictating strategies that emphasized defense, baserunning, and the hitters' ability to advance base runners without the long ball. In an article featured in this anthology, Simon and Simonoff (2002) focused on the modern (i.e., post-deadball) era, a period during which home runs had (and still have) a profound impact on strategy. Since 1903, the champions of the two leagues have played in a best of seven- or nine-game set referred to as the World Series for the MLB championship at the conclusion of the season. Prior to 1961, both leagues consisted of eight teams and no divisions; since 1961 both leagues have expanded several times, divided into divisions, and adopted various playoff systems.

A baseball game is played on a field. The layout of the field is shown in Figure P I.1. The area between the first base line and third base line is referred to as fair territory, and the remainder of the field (the yellow area in Figure P I.1) is referred to as foul territory.

In a baseball game, two teams consisting of nine players each play nine innings, each of which is divided into two halves. In the top half of each inning, one team plays offense (or is said to be *at bat*) and tries to score runs and other team plays defense (or is said to be *on the field*) and tries to prevent runs from being scored. The nine defensive positions are shown in Figure P I.2.

At the beginning of each half inning, the defensive players take their positions; each of these players wears a special glove on one hand to use to catch the ball. A player from the offensive

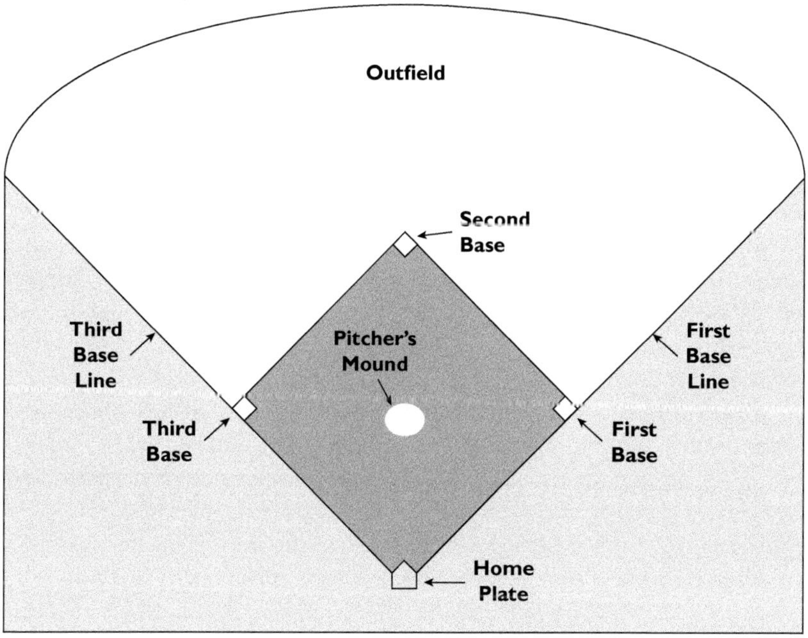

Figure P 1.1 Layout of a baseball field.

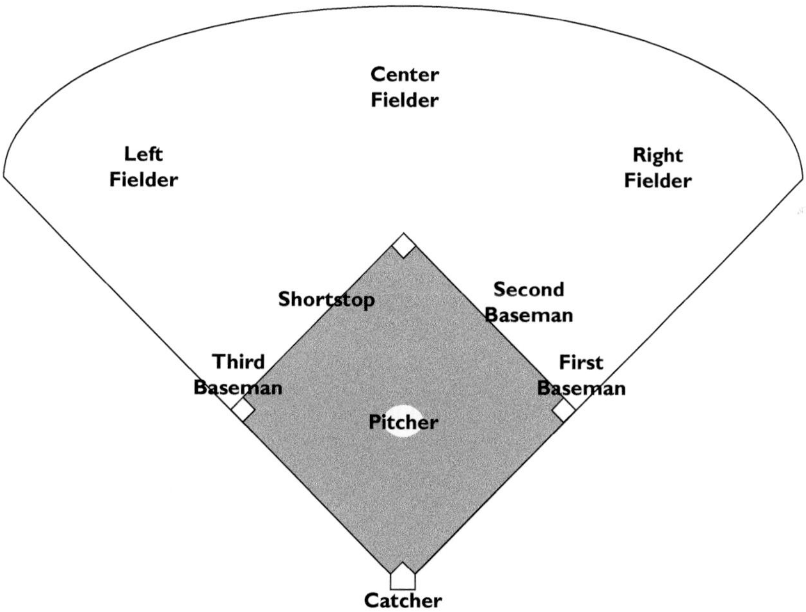

Figure P 1.2 Defensive positions.

team then assumes a position standing in foul territory along one side of home plate facing the pitcher; this player, who is called the batter, chooses which side of home plate to stand on, and this is determined by whether the batter is left handed or right handed. The pitcher for the defensive team stands on the mound and initiates play by throwing a baseball toward home plate. The batter tries to hit the thrown ball, or *pitch*, with a wooden or metal club called a bat. A pitch is a strike in such a case that the batter swings at the pitch and does not make contact with the ball, or others.

A pitch that is not a strike is a called a ball. Each batter is allocated three strikes and four balls. If three strikes occur during an at bat, the batter has made one out for his team and is said to have struck out; if four balls occur during an at bat, the batter is said to have walked and proceeds to first base.

If the batter hits the pitched ball into fair territory, the ball is said to be a fair ball; if the ball is hit into foul territory, it is called a foul ball. If the batter hits a fair ball, the batter proceeds to run to each of the bases in order and can stop at any base and occupy that base safely. In addition to striking out, a batter can also be put out if he hits the ball and such a case or others occurs that a defensive player catches the ball before it strikes any part of the field.

Once a fair ball is hit, the batter runs to each base in order until the batter elects to stop at a base or is put out. If a batter safely occupies first base, second base, or third base at the end of the at bat, the batter become a baserunner and will occupy that base until the next batter takes a turn. If the batter safely touches all bases and reaches home plate without being put out, the batter has scored a run for the team that is at bat. If a batter is put out or scores a run, the batter then leaves the field.

Baserunners only have to run to the next base if forced. This occurs when the current batter hits a fair ball, the batted ball is not caught by a defensive player before it strikes any part of the field, and all proceeding bases are occupied by baserunners. Otherwise baserunners proceed from one base to another at their own risk. A baserunner can be put out, for example, if the baserunner is forced to proceed to the next base and a defensive player in possession of a ball touches the base the baserunner is forced to proceed to before the baserunner safely reaches that base.

If a runner is able to reach each base including home plate without being put out, the baserunner has scored a run for the team that is at bat. If a baserunner is put out or scores a run, the baserunner then leaves the field.

The nine players on the offensive team take turns in a predetermined order, referred to as the batting order, until it has made three outs, so the obvious objective for a team is to score as many runs as it can before making its third out in an inning. At this point the teams exchange roles; the team that has been playing defense now is on offense, and the team that has been on offense now plays defense. An inning comprises one opportunity for each team to play offense. The team with the greater number of runs after nine innings is the winner. If the teams are tied after nine innings, extra innings are played until one team is ahead at the conclusion of an inning. Thus a baseball game has no absolute limit on innings; the game will continue until one team has won. on May 8, 1984, it took the Chicago White Sox 25 innings to defeat the Milwaukee Brewers! This game lasted 8 hours and 25 minutes; not long by cricket test match standards, but still quite a marathon.

Our summary of how baseball is played provides a broad overview of the sport; there are many other nuances and rules that cannot be covered in such a brief overview. For a more formal and detailed explanation of baseball rules and fundamentals and how baseball is played, please see one of a wide selection of books such as Shatzkin (1988), Hample (2008), and Mahony (2014).

We now provide a summary of major topics addressed by statistical research, showing how the articles selected in this chapter fit into these topics. Baseball is replete with measures of both offensive and defensive performance at both the individual player and team levels. These data provide ample opportunities for statistical analyses of a wide range of topics. In terms of evaluation of players, batters and pitchers are assessed traditionally by such standard measures as batting average (BA), computed by dividing the number of hits by the number of at bats, and earned run average (ERA), computed by multiplying nine times the ratio of the number of runs allowed to the number of innings pitched. Schall and Smith (2000) focused on these measures of performance in their analysis. In addition to these and other standard measures, a variety of advanced measures such as adjusted production (AP) have recently been developed. These measures are useful and readily available, as demonstrated by Cochran (2000) in his discussion of their use in statistics education. An approach to integrating a variety of such measures into a single measurement is the Player Game Percentage (PGP) proposed by Bennett and Flueck (1984), which measures a player's contribution to win probability.

In terms of evaluation of teams, win probability can be estimated with "The Pythagorean Method" proposed by Bill James. It provides a connection between a team's win probability and the number of runs it scores and allows, which can be estimated from the team players' abilities to score or allow runs. This method has been investigated and refined by several researchers and has even been extended to the National Hockey League, the National Football League, the National Basketball Association, and college basketball.

Recently, research for the evaluation of players and teams has shifted to the use of a multiple output approach. Koop (2002) used frontier analysis for comparing players' offensive performances by means of their efficiency; here efficiency of an individual player is calculated based on a comparison with a benchmark set by the best players. He developed Bayesian methods for statistical inference and estimated the efficiencies of 535 players in 1995–1999. As a result, various types of hitters are ranked highly such that hitters who get on base frequently (e.g., Chuck Knoblauch), hit for power (e.g., Barry Bonds), or some combination of these two skills (e.g., Edgar Martinez). There is also literature on data envelopment analysis to measure the efficiencies of players or teams, providing their relative ranking (e.g., Ruggiero, 2011).

Looking at the evaluation of true ability of players and teams, we can explore the underlying truth from the data by means of statistical approaches that help distinguish differences due to real effects from differences that can be explained by chance. Related to clichés such as sophomore slump and rookie-of-the-year jinx, Schall and Smith (2000) studied the underlying effect of regression toward the mean. These authors looked at the correlation coefficient for adjacent-season performance to demonstrate the predictability of a player's performance based on past performance.

In analyzing great achievements of players and teams, statistical approaches are also useful for assessing the rareness of an achievement. At the individual player level, Joe DiMaggio's 56-game hitting streak has inspired research on "streakiness" in baseball. For example, a variety of methods have been proposed for detecting streakiness for batting performance. At the team level, Albert (2004) has looked at the patterns of wins and losses for MLB teams. He focused on the Oakland Athletics's 20-game winning streak during the 2002 season, and tried to better understand the uniqueness of this streak by placing it in the context of the expected winning and losing streaks for all 30 MLB teams over several seasons.

The desire to compare players (or teams) across eras has led to further research in techniques to adjust for variations in the play over history. For example, Simon and Simonoff (2002) assess how the 1996–2000 New York Yankees compare to other MLB teams that have had outstanding performances over five consecutive seasons.

Another important area of baseball research is the situational effect and the related clutch effect. Many situational factors have been tested for significant difference from batting ability. This research has generally shown that batting abilities are enhanced by playing at home or facing against a pitcher of opposite handedness (i.e., left versus right or vice versa), and not by playing on grass or artificial turf, in day or night, or before or after the All-Star break. Clutch hitting (i.e., the abilities of batters to improve their performances in game critical situations) is known to be difficult to detect. One approach to this problem is integrating it into player contribution using PGP, as suggested by Albert and Bennett (2003).

Evaluation of players and teams are also suitable material for statistics education (e.g., Albert, 2003). Cochran (2000, 2002) uses the MLB Hall of Fame (HOF) data and MLB attendance data in the classroom. The MLB HOF dataset contains information about career totals not only for standard baseball statistics, but also for various sabermetric measures, with regard to the 1340 MLB players (retired prior to 1993). Cochran teaches descriptive statistics, classification, and discrimination by assigning students the task of modeling whether the players have been elected to the HOF. The downloadable HOF data provided with this article on the *Journal of Statistics Education* website (https://www.amstat.org/publications/jse/) has been used in research by many other statisticians such as Eksioglu et al. (2005), Kim et al. (2007), Loh et al. (2007), and Wang (2007). These data have also been used in statistics textbooks such as Simonoff (2003) and Watkins et al. (2011). The MLB Attendance data contains home game attendance and measures of team performance such as runs scored, runs allowed, wins, losses, and number of games behind the division leader for each major-league franchise for the 1969 through 2000 seasons. He uses these data to teach descriptive statistics, regression analysis, and econometrics by assigning students the task of modeling attendance by season and franchise.

As shown in this introduction, baseball is worthy of statistical study in its own right. In addition, it provides a variety of interesting contexts for statistics research and education.

REFERENCES

Albert, J. (2003). *Teaching Statistics Using Baseball*, Washington: Mathematical Association of America.

Albert, J. (2004). Streakiness in Team Performance. *Chance*, 17(3), 37–43.

Albert, J. and Bennett, J. (2003). *Curve Ball*, Copernicus Books, New York.

Bennett, J.M. and Flueck, J.A. (1984). Player Game Percentage. *Proceedings of the Social Statistics Section*, American Statistical Association, 378–380.

Cochran, J.J. (2000). Career Records for all Modern Position Players Eligible for the Major League Baseball Hall of Fame. *Journal of Statistics Education*, 8(2).

Cochran, J.J. (2002). Data Management, Exploratory Data Analysis, and Regression Analysis with 1969–2000 Major League Baseball Attendance. *Journal of Statistics Education*, 10(2).

Eksioglu, B., Demirer, R., and Capar, I. (2005). Subset Selection in Multiple Linear Regression: A New Mathematical Programming Approach. *Computers & Industrial Engineering*, 49(1), 155–167.

Hample, Z. (2008). *Watching Baseball Smarter: A Professional Fan's Guide for Beginners, Semi-Experts, and Deeply Serious Geeks*, Knopf Doubleday Publishing Group, New York.
Kim, H., Loh, W-Y, Shih, Y-S, and Chaudhuri, P. (2007). Visualizable and Interpretable Regression Models with Good Prediction Power. *IIE Transactions*, 39(6), 565–579.
Koop, G. (2002). Comparing the Performance of Baseball Players: A Multiple-Output Approach. *Journal of the American Statistical Association*, 97(459), 710–720.
Loh, W-Y, Chen, C-W, and Zheng, W. (2007). Extrapolation Errors in Linear Model Trees. *ACM Transactions on Knowledge Discovery from Data*, 1(2), 1–17.
Mahony, P. (2014), *Baseball Explained*, McFarland, Jefferson.
Ruggiero, J. (2011). *Frontiers in Major League Baseball: Nonparametric Analysis of Performance Using Data Envelopment Analysis*. Series: Sports Economics, Management and Policy, Vol. 1, Springer, New York.
Schall, T. and Smith, G. (2000). Do Baseball Players Regress Toward the Mean? *The American Statistician*, 54(4), 231–235.
Shatzkin, M. (1988). *Baseball Explained*, Pelham, New York.
Simon, G.A. and Simonoff, J.S. (2002). Were the 1996–2000 Yankees the Best Baseball Team Ever? *Chance*, 15(1), 23–29.
Simonoff, J.S. (2003). *Analyzing Categorical Data*, Springer-Verlag, NY.
Wang, S.C. (2007). Teaching Statistical Thinking Using the Baseball Hall of Fame. *Chance*, 20, 25–31.
Watkins, A.E., Scheaffer, R.L., and Cobb, G.W. (2011). *Statistics: From Data to Decision*, John Wiley and Sons, New York.

CHAPTER 1

Were the 1996–2000 Yankees the best baseball team ever?

GARY A. SIMON AND JEFFREY S. SIMONOFF

Introduction

In the aftermath of the New York Yankees' 2000 World Series victory over the New York Mets, four games to one, many fans, sportswriters, and baseball insiders focused attention on whether the Yankees' achievement of four World Series championships in five years was the most impressive team accomplishment in the history of the sport. In particular, how did it compare to the five championships in six years (including four consecutive championships) of the 1936–1941 Yankees or the six championships in seven years (including five consecutive championships) of the 1947–1953 Yankees?

A commonly expressed opinion was that of MSNBC sports columnist Bob Herzog, who wrote, "The greatest team in Yankees history is playing right now." The reason commonly given for this opinion is that, with two postseason playoff rounds before the World Series, there is the added difficulty of just getting to the World Series (let alone winning it). "The simple act of repeating is more difficult than ever" opined Fox Sports commentator Keith Olbermann, and Atlanta Braves manager Bobby Cox agreed, saying, "It's so hard to go all the way every year now. It's amazing to me what the Yankees have done."

Although it is certainly true that it is harder to win three series (one best three out of five games, two best four out of seven games) than to win one (best four out of seven games), that is only half the story. What Herzog, Olbermann, and Cox are ignoring is that it is apparently easier now to get into the postseason, when eight teams are eligible, than it was 50 years ago, when only two teams were eligible. In this article we examine both sides of the question and try to establish whether it really is harder to win the World Series now.

The first step is to acknowledge that we are not going to actually answer the question posed in the title. Although it might be a more impressive achievement to win three series rather than one to win a championship, that hardly implies that a team that only had to win one series is less good. After all, it isn't the fault of the 1927 Yankee team, which won more than 71% of

Chad Curtis celebrates his game-winning home run in the 1999 World Series

its regular season games and then swept the World Series from the Pittsburgh Pirates, that it only had to win four postseason games to win it all! Rather, we will attempt to quantify the "impressiveness" of a given team's achievement as the answer to the question, "How surprising would it be for a team to achieve a given level of regular season and postseason success?"

Measuring impressiveness

Answering that question involves several steps. First, we must define what we mean by "a given level of regular season and postseason success." Since the number of games played in a season has varied over time, regular season success should be assessed using winning percentage, rather than number of wins. If a team wins 100 out of 162 games, for example, that corresponds to a winning percentage of 0.617, and we will treat that event as having a winning percentage of at least 0.617. The intuition behind this is that baseball fans who refer to a team "winning 100 games" actually mean *at least* 100 games.

Postseason success corresponds to a two-part process. First, a team must earn a position in postseason play. Before 1969, only one team from each league (American and National) earned a spot in the World Series, the only postseason competition. From 1969 through 1993, each league was divided into two divisions, with the division winners in each league first playing each other in a League Championship Series and the league champions then meeting in the World Series. Since 1995 each league has been split into three divisions, and the three division winners, plus the team with the next best record (the wild card team), play in the Division Series, with the winners playing in the League Championship Series for the right to play in the World Series. There was no postseason play in 1994 because of a players' strike. These different systems mean that the impressiveness of a team's achievement in making the postseason was very different in 2000 than it was in 1950, with four times as many slots available.

Density Estimation

The distribution of a continuous random variable X, like income, weight, or (in the accompanying article) winning percentage, is usually described by its density function. The density function gives the relative likelihood of different values of X. It is usually denoted as $f(x)$; we require $f(x)$ to be greater than or equal to 0 and that $f(x)$ integrates to 1 over the range of possible values. In some cases researchers assume a particular form of density, for example, we might use a normal density to describe the distribution of weights in the population. In other cases, the goal is to estimate the density nonparametrically from observed values $x_1, \ldots x_n$.

There are many approaches to density estimation. The authors of the accompanying article use an approach known as local quadratic likelihood density estimation. This method is known to have good tail behavior, and the impressiveness measure depends heavily on density estimates in the upper tail. The procedure requires a kernel weight function $K(\cdot)$ and a smoothing parameter h (this controls how smooth the estimated density is). Then the estimate $\widehat{f}(x)$ is defined as the function that maximizes

$$\sum_{i=1}^{n} K\left(\frac{x - x_i}{h}\right) \log(f(x_i)) - n \int K\left(\frac{x-u}{h}\right) f(u)du$$

over all functions where $\log f(\cdot)$ is locally quadratic. Some density estimation references are provided below. They provide more details about the choice of $K(\cdot)$ and h. The local quadratic likelihood approach was described by Hjort and Jones (1996), and Loader (1996).

The second part of postseason success is simply how many series a team won. Although it could be argued that a four-games-to-zero World Series triumph is more impressive than four-games-to-three, ultimately all that is remembered is which team won.

A statistical approach

Formally measuring impressiveness requires quantifying the difficulty of achieving each of the three milestones required to win the World Series—that is, (1) the probability of earning a winning percentage x, (2) the probability of qualifying for the postseason given a winning percentage of x, and (3) the probability of winning the World Series given that one qualifies for postseason play. Our approach is described briefly here with empirical results provided next.

For the first step, we use a density estimation method (see sidebar) to construct a function $\widehat{f}(x)$ that estimates the density function for team winning percentage. The density function basically gives the likelihood of a team earning a winning percentage x. To reflect the changing pattern of winning percentages over time, separate density estimates are constructed for the three eras of the so-called "modern age" of baseball (1920–1968, 1969–1993, and 1995–2000).

For the second step, the probability of qualifying for postseason play, we use logistic regression (see sidebar). A separate model is fitted for each era, yielding an estimate of the probability that a team qualifies for postseason play given a winning percentage x, which we denote $\hat{P}(S|x)$.

For the final step, the probability of winning the World Series given that a team has qualified for postseason play, we begin with a simple coin-flip model. If the two teams in any postseason series are evenly matched, the probability of winning a series is simply 0.5; assuming independence of postseason series (a coin-tossing model), the probability of winning the World Series once the team has qualified for the postseason is 0.5 (pre-1969), 0.25 (1969–1993), and 0.125 (1995–2000), respectively. We denote this piece as $\hat{P}(S|x)$. It is possible to refine these figures by allowing the probability of winning a series to be a function of the quality of the two teams playing in the series; we examine this possibility later.

The three pieces are combined into a single measure of impressiveness as follows: According to the laws of probability, the likelihood that a team earns winning percentage x, qualifies for the postseason, and wins the World Series is the product $\hat{f}(x)\hat{P}(S|x)\hat{P}(W|S)$. The probability of achieving a winning percentage x exactly may not be representative since only certain winning percentages are possible in a standard 162-game season. Because of this (and our earlier discussion), we make the impressiveness measure, denoted I, the probability of achieving a winning percentage *greater than or equal to* x and then winning the World Series. I is obtained by computing the product $\hat{f}(x)\hat{P}(S|x)\hat{P}(W|S)$ for all winning percentages greater than or equal to x and combining the results.

Empirical results

The analyses presented here are based on data for all teams for the Major League seasons 1920 through 2000, excepting 1981 and 1994. We begin with 1920 because that is generally considered the beginning of the modern era of baseball (succeeding the "dead ball" era). Both 1981 and 1994 were disrupted by player strikes that affected the postseason (1981 was played as a split season, with first-half champions playing against second-half champions in the postseason, a one-of-a-kind arrangement, and thus not suited for inclusion in our analysis, while the 1994 postseason was canceled). This time period is split into three eras: Era 1, 1920 through 1968 (when one team per league played in one postseason playoff series, the World Series); Era 2, 1969 through 1993 (when two teams per league played in two postseason playoff series, the League Championship Series, and the World Series); and Era 3, 1995 through 2000 (when four teams per league have played in three playoff series, the Division Series, League Championship Series, and World Series).

Figure 1.1 gives density estimates for the winning percentages for all teams for the three eras (Era 1, solid line, Era 2, dotted line, Era 3, dashed line). The average winning percentage in any year is, of course, 0.500, but it is noticeable that in the earliest era the distribution of winning percentages is decidedly asymmetric (with mode at 0.520 and a long left tail) and exhibits higher variability (the standard deviation of winning percentages for this era is 0.090). The winning percentage distributions for the latter two eras are very similar, both being reasonably symmetric, peaking at 0.500, with less variability than for Era 1 (standard deviation of winning percentages 0.069 for both). These densities imply that competitive balance has been much greater in the last 30 years than in the previous 50 years. Although almost 15% of all Major League baseball teams won at least 60% of their games during Era 1, only about 7% did

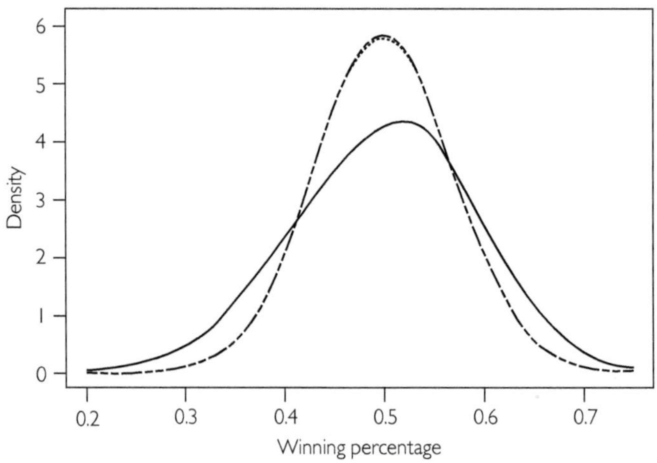

Figure 1.1 Density estimates of winning percentage by era: solid line, Era 1 (1920–1968); dotted line, Era 2 (1969–1993); dashed line, Era 3 (1995–2000).

in Eras 2 and 3; the percentages winning at least two-thirds of their games are 2.0%, 0.7%, and 1.1%, respectively. Unfortunately, this surfeit of apparent excellence in the earlier part of the century is balanced by excess incompetence, with 14.5% of Major League teams in Era 1 winning fewer than 40% of their games (compared with 8.5% and 4.6% in Eras 2 and 3, respectively) and 3.9% winning no more than one-third of their games (compared with 0.7% and 1.1% for Eras 2 and 3, respectively). This means that a winning percentage of 0.614 in Era 1 (the best 10% of performances) is equivalent to a winning percentage of 0.591 in the latter eras, which corresponds to almost four fewer wins in a 162-game schedule.

The paleontologist Stephen Jay Gould (1986) noted this same pattern of narrowing distributions over the years when examining batting averages. Rather than point to this as evidence of a decline of hitting ability, he attributed the pattern to an overall increase in quality of both pitchers and hitters. The same argument holds here: With fewer profoundly poor teams in baseball (such as the Philadelphia Phillies, which lost more than 65% of their games 11 times between 1921 and 1945, or the Boston Red Sox, which did so 9 out of 14 years from 1922 to 1935), it is now much more difficult for even truly excellent teams to have dominant seasons.

Figure 1.2 gives the three separate fitted logistic regressions used to model the relationship between the probability of getting into postseason play and winning percentage (one for each of the three eras). Each team's response (make playoffs or don't make playoffs) is modeled as independently Bernoulli distributed, ignoring the slight dependence that comes from the fixed number of postseason spots in a given year. As would be expected, the curves are shifted to the left in each successive era, corresponding to a higher probability of appearing in the postseason for any given winning percentage (since a higher percentage of teams make the playoffs). The winning percentages corresponding to a 50% chance of making the playoffs are 0.611, 0.577, and 0.543, respectively. As would be expected, this also translates into the potential for World Series winners with lower winning percentages than before, such as the 1987 Minnesota Twins (with a winning percentage of 0.525) and 2000 New York Yankees (with a winning percentage of 0.540).

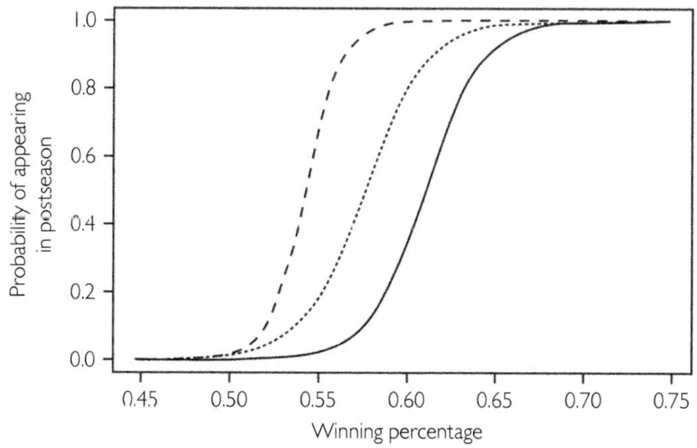

Figure 1.2 Estimates of the probability of qualifying for postseason play as a function of winning percentage by era: solid line, Era 1 (1920–1968); dotted line, Era 2 (1969–1993); dashed line, Era 3 (1995–2000).

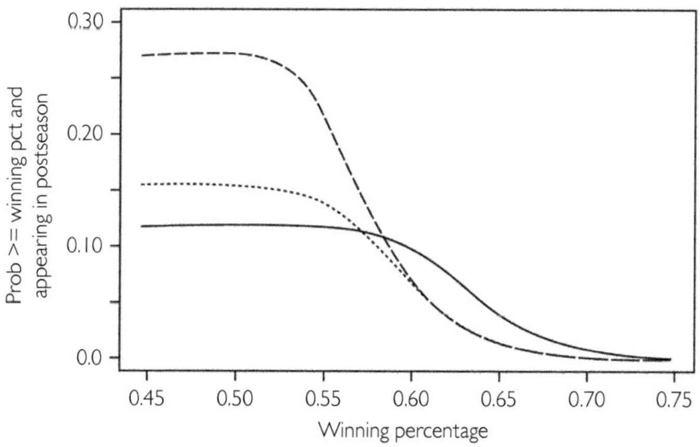

Figure 1.3 Estimates of the probability of achieving at least a given level of regular season success and qualifying for the postseason, by era: solid line, Era 1 (1920–1968); dotted line, Era 2 (1969–1993); dashed line, Era 3 (1995–2000).

The two factors of a narrowing distribution of winning percentage (Fig. 1.1) and increased probability of making the postseason (Fig. 1.2) combine to give Figure 1.3. This figure gives the probabilities of a team having at least a given winning percentage and making the postseason. The figure shows that the increased number of playoff spots recently is not the only story in evaluating the ease of making the postseason. For winning percentages up to 55–60%, it was indeed more difficult in Era 1 to achieve at least that level of regular season success and make the postseason (which, in that era, meant winning the league championship) because of the limited number of postseason spots (2 out of 16 teams qualified for the postseason). However, the much higher probability of achieving a winning percentage higher than 60% in the earliest era means that it was, in fact, *easier* to have a winning percentage over 60% and reach the postseason before 1969 than it has been since then.

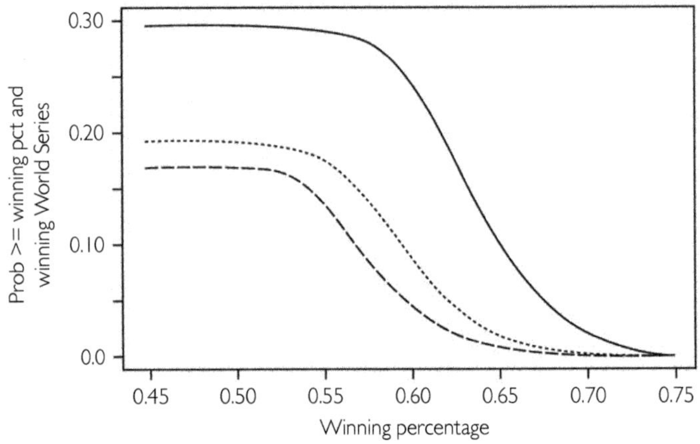

Figure 1.4 Estimates of the probability of achieving at least a given level of regular season success and winning the World Series, by era: solid line, Era 1 (1920–1968); dotted line, Era 2 (1969–1993); dashed line, Era 3 (1995–2000).

The final calculation needed is that of winning the World Series, given a team makes it to the postseason. We first consider the simple coin-tossing model. The values 0.5, 0.25, and 0.125, respectively, are multiplied by the values in Figure 1.3 to give the impressiveness measures presented in Figure 1.4. The picture here is very clear. For any given winning percentage, it was a more impressive feat to achieve that level of success and then win the World Series in era 2 than it was in era 1, and it is more impressive now than it was in era 2. When combined with the implication of increased overall quality from Figure 1.1, the ability to win championships with any consistency nowadays is quite remarkable.

The most impressive seasons

There are (at least) two ways that we could use these results to assess the most impressive teams ever (or, at least, since 1920). Table 1.1 lists the 10 most impressive championship performances since 1920. Recall that small values of the impressiveness measure I indicate unusually good, and hence most impressive, performances. Note that by requiring a championship performance, several teams with very impressive regular season performances (that might have appeared in the table), such as the 1931 Philadelphia Athletics, 1954 Cleveland Indians, 1969 Baltimore Orioles, and 1995 Cleveland Indians, are excluded; when considering one-year performance, this seems reasonable, since presumably none of the members of those teams would have considered the year as completely successful.

The most obvious message in Table 1.1 is that the performance of the 1998 New York Yankees was by far the most impressive performance of the last 80 years. Winning more than 70% of their regular season games in the current era is remarkable in itself, and to then add to that victory in three playoff series (with a record of 11–2) puts their impressiveness rating less than one-tenth the next best value. The next best performance is shared by three teams from Era 2 (1970 Baltimore Orioles, 1975 Cincinnati Reds, and 1986 New York Mets) that won two-thirds of their games and then a League Championship Series and World Series. The famous "Murderer's Row" 1927 New York Yankees of Babe Ruth and Lou Gehrig round out the top five,

Table 1.1 ***Most impressive championship teams since 1920 (the impressiveness measure I is based on assuming equiprobable winners in postseason series).***

Rank	Winning Team	Series Record	pct.	Won	I
1	1998 New York Yankees	114–48	0.704	3	0.000173
2	1975 Cincinnati Reds 1970 Baltimore Orioles 1986 New York Mets	108–54	0.667	2	0.001919
5	1927 New York Yankees	110–44	0.714	1	0.002229
6	1939 New York Yankees	106–45	0.702	1	0.003777
7	1995 Atlanta Braves	90–54	0.625	3	0.004340
8	1984 Detroit Tigers	104–58	0.642	2	0.004955
9	1932 Ne w York Yankees	107–47	0.695	1	0.004975
10	1929 Philadelphia Athletics	104–46	0.693	1	0.005363

based on the second-best regular season percentage of the last 80 years (the 1954 Indians had the best regular season percentage). As was noted earlier, this team reinforces the distinction between most impressive team and best team, since the 1927 Yankees did everything they were asked and would probably have beaten any other playoff opponent put in front of them. Still, Table 1.1 is comforting because it includes many teams that are commonly considered among the best of all-time. For example, Rob Neyer and Eddie Epstein (2000), in their book *Baseball Dynasties*, rated five of the top six teams in Table 1.1 as being the five best teams since 1910, the only exception being the 1986 Mets (which they put in the top 10 all-time).

Improving our measure

As we noted earlier, a simple coin-tossing (equiprobable) model is used to determine the probability of any team winning a postseason series in defining I. One would imagine that better estimates of the probability of winning a series could be derived based on the available data, such as the difference between the winning percentages of the two teams in the series.

Logistic Regression

Standard regression analysis is used to relate one or more predictor variables to a response variable of interest when the response is a continuous variable. In many cases of interest, the response variable is binary—for example, whether a patient dies or not in a medical study. In the accompanying article, the authors would like to predict whether a baseball team will qualify for postseason play or not using the team's winning percentage as a predictor. This can be done using logistic regression. If we take $P(S|x)$ to denote the probability that the event S (qualifying for the postseason) occurs given the predictor variable x, then the logistic regression model assumes

$$\log\left(\frac{P(S|x)}{1 - P(S|x)}\right) = \beta_0 + \beta_1 x$$

or equivalently

$$P(S|x) = \frac{e^{\beta_0+\beta_1 x}}{1 + e^{\beta_0+\beta_1 x}}$$

Parameters are estimated using the method of maximum likelihood. Most statistical software packages will perform logistic regression.

We explored this by using logistic regression to model the probability that the better team wins a series as a function of the difference in winning percentage of the two teams. This approach was used for each series of each era with the surprising result that the difference in winning percentage is not a statistically significant predictor in most series. Only the League Champion Series in Era 2 and the Division Series in Era 3 yield significant results at a 0.10 level. This can also be seen in Table 1.2 which gives the winning percentage of the "better" team in each series for each era. It is easy to imagine a psychological explanation for this: Although teams with lesser regular season records might not be very confident going into their first series against a good team, if they win it they feel that they have proved that they can compete and are much more self-assured (and therefore more successful) in later series.

We recomputed our impressiveness measure using the results of the significant (and near significant) regression models, but this gave only a small change to the ordering in Table 1.2. Given this finding, and the general lack of predictability of postseason series, we do not report further on this part of our work.

The most impressive teams

As noted in the introduction, however, it is not one-year performance that has led to the Yankee teams of the 1930s, 1950s, and 1990s being considered the best ever but rather sustained brilliance. Table 1.2 addresses this question. The four teams we consider include three versions of the New York Yankees—the 1936–1941 team, which won five titles in six years (including four in a row); the 1947–1953 team, which won six titles in seven years (including five in a row); and the 1996–2000 team, which won four titles in five years (including three in a row). Each of these teams includes a year in which the Yankees did not win the World Series, which must be addressed. The 1940 Yankees finished third in the American League with a record of 88–66, so its impressiveness measure only includes the probability of having a winning percentage of at least 0.571. Similarly, the 1948 Yankees had a record of 94–60, but finished third in the standings. The 1997 Yankees qualified for the postseason as a wild card with a record of 96–66 but lost in the division series to Cleveland, so their impressiveness does not include any postseason series victory effect. The impressiveness for any string of years is defined as the product of individual year's impressiveness. This is based on an independence assumption that is undoubtedly violated, but it is not apparent how to correct for this. To account for the years that did not end in World Series championships, the product obtained is then multiplied by the number of ways that a team could win five titles in six years $\left(\binom{6}{5}, \text{ or } 6\right)$, six titles in seven years (7), or four titles in five years (5), respectively. The fourth team is the 1971–1975

Table 1.2 ***Impressiveness measures for the teams that won at least three consecutive titles.***

Year	Status	Winning pct.	*I*
1936 Yankees	Champion	0.667	0.012731
1937	Champion	0.662	0.014729
1938	Champion	0.651	0.019866
1939	Champion	0.702	0.003777
1940	Third place	0.571	0.219362
1941	Champion	0.656	0.017402
			0.026184
1947 Yankees	Champion	0.630	0.032162
1948	Third place	0.610	0.107670
1949	Champion	0.630	0.032162
1950	Champion	0.636	0.028402
1951	Champion	0.636	0.028402
1952	Champion	0.617	0.040330
1953	Champion	0.656	0.017402
			0.046083
1971 A's	Won division	0.627	0.032701
1972	Champion	0.600	0.017327
1973	Champion	0.580	0.025735
1974	Champion	0.556	0.033606
1975	Won division	0.605	0.061332
			0.049612
1996 Yankees	Champion	0.568	0.019986
1997	Wild card	0.593	0.088588
1998	Champion	0.704	0.000173
1999	Champion	0.605	0.007971
2000	Champion	0.540	0.030457
			0.013008

Oakland A's. This is the only other team to win three consecutive World Series (1972–1974), and they also finished first in their division in 1971 and 1975. Note that their impressiveness for 1971–1975 is multiplied by $\binom{5}{3} = 10$, the number of ways to win three titles in five years.

We must be clear that we are ignoring the fact that these particular years were chosen after seeing their exceptional nature, and thus the final impressiveness measures cannot be viewed as prospective probabilities. There is also a difficulty in comparing strings of years of different lengths, since each additional year multiplies the impressiveness by a number less than 1, reducing it. We correct for this as follows: If the original measure is based on k years, the final measure reported is its kth root. If all of the years were championship years, this would correspond to the geometric mean of the annual impressiveness measures.

The verdict is very clear: The success of the 1996–2000 Yankees is the most impressive run of any team in the past 80 years, and this team has earned the right to be considered the best Yankee team ever. The 1949–1953 Yankees, which won five straight championships, suffer in comparison to the team of 1936–1941, since the earlier team won more than 100 games four times, while the later team never won more than 99. It might be surprising that the 1971–1975

A's performance is apparently comparable to the 1947–1953 Yankees, a team that won five straight championships. The A's are probably over-looked as a great team for several reasons. Their best regular season performances came in 1971 and 1975, when they lost in the American League Championship Series and didn't even appear in the World Series. They also suffer from the lack of recognition of the pattern in Figure 1.1, that a 0.600 winning percentage in 1972 was considerably more impressive than it would have been 20 or 30 years earlier (being so early in Era 2 would reinforce that problem). Finally, the additional playoff round made it more difficult to win the World Series (as they learned in 1971 and 1975!).

Having said that, the comparison in Table 1.2 is a bit unfair because of the uneven number of years. It is reasonable to just look at the 1949–1953 Yankees when comparing to the 1971–1975 A's and 1996–2000 Yankees, since then all three comparisons are based on five consecutive years. The overall impressiveness I of the 1949–1953 Yankees is 0.028318, much closer to that of the earlier Yankee team and considerably better than that of the A's.

Conclusion

In this article we have proposed a measure of how impressive a team's performance is, taking into account the competitive balance in the league and the playoff structure. Although this is not the same as measuring which team is best, it does provide one way of comparing teams from different eras, and impressiveness seems to correlate well with acknowledged "greatness" of a team. The performance of the 1996–2000 New York Yankees, and in particular the 1998 Yankees, stands out as being the most impressive of the past 80 years of major league baseball.

REFERENCES AND FURTHER READING

Berry, S. M. (2001), "Luck in sports," *Chance*, 14(1), 52–57.

Gould, S. J. (1986), "Entropic homogeneity isn't why no one hits 0.400 any more," *Discover*, August, 60–66.

Hjort, N. L., and Jones, M. C. (1996), "Locally parametric density estimation," *The Annals of Statistics*, 24, 1619–1647.

Loader, C. R. (1996), "Local likelihood density estimation," *The Annals of Statistics*, 24, 1602–1618.

Neyer, R., and Epstein, E. (2000), *Baseball Dynasties*, New York: W.W. Norton.

Simonoff, J. S. (1998), "Three sides of smoothing: categorical data smoothing, nonparametric regression, and density estimation," *International Statistical Review*, 66, 137–156.

Addendum

This paper was completed before the end of the 2001 baseball season, and is based on data through the 2000 season. The New York Yankees finished the 2001 regular season with a 0.594 winning percentage, and went on to defeat the Oakland A's and Seattle Mariners to win the American League championship, before falling to the Arizona Diamondbacks in seven games in the World Series. Including this year in a 1996–2001 treatment of the Yankees would raise the impressiveness measure to 0.016991.

CHAPTER 2

Comparing the performance of baseball players

A multiple-output approach

GARY KOOP

Abstract
This article extends ideas from the economics literature on multiple-output production and efficiency to develop methods for comparing baseball players that take into account the many dimensions to batting performance. A key part of this approach is the output aggregator. The weights in this output aggregator can be selected *a priori* (as is done with batting or slugging averages) or can be estimated statistically based on the performance of the best players in baseball. Once the output aggregator is obtained, an individual player can then be measured relative to the best, and a number between 0 and 1 characterizes his performance as a fraction of the best. The methods are applied to hitters using data from 1995–1999 on all regular players in baseball's major leagues.

Keywords: *Bayesian; Efficiency; Markov chain Monte Carlo; Stochastic frontier model.*

Introduction

Academics and sports fans alike are often interested in a statistical comparison of baseball players. This is hard to do, because baseball is fundamentally a multiple-output sport. For instance, some players are power hitters and excel in hitting home runs, whereas others hit more singles. To directly compare a power hitter like Mark McGwire with someone who hits for average like Tony Gwynn is difficult. In the sporting press, there are many different ways of aggregating performance in different offensive categories into a single-number. Examples of such *output aggregators* include batting average, on-base percentage (OBP), and slugging percentage. Such standard ways of combining different offensive categories into one number entail two related problems. First, they are all imperfect measures of offensive performance. For instance, batting averages do not correct for different player circumstances (e.g., Coors Park is notorious for being a hitter's park, and, accordingly, it is easier for Colorado Rockies players to hit well). Second, the weights in all of the output aggregators are somewhat *ad hoc*. For instance, slugging average weights home runs precisely four times as heavily as singles, batting average weights singles and home runs equally, and OBP weights walks and hits equally. All such weighting choices can be criticized. The purpose of this article is to use statistical methods to address these issues. All of the models used herein address the first issue by correcting for team, year, and league. Addressing the second issue is more complicated; several ways of estimating or calculating output aggregators for offensive performance of baseball players are considered. Individual players are then compared to the benchmark established by a given output aggregator.

Many of the statistical methods used in this article are adapted from the economics literature, and hence some economic terminology is used. That is, baseball players (like firms in an economic context) are viewed as producing outputs given firm characteristics (e.g., a batter produces the "output" hits, which depends on his situation, including what team he plays for, etc.). In an economics context, typically firms operate with different output mixes, just as in baseball batters have different mixes of hits (e.g., singles, doubles, home runs). By looking at the best firms in different regions of output space, the economist can trace out a *production possibilities curve*, which measures the maximum feasible combinations of outputs that can be produced. In baseball, the best batters can be considered in different regions of output space and trace out a comparable curve. For instance, Mark McGwire might heavily influence the production possibilities curve in the power dimensions, whereas Tony Gwynn might have influence in batting average dimensions. Once this production possibilities curve is estimated, individuals can be compared to it. Because the curve reflects best practice, an average player will lie inside the curve. As discussed later, a number between 0 and 1 can be calculated that reflects how far the player is from the nearest point on the curve. Following the economics literature, to this number is called *efficiency*. So, for instance, a certain player may have an efficiency of 0.8. This number has a simple, intuitive interpretation: The player under consideration is only 80% as productive as the best players with a comparable output mix. In other words, the proposed methodology enables comparison of any player with similar players and provides an easily interpretable, single-number summary of a player's performance. Furthermore, the methodology incorporates all of the outputs that a batter produces and corrects for the player's situation (e.g., playing in a particularly good or bad hitter's park).

The output aggregator in this first approach provides a sensible efficiency measure, *performance as percentage of best comparable player* (see the section on The model). However, it aggregates the individual outputs (e.g., home runs, singles) using weights estimated from the data based on the shape of the production possibility curve. These weights will not necessarily reflect the value of each output in team success. Questions relating to the value of individual outputs are quite difficult to address (e.g., is a home run four times as valuable as a single?) and an advantage of the previous approach is that such valuation is not required. Nevertheless, a second approach is developed, based on the contribution of individual outputs to team winning percentage, which estimates a value for each output. These values are then used as weights in the output aggregator. For comparison, a third approach is considered that simply weights all outputs equally. This means that efficiencies are calculated using OBP as a single output. A Markov chain Monte Carlo (MCMC) algorithm is derived for carrying out Bayesian estimation of player efficiencies. This algorithm is used to carry out statistical inference relating to the performance of all batters who played regularly during the years 1995–1999.

A voluminous literature uses statistical methods to analyze baseball data. However, apparently there has been no previous academic study of the multiple-output character of baseball that uses efficiency concepts to measure a player's performance. Many of the more influential works use data on the performance of the team as a whole (e.g., Barry and Hartigan 1993; Ferrall and Smith 1999; Kaigh 1995), focus on a particular measure or two of player performance (e.g., Albright 1993; Berry, Reese, and Larkey 1999; Lackritz 1996; Schall and Smith 2000), or discuss issues relating to salary determination (e.g., Chapman and Southwick 1991; Depken 2000; Hoaglin and Velleman 1995; Kahn 1993; Lackritz 1990; Scully 1974; Watnik 1998). The work of Lackritz (1990) is of particular relevance, because he explicitly develops a model for evaluating players in terms of their effects on team winning percentage. Here a similar (but simpler) model is used to construct values for outputs in one of the efficiency calculations. There are also many works that discuss and compare different ways of aggregating outputs. The article by Bennett and Flueck (1983) is a good introduction to this literature, because it compares ten different offensive performance indicators. A myriad of websites provide a range of statistics and nonacademic discussion of baseball issues (e.g., www.baseball1.com, www.sabr.com, www.baseball-reference.com, and www.stats.com/baseball.asp). The Elias Sporting Bureau (www.esb.com) is another popular nonacademic source that provides a ranking of players.

The literature on efficiency is also enormous, although very little of the statistical literature deals with the case in which multiple outputs must be treated as dependent variables. When data on costs of production, prices of inputs, or prices of outputs exist, then treatment of multiple outputs is relatively simple. However, when such data are not available (as in this article), the existence of multiple outputs greatly complicates things (see Fernandez, Koop, and Steel 2000 for details). There is also a largely nonstatistical literature that uses data envelopment analysis to address efficiency issues. The text of Kumbhakar and Lovell (2000) provides a good introduction to these literatures.

The data

All data used in this study were obtained from Sean Lahman's Baseball Archive (www.baseball1.com), which contains a myriad of statistics on every player who has ever played major league baseball. To focus the study, only performance indicators relating to batting are considered, although similar models could be developed to investigate pitching performance. Data limitations make it difficult to model defensive performance. Accordingly, all pitchers are omitted from the sample, as are all players with fewer than 200 at bats in a given year. (Having fewer than 200 at bats usually reflects injury or very irregular play.) Furthermore, Berry et al. (1999) provided convincing evidence that a player's ability can change substantially over the course of his career. To avoid explicitly modeling such aging effects, data from 1995–1999 are used, and a player's ability is assumed to be roughly constant over such a short time span.

The resulting dataset contains 1,492 observations on 535 different players. The following outputs are used:

- y_1 = number of singles and walks (including intentional walks and hits-by-pitch) per plate appearance

- y_2 = number of doubles plus triples per plate appearance
- y_3 = number of home runs per plate appearance.

The question of which outputs to include is fundamental to any efficiency analysis and requires careful judgment. It is important to include only outputs that are important and that are direct measures of individual performance. The importance of not including obscure outputs individually becomes clearer in the next section. Essentially, it is desirable to avoid cases where good performance in one obscure category yields a high efficiency score for an otherwise mediocre individual. In baseball, triples are very rare (the median number in the dataset used here is two per player per year) and hence a relatively unimportant part of the game. The aim is to rule out the possibility that a weak player hits a few fluke triples and suddenly appears to be a very efficient player by doing well in this one relatively unimportant category. Accordingly, doubles and triples are added together into one category. Another reason for adding doubles and triples together is that the preferred specification (see later) is a logged one and hence does not accommodate zero values for any of the outputs. Hence, all player years with zero values for any of the outputs are deleted. By adding doubles and triples together into one output, this procedure only involves the removal of 19 observations (i.e., player-years). Results do not substantively change if doubles and triples are included separately.

Similarly, the various types of walks (i.e., base-on-balls, hit-by-pitch, and intentional walks) and singles are combined into one category. This is motivated by the facts that these are similar in terms of their value in a game, and the aim is to reduce the risk that a player who walks an unusual amount is deemed an excellent hitter without actually hitting the ball. There is, of course, some cost to this, in that the value of doubles and triples and the value of singles and walks are not identical. However, the values are likely very similar, so that the costs of combining some outputs together are outweighed by the benefit of avoiding the situation where mediocre players appear good due to a lucky performance in one obscure category.

The omission of runs scored and runs batted in (RBIs) from the list of outputs is warranted by a desire to include only direct measures of performance. Runs scored and RBIs are not direct measures of batting performance, in that they contain little information not already contained in y_1, y_2, and y_3. Runs and RBIs reflect the ability of a player either to get on base or to drive home other players who are already on base; these are already measured by the other outputs. Runs and RBIs also depend on the performance of other players (which is already measured in their batting statistics) and on baserunning ability (which is not the focus of the present study). Furthermore, RBIs depend crucially on a player's location in the batting order and the present study does not have data on this. Accordingly, runs scored or RBIs are not included as separate output measures.

Two players with the same level of ability can have different outputs for several reasons. An attempt is made to control for this by including the following explanatory variables:

- x_1 = an intercept
- $x_2 - x_5$ = dummies for the years 1996, 1997, 1998, and 1999 (1995 is the omitted year)
- x_6 = dummy for league (= 1 for the American League)
- $x_7 - x_{35}$ = dummies for 29 of the 30 teams that existed for all or part of the 1995–1999 period (the Baltimore Orioles are the omitted team).

The league dummy is included to reflect differences across the American and National Leagues. A primary difference is the use of the designated hitter (i.e., a batter designated to bat in place of the pitcher) in the American League. However, differences also might occur due to umpires in the different leagues having different styles (e.g., different strike zones). The team dummies are included to reflect both physical effects (i.e., baseball parks have different characteristics that make some easier to hit in than others) and proxy for the effect of other players on a team on an individual's hitting performance (e.g., opposing pitchers can "pitch around" a good hitter on a weak team and not offer him any good pitches to hit, knowing that weaker hitters will follow him in the batting order).

Another possibly relevant issue is the treatment of outs. A player can get out in several ways (e.g., strikeout, ground out, fly out). Strikeouts are typically considered the worst way of getting out, in that runners cannot advance on a strikeout except in very unusual situations (although with strikeouts, the chances of a double play are small, in contrast to ground outs or even fly outs). The dataset used in this article contains data on strikeouts, but not on other ways of getting out. An early version of this article included strikeouts per at bat as an undesirable output in the model. Results for this model excessively penalized players with many strikeouts and did not seem reasonable (e.g., Mark McGwire was ranked as a very average player due to his high number of strikeouts). Perhaps if the data were available then including strikeouts, ground outs, and fly outs as three separate undesirable outputs would be sensible. But in the absence of complete data on these, it was decided to omit strikeouts altogether.

The model

The models used in this article are all multiple-output extensions of stochastic frontier models. Stochastic frontier models were first introduced by Aigner, Lovell, and Schmidt (1977) and Meeusen and van den Broeck (1977), and the first Bayesian analysis was provided by van den Broeck, Koop, Osiewalski, and Steel (1994). Recent surveys include those of Kumbhakar and Lovell (2000) and, with a focus on Bayesian methods, Koop and Steel (2000). Univariate stochastic frontier models posit a function relating a k-vector of explanatory variables, x, to an output, y,

$$y = h(x).$$

The function $h(x)$ is interpreted as reflecting best practice, and individuals will typically fall short of this benchmark. Adding a subscript i to indicate an individual and letting τ_i be a measure of this shortfall with $0 < \tau_i < 1$,

$$y_i = h(x_i)\tau_i$$

or

$$\tau_i = \frac{y_i}{h(x_i)}.$$

Here τ_i is the measure of individual i's efficiency. A statistical model is obtained by adding measurement error (usually assumed to be normal) choosing a particular function form for $h(\cdot)$ and a distribution for τ_i.

In the case of univariate output, the definition of efficiency is uncontroversial, as measuring actual output relative to the maximum possible. To extend these methods to deal with multiple outputs, an output aggregator is required (see Fernandez, Koop, and Steel 2000). If now y

is a p-vector of outputs and $\theta(y)$ is a scalar function that aggregates these outputs, then the previous equations can be modified as

$$\theta(y_i) = h(x_i)\tau_i \tag{2.1}$$

or

$$\tau_i = \frac{\theta(y_i)}{h(x_i)}. \tag{2.2}$$

In the baseball application, $\theta(y_i)$ is the aggregate output of player i. The function $h(x_i)$ can be interpreted as the maximum amount of output that can be produced by a player with characteristics given by x_i, where output is measured by the output aggregator $\theta(y)$. A typical player will produce less than the value implied by $\theta(y)$, and $0 < \tau_i < 1$ is the measure of this deviation.

The choice of $\theta(y)$ is of great importance, and two different approaches are considered. In the first of these, $\theta(y)$ does not depend on unknown parameters. In this case, standard univariate stochastic frontier methods can be used, with $\theta(y)$ being the single aggregate output (e.g., on-base percentage). The second approach involves allowing $\theta(y)$ to depend on unknown parameters. To illustrate the basic ideas behind this second approach, consider the case where baseball players produce two outputs, power and batting average. The outputs of six hypothetical baseball players are plotted in Figure 2.1, marked as #1–#6. Looking first at these points, there appears to be a trade-off between power and average. Players #1, #2, and #3 have relatively low values for power but relatively high batting averages. Players #4, #5, and #6 are power hitters. The statistical methods used here estimate a production possibilities curve of the best power/average combinations that are observed in the data and use this curve as $\theta(y)$. This curve, plotted on Figure 2.1, can be seen to be heavily influenced by players #1, #2, #5, and #6.

Direct comparisons of, say, player #1 to player #6 would be considered by many to be relatively meaningless. After all, there is no one better than either of these players at doing what he specializes in. However, player #4 is a power hitter and can be meaningfully compared to power hitters #5 or #6. Similarly, player #3 can be meaningfully compared to players #1 or #2. The statistical methods herein use the ratio of distances, $\frac{\text{O\#4}}{\text{OA}}$ as a measure of the performance

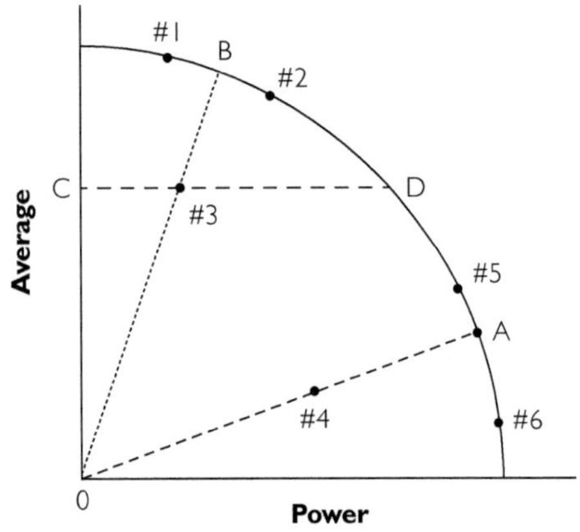

Figure 2.1

of player #4 and $\frac{\text{O\#3}}{\text{OB}}$ as a measure of the performance of player #3. It can be seen that these ratios are identical to τ_i as defined in (2.2). Note that players are always being measured relative to a curve largely defined by the best comparable players (e.g., power hitters are compared to power hitters). Because of this characteristic, efficiencies calculated in this way are called performance as a *proportion of the best comparable players'* (PBCP) efficiencies.

Figure 2.1 is a hypothetical illustration where hitters are either power or average hitters. Of course, in reality there will be a continuum of hitters and more outputs. Nevertheless, the same intuition holds. The performance of the best players for a given combination of outputs defines the production possibilities curve at this combination of outputs. Other players are compared to points on the curve largely defined by players with a similar mix of outputs.

The two approaches proposed herein have advantages and disadvantages. The first approach requires the choice of weights with which to aggregate outputs, and this choice is potentially ad hoc (e.g., OBP weights all outputs equally, whereas slugging percentage weights home runs four times as heavily as singles). It would be desirable to weight outputs proportionally to their values. In an economic application, such values would be proportional to the prices of each output (if such data were available), but in this baseball application the "value" of outputs is an elusive concept. Below, team data are used to estimate the influence of the outputs on team winning percentage. These estimates are then used as "values" to construct weights for the output aggregator. It can be argued that this is better than simply using OBP, batting average, or slugging percentage as an aggregate output. However, these estimates will be imperfect measures of the value of each output. In contrast, the second approach does not require that the statistician choose the weights in the output aggregator, because the production possibilities curve is used to define the output aggregator, $\theta(y)$. The interpretation of efficiencies obtained in this fashion is sensible (i.e., they reflect performance relative to the best comparable player); however, the weights used to aggregate outputs will not necessarily be proportional to the value of each output.

The previous paragraphs have sketched out the basic ideas of the proposed approach in an intuitive manner. The remainder of this section formalizes these ideas by assuming specific forms for $\theta(\cdot)$ and $h(\cdot)$, explicitly incorporating the longitudinal nature of the data, adding measurement error, and making additional assumptions necessary to achieve a valid sampling model.

Let $y_{(I,t,j)}$ be the amount of the jth output produced by, and let $x_{(i,t,l)}$ be the lth explanatory variable of, player i at time t, where $i = 1, \ldots, N$ and $t = 1, \ldots, T_i$. Note that this is an unbalanced panel where, for some players, data is missing for some values of t. To model the production possibility curve, an output aggregator is defined given by

$$\theta_{(i,t)} = \left(\sum_{j=1}^{p} \alpha_j^q y_{(i,t,j)}^q \right)^{1/q},$$

which depends on parameters $\alpha_j \in (0, 1)$ for all $j = 1, \ldots, p$, with $\sum_{j=1}^{p} \alpha_j = 1$ and $\alpha = (\alpha_1, \ldots, \alpha_p)'$. This is closely related to the commonly used constant elasticity of scale function. As described by many authors (e.g., Fernandez et al. 2000), this is a flexible form for the output aggregator that allows for a wide variety of shapes. The parameterization ensures invariance to the scaling of the jth output (e.g., measuring y_1 in terms of singles per 100 plate appearances instead of singles per plate appearance will only alter α_1 and not affect the measurement of efficiency). Here $\theta_{(i,t)}$ is interpreted as an aggregate output, which collapses the p-dimensional output vector into one dimension.

In the case where α and q are treated as known, two sets of parameter values are considered. The first is simply OBP, obtained by setting $q = 1$ and $\alpha_1 = \alpha_2 = \alpha_2 = 1/3$. The second sets $q = 1$ and uses a team-level regression to select a value for α. In particular, using data from 1995–1999, team winning percentage is regressed on team values for the three outputs, various measures of team pitching performance (i.e., ERA, saves, complete games, and shutouts), and team and year dummies. The coefficients on the three outputs can be interpreted as the value of each output in terms of its contribution to winning. Normalized to sum to 1, the coefficients imply weights of $\alpha_1 = 0.190$, $\alpha_2 = 0.277$, and $\alpha_3 = 0.533$. Interestingly, these weights are similar to those used to calculate slugging percentage, and hence this other common output aggregator is not considered separately. It is also interesting to note that in team-level regressions, disaggregating the outputs (e.g., breaking y_1 into singles and walks individually) has virtually no effect on the fit of the regression (as measured by R^2). Adding runs scored by a team increases the R^2 by only a small amount; that is, after taking into account team hitting, the number of runs actually scored by a team has little additional explanatory power for team winning percentage. These findings provide some additional evidence that the choice of outputs used for measuring player performance is reasonable.

Defining the NT-vector containing the logs of the aggregate outputs as

$$\log\theta = (\log\theta_{(1,1)}, \log\theta_{(1,2)}, \ldots, \log\theta_{(1,T_1)}, \ldots, \log\theta_{(N,T_N)})', \tag{2.3}$$

where $NT = \sum_{i=1}^{N} T_i$, $\log\theta$ is modeled as depending on an $NT \times k$ matrix of explanatory variables,

$$X = \begin{bmatrix} x_{(1,1,1)} & x_{(1,1,2)} & \cdot & x_{(1,1,k)} \\ x_{(1,2,1)} & & \cdot & \cdot \\ \cdot & & \cdot & \cdot \\ x_{(N,T_N,1)} & x_{(N,T_N,2)} & \cdot & x_{(N,T_N,k)} \end{bmatrix},$$

and an N-vector of individual effects, $z = (z_1, z_2, \ldots, z_N)'$, and a term is added to reflect measurement error,

$$\log\theta = X\beta - Dz + \varepsilon. \tag{2.4}$$

The $(NT \times N)$ matrix D is selected to match the individual effect with the appropriate individual, that is,

$$D = \begin{bmatrix} \iota_{T_1} & 0 & \cdot & 0 \\ 0 & \iota_{T_2} & \cdot & \cdot \\ \cdot & 0 & \cdot & 0 \\ 0 & 0 & \cdot & \iota_{T_N} \end{bmatrix},$$

where ι_S is an S-dimensional vector of 1s.

Some motivation for the specification in (2.4) is called for. One way of interpreting it is as a standard panel data stochastic frontier model (see, e.g., Schmidt and Sickles 1984) in terms of the aggregate output, θ. Adding the term zi is crucial to the analysis, because it measures the deviation of player i from best practice and thus relates to efficiency. The assumption that a player's ability is roughly constant over the short time span of the data implies that z_i is constant over time. Given the log specification in (2.4), it follows that player efficiency [see eq. (2.2) or Fig. 2.1] is given by

$$\tau_i = \exp(-z_i).$$

Because $0 < \tau_i < 1$, it follows that $z_i > 0$. To keep the number of parameters reasonably small relative to the number of observations, it is standard practice to make a distributional assumption for z. Such an assumption can be treated either as a hierarchical prior or as part of the likelihood function. Assume that

$$p(z \mid \lambda_0, \lambda_1) \propto \prod_{i=1}^{N} f_G(z_i \mid, \lambda_0, \lambda_1), \tag{2.5}$$

where $fG(z_i|a,\ b)$ denotes the pdf of a gamma distribution with mean a/b and variance a/b^2. Note that previous work with stochastic frontier models typically uses more restrictive distributions, such as the exponential or the half-normal. However, as stressed by Greene (1990, 2000), there may be substantial advantages in using the more flexible gamma. Here, use of the gamma is justified on the grounds that with the large amount of data that are available, it is appropriate to have a more flexible form to let the data determine the form of the efficiency distribution. Furthermore, histograms of the data indicate that the gamma is probably more suitable than the exponential or half-normal. For instance, histograms of $-\ln(y_{(i,t,j)})$ should give a rough idea of the shape of the distribution of z_i. These typically have an interior mode and are skewed to the right. Such properties can be easily accommodated using the gamma distribution, but not the exponential, the half-normal, or even the more general truncated normal distributions.

The normally distributed error term, ε, reflects the usual measurement error and model imperfections. If $f_N^R(\varepsilon|a, A)$ denotes the R-variate normal pdf with mean a and covariance matrix A, then assume that

$$p(\varepsilon) = f_N^{NT}(\varepsilon \mid 0, h^{-1} I_{NT}), \tag{2.6}$$

where h is the error precision and INT is the identity matrix.

For the case where α and q are treated as unknown parameters, it is important to note that eq. (2.4), plus the assumptions in (2.5) and (2.6), are not enough to fully characterize a likelihood function. For each player in each year, data are available on p outputs. A single equation such as (2.4) is not adequate to define a likelihood function for $p > 1$ endogenous variables. When α and q are known, (2.4) provides a valid likelihood function for $\log(\theta)$ but not the individual outputs that go into this aggregate output. Completing the specification of a likelihood function, requires some statement about the individual outputs. This is done so by defining the weighted output shares,

$$\eta_{(i,t,j)} = \frac{\alpha_j^q y_{(i,t,j)}^q}{\sum_{l=1}^{p} \alpha_l^q y_{(i,t,l)}^q}, \quad j = 1, \ldots, p, \tag{2.7}$$

which can be interpreted as the contribution of an individual output to the aggregate. Because $\sum_{j=1}^{p} \eta_{(i,t,j)} = 1$, (2.7) provides the extra $(p-1)$ dimensions necessary to produce a valid probability density for the p outputs. Because $0 \leq \eta_{(i,t,j)} \leq 1$, it is logical to group these shares as $\eta_{(i,t)} = (\eta_{(i,t,1)}, \ldots, \eta_{(i,t,p)})'$ and assume independent sampling from

$$p(\eta_{(i,t)}|s) = f_D^{p-1}(\eta_{(i,t)}|s), \tag{2.8}$$

where $s = (s_1, \ldots, s_p)'$ and $f_D^{p-1}(\cdot|s)$ is the pdf of a Dirichlet distribution with parameter vector s. Equations (2.7) and (2.8) are not added simply to complete the likelihood function. As discussed later, η has a simple intuitive interpretation relating to the existence of different types of hitters (e.g., those who hit for power and those who hit for average).

Define Y to be the $NT \times p$ matrix containing data on all outputs for all players in all years, with elements ordered in the same manner as $\log \theta$ in (2.3). Equations (2.4), (2.6), (2.7), and (2.8) imply the following likelihood function for Y:

$$p(Y|\beta, z, h, \alpha, q, s) = f_N^{NT}(\log \theta | X\beta - Dz, h^{-1} I_{NT}) \times \prod_{i,t} \left[f_D^{p-1}(\eta_{(i,t)}|s) \left(\prod_{j=1}^{p} q^{1-\frac{1}{p}} \frac{\eta_{(i,t,j)}}{y_{(i,t,j)}} \right) \right].$$

In this article a Bayesian approach is adopted. Accordingly, (2.5) is interpreted as a hierarchical prior. In addition, priors for the parameters (β, λ_0, λ_1, h, α, q, s) are needed. Details on the priors are given in the Appendix. Suffice it to note here that relatively noninformative priors are chosen that allow the data information to be predominant.

To carry out posterior inference, a Markov chain Monte Carlo (MCMC) algorithm is developed that involves sequentially drawing from $p(\beta|Y, z, \lambda_0, \lambda_1, h, \alpha, q, s)$, $p(h|Y, z, \beta, \lambda_0, \lambda_1, \alpha, q, s)$, $p(\alpha|Y, z, \beta, \lambda_0, \lambda_1, h, q, s)$, $p(q|Y, z, \beta, \lambda_0, \lambda_1, h, \alpha, s)$, $p(s|Y, z, \beta, \lambda_0, \lambda_1, h, \alpha, q)$, $p(\lambda_0|Y, z, \beta, \lambda_1, h, \alpha, q, s)$, $p(\lambda_1|Y, z, \beta, \lambda_0, h, \alpha, q, s)$, and $p(z|Y, \beta, \lambda_0, \lambda_1, h, \alpha, q, s)$. Details on the precise forms of these conditional posteriors, methods for drawing from them, and MCMC convergence are given in the Appendix. Note that this approach combines various techniques from Fernandez et al. (2000), Koop, Steel, and Osiewalski (1995), and Tsionas (2000).

Empirical results

The report of results begins with results relating to player efficiency based on three different choices of output aggregator, OBP efficiency (i.e., setting $q = 1$ and $\alpha_1 = \alpha_2 = \alpha_2 = \frac{1}{3}$), value-weighted efficiency (i.e., setting $q = 1$ and $\alpha_1 = 0.190$, $\alpha_2 = 0.277$, and $\alpha_3 = 0.533$), and PBCP efficiency (i.e., parameters estimated from the data and weights taken from production possibilities curve). Using the latter model (i.e., the one used to calculate PBCP efficiency), home run (HR) efficiency is also calculated. To motivate this measure, note that PBCP efficiency is a radial-oriented measure (i.e., it is based on distances along rays in Fig. 2.1). In words, such an efficiency measure relates to the question of how much can *all outputs be increased by the same proportion* without moving outside the production possibilities curve. However, output-oriented efficiency measures can also be developed (see Fernandez et al. 2001) that relate to the question of how much *output j only* can be increased without moving outside the production possibilities curve. In Figure 2.1, a power-oriented efficiency measure for player #3 is given by $\frac{C\#3}{CD}$. In this section, HR efficiency is calculated using such a HR-oriented measure. Such a measure might be of interest for the reader who feels that PBCP efficiency does not properly weight the outputs, but does not wish to provide explicit weights as in the value-weighted efficiency herein. HR efficiency takes the output that is undoubtedly of greatest value and measures inefficiency in terms of the shortfall of this output. So, as with PBCP efficiency, HR efficiency has the advantage that the weights in the output aggregator are estimated from the production possibility curve, but decreases the disadvantage of PBCP efficiency in that it apparently underweights home runs.

Because the focus here is on player efficiency, posterior results are not presented for all parameters in the model. Parameters that are common to all models have similar posteriors. In all cases, the posterior for λ_0 is concentrated in the region of 2.0, providing strong evidence for the importance of using the gamma for the distribution of the z_i's (as opposed to the exponential, which has $\lambda_0 = 1$). For the model that yields PBCP efficiency, q is near 1 (that is, its 95% HPDI includes 1), and posterior medians of the elements of α are $\alpha_1 = 0.433$, $\alpha_2 = 0.473$, and $\alpha_3 = 0.094$). For this model, only 5 of the 35 elements of β are significantly different from 0 (in the sense that the 95% HPDI does not include 0). However, a further 15 elements are possibly important in that they have 70% HPDIs that exclude 0. Furthermore, extensive experimentation suggests that posterior inference on the efficiencies is not greatly affected by the inclusion or exclusion of the explanatory variables that appear insignificant. Accordingly, the results presented in this section simply include all of the explanatory variables. Results for these explanatory variables are similar for all models and are sensible. For instance, the coefficients on team dummies for Boston, Cleveland, New York Yankees, Colorado, and Houston tend to be strongly positive in all specifications. Thus these teams either play in hitters' parks or have hitters' teams relative to the omitted team, Baltimore. A commonly used measure of model fit is the expected value of percentage of the total squared error that is due to measurement error (as opposed to efficiency). For the three specifications here, this measure of fit is at most 0.17, indicating that at least 83% of the squared deviation from the frontier is being picked up by the (time-invariant) efficiency component. This is interpreted as saying that the models are fitting well, in that the role of measurement error is quite small.

Space constraints preclude presenting efficiency results for all of the 535 players. Figure 2.2(a), a histogram of the posterior means of τ_i using PBCP efficiency, indicates a skewed efficiency distribution with an interior mode, but with relatively low dispersion (e.g., PBCP efficiency estimates run from 0.66 through 0.99, with most of the players clustered in the interval [0.8, 1.0]). Results for other efficiency definitions are similar. This is unsurprising given that in baseball, a batter who hits safely one time in four risks demotion to the minor leagues, whereas a batter who hits safely one time in three is close to winning a batting title. Table 2.1 presents posterior means and standard deviations for the 10 players with the highest posterior mean efficiencies when PBCP efficiency is used. Table 2.2 presents results for players with posterior means of PBCP efficiencies that are at the minimum, first quartile, median, and third quartile of the efficiency distribution. The last three columns of each of these tables is the rank of these players when OBP, value-weighted efficiency, and HR efficiency are used.

The results in Table 2.1 are quite reasonable, with many of the top names in baseball listed. It is also reassuring to see that various types of hitters are ranked highly using PBCP efficiency. The top 10 list contains hitters who are well known as getting on base a lot (e.g., Chuck

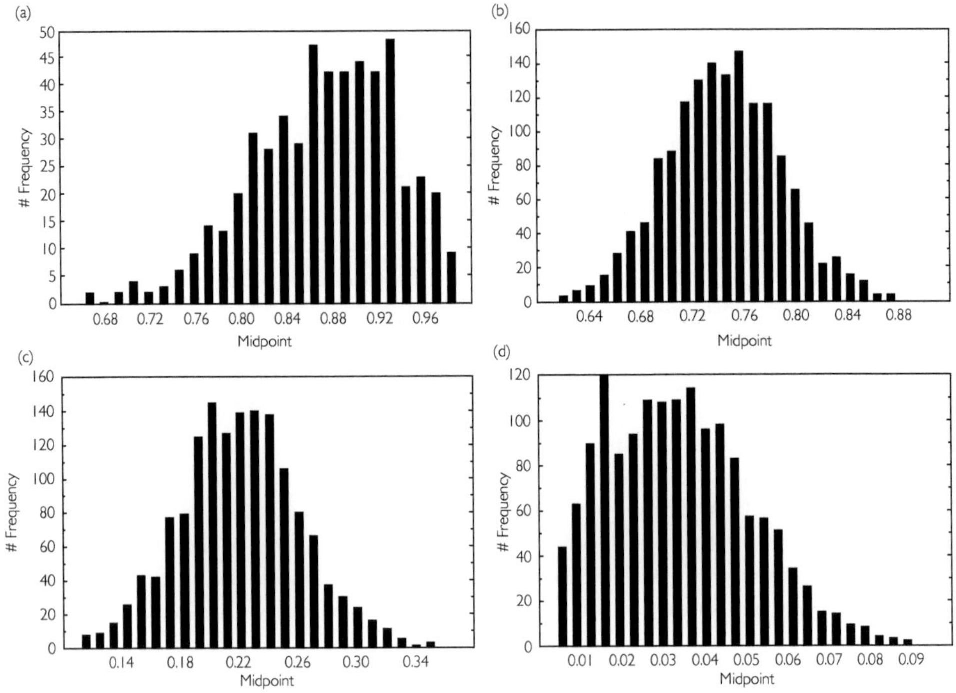

Figure 2.2 (a) Histogram of player efficiencies, (b) histogram of Eta (2.1), (c) histogram of Eta (2.2), and (d) histogram of Eta (2.3).

Table 2.1 ***Posterior results for Top 10 players ranked according to PBCP efficiency.***

PBCP rank	Name	PBCP efficiency (mean)	PBCP efficiency (standard deviation)	OBP rank	Value-weight rank	HR efficiency rank
1	E. Martinez	0.989	0.009	2	7	1
2	G. Sheffield	0.985	0.011	1	2	2
3	F. Thomas	0.985	0.012	4	5	4
4	M. Grace	0.982	0.013	10	29	12
5	J. Olerud	0.980	0.014	9	26	11
6	B. Bonds	0.980	0.014	5	3	5
7	J. Thome	0.977	0.016	6	9	6
8	J. Kendall	0.977	0.016	12	51	37
9	J. Bagwell	0.976	0.016	7	6	8
10	C. Knoblauch	0.973	0.013	16	35	32

Knoblauch), along with power hitters (e.g., Barry Bonds, Frank Thomas, Jim Thome) and hitters who combine some power with high batting average (e.g., Edgar Martinez, Gary Sheffield). The presence of explanatory variables means that two players with identical outputs on different teams can have different efficiencies. The fact that Larry Walker (with batting average of 0.344 over the 1995–1999 period) played in Colorado accounts for his absence from the top 10 list (although with an estimated efficiency of 0.967, he is not far off the list).

Other notable absences from this list are Mark McGwire and Sammy Sosa (record-setting home run hitters). Sammy Sosa's absence is explained by relatively poor performance in the early years of this study (1995–1997). In walks per plate appearance, in particular, he is significantly inferior to Mark McGwire. McGwire, with a posterior mean PBCP efficiency of 0.973, is just off the list, ranking 11th. However, it is worth noting that in terms of value-weighted and HR efficiencies, Mark McGwire is ranked number 1 and number 3. This shows the importance of the weights in the output aggregator. When the production possibility curve is used as an output aggregator (as stochastic frontier methods typically do), efficiency calculations are based on players with comparable output mixes. The implied weights can be quite different from the values of the outputs. Table 2.1 clearly shows that players who are not known for their power (i.e., Grace, Olerud, Kendall, and Knoblauch) score very highly with PBCP efficiency but somewhat lower with value-weighted efficiency. But of course, with value-weighted efficiency, these individuals are no longer being compared to players with a similar output mix. HR-oriented efficiency provides results that are somewhere between PBCP and value-weighted efficiency, in that Grace, Olerud, Kendall, and Knoblauch do worse than with PBCP efficiency, but better than with value-weighted efficiency. HR efficiency may thus be a reasonable compromise for the researcher who does not want to choose values for taking a weighted average of outputs, but wants to incorporate the concept that home runs are worth more than other outputs.

The players listed in Table 2.2 are less well known than those in Table 2.1 (e.g., J. R. Phillips played only in 1995 and even then only in 92 games). Nevertheless, examining the data, it does seem as though the proposed model provides a reasonable ranking of efficiency. Furthermore, the four different efficiency measures give similar results. It is worth noting that calculating the correlation matrix for the point estimates of the four different efficiency measures yields all correlations between 0.71 and 0.96. The most weakly correlated measures are PBCP and HR efficiency, whereas the most strongly correlated measures are PBCP and OBP efficiency.

Table 2.2 ***Posterior results for quartile players ranked according to PBCP efficiency.***

PBCP rank	Name	PBCP efficiency (mean)	PBCP efficiency (standard deviation)	OBP rank	Value-weight rank	HR efficiency rank
Minimum	J. R. Phillips	0.661	0.054	533	502	434
First quartile	S. Martinez	0.829	0.060	438	465	427
Median	O. Cabrera	0.877	0.046	344	392	339
Third quartile	J. Reed	0.919	0.037	103	140	100

The discussion to this point has focused on point estimates of efficiencies. One advantage of the Bayesian approach over non-Bayesian alternatives is that it is simple to derive posterior standard deviations or HPDIs for player-specific efficiencies to see whether rankings based on point estimates are truly statistically reliable. Statistically reliable means that if player A is ranked above player B (based on point estimates), then the posterior probability that player A truly is more efficient than player B is high (see Fernandez et al. 2001 for a detailed discussion of this issue). An examination of Tables 2.1 and 2.2 indicate that posterior standard deviations for PBCP efficiencies can be fairly large, indicating that caution must be taken in comparing individual players. Standard deviations for value-weighted and OBP efficiencies are of a slightly smaller magnitude, which is sensible, because fewer parameters are being estimated. The investigation of the statistical validity of rankings based on posterior means focuses on PBCP efficiency, because if the rankings are meaningful for this case, they will also be meaningful for OBP and value-weighted efficiencies.

The estimated efficiencies of the top 10 players are usually within one standard deviation of one another. A naive interpretation of this finding is that the top 10 ranking is pretty meaningless. However, correlations between the efficiencies of different players imply that posterior standard deviations may not tell the full story. To shed additional light on this issue, various posterior probabilities addressing the question of the probability that player A is more efficient than player B can be calculated. Of course, a myriad of possible comparisons could be made. As a general rule, it is not easy to distinguish between players ranked very close to each other in terms of point estimates, but it is possible to distinguish between players farther apart in the ranking. The following comparisons illustrate these points.

One finding is that there is a 0.29 probability that Mark McGwire (ranked 11th, with an estimated efficiency of 0.960) should be on the top 10 list (i.e., in 29% of the MCMC draws, he falls in the top 10 in terms of efficiency). Other findings include a 0.155 probability that Edgard Martinez truly is the most PBCP-efficient player and a 0.701 probability that he should be ranked in the top 10. Hence, although the results do not conclusively show that Martinez should be at the top of Table 2.1, but they do indicate that he belongs in this top 10 table.

Tables 2.3 and 2.4 offer more systematic comparisons. Table 2.3 takes the players ranked first, fifth, and 10th and compares them. In particular, the element in the ith column and jth row in Table 2.3 is the probability that player i is more efficient than player j. Table 2.4 is of the same form, but takes the five quartile players (i.e., minimum, first quartile, median, third quartile, and maximum efficient players). The results in these tables are moderately encouraging. Of course, a fair bit of statistical uncertainty is associated with estimating the large number of parameters in the model, as indicated in Table 2.3. However, Table 2.4 indicates

Table 2.3 ***Probability that a player in a column is more efficient than that player in a row.***

	E. Martinez	J. Olerud	C. Knoblauch
E. Martinez			
J. Olerud	0.705		
C. Knoblauch	0.794	0.611	

Table 2.4 ***Probability that a player in a column is more efficient than that player in a row.***

	E. Martinez	J. Reed	O. Cabrera	S. Martinez	J. R. Phillips
E. Martinez					
J. Reed	0.985				
O. Cabrera	1.000	0.733			
S. Martinez	1.000	0.974	0.793		
J. R. Phillips	1.000	1.000	0.997	0.996	

that the proposed approach can reliably say that players ranked in different quartiles from one another have different efficiencies. Table 2.3 compares players ranked roughly one percentile apart, and thus it is unsurprising that results are not as strong. Nevertheless, there is a good chance that the top 10 rankings are at least roughly sensible.

Player-specific efficiencies are the focus of the present study. Nevertheless, it is worthwhile to discuss the $\eta_{(i,t,j)}$s, because these are innovative parts of the model used to calculate PBCP and HR efficiency. Remember that they can be interpreted as being the share of the aggregate output attributed to output j for player i in time t [see eq. (2.7)]. Figures 2.2(b)–(d) present histograms using $E(\eta_{(i,t,j)}|Data)$ for every i, t, and j. These imply, for instance, that the output singles plus walks receive most of the weight in the aggregate output. The weights reflect both the shape of the production possibility curve (i.e., α and q, which are estimated from the data) as well as the actual value of the output in each category (i.e., y_1, y_2, y_3). The shapes of the histograms in Figures 2.2(b)–(d) are, with one exception, roughly normal. The exception is home runs, which is skewed and has a few interesting bumps in the right tail. The bins furthest are Mark McGwire and Sammy Sosa in the years 1998 and 1999.

Figures 2.2(b)–(d) illustrate a relatively high variability across players (e.g., the share of singles plus walks varies from 0.626 to 0.865, the share of home runs varies from 0.005 to 0.079). In other words, there are many different types of players with very differing performance in terms of each of the individual outputs. This indicates that any ranking of players based on a single output could be very misleading and emphasizes the value of finding a statistically justifiable way of aggregating outputs.

This article has described various methods of aggregating outputs that allows for comparison of different types of players. However, some readers may not like such comparisons. Such a reader may argue that a player who is producing 80% of what Mark McGwire produces is worth more than a player producing 80% of what Chuck Knoblauch produces, but PBCP efficiency might give both these players efficiencies of roughly 0.80. Furthermore, this reader might argue that trying to measure the "value" of outputs in baseball is too difficult and hence may not like our measure of value-weighted efficiency. For this reader, the $\eta_{(i,t,j)}$'s should be of particular interest, because they allow for a very sensible way of determining which players are comparable in terms of their mix of outputs. That is, one may wish to divide players into "power hitters" and "non-power hitters," arguing that comparisons involving players in different categories are hard to interpret. But this begs the question of how to group players into these two categories. The $\eta_{(i,t,j)}$'s offer a simple and sensible way answering this question.

Table 2.5 ***Top 10 players for different categories ranked according to PBCP efficiency.***

PBCP rank	Power hitters	Hitters with some power	Non-power hitters
1	G. Sheffield	E. Martinez	J. Kendall
2	F. Thomas	M. Grace	C. Knoblauch
3	B. Bonds	J. Olerud	T. Gwynn
4	J. Thome	B. Larkin	R. Henderson
5	J. Bagwell	W. Clark	J. Offerman
6	M. McGwire	B. Williams	D. Magadan
7	M. Salmon	R. Greer	W. Boggs
8	M Piazza	R. Alomar	K. Seitzer
9	L. Walker	J. Cirillo	K. Lofton
10	H. Baines	B. Abreu	E. Young

Table 2.5 presents rankings for a grouping based on the share of home runs in the output aggregator, $\eta_{(i,t,3)}$. In particular, for each player, $E(\eta_{(i,3)}|Data) = \frac{\sum_{t=1}^{T_i} E(\eta_{(i,t,j)}|Data)}{T_i}$ is calculated (i.e., average over time to get one value per player), and then the players are divided into three equal groups. These three groups have the lowest, middle, and highest values, based on this average share of home runs. These three groups of players are called non–power hitters, hitters with some power, and power hitters. Because shares sum to 1, there is a high degree of correlation between the various shares. A division of players into groups based on $\eta_{(i,t,1)}$ or $\eta_{(i,t,2)}$ would be similar to the one used here.

It is important to stress that although column 2 of Table 2.5 enables a comparison of power hitters, they are not being compared solely in terms of their home runs. This accounts for the fact that McGwire is not ranked first, despite the fact that his number of home runs per plate appearance is the highest in the sample. Furthermore, as with all of the efficiency measures presented in this article, different player characteristics are corrected through a vector of explanatory variables [see eq. (2.1)], which means that the rankings can differ from uncorrected rankings of, for example, slugging or on base percentage. It can be argued that both of these features of the comparisons are desirable.

Both the division of players into categories and the rankings are very sensible. The list of power hitters contains virtually all of the top home run hitters in the game (the reason for Sammy Sosa's omission was discussed earlier). The players in the middle category all are excellent players who have some power but are not known as pure home run hitters. The list of non–power hitters includes some great leadoff hitters and batting title winners. Thus, should one wish to separately compare players with different output mixes, this statistical framework allows for a sensible way of doing this based on the $\eta_{(i,t,j)}$'s.

Conclusion

This article has developed statistical methods for comparing the offensive performance of individual baseball players that explicitly take into account the many outputs that players produce. A key component in the method is an output aggregator. The weights in this output aggregator can be selected either *a priori* (e.g., OBP weights all outputs equally) or based on some measure of the value of each output (e.g., from a regression using team data that estimate the effect of each output on team winning percentage), or they can be estimated based on the performance of the best players in the various output dimensions. Considered are three ways of aggregating outputs, which provide three different measures: OBP, value-weighted, and PBCP efficiency. The efficiencies of individual players are based on a comparison with a benchmark set by the best players. The result is a number between 0 and 1 that has a simple, intuitive interpretation. For instance, in the case of PBCP efficiency, if a player's efficiency estimate is 0.75, then it can be said that he is producing 75% of the output of the best comparable players. Such a number, it is argued, has advantages over traditional measures of batting performance in that it is simple to interpret and does not require the choice of *ad hoc* weights to create an aggregate output. Bayesian methods for statistical inference were developed and used to estimate the efficiencies of 535 players in the years 1995–1999. Results are found to be sensible. For instance, the players that are highly ranked would also undoubtedly be highly ranked by students of baseball.

Given the great, and increasing, importance of sport in economic and social life, developing such techniques is of direct interest. The techniques used here can be applied to any sport that has good measures of player outputs. Unfortunately, for many sports (e.g., football and soccer), it may be difficult to find suitable outputs (e.g., in soccer, goals scored is a flawed measure of individual performance, and few other measures are available). However, for other sports (e.g., cricket, hockey, and basketball) the large (and increasing) amount of data collected on individual players should make them well suited for an application using the techniques discussed in this article. Furthermore, in many social and biological sciences, cases exist where the researcher is interested in evaluating the performance of an individual against a standard set by others. In such cases, models similar to the one used in this article are of potential interest.

Appendix: Prior and MCMC algorithm

A prior of the form $p(\beta,\ \lambda_0,\ \lambda_1,\ h,\ \alpha,\ q,\ s) = p(\beta)p(\lambda_0)p(\lambda_1) \times p(h)p(\alpha)p(q)p(s)$ is used. For β, assume that $p(\beta) \propto f_N^k(\beta|b_0, H_0^{-1})$ and make the noninformative choices of $b_0 = 0_k$ and $H_0 = 10^{-4} \times I_k$. Next assume that $p(\lambda_0) = fG(\lambda_0|\underline{\lambda}_{00}, \underline{\lambda}_{01})$ and choose $\mathrm{E}(\lambda_0) = 3$ and $\mathrm{var}(\lambda_0) = 1$, and hence $\underline{\lambda}_{00} = 9$ and $\underline{\lambda}_{01} = 3$. Furthermore, $p(\lambda_1) = fG(\lambda_1|\underline{\lambda}_{10}, \underline{\lambda}_{11})$ with $\underline{\lambda}_{10} = 3$ and $\underline{\lambda}_{11} = -\ln(\tau^*)$. If $\lambda_0 = 3$ (i.e., its prior mean), then these values would imply that the prior median efficiency is τ^*, which is a natural quantity to elicit. Set $\tau^* = 0.70$. For h, take $p(h) = fG(h|n_0/2,\ a_0/2)$ and set $n_0/2 = 1$ (which leads to an exponential prior for h) and $a_0/2 = 10^{-6}$. These values imply large prior uncertainty. Because the components of α are all in the interval (0, 1) and sum to 1, an obvious choice of a prior distribution is a Dirichlet with pdf $p(\alpha) = f_D^{p-1}(\alpha|a)$ with hyperparameter vector $a = (a_1, \ldots, a_p)'$. Using the diffuse choice of $a = \iota p$ makes the prior uniform over the $(p—1)$-dimensional unit simplex. For q, take an exponential prior $p(q) \propto fG(q|1,\ d)$. Adopt the diffuse choice of $d = 10^{-6}$. Assume p independent gamma distributions for the components of s, $p(s) = \prod_{j=1}^{p} p(s_j) = \prod_{j=1}^{p} f_G(s_j|b_j, c_j)$. Make a diffuse choice of $b_j = 1$ and $c_j = 10^{-6}$ for all j.

The MCMC algorithm here involves sequentially drawing from the posterior conditionals described in the body of the text. The first of these is $p(z|Y, \beta, \lambda_0, \lambda_1, h, \alpha, q, s) \propto \prod_{i=1}^{N} z_i^{\lambda_0-1} f_N^1(z_i|m_i, R_i) \times I(z_i \geq 0)$, where mi is the ith element of the N-vector $m =$

$(D'D)^{-1}$ $\{D'(X\beta - \log(\theta)) - h^{-1}\lambda_1\}$ and Ri is the ith diagonal element of the $N \times N$ diagonal matrix $R = h^{-1}(D'D)^{-1}$. Because the elements of z are conditionally independent, they can be drawn separately. In practice, a modified version of the acceptance algorithm of Koop (1995) is used to draw from this nonstandard density.

The pdf of the posterior conditional distribution of λ_0 is given by

$$p(\lambda_0|Y,z,\beta,h,\alpha,q,s,\lambda_1)$$
$$\propto \lambda_1^{N\lambda_0}\lambda_0^{\lambda 00-1}\,\Gamma(\lambda_0)^{-N}\exp\left[\left(\sum_{i=1}^{N}\ln(z_i)-\lambda_{01}\right)\lambda_0\right].$$

This distribution is nonstandard, but a simple acceptance algorithm can be derived following Tsionas (2000). Candidate draws, λ_0^{can}, are taken from $f_G(\lambda_0|1,\ A)$, where A is chosen to maximize the acceptance probability.

The conditional posterior distribution of λ_1 is $p(\lambda_1|Y,z,\beta,h,q,s,\lambda_0) = f_G(\lambda_1|\underline{\lambda}_{10}+N\lambda_0,\underline{\lambda}_{11}+\sum_{i=1}^{N}z_i)$, and the conditional posterior distribution of β is $p(\beta|Y,z,\lambda_0,\lambda_1,h,\alpha,q,s) = f_N^k(\beta|b_*,H_*^{-1})$, where $H_* = H_0 + hX'X$ and $b_* = H_*^{-1}\{H_0b_0 + hX'(\log(\theta)+Dz)\}$.

For h,

$$p(h|Y,z,\lambda_0,\lambda_1,\beta,\alpha,q,s)$$
$$= f_G\left(h|\frac{n_0+NT}{2},\frac{a_0+(\log(\theta)-X\beta+Dz)'(\log(\theta)-X\beta+Dz)}{2}\right).$$

For α, the following nonstandard pdf is obtained:

$$p(\alpha|Y,z,\lambda_0,\lambda_1,\beta,h,q,s)$$
$$\propto \prod_j \alpha_j^{a_j+s_jqNT-1}\prod_{i,t}\left(\sum_j \alpha_j^q y_{(i,t,j)}^q\right)^{-\sum_j sj}$$
$$\times\exp\left\{-\frac{1}{2\sigma^2}(\log(\theta)-X\beta+Dz)'(\log(\theta)-X\beta+Dz)\right\}.$$

The conditional posterior for q is also nonstandard and is given by

$$p(q|Y,z,\beta,\lambda_0,\lambda_1,h,\alpha,s)\propto q^{NT(p-1)}\exp(-dq)\exp\{-A(q)\},$$

where

$$A(q) = \frac{h}{2}(\log(\theta)-X\beta+Dz)'(\log(\theta)-X\beta+Dz)$$
$$+\sum_{i,t,j}s_j\log\left(\frac{\sum_{l=1}^{p}\alpha_l^q y_{(i,t,l)}^q}{\alpha_j^q y_{(i,t,j)}^q}\right).$$

Each of the p components of s is drawn separately. The conditional posterior distribution of $s_j, j = 1, \ldots, p$ (also given the remaining components of s) has pdf on $(0, \infty)$,

$$p(s_j|Y, z, \lambda_0, \lambda_1, \beta, h, \alpha, q, \{s_h : h \neq j\})$$
$$\propto \frac{\Gamma\left(\sum_l s_l\right)^{NT}}{\Gamma(s_j)^{NT}} s_j^{b_j-1} \exp\left[-s_j\left\{c_j + \sum_{i,t} \log\left(\frac{\sum_l \alpha_l^q y_{(i,t,l)}^q}{\alpha_j^q y_{(i,t,j)}^q}\right)\right\}\right].$$

For the last three distributions, random-walk Metropolis–Hastings algorithms involving normal candidate-generating densities with variance calibrated in preliminary MCMC runs are used, so that the acceptance probability is in the region of 5%–50%.

The results of Fernandez et al. (1997) and Fernandez et al. (2000) imply that the posterior is proper and the MCMC algorithm will converge. However, proof has yet to be found that posterior means and higher moments exist for parameters other than the efficiencies (although numerical evidence suggests that higher moments of all parameters do exist). Hence, except for the efficiencies, this work discusses only HPDIs (which do exist).

The final results are based on 260,000 MCMC replications. The first 10,000 are discarded to mitigate startup effects. Of the remaining 250,000, every 10th replication is retained to break serial correlation in the draws (and minimize storage costs). Thus the posterior properties reported are based on 25,000 replications. MCMC convergence is monitored using diagnostics suggested by Geweke (1992) and Gelman and Rubin (1992).

REFERENCES

Aigner, D., Lovell, C. A. K., and Schmidt, P. (1977), "Formulation and estimation of stochastic frontier production function models," *Journal of Econometrics*, 6, 21–37.

Albright, S. C. (1993), "A statistical analysis of hitting streaks in baseball" (with discussion), *Journal of the American Statistical Association*, 88, 1175–1196.

Barry, D., and Hartigan, J. A. (1993), "Choice models for predicting divisional winners in baseball," *Journal of the American Statistical Association*, 88, 766–774.

Bennett, J., and Flueck, J. (1983), "An evaluation of major league baseball offensive performance models," *The American Statistician*, 37, 76–82.

Berry, S., Reese, C. S., and Larkey, P. (1999), "Bridging different eras in sport" (with discussion), *Journal of the American Statistical Association*, 94, 661–686.

Chapman, K. (1991), "Testing the matching hypothesis: the case of major league baseball," *American Economic Review*, 81, 1352–1360.

Depken, C. A. II (2000), "Wage disparity and team productivity: evidence from major league baseball," *Economics Letters*, 67, 87–92.

Fernandez, C., Koop, G., and Steel, M. F. J. (2000), "A Bayesian analysis of multiple-output production functions," *Journal of Econometrics*, 98, 47–79.

—— (2002), "Alternative efficiency measures for multiple output production," *Journal of the American Statistical Association*, 97, 432–442.

Ferrall, C., and Smith, A. (1999), "A sequential game model of sports championship series: theory and estimation," *Review of Economics and Statistics*, LXXXI, 704–719.

Gelman, A., and Rubin, D. (1992), "Inference from iterative simulation using multiple sequences," *Statistical Science*, 7, 457–511.

Geweke, J. (1992), "Evaluating the accuracy of sampling-based approaches to the calculation of posterior moments," in *Proceedings of the Fourth Valencia International Meeting on Bayesian Statistics*, eds. J. O. Berger, J. M. Bernardo, A. P. Dawid, and A. F. M. Smith, Oxford, U.K.: Oxford University Press, pp. 169–194.

Greene, W. (1990), "A gamma distributed stochastic frontier model," *Journal of Econometrics*, 46, 141–164.

—— (2000), "Simulated likelihood estimation of the normal-gamma stochastic frontier model," unpublished manuscript, available at www.stern.nyu.edu/~wgreene/.

Hoaglin, D., and Velleman, P. (1995), "A critical look at some analyses of major league baseball salaries," *The American Statistician*, 49, 277–284.

Kahn, L. (1993), "Free agency, long-term contracts and compensation in major league baseball: estimates from panel data," *Review of Economics and Statistics*, 75, 157–164.

Kaigh, W. (1995), "Forecasting baseball games," *Chance*, 8, 33–37.

Koop, G., Osiewalski, J., and Steel, M. F. J. (1997), "Bayesian efficiency analysis through individual effects: hospital cost frontiers," *Journal of Econometrics*, 76, 77–105.

Koop, G., and Steel, M. F. J. (2000), "Bayesian analysis of stochastic frontier models," in *Companion to Theoretical Econometrics*, ed. B. Baltagi, Oxford, U.K.: Basil Blackwell, pp. 520–537.

Kumbhakar, S., and Lovell, C. A. K. (2000), *Stochastic Frontier Analysis*, Cambridge, U.K.: Cambridge University Press.

Lackritz, J. (1990), "Salary evaluation for professional baseball players," *The American Statistician*, 44, 4–8.

—— (1996), "Two of baseball's great marks: can they ever be broken?," *Chance*, 9, 12–18.

Meeusen, W., and van den Broeck, J. (1977), "Efficiency estimation from Cobb–Douglas production functions with composed errors," *International Economic Review*, 8, 435–444.

Schall, T., and Smith, G. (2000), "Do baseball players regress toward the mean?" *The American Statistician*, 54, 231–235.

Schmidt, P., and Sickles, R. (1984), "Production frontiers and panel data," *Journal of Business and Economic Statistics*, 2, 367–374.

Scully, G. (1974), "Pay and performance in major league baseball," *American Economic Review*, 64, 915–930.

Tsionas, E. (2000), "Full likelihood inference in normal-gamma stochastic frontier models," *Journal of Productivity Analysis*, 13, 183–205.

van den Broeck, J., Koop, G., Osiewalski, J., and Steel, M. F. J. (1994), "Stochastic frontier models: a Bayesian perspective," *Journal of Econometrics*, 61, 273–303.

Watnik, M. (1998), "Pay for play: are baseball salaries based on performance?" *Journal of Statistics Education*, 6, 1–5.

CHAPTER 3

Career records for all modern position players eligible for the Major League Baseball Hall of Fame

JAMES J. COCHRAN

Abstract

The dataset "Career Records for All Modern Position Players Eligible for The Major League Baseball Hall of Fame" contains information for the 1340 major league baseball players who had retired prior to the 1993 season and who were eligible for the Major League Baseball Hall of Fame (had played in at least ten seasons). Traditional performance measures included are number of seasons played, games played, official at-bats (AB), runs scored, hits (H), doubles (2B), triples (3B), home runs (HR), runs batted in (RBI), walks (BB), strikeouts (SO), batting average (BA), on base percentage (OBP), slugging percentage (SLG), stolen bases (SB), times caught stealing (CS), fielding average (FA), and primary position played (POS). In addition, the following composite measures are included: adjusted production (AP), batting runs (BR), adjusted batting runs (ABR), runs created (RC), stolen base runs (SBR), fielding runs (FR), and total player rating (TPR). Finally, the dataset includes an indication of whether or not each player has been admitted into the Major League Baseball Hall of Fame and, if so, under what set of rules he was admitted.

Keywords: *classification, classroom data, descriptive statistics, discrimination, multivariate statistics*

Introduction

1 In Major League Baseball, a wide variety of performance measures are collected for individual players. These measurements, which stimulate interest in both casual and serious baseball fans, are widely disseminated and at least reasonably well understood

by a large proportion of the U.S. population. Indeed, baseball phrases such as *hitting a home run, striking out*, and *play ball!* are part of our nation's collective lexicon. Because of this broad interest, measurements of Major League Baseball player performances are particularly useful in classroom demonstrations of various statistical concepts.

2 One of the most controversial topics of debate among baseball fans is the question of which players have earned, by virtue of their career performances, membership in the Major League Baseball Hall of Fame. There are two ways in which membership may be conferred upon a player. Five years after his retirement, the player becomes eligible for election by the Baseball Writers' Association of America (the BBWAA). At this stage a player is inducted if he can, in an annual election, gain the support of at least 75% of all writers who have belonged to the BBWAA for at least ten years. If a player is not elected by the BBWAA within twenty years of his retirement, the Veterans Committee may then consider him for induction. This is a small committee comprised of retired players, managers, coaches, and administrators with many years of experience in Major League Baseball. The Veterans Committee also meets annually and may choose to induct any player no longer eligible for election by the BBWAA. This committee's purpose is somewhat similar to that of the Old Timers Committee, which was disbanded a few years after the establishment of the Major League Baseball Hall of Fame because it was felt at that time that the committee had outlived its usefulness. Finally, a player must have played at least ten major league seasons to be eligible for membership in the Major League Baseball Hall of Fame.

3 Because the variable of primary interest (*Major League Baseball Hall of Fame Membership*) in this dataset is categorical, the dataset is appropriate for use in a course on generalized linear models, classification, or discrimination. The career performance measures for the players are suitable independent variables, and each player included in the dataset constitutes one record/observation. Additionally, the data can be (and have been) used to demonstrate concepts in descriptive statistics, basic inference, and multivariate statistics.

The dataset

4 These data appear in *Total Baseball* (Thorn and Palmer 1993) and *The Baseball Encyclopedia* (Reichler 1993). The data include the traditional performance measures *Number of Seasons Played, Games Played, Official At-Bats, Runs Scored, Hits, Doubles, Triples, Home Runs, Runs Batted In, Walks, Strikeouts, Batting Average, On Base Percentage, Slugging Percentage, Stolen Bases, Caught Stealing*, and *Fielding Average* for each modern (post-1900) position player with at least ten years of major league experience. The *Primary Position Played by the Player* throughout his career, defined as the position at which the player played the most games in his career, is also included. This variable is coded in the following manner:

C = Catcher,
1 = First Base,
2 = Second Base,
3 = Third Base,
S = Shortstop,

O = Outfield, and
D = Designated Hitter.

5 Additionally, an indication of the *Player's Hall of Fame Membership Status* is included. This variable is coded in the following manner:
0 = Not a member,
1 = Elected by the BBWAA, and
2 = Chosen by the Old Timers Committee or Veterans Committee.

6 The Old Timers Committee, which existed for only a brief period early in the life of the Hall of Fame, was charged with selecting players whose careers were in the nineteenth century. After the demise of the Old Timers Committee, the Veterans Committee was formed and given the responsibility of selecting players overlooked by the BBWAA. Understandably, their selections have often been the source of great controversy (by definition, a player under consideration by the Veterans Committee has already been deemed undeserving by baseball writers).

7 Finally, various composite measures developed by sabermatricians including John Thorn and Pete Palmer (1984, 1993) and Bill James (1982) are included. Definitions and explanations of the composite measures follow.

Adjusted production

Also called *Production Plus*, this measure equals the sum of *On Base Percentage* and *Slugging Percentage*, each indexed by the league average (Thorn and Palmer 1993). This is calculated as

Batting runs

The difference between the linear weights-estimated number of runs generated by the player through his plate appearances and the league average (Thorn and Palmer 1984). Calculated as

$$\begin{aligned} BR = {} & 0.47(1B) + 0.78(2B) + 1.09(3B) + 1.40(HR) + 0.33(BB + HBP) \\ & -0.25(AB - H) - 0.50(OOB), \end{aligned}$$

where 1B represents *Singles*, HBP represents *Hit By Pitches*, and OOB represents *Outs On Base.*

Adjusted batting runs

Batting Runs adjusted for differences in home park and indexed against the league average (Thorn and Palmer 1984).

Runs created

A composite estimate of run contribution from a variety of batting and base running events (James 1982). Calculated as

$$RC = \frac{(H + BB)(TB)}{AB + BB},$$

where TB represents *Total Bases* or 1B + 2(2B) + 3(3B) + 4(HR).

Stolen base runs

The difference between the linear weights-estimated number of runs generated by the player through his stolen base attempts and the empirically determined historical break-even success rate of 0.67 (Thorn and Palmer 1984). Calculated as

$$SBR = 0.30(SB) - 0.60(CS).$$

Fielding runs

The difference between the linear weights-estimated number of runs prevented by the player through his fielding chances and the league average at his position (Thorn and Palmer 1984).

Total player rating

The sum of the player's *Adjusted Batting Runs, Fielding Runs*, and *Stolen Base Runs* minus a positional adjustment, then divided by the estimated number of runs a player of that era must generate to create an additional win for his team (Thorn and Palmer 1984).

8 The traditional performance measures are readily available and widely disseminated, and so they are most likely the primary objective criteria used by BBWAA and Veterans Committee members in deciding for whom they will vote. However, the composite measures have been developed in response to perceived inadequacies in the traditional measures. Although the composite measures have gained exposure in recent years, they still are not widely accepted and so are not likely to be considered explicitly by BBWAA and Veterans Committee members. In fact, some eschew the use of these measures because they believe the measures are contrived and ill conceived. However, their inclusion in this dataset is interesting because it allows for some assessment of their validity and consistency with traditional measures.

Classroom use

9 I regularly teach a one-quarter (ten-week) standard introductory business statistics course. In this course I cover basic descriptive statistical methods (numerical, graphical, and tabular), probability, sampling distributions, and confidence intervals. I have used these data extensively in this course to demonstrate descriptive statistics. For example, the data can be used to illustrate why bar charts based on relative or percentage frequencies are more meaningful than bar charts based on absolute frequencies when comparing two groups of different sizes. Back-to-back bar charts of *Primary Position Played* for Hall of Fame members and non-members, based on absolute frequencies, are given in Figure 3.1.

10 It is difficult, based on this graph, to discern if there is a positional bias in Hall of Fame membership. The large discrepancy in the sizes of the two groups (the dataset consists of 124 position players who are members of the Hall of Fame and 1216 who are not members) masks relative differences. However, a similar chart based on relative frequencies, such as the chart provided in Figure 3.2, makes the bias much more evident.

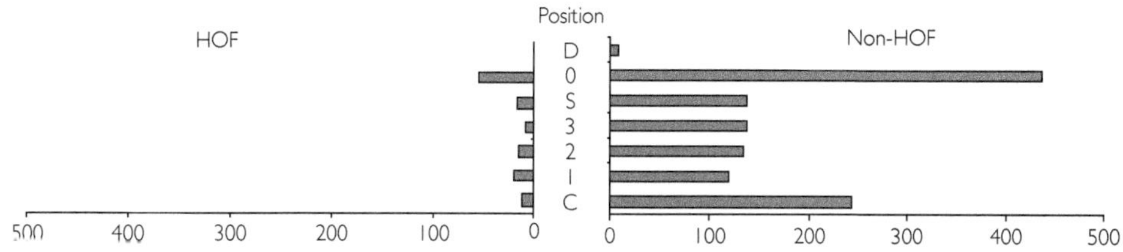

Figure 3.1 Absolute frequencies by primary position played.

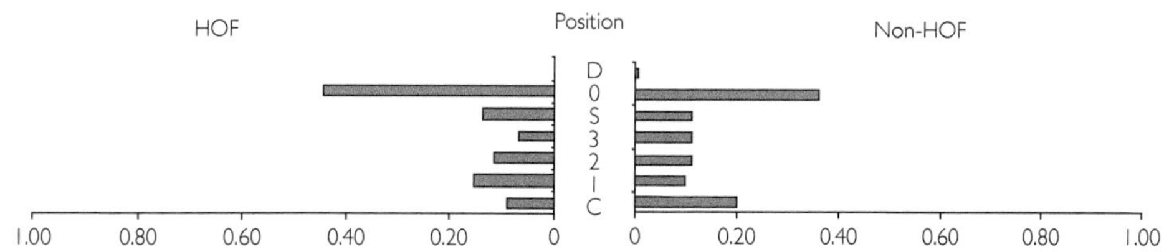

Figure 3.2 Relative frequencies by primary position played.

11 Using total number of players eligible as our basis, it is easy to see that outfielders and first basemen enjoy overrepresentation in the Hall of Fame, and third basemen and catchers suffer from under-representation. At this point, I also discuss the importance of the basis of comparison—if we used at-bats or games played as our basis, catchers may not appear to be underrepresented. This is, of course, because the demands of the position generally limit catchers to shorter careers.

12 I also provide my students with a similar example using *Total Player Rating* by *Hall of Fame Status* (eligible but not elected, elected by the BBWAA, elected by the Veterans Committee). A sample side-by-side histogram is provided in Figure 3.3.

13 This side-by-side histogram does suggest that the great majority of eligible players with relatively low *Total Player Ratings* are not in the Hall of Fame. However, the broad scale of the *y*-axis, necessitated by the large discrepancy in the size of the three groups (1216 eligible players who are not members of the Hall of Fame, 57 position players who have been elected by the BBWAA to the Hall of Fame, and 67 position players who have been elected by the Veterans Committee to the Hall of Fame) makes it difficult to draw further conclusions. Again, a similar chart based on percentage frequencies, such as the chart provided in Figure 3.4, is more revealing.

14 From this chart we can conclude that the BBWAA generally elects players with the highest *Total Player Ratings*. Furthermore, the Veterans Committee, choosing from players who have been passed over by the BBWAA, generally elects players from the remaining pool with the highest *Total Player Ratings*. Thus we observe some agreement between the composite measurement *Total Player Rating* and the behavior of the BBWAA and Veterans Committee. We also see potential explanations for the controversy regarding some of the selections made by the Veterans Committee.

15 Frequently I also teach a one-quarter follow-up introductory course on inferential methods for business. In this course I introduce one-sample and two-sample hypothesis testing techniques, chi-square tests of independence and equality of

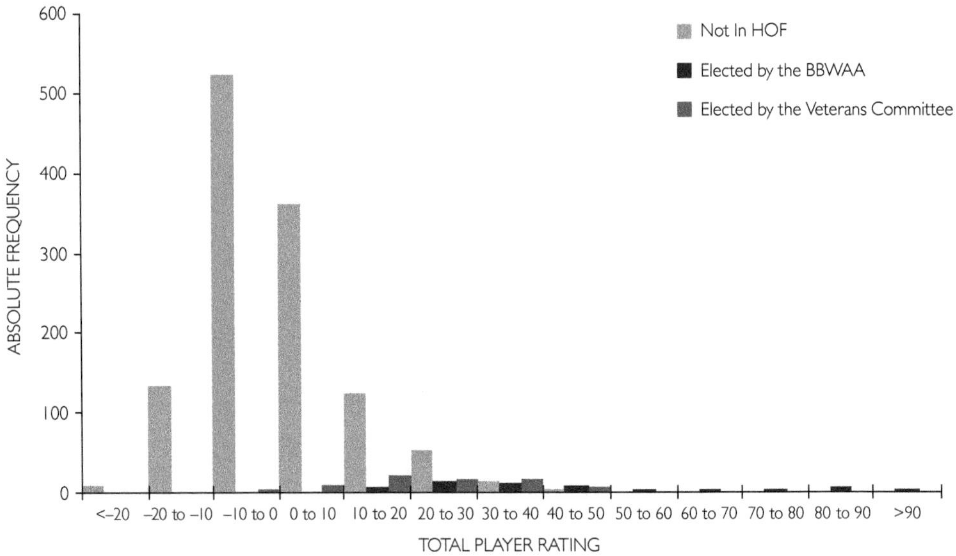

Figure 3.3 Absolute frequencies for hall of fame membership by total player rating.

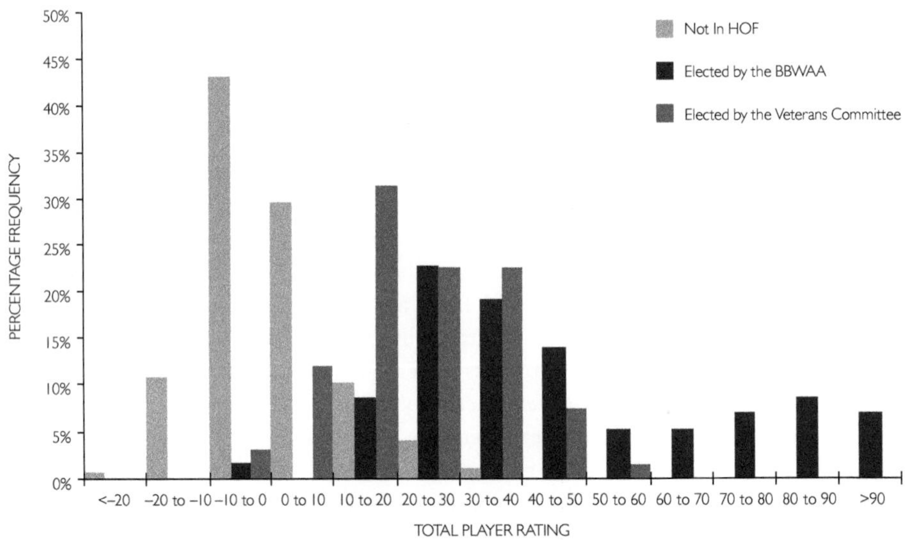

Figure 3.4 Percentage frequencies for Hall of Fame Membership by total player rating.

proportions, one-factor and two-factor analysis of variance, and simple and multiple linear regression analysis. I have used these data extensively in this course as well. For example, I ask students to consider independence of total career *Strikeouts* and *Hall of Fame Membership Status*. The resulting summary of the data could look like Table 3.1.

Table 3.1 ***Frequencies of total career strikeouts.***

Total career strikeouts	Not in the Hall of Fame	In the Hall of Fame	Total
0–250	381	27	408
250–500	429	38	467
500–750	225	25	250
750–1000	98	8	106
1000–1250	42	9	51
1250–1500	15	6	21
> 1500	7	10	17
?	19	1	20
Total	1216	124	1340

Table 3.2 ***Cell values for total career strikeouts.***

Total career strikeouts	Not in the Hall of Fame	In the Hall of Fame	Total
0–250	0.328	3.193	3.521
250–500	0.072	0.699	0.771
500–750	0.013	0.125	0.138
750–1000	0.037	0.357	0.393
1000–1250	0.390	3.797	4.187
1250–1500	0.858	8.354	9.212
> 1500	4.594	44.712	49.306
Total	6.292	61.236	67.529

16 Strikeouts were not recorded in early professional baseball, so career strikeout totals are unavailable for twenty players. Ignoring these twenty players, we obtain the chi-square values in Table 3.2.

17 The p-value associated with the corresponding value (67.529) for six degrees of freedom is less than 0.0001, supporting rejection of the hypothesis of independence between total career *Strikeouts* and *Hall of Fame Membership Status*. A further review of the individual cell values reveals that the actual number of Hall of Fame members in the high strikeout categories (1000–1250, 1250–1500, 1500) greatly exceeds what would be expected if total career *Strikeouts* and *Hall of Fame Membership Status* were independent. A classroom discussion on possible explanations (length of career, relative propensity for home run hitters to strike out, etc.) for this seemingly paradoxical relationship ensues.

18 These data are also useful in multivariate statistics courses. Factor analysis of the traditional performance measurements for all players included in this dataset can be used to derive latent patterns that may represent "prototypes" of baseball players with long careers. Canonical correlation analysis of the traditional performance measurements against the composite performance measurements may yield interesting insight into the reliability of the composite measures. Finally, the dataset is well suited for demonstration of generalized linear models, as well as various methods for classification and discrimination. In fact, Cochran and Levy (2000) have applied regression tree methodology to these data in an attempt to explain Hall of Fame membership.

19 Because of the general level of interest in baseball, these data are extremely valuable in providing examples that lend insight and promote enthusiasm among students. Those students who are fans of baseball are more engaged and motivated by such examples. While the remaining students might not share their classmates' enthusiasm, they cannot help but be impressed by the usefulness of statistics when applied to something as "mundane" as baseball.

Getting the data

20 The file MLBHOF.new.dat.txt contains the raw data. The file MLBHOF-tab.new.dat.txt contains the data in tab-delimited format, and MLBHOF.new.xls is an Excel file of the data. The file MLBHOF.new.txt is a documentation file containing a brief description of the dataset.

Appendix—Key to variables in MLBHOF.new.dat.txt

Columns

1–19 Name
20–21 Number of Seasons Played
27–30 Games Played
32–36 Official At-Bats
38–41 Runs Scored
43–46 Hits
48–50 Doubles
52–54 Triples
56–58 Home Runs
60–63 Runs Batted In
65–68 Walks
70–73 Strikeouts
76–79 Batting Average
82–85 On Base Percentage
88–91 Slugging Percentage
93–95 Adjusted Production
97–100 Batting Runs
102–105 Adjusted Batting Runs
107–110 Runs Created
112–115 Stolen Bases

117–119 Caught Stealing
121–124 Stolen Base Runs
128–131 Fielding Average
133–136 Fielding Runs
139 Primary Position Played
141–146 Total Player Rating
149 Hall of Fame Membership

Values are column-aligned. The few missing values occur only where the data were either not collected or are unavailable. Measures such as *Caught Stealing*, for example, have been collected sporadically throughout Major League Baseball's history.

REFERENCES

Cochran, J. J., and Levy, M. S. (2000), "Who 'Deserves' to be in the Major League Baseball Hall of Fame?" University of Cincinnati Working Paper #2000–01.

James, B. (1982), *The Bill James Baseball Abstract 1982*, New York: Ballantine Books.

Reichler, J. L. (ed.) (1993), *The Baseball Encyclopedia*, New York: MacMillan Publishing Company.

Thorn, J., and Palmer, P. (1984), *The Hidden Game of Baseball: A Revolutionary Approach to Baseball and Its Statistics*, New York: Doubleday.

—— (1993), *Total Baseball*, New York: Harper Collins Publishers.

CHAPTER 4

Streakiness in team performance: How often would we expect to see a 20-game winning streak like that of the Oakland Athletics in 2002?

JIM ALBERT

Baseball fans and the media are fascinated with the occurrence of streaky behavior among players and teams. In the 2002 baseball season, the Oakland Athletics won 20 games in a row that was the longest winning streak since the Chicago Cubs' winning streak of 21 games in 1935. In the 2003 season, the Detroit Tigers lost 11 games in a row, and Albert Pujois hit successively in 30 consecutive games.

Based on our experience in watching baseball, it is clear that we will continue to see streaky behavior among teams and players in future seasons. But the important question is "How do we interpret this streakiness?" Is this observed streakiness just a manifestation of chance variability inherent in day-to-day baseball performance of teams and players? Or does this streaky performance say something about the ability or quality of the team or player to be truly streaky? The intent of this paper is to use a simple probability model to understand the pattern of streaky performance of baseball teams that can be attributable to chance. Once we see the pattern of long streaks predicted by this model, we can better understand the significance of a particular team's long streak, such as Oakland's 2002 winning streak.

First, we should clarify what we mean by streaky ability. There is no doubt that teams (and players) go through good and bad stretches of performance, and we'll observe streaky behavior.

But streaky ability means that the ability of the team to win games also goes through high and low periods. We can quantify a team's ability by its probability of winning a game. If a team's win probability for a particular game depends on its success (or lack of success) in recent games, then the team is truly streaky. The tricky part, from a statistical perspective, is how we learn about a team's true streakiness given its observed streaky behavior.

It is easiest to talk about true streakiness by first talking about the opposite type of team—the truly consistent team. This team will win every game with the same win probability p and the result of a game will not depend on the results of previous games. In more formal language, we are assuming that the game results (win or loss) are independent Bernoulli outcomes with constant probability of success p.

Do we believe that this consistent model is true for a baseball team? Actually, no. I would suspect that the probability that a team wins a game depends somewhat on the starting pitcher, the opposing team, the health status of the players, and the month of the season. But the consistent model may be a reasonable approximate representation for win/loss data in that it gives predictions similar to actual game results.

The usual approach for detecting true streakiness of a particular team is to first assume this consistent model is true with an estimated value of the winning probability p; second, focus on some interesting streaky statistic from the game results of the team, and finally compute the p value that is the probability that the streaky statistic is at least as extreme as the observed value given the consistent model. If this p value is small (say, under 5%), then this is considered sufficient evidence that the consistent model is false and the team is displaying some streaky ability.

Let's illustrate this approach with Oakland's 2002 winning streak of 20 games. In this season, Oakland's winning percentage was 64.8. We assume that Oakland's game outcomes are independent Bernoulli trials with win probability $p = 0.648$ and compute the probability that Oakland gets a winning or losing streak of 20 games or longer. This computation is easy to perform by simulation, and we obtain

$$\text{Prob}\left(\text{winning streak} \geq 20|\text{consistent model with } p = 0.648\right) = 0.0068$$

This p value is small, indicating that the consistent model is false and Oakland displayed some true streakiness this particular season.

There are a number of problems with this approach. First, there is the issue of selection bias. We focused on the streakiness quality of Oakland, since they had won a large number of consecutive games. Since we are deciding to perform inference on an extreme team, it will appear to be streakier than if we had instead explored the pattern of winning and losing of a "random" team. Second, there is the related issue of multiplicity. Since there are currently 30 major teams and many seasons of team performance in our recent memory, it may be very likely to see a team win 20 consecutive games in this large group of data. To decide on the significance of Oakland's winning streak, we should view it in the context of all teams that have played baseball in recent seasons. Another issue is that there might be a relationship between a team's ability and its tendency to be streaky. Since Oakland won a much higher percentage of games than a mediocre 2002 team such as Philadelphia, it might be more likely for Oakland to have a long winning streak than Philadelphia. Last, how do we interpret this small p value value? Maybe we have demonstrated that Oakland has demonstrated streaky behavior beyond what could be explained by chance, but it is not clear how to interpret this pattern of streakiness by means of true streaky behavior.

Previous work

There has been substantial interest in the detection of true streakiness since Gilovich et al. (1985) who claimed that any observed streakiness in sports data is simply people's misperception of the streaky patterns inherent in random data. Stern (1997) has a general discussion on the statistical detection of streakiness and Alan Reifman has a web site (www.hs.ttu.edu/hdfs3390/hothand.htm) devoted to hot-hand research.

Albright (1993) did an extensive analysis of streakiness of hitting data for many major league baseball players. By incorporating "streakiness" parameters in a regression model, he determined that these parameters were statistically significant for particular players. Albert (1993), in his discussion of Albright's paper, introduced a streaky model for hitting performance. He assumed that a baseball hitter had two possible ability states, low and high, with corresponding hitting probabilities p_{LO} and p_{HI}. A player would move between the two ability states according to a Markov Chain with a given transition matrix. One could then measure true streakiness for a given player by estimating the difference in hot and cold hitting probabilities $p_{HI} - p_{LO}$. (We will return to this model later.)

Once a streaky model has been defined, then one can investigate the power of different statistics, such as the longest streak of wins, in detecting values of parameters of the streaky model. Albert and Williamson (2001), using a Bayesian viewpoint, showed that a number of streaky statistics such as the longest streak and the total number of streaks were not very powerful in detecting true streakiness.

Albert and Bennett (2003), Chapter 5, show how p value calculations can be used to assess the true streakiness of a given player or team. As above, one begins by computing the p value of a streaky statistic of a team assuming the consistent model. By also computing the p value of this statistic under a competing streaky model, one can assess the probability that the given team is streaky given the observed data.

Since the intent here is to look at the win/loss patterns for all major league teams, it is relevant to look at one model for baseball competition discussed in Chapter 12 of Albert and Bennett (2003). The model assumes that teams have underlying talents that are drawn from a normal population and the baseball game results are Bernoulli outcomes where the win probabilities that are functions of the talents. In Albert and Bennett's analysis, the goal is to relate a team's ability with its performance and they show the prominent role of chance in the determination of the winner of the World Series.

Empirical work

Streakiness of teams during the 2002 season

We begin by exploring the streaky behavior of the teams for the 2002 season. For each team, we observe the sequence of wins and losses for the season. There are many ways of measuring the streakiness or lack of consistency of the team's performance. We can focus on long streaks. For example, we can compute the length of the longest winning or losing streak, or count the number of streaks (consecutive runs of winning or losing games) that exceed a specific length. Alternatively, we can compute the number of streaks in the sequence. Teams that go through extreme high and extreme low performances will tend to have a small number of streaks. An equivalent measure to the number of streaks is the number of switches (either win to loss or loss to win) during a season.

Albert and Bennett (2003) introduce a measure of consistency of a team's performance based on moving averages of a team's performance. To study a team's performance during short time periods, Albert and Bennett define a particular window, say 12 games, and compute the winning fraction of the team for games 1–12, games 2–13, games 3–14, and so on. (These winning fractions are called moving averages.) We can observe the short-term performance of a team by plotting these moving averages against the game number. This moving average plot for the 2002 Oakland Athletics is displayed in Figure 4.1. If the moving averages fall close to the season winning proportion (shown by a horizontal line), then the team is very consistent. In contrast, if the moving averages show much variation about this line, then the team is performing streaky. We can measure the consistency of a team by averaging the distances of the moving averages from the overall winning proportion. If $y_1, \ldots, y_m$ denote the sequence of moving averages, then this statistic, called BLACK, is given by

$$BLACK = \sum_{i=1}^{m} |y_i - \bar{y}|/m$$

This statistic is called BLACK since it is proportional to the shaded area between the moving averages and the season winning fraction. In Figure 4.1, we see that Oakland had a significant

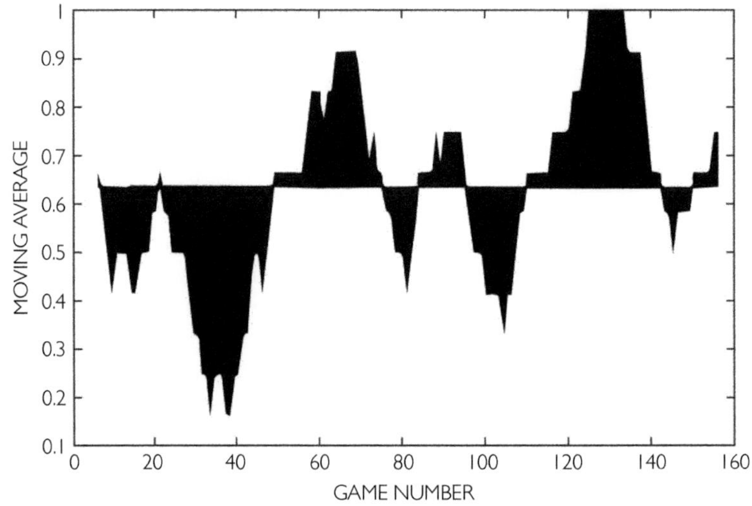

Figure 4.1 Moving average plot of game results of Oakland during 2002 season using a window of 12 games. The shaded region represents the average distance of the moving averages from the season winning fraction.

cold spell early in the season and a large hot spell (that included the winning streak of 20 games) towards the end of the season.

In this paper we focus on two streaky statistics, the length of the longest streak and the BLACK statistic. They represent different aspects of a team's performance. The length of the longest streak measures a team's extreme performance (with regard to streakiness) and the BLACK statistic is an overall measure of a team's consistency over the entire season.

As described earlier, the streakiness of a team is confounded with its success or lack of success. This is demonstrated in Figure 4.2 that plots (for the 2002 teams) the length of the longest

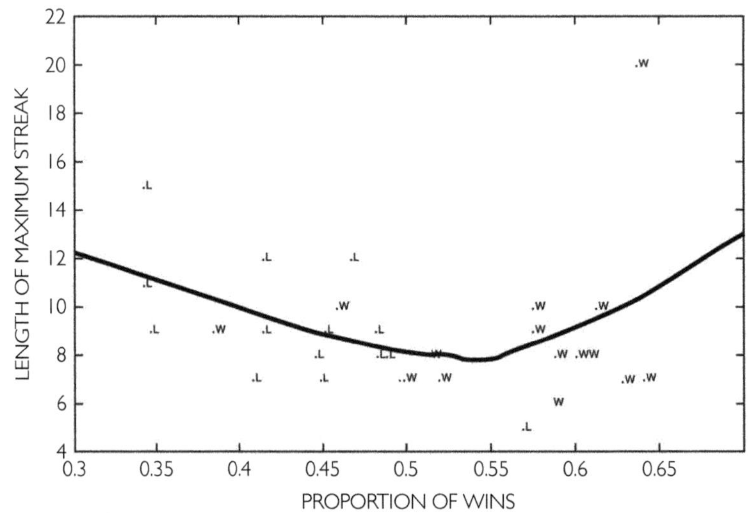

Figure 4.2 Scatterplot of proportion of wins and length of longest streak for 2002 teams. The solid curve is a loess smoother, to the scatterplot.

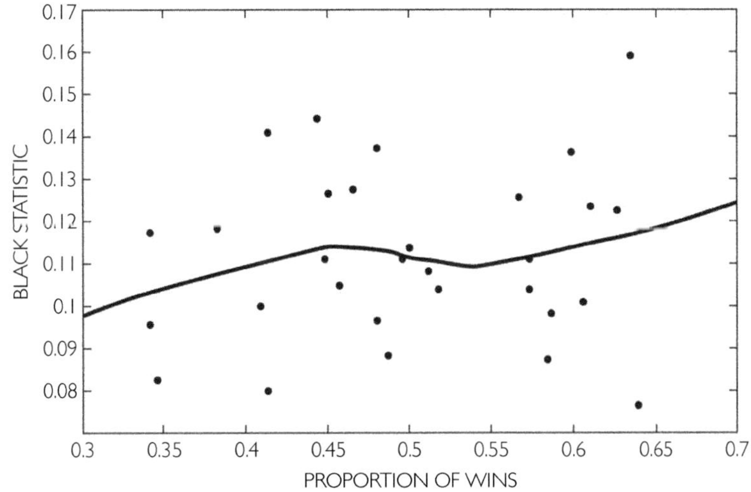

Figure 4.3 Scatterplot of proportion of wins and BLACK statistic for 2002 teams. The solid curve is a loess smoother to the scatterplot.

streak against the team's season winning fraction. Each point in the graph is labeled—a "W" denotes that this streak was a winning streak, "L" indicates it was a losing streak, and "B" indicates that the team had both a winning and a losing streak of this length.

A loess smooth is placed on the scatterplot indicating the basic pattern in the graph. We see from the figure that long winning and losing streaks tend to correspond to teams with extreme (high or low) season performances.

Figure 4.3 shows a scatterplot of the BLACK statistic and the winning fraction for the 2002 teams. It is interesting to note that, unlike Figure 4.1, there is no general pattern to this scatterplot, indicating that there is little relationship between a team's BLACK measure and its general success in winning games.

Winning fractions of teams through the years 1973–2002

Since the streakiness performance of a team appears to be related to its strength, we look closer at the pattern of winning fractions of baseball teams. Figure 4.4 displays a dot-plot of the winning fractions of the 2002 teams. We notice a large spread in these fractions from 0.34 to 0.64. We can measure the general level of competition of baseball teams by the standard deviation of these fractions. If $p_1, \ldots, p_{30}$ denote the winning fractions of the 30 teams during year y, then this standard deviation is given by

$$s_y = \sqrt{\frac{1}{29} \sum_{i=1}^{30} (p_i - \bar{p})^2}.$$

If this standard deviation is large relative to previous years, then that would suggest that baseball teams are not as competitive as they were in the past.

To look at this question further, we compute the standard deviation of the winning fractions for the teams for each of the previous 30 seasons (1973–2002). Figure 4.5 plots these standard

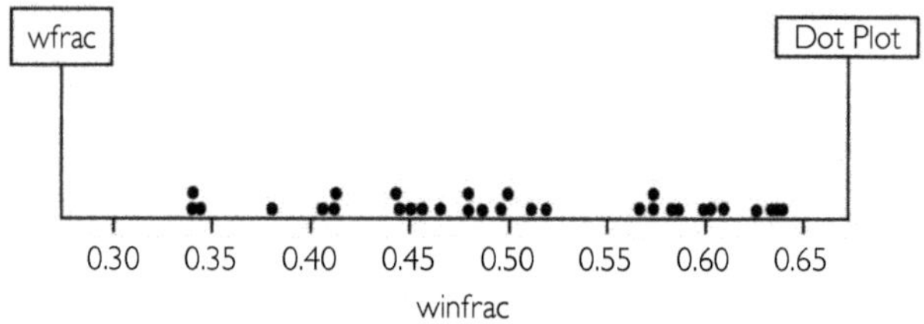

Figure 4.4 Dotplot of winning proportions for 2002 baseball teams.

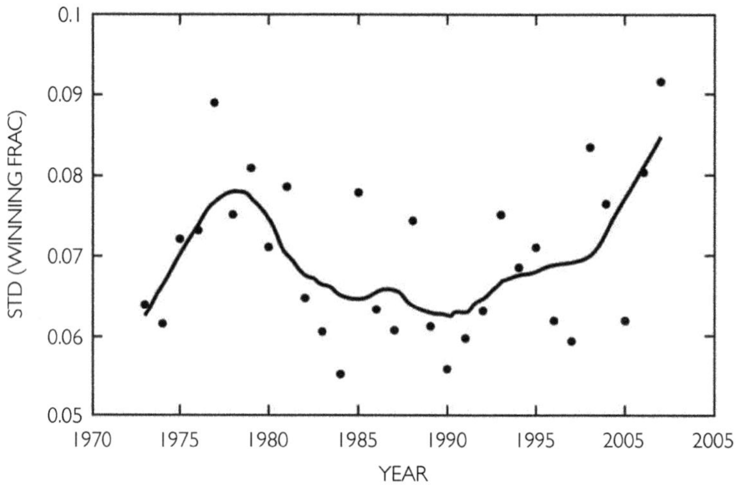

Figure 4.5 Time series plot of standard deviation of team winning percentages for the last 30 baseball seasons.

deviations against season and places a lowess smoothing curve on top. The smooth has an interesting pattern. It appears that the teams were less competitive in the late 1970s, more competitive in the 1980s and early 1990s, and there has a lack of competitive balance in recent years.

We can measure the variation in these standard deviations by a standard deviation that is denoted by

$$S30 = \sqrt{\frac{1}{29} \sum_{i=1}^{30} (s_y - \bar{s})^2}.$$

Here, the value of $S30$ is 0.0097—it measures the variability in the consistency of the teams over the 30 seasons.

Model for team abilities

We are interested in constructing a relatively simple model that can accurately predict the pattern of team winning fractions in recent years. A good, simple model for team competition is the Bradley–Terry model described in Chapter 12 of Albert and Bennett (2003).

During a particular season, the ith team has a true talent θ_i. If team i plays team j, then the team i will win with probability given by the logistic form

$$\text{Prob(team i defeats team j)} = \exp(\theta_i)/(\exp(\theta_i) + \exp(\theta_j)).$$

To complete the model, the talents for the 30 teams, $\theta_1, \ldots, \theta_{30}$ are assumed to be a sample from a normal distribution with mean 0 and standard deviation τ.

One can use the observed win/loss data for a season to estimate the team talents $\{\theta_i\}$ and the spread of the population talent distribution τ. The focus here is on the estimation of the standard deviation parameter τ that reflects the competitive balance of the teams. If $\tau = 0$, then the teams all have the same ability and the outcomes of games are equivalent to the results of tossing a fair coin.

An easy way of estimating τ is based on the model/data simulation algorithm described by Albert and Williamson (2001). It is an approximate Bayesian method for performing inference on a parameter. Suppose we think, before looking at any data, τ that is equally likely to be any value on the interval (0, 1). To run this model/data simulation, we

(1) simulate a value of τ uniformly from (0, 1)
(2) simulate values of the talents from a N(0, τ) distribution
(3) using the Bradley–Terry model, simulate a full season of baseball games (using the 2002 schedule) and compute the team winning fractions
(4) compute the standard deviation s_y of the team winning fractions

We repeat this procedure many times, each time collecting the value of the talent dispersion parameter τ and the value of the standard deviation s_y from the game results.

Figure 4.6 constructs a scatterplot of the (τ, s_y) pairs. As expected, there is a positive relationship in the scatterplot—as the abilities of the teams get more dispersed, the team performances will also get more spread out.

Once this simulation is run, it is easy to perform inference about τ for an observed value of s_y. To illustrate, if the standard deviation of the winning fractions for a particular year is $s_y = 0.07$, then we focus on the simulated values of τ where s_y falls in a small neighborhood, say (s_y—0.002, s_y + 0.002), about s_y. Figure 4.7 shows a histogram of the simulated values of the ability standard deviation τ. Based on this histogram, a reasonable estimate at τ would be 0.25 which is approximately the modal value.

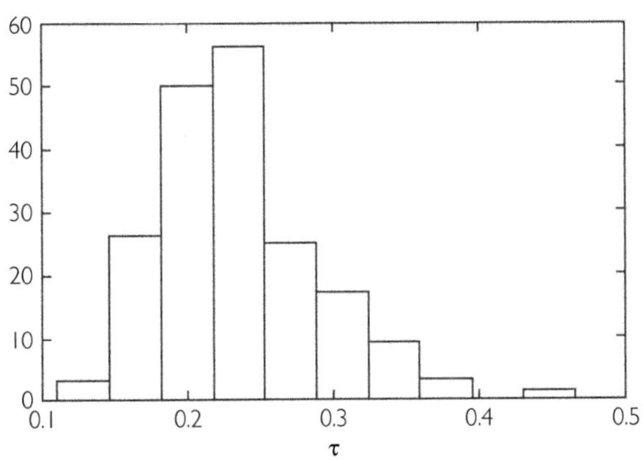

Figure 4.6 Scatterplot of values of the ability dispersion parameter t and the standard deviation s_y of the winning fractions for 10,000 simulations.

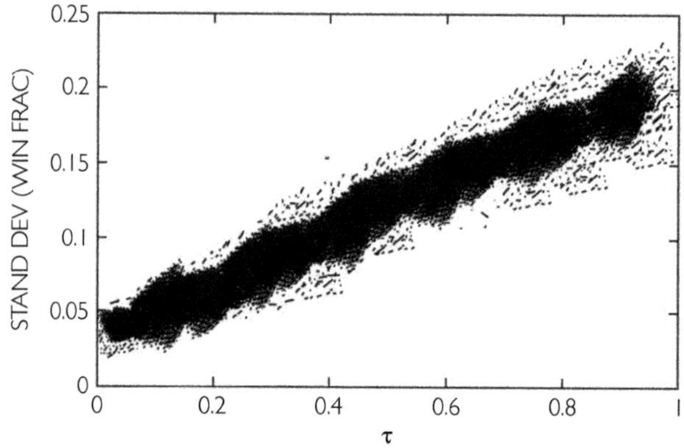

Figure 4.7 Simulated values of the talent dispersion parameter for values of s_y in the range $(0.07 - 0.002, 0.01 + 0.002)$.

Actually, from Figure 4.5, we see that $s_y = 0.07$ is a typical measure of spread of the winning fractions of teams for a season in the past 30 years. So the value $\tau = 0.25$ would be a reasonable estimate at the spread of the talent distribution for recent years.

To see if the Bradley–Terry model with the abilities distributed according to N(0, 0.25) is a good model, we simulated 30 years of team competition using this model. For each season, we computed a standard deviation s_y, and then we summarized the spread of the 30 values of s_y (for the 30 seasons) by a standard deviation. As in the previous section, we denote the latter measure by $S30$—it represents the variability of the competitive balance of the teams over the 30-year period. We then repeated this 30-year simulation 1000 times, collecting a value of the across-seasons standard deviation $S30$ each simulation. At the end of the simulation, we computed the mean of the $S30$ values, and it was very close in value to the actual value of $S30$ for the seasons 1973–2002. So this Bradley–Terry model, with the N(0, 0.25) talent distribution, seems to be a good representation of the season-to-season performances of major league teams.

Team streakiness from random effects model

Now that we have a good model for team performances, we use this model to simulate game-to-game performances of teams during a 162-game season and from this simulated data, we compute various streaky statistics of interest.

To simulate data for one season, we first simulate a group of talents from the N(0, 0.25) distribution. Then using these team talents, we simulate a full season of game results using the Bradley–Terry model. For each team, we collect the game-to-game win/loss records, and compute two streaky statistics: the length of the longest streak, and the BLACK statistic using a window of 12 games. We repeat this simulation process for 100 seasons, and obtain values of the streaky statistics for the 30 teams for the 100 seasons.

Our objective is to compare the pattern of streakiness in the game results for the 2002 teams with the streakiness in the simulated data. Let's first look at the length of the longest run. The longest streaks for each of the 30 teams in the 2002 season, in ascending order, were

5 6 7 7 7 7 7 7 7 8 8 8 8 8 8
8 9 9 9 9 9 9 10 10 10 11 12 12
15 20.

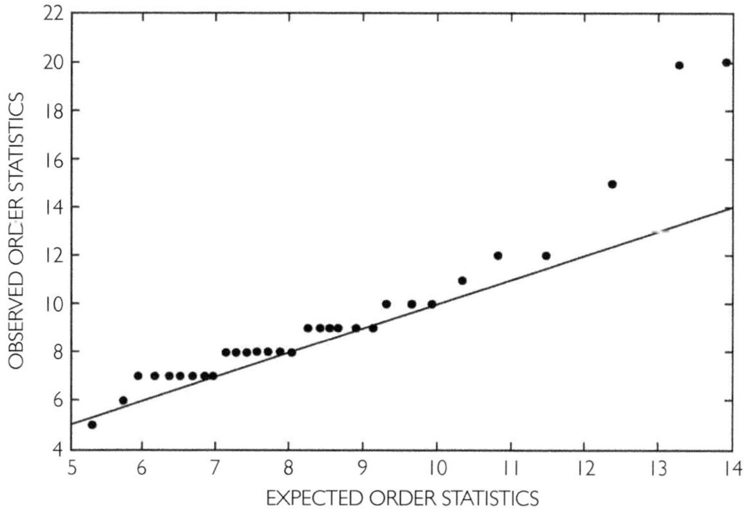

Figure 4.8 Scatterplot of expected ordered streak lengths against actual 2002 streak lengths.

To see if these lengths are unusual, we compare them to the averages of the ordered lengths (called the expected order statistics) in our simulation. Figure 4.8 constructs a scatterplot of the average ordered lengths against the 2002 ordered lengths and places the line $y = x$ on the plot for comparison. Generally, the two 2002 streaks that stand out in this figure correspond to the two points above the line. These points correspond to Oakland's winning streak of 20 games and Tampa Bay's losing streak of 15 games.

Let's focus on Oakland's 20-game win streak. Figure 4.8 indicates that this length is larger than one would expect by chance using our probability model. To better judge the extreme

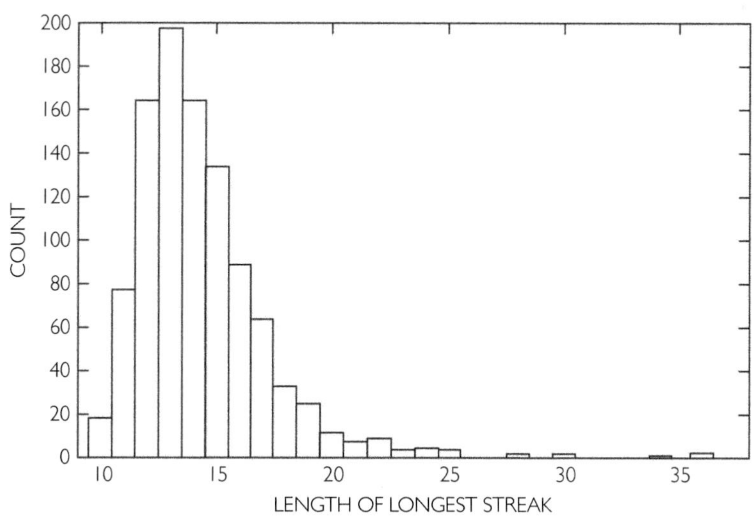

Figure 4.9 Histogram of length of the longest streak for the 100 simulated seasons.

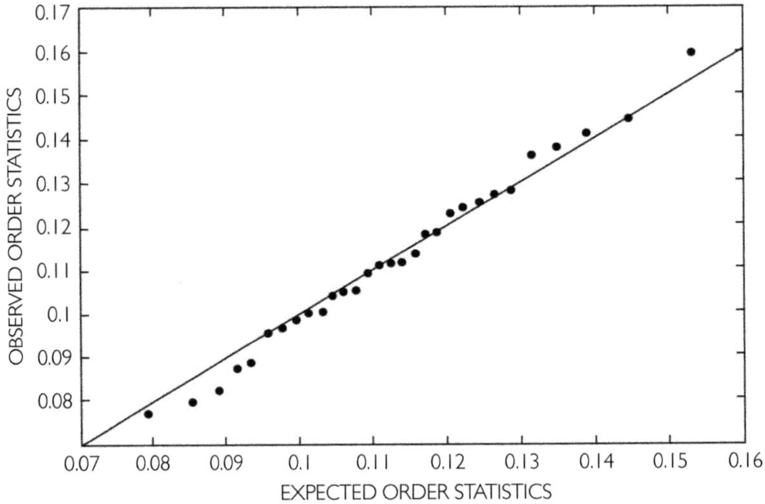

Figure 4.10 Scatterplot of expected ordered BLACK values against actual 2002 BLACK values.

nature of this streak, we repeat the simulation for 1000 seasons and collect the length of the longest streak for each season. Figure 4.9 displays a histogram of these lengths. We see that the longest streak for a season is typically in the 12–16 game range, but there were a number of seasons where we observed streaks of length 20 or more—the proportion of simulated seasons with a streak length of 20 or more was 0.039.

Figure 4.10 compares the ordered values of the BLACK statistic for the 2002 season with the expected values of the ordered values of BLACK in the 100-year simulation. It is interesting to note how close the points fall on the line. This means the consistency behavior of the 2002 teams is close to what one might expect by chance. Figure 4.11 displays a histogram of the

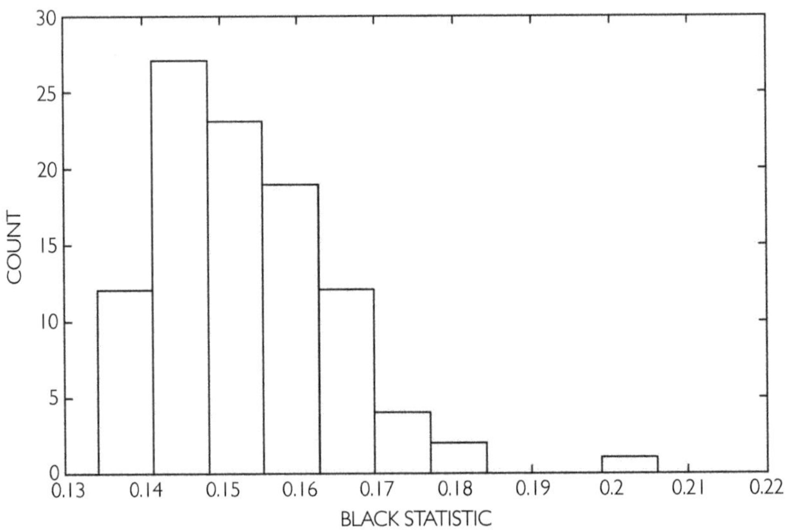

Figure 4.11 Histogram of largest BLACK values for the 100 simulated seasons.

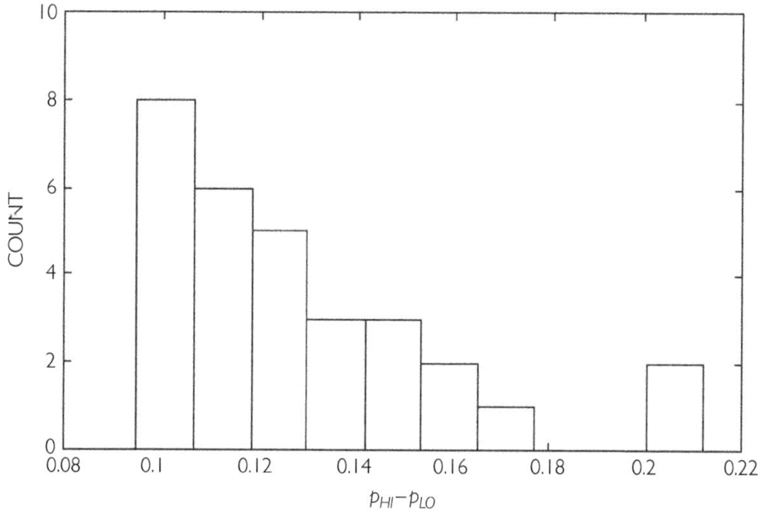

Figure 4.12 Histogram of the posterior means of the streakiness parameter $p_{HI}-p_{LO}$ for the 30 teams in the 2002 season.

largest value of BLACK in the 100 seasons. The largest BLACK value in 2002 was Oakland's value of 0.1595—in contrast to the long streak of 20, we can see from Figure 4.12 that a BLACK value of 0.1595 is not that extreme a value.

Measuring true streakiness for the 2002 teams

In the analysis described above, we focused on Oakland's long streak of wins and investigated how extreme this streaky statistic would be if win/loss data were simulated from a "nonstreaky" random effects model for team competition. Here we look at Oakland's streaky behavior from an inferential perspective. Given Oakland's pattern of wins and losses for the 2002 season, can we estimate a parameter that measures true streakiness?

One convenient model for representing true streakiness was proposed by Albert (1993) and Albert and Bennett (2003). Suppose a team has two states during the season, called "hot" and "cold" states, where the team wins games with probabilities p_{LO} and p_{HI}, respectively, where $p_{LO} < p_{HI}$. Moreover, during games of the season, the team will move between the two states according to a Markov Chain with transition matrix P, where the probability of staying in a hot (cold) state given that the team is currently in a hot (cold) state is ρ_{STAY}. (The probability that a team will change states, either hot to cold, or cold to hot is 1-ρ_{STAY}.) The model is parameterized by the hot and cold probabilities p_{LO}, p_{HI}, and the staying probability ρ_{STAY}. True streakiness can be measured by the difference in hot and cold probabilities $p_{HI}-p_{LO}$. Due to the high correlation between the transition probabilities and (p_{LO}, p_{HI}) in the model fitting, it is convenient to fix ρ_{STAY} at a large value (here $\rho_{STAY} = 0.9$) which reflects our belief that if a team is in a particular state, it is likely to remain in that same state in the next game.

The introduction of latent data and Gibbs sampling can be used to fit this model to the sequence of wins and losses for a particular team. We summarize the fit by computing the posterior mean of the win minus cold probabilities $p_{HI}-p_{LO}$. One nice feature of this fit is that one is using this streaky model to detect streakiness in the team's overall season pattern of wins and losses instead of limiting our attention on a single long winning streak.

This model was fit to each of the win/loss sequences for the 30 teams in the 2002 baseball season. Figure 4.12 displays a histogram of the posterior means of the streakiness parameters $\{p_{HI}-p_{LO}\}$ for the 30 teams. Note that most of the estimates are in the 0.10 to 0.16 range, and two teams, Toronto and Oakland, had streaky estimates that are set apart from the rest. It is interesting that the streakiest team in the 2002 season, using this measure, was not Oakland but Toronto.

This model is useful in measuring the size of true streakiness inherent in a team's performance. But this analysis begs the question—should we be particularly surprised by estimated streaky parameters over 0.2? To answer this question, we simulate data from the random effects model described earlier. For each of 20 simulated baseball seasons,

- we simulated game results for 30 teams using the random effects model with $\tau = 0.25$
- for each team's win/loss sequence, we fit the above streaky model
- we collected the estimates of the streaky parameters $\{P_{HI}-p_{LO}\}$ for the 30 teams

For each season, we order the streaky parameter estimates—call them $\theta_1, \ldots, \theta_{30}$. Figure 4.13 displays boxplots of the ordered estimates—the values of the streaky estimates from the 2002 season team data are drawn on top of the boxplots by a connected line. Note that the 2002 estimates seem pretty consistent with the streaky estimates from the simulated data. To summarize, this streaky model is useful for detecting true streakiness in Oakland's (and Toronto's) performances, but one would observe similar streaky estimates from simulated data based on a coin-tossing model with varying team abilities.

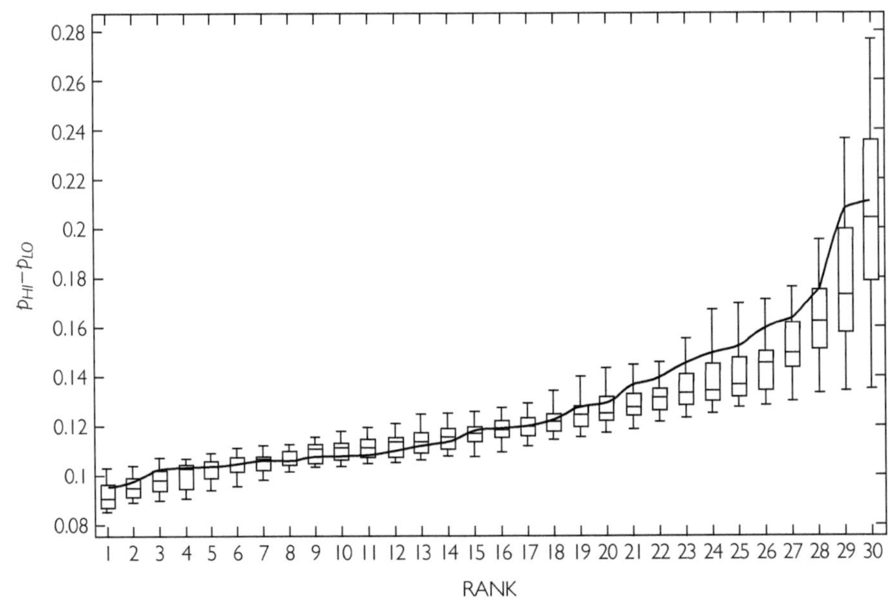

Figure 4.13 Boxplots of ordered streaky estimates from 20 seasons of simulated data assuming random effects model. The ordered streaky estimates from the 2002 team data are shown with the solid line.

Conclusion

Most baseball fans were surprised with Oakland's 20-game winning streak and it is natural to attempt to compute the likelihood of this particular event. Rob Neyer, in his espn.com column, reports that a Topps baseball card described the streak as "truly one in a million" and also reports on the computed odds of this event among 13 readers of his column. But we have indicated a number of problems inherent in these types of probability calculations. Generally one focuses on the significance of a single extreme result among many seasons of baseball. Since we are ignoring the fact that this extreme result was taken from many seasons of baseball, the significance of this streak should be smaller than implied by the reported p value of 0.0068. In this article, we have tried to better understand the significance of this streak by placing it in the context of winning and losing streaks for 30 teams in many years of baseball competition. Using a relatively simple, but realistic model for team competition earlier, we saw that long streaks of wins and losses are relatively common, and in fact 3.9% of the seasons in our simulation actually had a streak of length 20 or longer. Oakland did have a long winning streak in the 2002 season, but we should expected to see strengths of similar lengths about 4% of the time, or about once every 25 years of baseball.

REFERENCES

Albert, J. 1993. Comment on: A statistical analysis of hitting streaks in baseball. *Journal of the American Statistical Association*, 88: 1184–1188.

Albert, J. and Bennett, J. *Curve Ball*, New York: Copernicus Books, 2003.

Albert, J., and Williamson, P. 2001. Using model/data simulations to detect streakiness. *The American Statistician*, 55(1): 41–50.

Albright, C. 1993. A statistical analysis of hitting streaks in baseball. *Journal of the American Statistical Association*, 88: 1175–1183.

Gilovich, T., Vallone, R. and Tversky, A. 1985. The hot hand in basketball: on the misperception of random sequences. *Cognitive Psychology*, 17: 295–314.

Stern, H. S. 1997. Judging who's hot and who's not. *Chance*, 10: 40–43.

CHAPTER 5

Statistical practice

Do baseball players regress toward the mean?

TEDDY SCHALL AND GARY SMITH

Abstract
Baseball performances are an imperfect measure of baseball abilities, and consequently exaggerate differences in abilities. Predictions of relative batting averages and earned run averages can be improved substantially by using correlation coefficients estimated from earlier seasons to shrink performances toward the mean.

Keywords: *least squares, predictions, Sports*

Introduction

Baseball players who win coveted awards and sign breathtaking contracts often disappoint fans, managers, and owners. These disappointments might be explained by regression toward the mean, which occurs when real phenomena are measured imperfectly, causing extreme measurements to exaggerate differences among the underlying phenomena. The degree of exaggeration depends on the correlation between the measurements and the real phenomena. In baseball, the correlation between performance and skill is far from perfect and, as a consequence, observed performance differences substantially overstate skill differences. Players who do exceptionally well in any particular season typically do not do as well the subsequent season—they regress toward the mean. We can improve our forecasts by adjusting our predictions accordingly.

Regression toward the mean

Galton (1886) observed regression toward the mean in his seminal study of the relationship between the heights of parents and their adult children. Because heights are affected by diet, exercise, and other environmental factors, observed heights are an imperfect measure of the genetic influences that we inherit from our parents and pass on to our children. A person who is 78 inches tall may have been pulled above a somewhat shorter genetically predicted height by positive environmental influences or may have been pulled below a somewhat taller genetic height by negative environmental factors. The former is more likely because there are many more people with genetically predicted heights below 78 inches than with genetic heights above 78 inches. Thus, the observed heights of unusually tall parents usually overstate the genetic heights that they pass on to their children.

Regression toward the mean can be seen in sports, where observed performance is an imperfect measure of skill. Of the last 20 major league baseball champions from 1979 through 1998, only two repeated the following year. Of those major league baseball teams that win more than 100 games in a season, 90% do not do as well the next season (James 1981). Outcomes depend on luck as well as skill, and those teams that do unusually well are more likely to have experienced good luck than bad. Few teams are so far superior to their opponents that they can win a championship in an off year. Thus, the performance of most champions exaggerates their skill and, because good luck cannot be counted on repeatedly, most champions regress to the mean.

The same is true of individual players. In 1989, *Sports Illustrated* reported that of those baseball players who hit more than 20 home runs in the first half of the season, 90% hit fewer than 20 during the second half. The writer concluded that there was a "second-half power outage" (Gammons 1989, p. 68). The regression-toward-the-mean explanation is that their skills did not deteriorate, but rather that their unusually good performances during the first half exaggerated their skills. Regression toward the mean can also explain such clichés as the Cy Young jinx, sophomore slump, rookie-of-the-year jinx, and *Sports Illustrated* cover jinx. Baseball players have good and bad years, and it would be extraordinary for a player to be the best in the league while having a bad year by his own standards. Most players who do much better than their peers are also performing better than their own career averages.

Tables 5.1 and 5.2 show the players who had the ten best batting averages (BAs) and earned-run averages (ERAs) in 1998 among those players who had at least 50 at bats or 25 innings pitched in 1997, 1998, and 1999. The mean batting average was approximately 30 points lower in the adjacent seasons than in the top-10 season; the mean earned run average was roughly 1 run higher. Thirteen of these 20 players had worse records in both 1997 and 1999 than in 1998.

More generally, Figures 5.1 and 5.2 compare the 1998 and 1999 BAs and ERAs of all major league players who had at least 50 at bats or 25 innings pitched in each of these years. [These figures also show the locally weighted scatterplot smoothing (LOWESS) lines with a bandwidth of 0.5. Ramsey regression specification error tests for second-, third-, or fourth-order terms have p values of 0.305 for batting averages and 0.172 for earned run averages.]

These graphs have two striking characteristics. First, although the relationships are highly statistically significant, the correlations are modest. For the 381 batters, the two-sided p value is 95.3×10^{-13} and the correlation coefficient is 0.36; for the 300 pitchers, the two-sided p value is 0.00000004 and the correlation coefficient is 0.31. Although the 1998 performances are statistically helpful in predicting 1999 performance, the predictions are far from perfect.

Second, the slopes of the least-squares lines are less than 1, indicating that performance regresses to the mean. Because the least-squares line goes through the average values of both

Table 5.1 ***How the ten players with the highest BAs in 1998 did in 1997 and 1999.***

	1997	1998	1999
Larry Walker	0.366	0.363	0.379
John Olerud	0.294	0.354	0.298
Bernie Williams	0.328	0.339	0.342
Mo Vaughn	0.315	0.337	0.281
Eddie Perez	0.215	0.336	0.249
Dante Bichette	0.308	0.331	0.298
Albert Belle	0.274	0.328	0.297
Mike Piazza	0.362	0.328	0.303
Eric Davis	0.304	0.327	0.257
Jason Kendall	0.294	0.327	0.332
Average	0.306	0.337	0.304

Table 5.2 ***How the ten players with the lowest ERAs in 1998 did in 1997 and 1999.***

	1997	1998	1999
Ugueth Urbina	3.78	1.30	3.69
Trevor Hoffman	2.66	1.48	2.14
Robb Nen	3.89	1.52	3.98
Mike Jackson	3.24	1.55	4.06
Graeme Lloyd	3.31	1.67	3.63
Mariano Rivera	1.88	1.91	1.83
John Wetteland	1.94	2.03	3.68
Jeff Shaw	2.38	2.12	2.78
Greg Maddux	2.21	2.22	3.57
Kevin Brown	2.69	2.38	3.00
Average	2.80	1.82	3.24

variables, the 0.378 slope for batters means that a player whose BA was 0.050 above (or below) the 1998 mean is predicted to have a 1999 batting average that is only 0.019 from the 1999 mean. The 0.340 slope for pitchers implies that a player whose ERA was 1.000 from the 1998 mean ERA is predicted to have a 1999 ERA that is only 0.340 from the 1998 mean.

For another suggestive indicator of regression toward the mean, we looked at all major league players since 1901 who had at least 50 at bats or 25 innings pitched in two consecutive seasons. Of 4,026 players who had BAs of 0.300 or higher in any season, 3,210 (79.7%) did worse the following season. Of 3,849 players who had ERAs of 3.00 or lower in any season, 3,077 (79.9%) did worse the following season. Clearly, baseball players regress toward the mean. To formalize these observations, we use the following model.

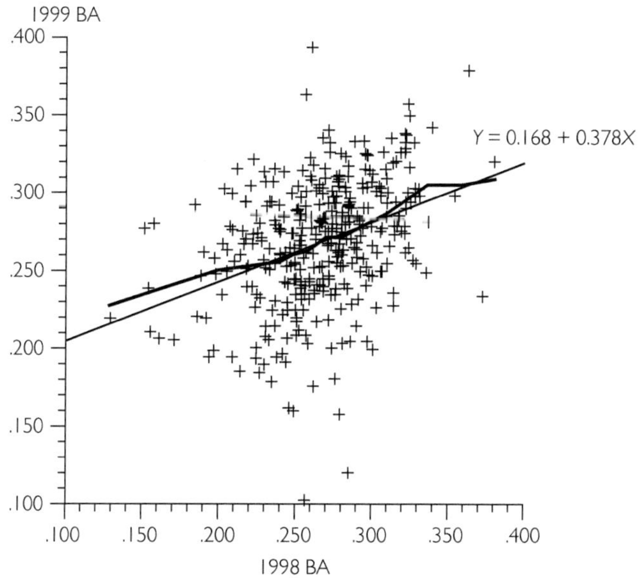

Figure 5.1 1998 and 1999 batting averages.

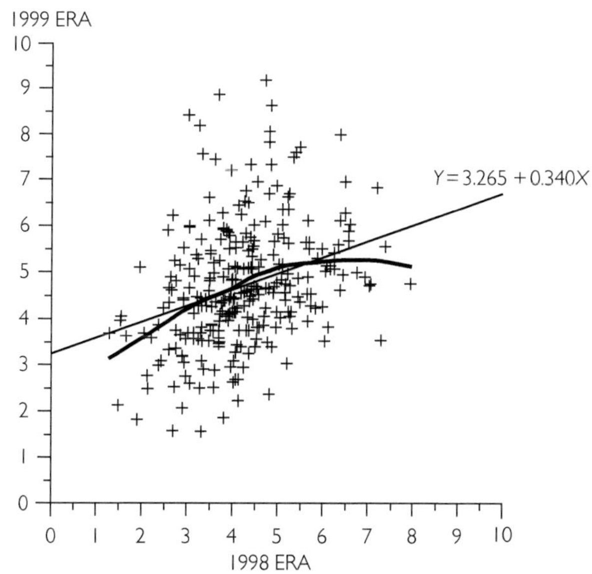

Figure 5.2 1998 and 1999 earned run averages.

A model

Let Y be a statistical measure of a player's performance (batting average for batters, earned run average for pitchers). To compare players from different seasons, we standardize performance by taking the difference between the player's performance in any given season and the mean performance for all players that season, and dividing this difference by the standard deviation of performance across players that season. Thus, Ted Williams' 0.406 batting average in 1941 was a standardized batting average of $Z = 3.14$; that is, 3.14 standard deviations above the mean that year. Sandy Koufax's 1.73 earned run average in 1966 was standardized to be $Z = -1.93$, or 1.93 standard deviations lower than the mean ERA that year.

A player's performance Z in any year is related to an expected value μ, which we can think of as this player's true ability (what his average performance would be over many similar seasons). The player's actual performance in any particular season differs from his ability by a random term ε that we assume has an expected value of zero and is independent of ability and of the value of the random term in other seasons:

$$Z = \mu + \varepsilon. \tag{5.1}$$

There is a distribution of abilities across players. If ε is independent of μ, then the variance of Z is equal to the variance of μ plus the variance of ε, and is therefore larger than the variance of μ: $\text{var}[Z] = \text{var}[\mu] + \text{var}[\varepsilon]$. Thus, the variation of actual batting averages (or earned run averages) across players in a season is larger than the variation of abilities. An extreme performance typically overstates how far this player's ability is from the mean ability.

If we knew the value of a player's ability μ, we could use μ to make an unbiased prediction of his performance in a season. However, players, fans, managers, and owners are interested in the reverse question—using performance to predict ability—and unbiased predictions require an adjustment for regression toward the mean.

Imagine that we had data on each person's ability μ and his performance Z in a season. The least-squares equation for predicting performance from ability is

$$\hat{Z} = r\mu, \tag{5.2}$$

where r is the correlation between Z and μ. The least-squares equation for predicting ability from performance is

$$\hat{\mu} = rZ. \tag{5.3}$$

Unless the correlation coefficient is 1 or -1, these two equations are not simply the reverse of each other.

Because they cannot use the unknown value of μ to predict performance Z in any given season, fans, managers, and owners might use the value $Z(-1)$ in the preceding season:

$$Z = \alpha + \beta Z(-1) + \gamma.$$

Using Equation (5.1) and assuming abilities are constant in adjacent seasons, the true relationship is

$$Z = Z(-1) + (\varepsilon - \varepsilon(-1)).$$

If $\text{var}[\varepsilon(-1)] = \text{var}[\varepsilon]$, then the correlation coefficient between Z and $Z(-1)$ is a simple function of the variation in abilities and the variation in performance about ability:

$$\rho = \frac{\text{var}[\mu]}{\text{var}[\mu] + \text{var}[\varepsilon]}.$$

The greater the variation in performance about abilities, the smaller is the correlation between adjacent-season performances and the more we should shrink predicted performances toward the mean.

Data

We looked at season batting averages and earned run averages for all major league baseball batters and pitchers from 1901 through 1999 (Thorn, Gershman, Palmer, and Pietrusza 1999). To reduce the influence of outliers, we excluded players who had fewer than 50 official times at bat or 25 innings pitched. This gave us a database of 29,310 seasons for 5,262 batters and 20,385 seasons for 4,212 pitchers. To compare players from different years, each player's performance each year was standardized by subtracting the mean for all players that year and dividing this difference by the standard deviation of performance across players that year. In addition, in a competitive sport like baseball, relative performance is what matters—not whether a player bats 0.290, but how many standard deviations the player is above or below the mean that year.

Performance correlations

We calculated correlation coefficients in two ways. For the most-recent-season calculation, we compared a player's performance in a given year with his performance in his most recent preceding season; for the more restrictive adjacent-season comparison, we only considered cases where the most recent season immediately preceded the current season.

For batting averages, a total of 24,047 most-recent-season observations have a correlation coefficient of 0.37 and 22,243 adjacent-season observations have a correlation coefficient of 0.38. For earned run averages, a total of 16,172 most-recent-season observations have a correlation coefficient of 0.22 and 14,714 adjacent-season observations have a correlation coefficient of 0.24. In all cases, these correlations are highly statistically significant, decisively demonstrating that performance is not uncorrelated across seasons. However, the correlations are far from perfect.

These results are not an artifact of a few outliers who did extraordinarily well or poorly one season and more nearly average the next season. Excluding batters who were more than two standard deviations from the mean in the most recent or adjacent season, the correlation coefficients drop slightly, to 0.34 and 0.35. Excluding pitchers who were more than two standard deviations from the mean, the respective correlation coefficients remain 0.22 and 0.24.

Looking at individual seasons, the BA correlation coefficients for adjacent seasons ranged from 0.18 in 1991 and 0.25 in 1940 to 0.63 in 1906 and 0.59 in 1931; the average was 0.39. Since World War II, the highest correlation coefficient was 0.46 in 1957. The ERA correlation coefficients for adjacent seasons ranged from 0.00 in 1913 and 1948 to 0.55 in 1939 and 0.48 in 1932; the average was 0.25. Since World War II, the lowest correlation coefficient was 0.05 in 1981 and the highest was 0.38 in 1977.

Predicting relative performance

Our model suggests that instead of using this year's relative performance to predict next season's relative performance, more accurate predictions can be made by shrinking each player's performance toward the mean; that is, predicting a player's $Z(+1)$ from rZ rather than Z. To see whether this is so, we considered batters and pitchers who played in two adjacent seasons, say 1998 and 1999. The unadjusted prediction of each player's 1999 Z value is simply his 1998 Z value. For the adjusted predictions, we used the data for all persons who played in both 1997 and 1998 to estimate the correlation coefficient between adjacent-season performance; we then predicted the 1999 Z value of each person who played in 1998 and 1999 by multiplying his 1998 Z value by this correlation coefficient.

We did this for every season and measured the overall accuracy of the predictions by the root mean squared error (RMSE) each year and for the entire time period. The first rows of Tables 5.3 and 5.4 show these results. For both batters and pitchers, the RMSE was improved in every one of the 97 years by shrinking the performances toward the mean. For the entire time period, the RMSE was reduced 16% for batters and 18% for pitchers.

For those who have played more than one year, more accurate predictions might be expected from an averaging of these previous seasons' performances. Our model suggests that these averages, too, should be shrunk towards the mean. To investigate this question, we considered batters and pitchers who had played in $2n$ adjacent seasons, using the average Z value for the

Table 5.3 ***Predicting standardized batting average from earlier Z and rZ.***

Years ahead	Number of predictions	Average correlation	Years with Lower RMSE		Cumulative RMSE	
			Z	rZ	Z	rZ
1	22,107	0.392	0	97	1.048	0.884
2	14,422	0.553	0	94	0.723	0.644
3	9,546	0.633	5	86	0.607	0.549
4	5,883	0.678	4	79	0.546	0.493
5	3,223	0.715	7	59	0.518	0.464
6	1,069	0.675	3	22	0.511	0.449
7	158	0.650	0	8	0.501	0.410

Table 5.4 ***Predicting standardized earned run average from earlier Z and rZ.***

Years ahead	Number of predictions	Average correlation	Years with Lower RMSE		Cumulative RMSE	
			Z	rZ	Z	rZ
1	14,649	0.245	0	97	1.129	0.923
2	8,806	0.290	2	92	0.803	0.677
3	5,190	0.319	11	73	0.667	0.577
4	2,508	0.294	6	44	0.610	0.537
5	1,088	0.223	11	17	0.581	0.548
6	182	0.374	1	7	0.531	0.483

n earlier years to predict the average Z value for the n subsequent years. For example, with $n = 2$, we looked at persons who had played in 1995, 1996, 1997, and 1998. The unadjusted prediction of each player's average Z value for 1997 and 1998 is his average Z value for 1995 and 1996. For the adjusted predictions, we calculated the correlation coefficient between average 1993–1994 Z values and average 1995–1996 Z values for all persons who played these four seasons. We then predicted each player's 1997–1998 Z by multiplying his 1995–1996 Z by this correlation coefficient. We only made predictions when data for at least 25 players were available for estimating the correlation coefficient. There were consequently no BA predictions for horizons longer than seven years and no ERA predictions for horizons longer than six years.

Tables 5.3 and 5.4 show the results. As the horizon lengthened and the Z values were averaged over more seasons, there is typically an increase in the correlation coefficient and a decline in the RMSE. For every horizon, the predictions were substantially improved by shrinking each player's relative performance toward the mean.

These results are robust with respect to the minimum number of at bats and innings pitched. Tables 5.5 and 5.6 show the predictive accuracy for next-season forecasts with the minimum number of at bats ranging from 50 to 400 and the minimum number of innings pitched ranging from 25 to 200. These increases in the minimum numbers raise the average correlation coefficient, but do not alter our conclusion that shrinkage gives more accurate predictions of relative performance.

At the suggestion of a referee, we also looked at this performance measure for batters: on-base average plus slugging average, where a player's on-base average is the sum of his hits, bases

Table 5.5 ***Predicting year-ahead standardized batting average.***

Times at bat	Number of predictions	Average correlation	Years with lower RMSE		Cumulative RMSE	
			Z	rZ	Z	rZ
50	22,107	0.392	0	97	1.048	0.884
100	19,025	0.422	0	97	1.048	0.895
200	14,594	0.460	0	97	1.033	0.893
400	7,250	0.503	2	93	1.007	0.881

Table 5.6 ***Predicting year-ahead standardized earned run average.***

Innings pitched	Number of predictions	Average correlation	Years with lower RMSE		Cumulative RMSE	
			Z	rZ	Z	rZ
25	14,649	0.245	0	97	1.129	0.923
50	12,159	0.269	0	97	1.139	0.945
100	7,540	0.297	0	97	1.134	0.937
200	1,798	0.317	4	55	1.158	0.968

Table 5.7 ***Predicting standardized on base average plus slugging average.***

Years ahead	Number of predictions	Average correlation	Years with lower RMSE		Cumulative RMSE	
			Z	*rZ*	*Z*	*rZ*
1	22,107	0.555	0	97	0.926	0.822
2	14,422	0.696	4	90	0.667	0.617
3	9,546	0.745	10	81	0.586	0.547
4	5,883	0.773	5	78	0.548	0.511
5	3,223	0.790	8	58	0.541	0.501
6	1,069	0.746	1	24	0.547	0.495
7	158	0.659	2	6	0.515	0.453

on balls, and times hit by pitches divided by the sum of his times at bat, bases on balls, times hit by pitches, and sacrifice flies; a player's slugging average is his total bases (one base for a single, two for a double, three for a triple, and four for a home run) divided by his times at bat. Table 5.7 shows that the correlations between performances in different years are somewhat higher, but that the results are otherwise very similar to those for batting averages.

Conclusion

Because baseball performances are an imperfect measure of underlying abilities, batting averages and earned run averages regress toward the mean. Outstanding performances exaggerate player skills and are typically followed by more mediocre performances. The average correlation coefficient for adjacent-season performance is 0.39 for batting averages and 0.25 for earned run averages. Predictions of standardized batting averages and earned run averages can be improved consistently and substantially by using correlation coefficients estimated from earlier seasons to shrink performances toward the mean.

REFERENCES

Galton, F. (1886), "Regression towards mediocrity in hereditary stature," *Journal of the Anthropological Institute*, 15, 246–263.

Gammons, P. (1989), "Inside baseball," *Sports Illustrated*, 103, 68.

James, B. (1981), "Esquire's 1981 baseball forecast," *Esquire*, 95, 106–113.

Thorn, J., Gershman, M., Palmer, P., and Pietrusza, D. (eds.) (1999), *Total Baseball* (6th ed.), New York: Total Sports; the latest data are available at www.baseball.com.

PART II

Olympics/Track & Field

Introduction

GERARD SIERKSMA

With decreasing differences in performance of competing top athletes, sometimes even within the error margins of the measuring systems, comparing performances, and deciding on winners is becoming a perilous affair. Yet these tasks are of utmost importance: top sport is a multi-billion-dollar industry, and wrong decisions of referees may have disproportionate consequences. Was the award of an Olympic gold medal to Michael Phelps in 2008 indeed correct? With the increasing number of top soccer matches ending in not more than one goal, the winners of these matches are frequently determined by referee mistakes.

The problem of "fair" performance comparisons becomes even more relevant when performances across different eras are compared. One may, for example ask: Who is the best ice hockey player of all time? Needless to say, for the latter question many performance-enhancing aspects need to be taken into account. In this respect, technological innovations and growing pools of top athletes seem to be the most relevant ones.

What aspects are responsible for the increasing performance densities at the top? What other trends can be observed?

In the case of Olympic speed skating, the results are nowadays given in 1/1000 of a second, while the error margin at the finish is widely agreed to exceed 3/1000 of a second. This silly rule caused the Dutch 1500-m specialist, Koen Verwey, to lose a gold medal at the 2014 Olympics in Sotchi: he was "beaten" by 3/1000 of a second according to the official measuring system, while Verwey and the gold winner were skating in different pairs (so there was no common photo finish). This picture shows two bike racers with the top one being declared the winner. But, what is the error margin in drawing the finish line? On the other hand, the two downhill skiers, Tina Maze from Slovenia and Dominique Gisin from Switzerland, both received the gold medal with a difference of 0.05 seconds!

Moreover, one may expect that selection procedures have become more and more precarious, certainly when there is a limit on the number of participants and the pool of competitive athletes is large. Basing selections on past performances, which are usually achieved under diverging circumstances, provides additional complications; statistical corrections for these circumstances also influence the error margins.

While the level of competition has increased over the years, for many sports, (i) the variation in performance, especially at the top, has steadily decreased, and (ii) extreme events are becoming rarer. This phenomenon is called the Gould effect (Talsma, 2013). The usual explanation is that, due to limitations of the human body, there is an absolute limit to sports performances, which is approached by the performances of the very best athletes. In Gould (1996) the author shows how the set of all performances in one year in a given sport can be represented as a statistical distribution in which the best performances lie at the right tail of this distribution. The distribution is bounded on the right-hand side. While the performance level of athletes increases, more and more athletes come closer to this boundary, and the difference between top and average athletes narrows down. Gould worked with baseball data, but similar results are found in basketball, and in golf, although in basketball a number of rule changes have kept the variability at a higher level than observed in baseball. Gould's paper discusses the disappearance of 0.400 baseball hitters, that is, of baseball players that are able to hit an average of over 40% of the balls during one season. Such high hitting averages were commonplace in the 1920s and 1930s of the last century, whereas such averages have completely disappeared since 1941.

For a long time, this phenomenon was attributed to the fact that the performance level of elite baseball players has gone down. However, Gould realized that in fact the opposite holds true. He claimed that the level has gone up, and that the disappearance of 0.400 hitters is due to the fact that the performance of the pitchers and of the field players has improved along with that of the batters. The hitting averages, averaged over all players, have been more or less constant, but because the average performances of both hitters and pitchers have increased over time, it has become harder and harder for the top hitters to achieve such high hitting averages and to become so much better than their opponents. It is precisely this phenomenon of growing small performance differences that brings us to the question of fairness in the judgment of elite sport performance.

The first of the four papers of this section, written by Scott M. Berry (2002A), is "One Modern Man or 15 Tarzans?" Berry presents a possible explanation for the ongoing performance improvements of elite athletes. The paper concentrates on seven Olympic events between 1900 and 2000, namely the 100 meters running, the marathon, the 100 and the 1500 meters freestyle swimming, the 500 and the 5000 meters speed skating, and the shot-put. For each of these seven disciplines, the gold medal performances were compared with the world population in the corresponding years. The purpose of the paper is to estimate the effect of the world population growth on the respective Olympic performances. To be precise, the expected Olympic winning performances, based on a population growth model, are related to the corresponding actual winning performances: Is a larger talent pool a plausible explanation of the performance progress? Note that both the winning results and the population values are time-dependent. The results of the paper show that the six Olympic winning times are predicted very well: population grows explains 80 to 95% of the variability in these performances. In the case of shot-put, there are clear deviations between the predicted and the actual results. Berry suggests that the lack of fit is caused by the "steroids effect followed by a testing-for-steroids effect." It may be interesting to collect and use the data after the year 2000. One may expect that the explanation power of the population growth model decreases, because the world population is still growing exponentially, while the rate of growth of the winning performance values may decrease, due to, among others, the Gould effect.

The second paper, by Trevor F. Cox and Ryan T. Dunn (2002), is "An Analysis of Decathlon Data." This paper deals with fairness of the order of the ten events of the decathlon. For example, an athlete who excels at the 100-meter race should not be more likely to win the decathlon than an athlete who excels at the discus. The tentative conclusion of the analyses of this paper is that the current organization of the decathlon favors those athletes who perform relatively well at the field events. The analysis uses data collected at the five IAAF World Athletics Championships between 1991 and 1999. The authors first carried out several cluster analyses, leading to several interesting observations. For instance, "discus" and "shot-put" are in one cluster for all linkage cluster methods used by the authors. Using classical scaling techniques, it turned out that the "javelin" is quite different from the other events. According to the authors, the order in which the events take place during the competition should be considered fair, if the population standard deviations of the various events are equal. Interesting is the use of CUSUM, which is used to test whether athletes who finish high in the rankings have gained advantage from a particular event. It turns out that no one event tends to benefit the more successful athletes.

The third paper, written by Samuel S. Wu and Mark C.K. Yang (2004), is titled "Evaluation of the Current Decision Rule in Figure Skating and Possible Improvements." This paper was inspired by the pair-skating controversy during the Salt Lake City Winter Games in 2002. The original silver medal was changed into a gold one, because one of judges was found guilty of vote fixing. The authors present a rule (that converts all jury scores of a skater into a final ranking of that skater) "... that is 'fair' and 'robust' against manipulation in most cases." Their new rule, called the NR, is based on, among others, the number of wins in pairwise comparisons of jury ranks, the median of the jury ranks, the number of ranks larger than the median, and a test that removes outliers produced by manipulators based on rank sums. Unlike many existing rules, the NR satisfies the obvious property that if one individual judge raises a skater's score, then this change does not decrease this skater's final ranking position. According to the authors, one additional strong point of the NR is the fact that it produces fewer ties than the other rules considered. Interesting is of course the author's application of the NR to the 2002 controversy. Unfortunately, the new rule does not change the initial results, due to the fact that the top two pairs dominated ranks one and two. But, in the light of the Gould effect, should we not expect more and more (almost) ties and dominating pairs? This, of course, raises an important open problem: How to keep the judgment fair and interesting for the public?

The last paper, also written by Scott M. Berry (2002B), is "Gold Medals, Protests, and Trading Ratings: The 2002 Winter Olympics Figure Skating." Remarkable about figure skating is the fact that it is the only sport where the rankings are determined entirely by a jury. No wonder that figure skating has a long history of judging controversies, especially when the jury scores are close. In 2002, two controversies shocked the public. The first one was in the pairs event, where the second ranked Canadian pair, after a loud booing from the supporting Canadian crowd, was awarded a gold medal, together with the Russian pair. The other controversy concerned the underdog American figure skater in the women's single event. After the "short program," she jumped from the fourth place to the Gold Medal. In this paper, Berry discusses in this respect the "homer effect," where judges favor their own compatriots, and concludes that there is strong evidence for homer effect in at least the case of the pairs figure skating.

REFERENCES

Berry, S.M. (2002A). One Modern Man or 15 Tarzans? *Chance*, 15(2), 49–53.

Berry, S.M. (2002B). Gold Medals, Protests, and Trading Ratings: The 2002 Winter Olympics Figure Skating, *Chance*, 15: 2, 14–18.

Cox, T.F. and Dunn, R.T. (2002). An Analysis of Decathlon Data. *Journal of the Royal Statistical Society. Series D (The Statistician)*, 51(2), 179–187.

Gould, S.J. (1996). *Full House: The Spread of Excellence from Plato to Darwin*, Three Rivers Press, New York.

Talsma, B.G. (2013). *Performance Analysis in Elite Sports*. Thesis, University of Groningen.

Wu, S.S. and Yang, M.C.K. (2004). Evaluation of the Current Decision Rule in Figure Skating and Possible Improvements. *The American Statistician*, 58(1), 46–54.

CHAPTER 6

Gold medals, protests, and trading ratings the 2002 Winter Olympics Figure Skating

Does the data show evidence of wrongdoing in figure skating judging at the 2002 Winter Olympics?

SCOTT M. BERRY

Introduction

Figure skating is probably the most popular Winter Olympics event. The names of Peggy Fleming, Dorothy Hamill, and Scott Hamilton still evoke thoughts of wonderful all-American champions. Perhaps the popularity of figure skating derives from its blend of athleticism and artistry. The competitors are supreme athletes that spend childhood perfecting their mastery and presentation. Winning in their sport is not simply jumping highest, skating fastest, or directly scoring against an opponent. The winners of figure-skating events are determined entirely by pleasing judges. Other sports, such as gymnastics, ski jumping, and snow boarding are determined, at least in part, by judges, yet there is much less controversy surrounding the judging of those events. Figure skating has a history of judging controversies and a general mistrust of the judging methodology. In the 2002 Winter Olympics the judging of figure skating took center stage when a judge admitted striking a deal to grade the Canadian skating pair lower. Later, the Russians protested the judging of the ladies individual event, claiming that their skater was unfairly judged.

Bring and Carling discussed some of the faults of figure skating judging in *Chance* in 1994. Since their article the scoring system has been altered, but many of their complaints still hold. First, they discuss what they call the principle of *independence of irrelevant alternatives.* They presented an example in which the results of one pair of skaters can switch the ordering of other pairs. This violation of the principle of independence of irrelevant alternatives occurred again in the 2002 ladies singles skating event. This is not what caused the major controversies in the 2002 Winter Olympics, but it still raises questions about scoring validity. Bring and Carling also mentioned evidence of judges favoring their own "compatriots." There is evidence of this "homer effect" in the 2002 Winter Olympics too. These occurrences cast doubt on the judging within figure skating—turning a beautiful, athletic, and popular sport into a quagmire of controversy.

The Ladies Singles Gold Medal controversy

Judging within figure skating has always been controversial. It is not shocking for someone to point out the faults of the judging system. Indeed many sports dignitaries point out these faults regularly. However, there were two figure skating controversies at the 2002 Winter Olympics that did shock the public. The first was in the pairs event. Originally the Russian pair of Elena Berezhnaya and Anton Sikharulidze were awarded the Gold Medal and Jamie Sale and David Pelletier, from Canada, were awarded the Silver Medal. As the scores went up for the Canadian pair the crowd booed loudly. The American announcers were shocked that the Russians had beaten the Canadians. There was widely held sentiment that the Canadians had skated better. In subsequent investigations the International Skating Union determined that the French judge had rated the Russians higher as part of a deal, and so the Canadian pair was also awarded a Gold Medal. In the ladies singles event, Sarah Hughes, an underdog American skater pulled a shocking upset by jumping from fourth place after the short program to the Gold Medal. The Russians protested that their skater, Irina Slutskaya, should have been awarded the Gold Medal—and at least they should both be given a Gold Medal. The Russians even threatened a boycott of the remaining events because of a number of rulings they felt were unjust.

I address the ladies singles event first. Within that event, understanding the scoring system is crucial to understanding the controversy. There are two skating programs for each competitor: the short program and the free program. Within each program, nine judges grade each competitor. They get grades, called marks, on *technical merit* and *presentation*, on a scale from 1 (poor) to 6 (perfect). For each competitor, and each judge, the marks for technical merit and presentation are added to form a score. Within each judge the competitors are ranked from 1 to n, where n is the number of competitors. A tie for two competitors within a judge is resolved differently for the short and free programs. In the short program they both receive the same rank. For example, if they tied for first they would both receive ranks of 1 and the next competitor would receive a rank of 3. In the free program the marks for presentation are a tie-breaker. If they are still tied then they receive the same rank. These ranks, or in figure skating terminology, *ordinals*, are used to determine the placing of each competitor.

Each competitor's ordinals are compared to every other competitor's ordinals. They receive a *win* for each competitor that they defeat head-to-head by winning on the most judges ordinals. The competitors are then ordered by the number of wins. Their position in the order is their place for that program. Half the place for the short program and the place for the free program are added to determine the final score for each participant. If two competitors' scores are identical the one with the better free program is the winner. If they receive identical free program ranks, then the result is a tie.

What happened in the ladies singles at the 2002 Winter Olympics? Table 6.1 summarizes the relevant scores for the top three competitors. After the short program Kwan was first, Slutskaya second, Cohen third, and Hughes fourth with respective half-place scores of 0.5, 1, 1.5 and 2. The chronology of subsequent events was quite interesting. Hughes was the first to skate in the free program, Kwan was second to last, and Slutskaya was last. Hughes skated the free program very well. Kwan fell once and struggled in her routine, but she still received decent marks. After Kwan's skate, Kwan and Hughes had both beaten every other competitor in head-to-head. In their own head-to-head comparison, Hughes beat Kwan on six of the nine judges and therefore was in first place in the free program.

Slutskaya was last to skate the free program, and depending on how she skated the order of Kwan and Hughes would reverse. If Slutskaya skated worse than both Hughes and Kwan, then Kwan who placed first in the short program and second in the free progam would win the Gold Medal with a final score of $0.5 + 2.0 = 2.5$. Hughes would go home with a final score of $2.0 + 1.0 = 3.0$ and a Silver Medal. If Slutskaya scored between Kwan and Hughes, Hughes would win the Gold Medal in a tie-breaker, and Slutskaya would win Silver, and Kwan the Bronze. If Slutskaya performed very well, beating Hughes and Kwan, then she would win the Gold Medal, Kwan the Silver Medal, and Hughes the Bronze Medal.

For Slutskaya to finish between Kwan and Hughes she would have to beat Kwan on at least five of the judges, but yet lose to Hughes on at least 5 of the judges. Slutskaya skated okay, with a few small mistakes (at least that is my impression from Scott Hamilton's commentary). She beat Kwan on six of the nine judges. Hughes beat Slutskaya on five of the nine judges. The final scores were Hughes $2.0 + 1.0 = 3.0$, Slutskaya $1.0 + 2.0 = 3.0$, and Kwan $0.5 + 3.0 = 3.5$. Hughes beat Slutskaya because the first tie breaker is the free program—which Hughes won.

Table 6.1 ***The marks for each of the three medal winners in the 2002 Winter Olympic ladies singles event. The Ordinals are the rank within all the competitors for each individual program.***

Competitor		Judge's Country								
		Ger	Rus	Slv	Den	Ita	Bul	Fin	Can	USA
Sarah Hughes	Tech	5.2	5.1	5.6	5.3	5.3	5.5	5.2	5.6	5.5
Short	Pres	5.7	5.5	5.7	5.5	5.5	5.7	5.6	5.7	5.7
	Ordinals	6	10	4	5	5	5	5	4	4
	Tech	5.7	5.8	5.8	5.8	5.8	5.8	5.7	5.8	5.8
Free	Pres	5.7	5.7	5.8	5.6	5.8	5.8	5.8	5.8	5.8
	Ordinals	1	4	3	4	1	2	1	1	1
Irina Slutskaya	Tech	5.8	5.8	5.8	5.7	5.6	5.8	5.8	5.8	5.7
Short	Pres	5.7	5.8	5.9	5.8	5.6	5.8	5.9	5.8	5.6
	Ordinals	3	1	2	2	3	1	1	1	3
	Tech	5.7	5.8	5.9	5.8	5.8	5.8	5.8	5.7	5.8
Free	Pres	5.6	5.9	5.9	5.8	5.6	5.9	5.7	5.7	5.7
	Ordinals	3	1	1	1	4	1	2	3	2
Michelle Kwan	Tech	5.8	5.7	5.9	5.7	5.5	5.7	5.7	5.7	5.6
	Pres	5.9	5.9	5.9	5.9	5.9	5.9	5.9	5.9	5.9
Short	Ordinals	1	2	1	1	1	2	2	2	1
	Tech	5.6	5.7	5.8	5.7	5.6	5.6	5.6	5.7	5.6
Free	Pres	5.7	5.8	5.9	5.8	5.8	5.9	5.8	5.8	5.8
	Ordinals	2	3	2	2	2	3	3	2	3

The scoring system is not the point to this article, but I cannot pass up the chance to criticize such a system. Combining ranks are a notoriously bad way to rank athletes. The PGA Tour uses a sum of ranks for their all-around statistic—and it is a terrible measure. Competitors are not fully rewarded for great performances, and a competitor can easily drop many ranks for small differences. How can Kwan be better than Hughes, unless Slutskaya performs between them?

Is there any evidence of bias against Slutskaya, as the Russians claim? Such a claim would be nearly impossible to back-up from the numbers—because they were all clearly close. Interestingly, only two judges, the USA and Finland, had the top three contestants ordered in the free program as they finished. Another interesting note is that Slutskaya and Hughes each received at least one rank of fourth—while Kwan received none. In the analysis of the data I consider only the ordinals each competitor received from each judge—since this is the only relevant information—in statistical terminology, they are sufficient. There were 27 competitors in the short program and 23 in the free program. The ordinals within judge for each of the competitors in the free program are shown in Table 6.2.

Table 6.2 ***The ordinals for each of the judges for the ladies singles free program. The "ALL" column is the overall placing for the free program.***

Skater	ALL	GER	RUS	SVK	DEN	ITA	BUL	FIN	CAN	USA
Hughes (USA)	1	1	4	3	4	1	2	1	1	1
Slutskaya (RUS)	2	3	1	1	1	4	1	2	3	2
Kwan (USA)	3	2	3	2	2	2	3	3	2	3
Cohen (USA)	4	5	2	4	3	3	4	4	4	4
Suguri (JPN)	5	4	8	5	5	5	7	5	5	5
Butyrskaya (RUS)	6	6	5	8	7	12	5	8	7	6
Robinson (CAN)	7	7	7	7	9	6	8	10	6	7
Sebestyen (HUN)	8	8	10	12	8	7	6	12	8	8
Kettunen (FIN)	9	9	9	13	6	12	10	7	11	14
Volchkovav (RUS)	10	10	6	14	11	10	12	6	9	15
Maniachenko (UKR)	11	13	12	11	12	16	11	11	10	9
Fontana (ITA)	12	14	11	18	16	9	15	9	12	10
Liashenko (UKR)	13	15	13	6	10	8	14	13	14	16
Onda (JPN)	14	11	14	10	15	15	13	15	13	11
Hubert (FRA)	15	12	17	17	13	11	16	14	15	13
Meier (SUI)	16	16	16	9	14	14	9	16	16	12
Gusmeroli (FRA)	17	17	15	15	17	17	18	17	17	17
Soldatova (BLR)	18	19	18	22	20	21	17	18	18	19
Hegel (CRO)	19	20	21	16	22	18	19	21	19	18
Giunchi (ITA)	20	18	19	20	21	19	20	20	20	20
Babiakova (SVK)	21	22	20	19	19	20	21	19	22	22
Kopac (SLO)	22	21	22	23	18	22	22	22	21	21
Lucar (ROM)	23	23	23	21	23	23	23	23	23	23

There were 22 instances in which a judge ranked a competitor from their "home" country. Five times the judge ranked the competitor the same as their place in that program—all from the USA judge. In 15 of the remaining 17 instances the judge ranked the competitor more favorably than their placing. The exceptions were Finland and Canada in the short program. To be fair, the USA competitors did very well, and thus it was difficult to rank them any higher. Of the six instances where the USA judge graded a USA skater, there were five agreements with the overall ranking and in the other case Sasha Cohen of the USA was ranked ahead of Slutskaya in the short program. The empirical evidence for a home country bias is strong. There is only a 0.0011 probability that 15 or more of the 17 gradings would be favorable for the home country if each instance of disagreement was equally likely to be positive or negative. It seems quite likely there is a home-country bias, but each of the top skaters had a home-country judge on the panel. The three Eastern European judges (Russia, Slovakia, and

Bulgaria) favored Slutskaya, while the two North American judges (USA and Canada) favored Hughes. The European judges fell three (Finland, Germany, Italy) to one (Denmark) in favor of Hughes.

If the three Eastern European judges were the only three judges on the panel then the short program would have finished: Slutskaya, Kwan, Butyrskaya (RUS), Cohen, and Hughes. The free program would have been: Slutskaya, Kwan, Hughes, Cohen, and Butyrskaya. The top five overall finishers would have been the same order as the free program.

The statistical problems with claiming bias are many. First we do not know the correct ordering. There is no answer key by which to compare judges. We do have the order that skaters finished, but perhaps the geographic balance of judges on the panel determined this order. Also, half the score for figure skating is presentation, so it may be that there are geographically-related styles which are judged more favorably by judges from the locales in which these styles predominate. In that case there will be geographic correlation but no bias. The number of data points is small enough to make investigating bias in this one Olympics challenging if not impossible. Despite the small sample size, I believe there is strong evidence that judges vote skaters from their own country higher—for whatever reason, be it bias or style preference. However, it would be silly to overturn the results on such a protest. You would need something much more concrete, like, say, a judge admitting they voted one contestant higher as part of a deal. But, that would never happen, would it?

The Pairs Gold Medal(s) Controversy

That brings us to the pairs competition. The rules for scoring are identical to the ladies singles competition. Table 6.3 shows the ordinals for the top three finishing pairs. The Russian pair of Berezhnaya–Sikharulidze finished first in the short program. They were first in the short program on seven of the nine judges. The Canadian and German judges were the only ones to select the Canadian pair of Sale-Pelletier. The Chinese pair of Shen–Zhao were selected third for all nine of the judges. In the free program the Chinese pair was also selected third by all the judges. Five of the judges selected the Russian pair best in the free program, with the remaining four selecting the Canadian pair. Whichever pair won the free program would have won the Gold Medal. Immediately following the disclosure of the Canadians' marks, the crowd started booing. The American announcers on the American telecast harshly criticized the marks—I am paraphrasing, but they referred to it as a travesty. If one of the judges who selected the Russian pair over the Canadian pair had reversed their order the Canadian pair would have won the Gold Medal.

In the following days the French judge admitted to having struck a deal with the Russian judge to mark the Russian pair higher, in exchange for favorable marks for a French skater(s). The French ice dancing team won a Gold Medal, and they were the only French competitors close to medaling. I could not find any verification of whether this was where the trade-off was to occur. A few days later the French judge, Marie Reine Le Gougne, claimed that there were no such deals, and that it was the American judge who was pushing for deals. The International Olympic Committee (IOC) in conjunction with the International Skating Union (ISU) decided to award both the Russian and the Canadian pairs Gold Medals.

The ordinals for the free program are shown in Table 6.4. There is more agreement between the judges and the overall rankings in the pairs free program than in the ladies singles free

Table 6.3 ***The ordinals for the top three teams in the pairs free program. The five to four advantage in the free program for the Russian pair over the Canadian pair earned them the Gold Medal.***

Pair	Program	RUS	CHN	USA	FRA	POL	CAN	UKR	GER	JPN
Berezhnaya-	Short	1	1	1	1	1	2	1	2	1
Sikharulidze	Free	1	1	2	1	1	2	1	2	2
Sale-Pelletier	Short	2	2	2	2	2	1	2	1	2
	Free	2	2	1	2	2	1	2	1	1
Shen-Zhao	Short	3	3	3	3	3	3	3	3	3
	Free	3	3	3	3	3	3	3	3	3

Table 6.4 ***The ordinals for each of the judges for the pairs free program. The "ALL" column is the overall placing for the free program.***

Pair Country	ALL	RUS	CHN	USA	FRA	POL	CAN	UKR	GER	JPN
RUS	1	1	1	2	1	1	2	1	2	2
CAN	2	2	2	1	2	2	1	2	1	1
CHN	3	3	3	3	3	3	3	3	3	3
RUS	4	4	5	5	4	4	7	4	5	4
USA	5	6	4	4	5	5	4	5	4	5
RUS	6	5	7	7	6	6	5	6	6	7
POL	7	7	6	6	7	7	6	7	7	6
CZE	8	9	8	8	9	8	8	8	8	8
CHN	9	8	9	9	8	10	10	9	9	9
CAN	10	10	10	12	10	9	9	10	11	12
CAN	11	12	11	10	11	11	11	11	10	11
CHN	12	11	13	13	12	12	12	12	12	13
USA	13	13	12	11	13	14	13	15	13	10
UKR	14	14	15	14	14	13	14	14	14	15
GER	15	15	14	15	15	15	15	13	15	14
UKR	16	16	16	16	16	16	16	16	16	16
SVK	17	18	17	17	17	17	18	17	18	17
UZB	18	17	18	18	18	18	17	19	17	19
ITA	19	19	19	19	19	19	19	18	19	18
ARM	20	20	20	20	20	20	20	20	20	20

program. The correlation between each judge's order and the overall order is larger than 0.98 for each of the nine judges in the pairs free program. In the ladies singles free program none of the judges had a correlation larger than 0.98. An interesting aspect of the controversy is that the French judge had an almost identical ranking to the overall ranking. She had the Chinese and the Czech Republic pairs reversed. In the free program, when judges had to rate competitors from their home country they agreed exactly with the overall placing on 9 of the 15 cases. Of the remaining 6 cases, 5 of them were rated more favorably. In the short program there were agreements in 6 of the 15 cases, with 5 rated higher and 4 rated lower. Overall, in only 10 of the 30 cases were home pairs rated higher. The only real disagreement in the pairs came at the very top. The Russian, Chinese, French, Polish, and Ukrainian judges selected the Russian pair. The judges from the USA, Canada, Germany, and Japan selected the Canadian pair. Again there is a hint of geographic bias in this case, but again there are far too few data points to discern a distinct pattern. In looking at the data the only concern this example brings may be the lack of variability. The agreement across the judges and the 20 pairs is rather amazing. This could arise because they are very good at marking pairs—more so than singles—or possibly a predetermined order?

We do not know what evidence or testimony was uncovered in the pairs case, but in neither of the cases is there empirical evidence of wrongdoing. Yes, there is evidence—and I believe strong evidence—that there is a "homer" effect. The fact that there is a homer effect presents the possibility of larger more wide reaching geographic bias—which in close competitions can clearly affect who ends up on the Wheaties box.

ADDITIONAL READING

Bring, J. and Carling, K. (1994), "A paradox in the ranking of figure skaters," *Chance*, 7, 4, 34–37.

CHAPTER 7

A statistician reads the sports pages

SCOTT M. BERRY

One modern man or 15 Tarzans?

One of my favorite statistical projects compared the abilities of athletes across eras. Shane Reese, Pat Larkey, and I (*Journal of the American Statistical Association*, September 1999) wondered whether Mark McGwire was a better home-run hitter than Babe Ruth, whether Wayne Gretzky was a better hockey scorer than Gordie Howe, and whether Tiger Woods was a better golfer than Jack Nicklaus. One of the challenges in answering these questions in baseball, hockey, and golf is that success is affected by the quality of your opponents. There is no clock or yard stick that gives an objective measure of a player's ability. Babe Ruth and Mark McGwire played against different pitchers, in different ballparks, and with different equipment. Hitting 60 home runs in 1927 is not equivalent to hitting 60 home runs in 2002. Like baseball, hockey has changed, and there is no objective measure of a hockey player's ability. While golf does have individual scores, the courses, equipment, and conditions changed with time. Tiger Woods's 68 in the 2002 Masters tournament may be a very different score from Jack Nicklaus's 68 in the 1972 Masters tournament.

Some sporting events have more objective measures of an athlete's ability. In many track events, the object is to run a distance as quickly as possible. A sprinter's time in the 100-meter dash in 1920 can be compared directly with a sprinter's time in 2002. Sure, training methods and the rewards have improved, but the nature of the event allows more direct comparison of sprinter speed in 1920 and 2002. In these events athletes are clearly getting better through time.

One of the more amazing stories demonstrating athletic improvement is that today's best swimmers race 1500 meters at a pace faster than Johnny Weissmuller's 1924 100-meter pace. Johnny Weissmuller was the 100-meter freestyle champion in the 1924 and 1928 Olympics. He is even more famous for playing Tarzan in a long-running TV series. His 1924 gold medal time in the 100-meter freestyle was 59.0 seconds. At this pace, Weissmuller would complete 1500 meters in 14:45 (14 minutes, 45 seconds). The Olympic record in the 1500-meter freestyle is 14:43.48, set by Kieren Perkins (Australia) in 1992. Imagine telling someone who witnessed Weissmuller's amazing, under-one-minute, 100-meter victory that the future will bring a swimmer beating this pace for 1500 meters!

The same result holds out of the water. The 1924 speed-skating champion in the 500 meters, Charles Jewtraw (USA), won the gold with a time of 44.0 seconds. This pace would give Jewtraw a time of 14:40 in the 10,000 meters. The 1998 Olympic champion, Gianni Romme (Netherlands), had a 10,000-meter time of 13:15.33. Jesse Owens is famous for his amazing 1936 Olympic sprinting performance. In Germany, in the face of Adolf Hitler, Jesse Owens won four gold medals. He set a world record of 10.3 seconds in the 100 meters. Jesse Owens could not compete with modem sprinters. His time of 10.3 would put him 4.4 meters behind Maurice Green at the finish line of the 2000 Olympics. Owens would hardly be considered a world-class sprinter in 2002. This is not to say that Owens wasn't a great athlete or champion. His accomplishments exceed those of Maurice Green, but he was undisputably not as fast as today's sprinters.

Modern strength-based athletes are also much better than those of previous eras. The 1994 Olympic shot put gold medalist, Bud Houser (USA), threw the shot put 49 feet, 2.25 inches. The current Olympic record (1988, Ulf Timmermann, East Germany) is 73 feet, 8.75 inches.

Why are today's athletes stronger and faster? Many reasons can be hypothesized, including better training methods and equipment, better diets, better medical treatments, performance-enhancing drugs, and much better financial incentives. But perhaps the most important reason is simple—the world population has more than tripled. The Olympic champion is now the best athlete of 6 billion people, while the 1924 Olympic champion was the best of 1.86 billion people.

So does population growth explain the improvement in our elite athletes? In this issue of *Chance*, I compare the expected Olympic-winning performances based on a population-growth model with the actual winning performances observed over the past century. The compatibility of the predicted and actual performances will help determine whether having a larger talent pool is a plausible explanation of our progress.

Olympic and world population data

I focus attention on seven Olympic events: two running races, the 100 meters and the marathon; two free-style swimming races, the 100 meters and the 1500 meters; two speed-skating races, the 500 meters and the 5000 meters; and one field event, the shot put. Figures 7.1, 7.2, 7.3, and 7.4 show the gold-medal-winning performances for each year the event was held. The figures show winning performances improving through time.

Figure 7.5 shows the world population (in billions) across the same time period. These numbers were reported by the U.S. Bureau of the Census and the United Nations (www.census.gov/ipc/www/worldhis.html). I was unable to find totals for each year, so I interpolated between data points using an exponential growth model. The fit of the following exponential model is very good.

$$\text{Population in year } t = 1.60e^{(t-1900)*0.0088} \text{ billion}$$

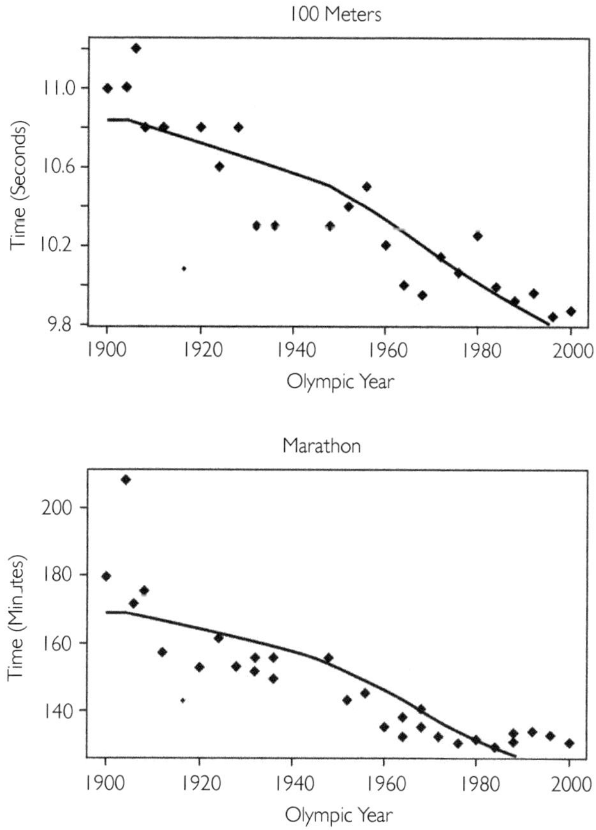

Figure 7.1 The Gold Medal–winning time in track for each year for the 100 meters and the marathon. The curves represent the predicted winning times from the population model.

Modeling an Olympic Winner

If we think of qualifying for the Olympics as the first-round of competition, nearly everyone on the planet is eliminated in the first-round. If there are 2 billion people on the planet, half male and half female, then there are 1 billion in each competition. I model the distribution of ability—for example, time in the 100 meters—as a normal distribution. The population has a normally distributed 100-meter time, with a mean of μ and a standard deviation of σ. This distribution is constant through time—thus athletes are not getting better, nor are they better trained—there are just more of them. The winner is the best of 1 billion people—the minimum of a normal distribution. I use a simple rule for this extreme. The value of a standard-normal distribution with 1 billionth of the probability less than it (1-billionth quantile) is −5.9978. For a normal distribution with a mean of μ and a standard deviation of σ the one-billionth quantile is $\sigma(-5.9978) + \mu$. For a population of n people this is simply $-\sigma Z_{1/n} + \mu$, where $Z_{1/n}$ is the 1/nth quantile of a standard-normal distribution. For the population of 0.875 billion men in 1900, the value of $Z_{1/n}$ is −5.98. The value of $Z_{1/n}$ for the 3.04 billion males in 2000 is −6.17.

In order to estimate the effect of population growth on Olympic performances, I find the least squares estimates of μ and σ for each event. Thus, I find the values of μ and σ that

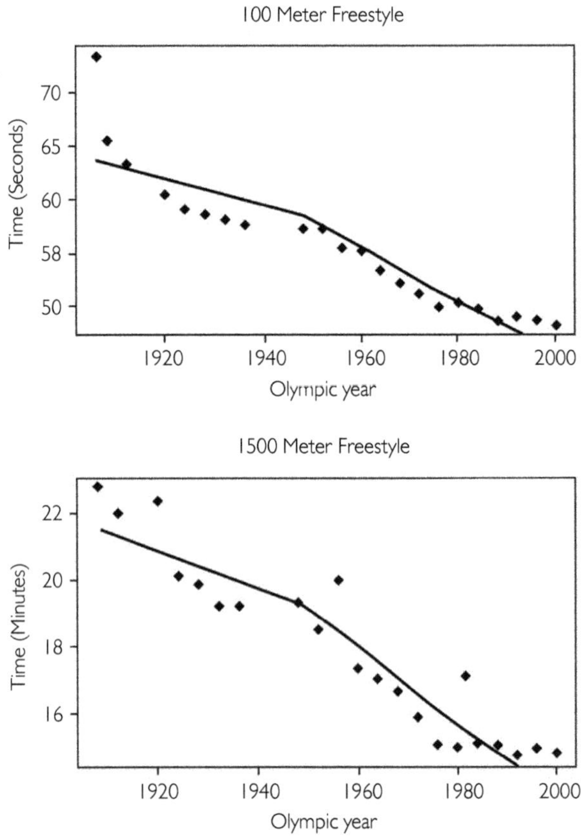

Figure 7.2 The Gold Medal–winning time for swimming in each year for the 100 meters and the 1500 meters. The curves represent the predicted winning times from the population model.

minimize the squared difference between the predicted Olympic winning time—using the population in that year—and the actual Olympic winning time (or distance). For males running the 100-meter race, the least squares estimates of μ and σ are 41.86 and 5.20, respectively. Taken literally this implies that the male population has an intrinsic mean 100-meter time of 41.86 seconds with a standard deviation of 5.20 seconds. Understanding that this includes all males regardless of athletic ability, this is a plausible estimate, but I think it is probably too high. Moreover the population is probably not normal—but I think the normal approximation is not too bad. The problem is essentially one of extrapolation—we are estimating the mean and standard deviation of a 1,000,000,000-member population from its smallest member. While the model will work reasonably well for the extremes, interpreting the mean and standard deviation can be fraught with disaster. I refer to this model as the "population model."

The estimate for the minimum time, using the population model, is shown by the curve on Figure 7.1. The fit is pretty good. A very convenient statistical measure of fit in this scenario is a summary referred to as R^2. It represents the percentage of the variability in the Olympic winning score that can be accounted for by the population model. If R^2 was 100% it would mean that the Olympic winning score was predicted perfectly from population size. If this percentage is small it means that the size of the population is a poor estimator or the functional form of my model is wrong. For the 100 meters the R^2 is 81.3%. I was pleasantly surprised that the R^2 was this large, as I was expecting a value of 50% to 70%. This means that 81.3% of the variation in Olympic winning times can be explained by the size of the male population. Table 7.1 presents the least squares estimates of μ and σ for each of the events, along with the R^2. The curves on Figures 7.1, 7.2, 7.3, and 7.4 show the estimated winning Olympic score using the population model. Of all seven events, the R^2 is highest, 95.4%, for the 500-meter speed skate and is the least, 68.7%, for the marathon. Based on the progress made in marathon times during the first 80 years, more progress was predicted for the past 20 years, but the winning score has stayed constant. The standard deviation was probably overestimated—thus predicting a bigger change in the Olympic scores with the rapid population growth of the past 20 years. Clearly, something other than population growth has affected marathon scores over the past 100 years.

The population model fits both swimming scenarios reasonably well. The R^2 for the 100 and the 1,500 meters is 83.8% and 90.6%, respectively. The effect seen in the marathon appears in both swimming events. The population model predicts a larger decrease in the times than actually occurred. Yet, the model does fit surprisingly well. In both swimming events there is a sharp decrease in the times in the mid-1950s in tandem with a huge population increase.

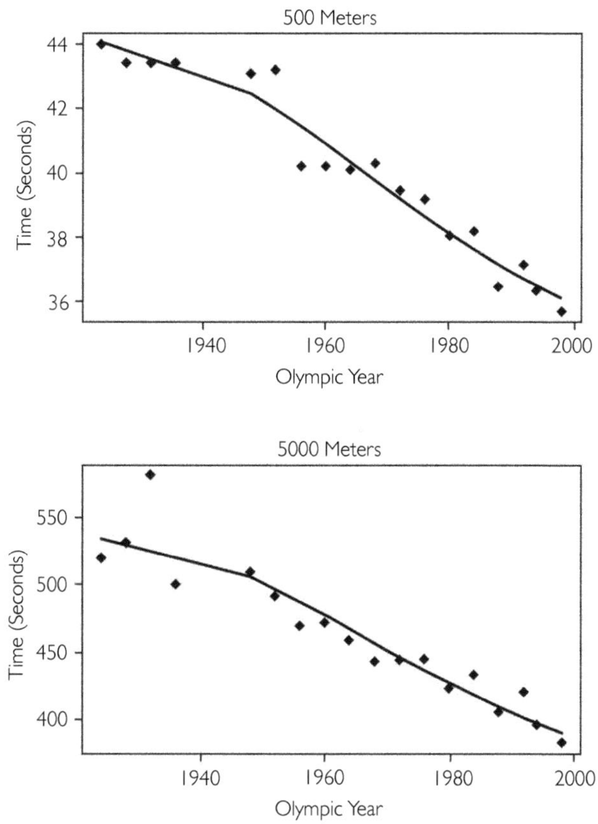

Figure 7.3 The Gold Medal–winning time in speed skating for each year for the 500 meters and the 5,000 meters. The curves represent the predicted winning times from the population model.

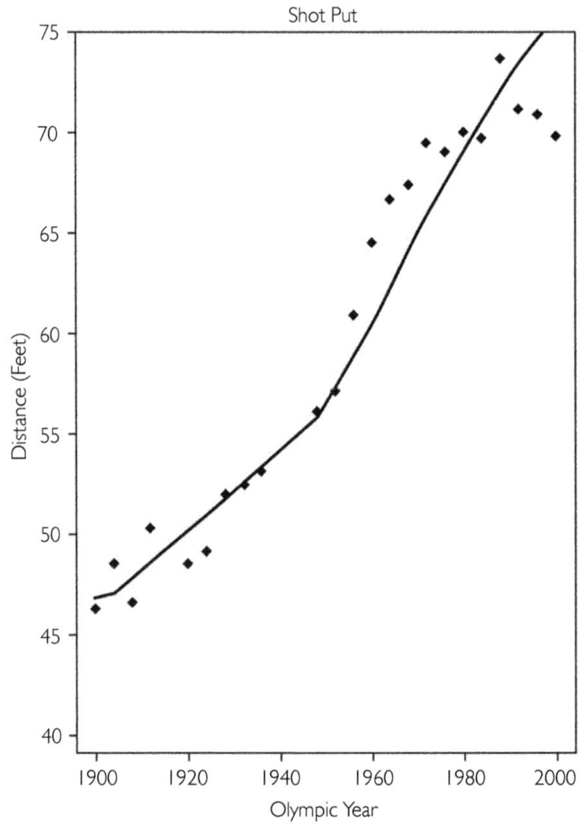

Figure 7.4 The Gold Medal–winning distance in each year for the shot put. The curve represents the predicted winning distance from the population model.

The population model fits the two skating events extremely well. The R^2s, 95.4% and 88.6%, are very high. Unlike the marathon and swimming events, the curves provide a very nice fit, especially over the past 20 years. Is speed skating inherently different from running and swimming? Is the impact of training and technique lower in speed skating? Why does the population model provide a better fit to the speed-skating Olympic records? While I don't know the answers, the conjectures implied by the first two questions seem possible.

The model was not constructed to predict future Olympic winning times, but as I write the 2002 Winter Olympics are a month away, and given the model's excellent fit, I wonder what it would predict for the winning 2002 Olympic speed-skating scores. With a 2002 population of 6.23 billion people (3.11 billion males), the predicted winning time in the 500 meters is 317.2—(6.18)(45.6) = 35.39 seconds and in the 1500 meters is 5458—(6.18)(821) = 384.22 seconds (6:24.22).

The shot put event has the second-highest R^2 of 92.1%, but there are clear deviations between the predicted and actual. There was a sharper increase in shot put distances in the 1960s and 1970s than predicted and a smaller increase than expected over the past 25 years. The shot put differs from the other events examined here in that strength rather than speed is the crucial aspect. I would think the shot put is a sport in which training, diet, and

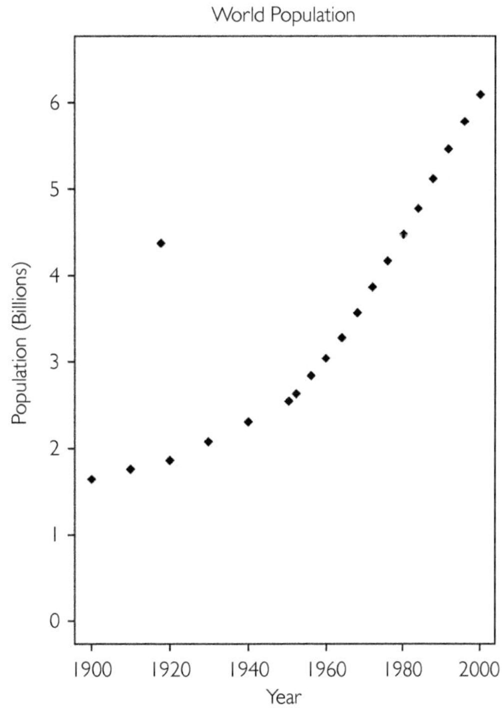

Figure 7.5 World population data with missing points interpolated from the exponential growth model.

Table 7.1 ***The estimated mean, μ, and standard deviation, σ, based on the least squares algorithm for each event. The R^2 represents the percentage of variation in winning performance that is explained by the population model.***

Event	$\hat{\mu}$	$\hat{\sigma}$	R^2
100 meters	41.86	5.20	81.3%
Marathon	1558	233	68.7%
Swim 100 meters	571.9	85.1	83.3%
Swim 1500 meters	253.0	38.7	90.6%
Skate 500 meters	317.2	45.6	95.4%
Skate 5000 meters	5458	821	88.6%
Shot Put	−780	139	92.2%

performance-enhancing drugs would play a very important role. Could the lack of fit over the past 40 years be due to a steroid effect followed by a testing-for-steroids effect?

Conclusion

The model used in this article assumes abilities follow a normal distribution. It assumes people have the same distribution of ability regardless of when they were born. Winning Olympic scores are predicted using the $1/n$th quantile, which is a function of the world population. This model—using only the world population—predicts Olympic winning times very well. While clearly there is more going on than just population growth, population growth explains 80 to 95% of the variability in Olympic winning performances in six of the seven events.

I guess it is a fallacy of each era to believe athletes are as good as they will ever get. Will an athlete in 2050 beat Ian Thorpe's pace in the pool in a race 15 times longer? Will an athlete come along who shatters Carl Lewis's times on the track? Will my memories of Eric Heiden dominating the 1980 Olympics be erased by athletes I can't imagine? While it doesn't seem possible, I am sure no one in 1924 thought 1992 would bring a man 15 times that of Tarzan.

Casey FitzRandolph (left) and Kip Carpenter (right) compete in the 500-meter speed skate at the 2002 Winter Olympics

Addendum

This paper was completed before the 2002 Salt Lake Winter Olympics began. The winner of the men's 5,000-meter speed-skating event in 2002 was Jochem Uytdehaage, of the Netherlands, and his time of 6:14.66 was a world record—shattering the old world record and the population model estimate of 6:24.22. The Olympics now records two times for 500-meter events, outside and inside lane times, which are averaged. The Gold Medal winner in the men's 500 meters, Casey Fitzrandolph of the USA, averaged 34.62 seconds in his two times. His inside-track time of 34.42 seconds was an Olympic record. In the 10-speed skating events, 8 world records were set, many were shattered. What happened? The Utah Olympic Oval is believed to be the fastest skating ice in the world. Because of the altitude and de-ionizing of the water, they were able to create the "hardest" ice in the world. It appears technology lowered times significantly. Such technological advances are not accounted for in the simple population-based model. Further details of the speed of the Utah Olympic Oval can be found at www.nbcolympics.com.

CHAPTER 8

Evaluation of the current decision rule in figure skating and possible improvements

SAMUEL S. WU AND MARK C. K. YANG

Abstract

The scandal in the 2002 Winter Olympic figure skating judging forced the International Skating Union (ISU) to address the vulnerability of the current nine-judge placement rule. The ISU proposed a stricter ethics code for the judges and changes to the decision rule. Unfortunately, the new decision rule is no better than the previous rule in countering manipulation. In this article, we construct an alternative rule and show that, under reasonable assumptions, our rule is more robust against manipulation than existing rules. When other properties such as Condorcet efficiency, frequency of ties, and violation of monotonicity are compared, our rule is also more desirable. In addition, a commonly used rank sum rule is examined and shown to be even worse than the other rules. These results are applicable to any decision rule based on the relative ranks of proposals or contestants.

Keywords: *median rank placement, rank sum decision rule, voting*

Introduction

Figure skating is one of the most beautiful and graceful sports. It received very high visibility in the Winter Olympic games. However, to determine the winner in a game is not an easy matter. Not only is it subjective, but it may also be vulnerable to manipulation. When national prestige is at stake, block voting and deal-making may occur. Recently, a judging controversy occurred in the pairs figure skating in the 2002 Salt Lake Winter Olympics when the French judge, Marie

Reine Le Gougne, was accused of fixing her vote as a result of the pressure from her federation president. Based on the evidence, Ms. Le Gougne was suspended as a future judge and the original decision was changed to award the silver medalists a gold medal (*New York Times*, Feb. 12–15, 2002; Berry 2002). Afterward, the International Skating Union (ISU) searched for ways to prevent improprieties and a tentative new rule was announced on June 7, 2002.

Even before the announcement of the new rule, prevention of manipulation had already been put into the rules for determining the winner. For example, the rule of giving scores to an individual was quite strict. Judges first watched the skaters practice and assigned a "base mark" for each skater. During the competition, the judges scrutinized each required element and deducted points from the base mark if the skater left out an element, made a mistake, or added extra elements. At the international level, the Technical Committee monitored the judges' scoring looking for such problems as block voting, and asking a judge to submit a letter to justify a score that did not agree with the panel. Thus, it was not easy for a judge to put too much "subjectivity" into his/her score. However, the rule could not be too strict to hinder a judge's sincere rating.

Because the new rule is still in the experimental stage, we will concentrate on the rules applied before June 7, 2002, and propose a different rule for determining the winner. The rules used at the 2002 ISU games are posted on the Web site http://www.frogsonice.com/skateweb. The rules used before 2002—presented in *The Official Book of Figure Skating* (1998)—had received statisticians' attention (Bassett and Persky 1994) and that rule is still being used by the U.S. Figure Skating Association (USFSA). Hence, both rules are worth evaluating. In recent Olympic figure skating, skaters are ranked based on the short program (1/3) and the free style skate (2/3) in singles and pairs skating; in ice dancing, there are two compulsory dances (10% each), an original dance (30%), and a free style dance (50%). For each component, a judge assigns two scores to each player on a 6.0 scale; one for presentation and the other for required elements or technical merits. The sum of these two scores is converted to an ordinal ranking among all players for each judge. If a judge gives the same total score to two skaters, the skater with the higher presentation score receives the higher rank. A placement of players is then derived based on all judges' (usually nine) ordinal ranks for this component. Finally, a competitor's overall placement is determined by the weighted average of the component placements.

Because the last stage is a simple arithmetic average, the fundamental ranking is established at each component. In this article we study the fairness in ranking players within each component. Let the sum of two within-component scores of skater i by judge j be denoted by x_{ij}, where, $i = 1, 2, \ldots, n$ and $j = 1, 2, \ldots, m$. It is apparent that the score sum $\sum_{j=1}^{m} x_{ij}$ should not be used to determine the winner because it gives the dominant power to the judge who uses a wider range. For example, if most judges use the range of 5.7 to 5.9 for good performances, then the judge who uses the range from 4 to 6 may have a larger decisive power in the final score. To give all the judges equal power in the final decision, it is fairer to base a rule on the rank within each judge. That is, we should consider only

$$r_{ij} = \text{the rank of skater } i \text{ by judge } j \text{ among all the skaters.}$$

Note that, for fixed j, r_{ij} are n distinct integers from the set $\{1, 2, \ldots, n\}$ with 1 as the best performance. The rank sum, or equivalently the Borda count in the public choice literature, is a reasonable choice for the final score. The name is to honor an equivalent election rule proposed

Table 8.1 ***A hypothetical rankings among 10 contestants with five judges.***

Skaters	Judges					Rank sum
	1	2	3	4	5	
A	1	2	2	3	10	18
B	2	5	3	4	1	15
C	5	1	1	6	2	15
D	3	3	5	5	5	21
E	4	4	4	9	9	30
...	...	...	...	...	...	...
J	...	...	...	...	...	...

by Borda (1781). Unfortunately, using the rank sum is not very robust because of the wide range of ranks when n is large. Table 8.1 shows a simple example where $m = 5$ judges are used to rank $n = 10$ contestants. Most people will agree that we should not consider skater B better than skater A, because B is ranked worse than A by four of the five judges, but skater B has the smaller (superior) rank sum. From this table it can also be seen that the rank sum is vulnerable to manipulation. In the case of candidates A and B in Table 8.1, a majority rule seems to be more reasonable when A is considered the winner if candidate A is rated better than candidate B by the majority of judges. Unfortunately, a pairwise majority rule is not transitive. If we let $A \succ B$ mean that A is considered better than B, then we can easily construct examples for three candidates A, B, and C such that $A \succ B$, $B \succ C$, and $C \succ A$ as demonstrated in Table 8.1. This nontransitivity is the basic difficulty encountered when attempting to determine an overall (public) consensus from individual rankings. The well-known Arrow's impossibility theorem states that under a set of reasonable requirements, there is no rule to draw a public consensus from individual rankings if $n > 2$ and $m > 1$ (Arrow 1963). Because the figure skating rule is a special case of the impossibility theorem, there is no way to determine the winner that will satisfy all rational persons.

However, a special case of the pairwise winner rule is less controversial when one wins pairwise over all the other contestants. This candidate is called the Condorcet winner in memory of its discoverer in the late eighteenth century (Condorcet 1785). A survey of his contribution can be found in Sommerlad and Me-Lean (1989) and Gehrlein (1997). There is no controversy of nontransitivity if we declare this candidate the winner when it exists.

Since voting is one of the fundamental procedures in democracy, there is a vast literature. A journal, *Public Choice*, is totally devoted to voting research. Gibbard (1973) and Satterthwaite (1975) showed that all reasonable rules can be manipulated. The hope is to choose a rule that is "fair" and "robust" against manipulation in most cases.

New and existing rules

Given m ranks $(r_{i1}, r_{i2}, \ldots, r_{im})$ received by contestant i, define S as the total sum of the ranks, let M be the median of the ranks, and let L be the number (*not sum*) of ranks that are greater than the median. Let W be the number of wins in pairwise comparisons. An example of how W is computed is given in Table 8.2. To avoid confusion between ranks by the judges and the final ranks of the skaters, we call the final ranks the (overall) placement in the game. Since the ranks are integers, one of the main difficulty in rule construction is how to handle ties. The rule proposed in 2002 by the ISU is called the one-by-one (OBO) rule which consists of two steps:

OBO1. The candidate with the largest value of W wins.

OBO2. When there is a tie in value of W, the one that has the smaller value of S is ranked higher. (The term "judges in favor" is used in the manual.)

In other words, the OBO rule is to order the candidates sequentially by W and S. The 1998 ISU and the current USFSA rule for ranking the candidates is as follows. We called this rule the median rank (MR) rule because the median rank plays a crucial role.

MR1. The contestants are ordered by the median rank M.

MR2. When there are ties in M, the one with a smaller value of L wins.

MR3. If the L's tie again, the one with smaller value of S wins. If the candidates still tie at this stage, they remain tied in the final placement.

To put the median rank as the first priority satisfies our intuition in one special case. If some skater has been ranked first with an absolute majority (e.g., five out of nine), that is, $M = 1$, then this candidate is the winner. When no skater has $M = 1$ and some skaters are tied at $M = 2$, the MR2 step says that the skater with the most first- and second-place ranks wins first place. This is called Condorcet's practical method which has been used for more than 100 years in Geneva (Nanson 1882). The International Skating Union and the United States Figure Skating Association defended the median rank rule, citing the following three desirable properties.

P1. If one individual judge raises a skater's score, then the change will not decrease the skater's placement.

P2. A skater ranked as the best by a majority will be placed as the best.

P3. This rule is not very sensitive to manipulation.

Bassett and Persky (1994) proved that this rule is the only one that satisfies properties P1 and the following modified P2′.

P2′. Suppose skater A has rank r_1 from at least half of the judges and skater B has rank r_2 from at least half of the judges. Then $r_1 < r_2$ implies $A \succ B$.

In other words, they proved the following theorem:

Theorem 8.1 (Bassett and Persky 1994). *MR is the only rule that satisfies properties P1 and P2′.*

This theorem gave the MR good justification, but it did not imply that there was no room for improvement. One concern about the MR rule is that the P2′ requirement cannot be met in many cases. For example, none of the first three contestants in Table 8.1 has the same rank from the majority of judges. In addition, P2′ can be quite unreasonable if, for example, candidate B is considered to be better than A in Table 8.2 by P2′, although A is voted better than B by eight out of the nine judges. Moreover, claim P3 was not quantified. We felt intuitively that a more complex rule that can still maintain the properties required by a good rule should be harder to

Table 8.2 ***An example illustrating the decision procedures of the four rules.***

Skaters	Judges														Results from			
	1	2	3	4	5	6	7	8	9	W	M	L	P	S	RS	MR	OBO	NR
A	1	1	1	1	3	3	3	3	3	4	3	0	7	19	1	2	1	1
B	2	2	2	2	4	4	4	4	2	3	2	4	10	26	3	1	2	2
C	3	3	3	3	2	2	2	1	4	2	3	1	10	23	2	3	3	3
D	4	4	4	4	1	5	5	2	1	1	4	2	12	30	4	4	4	4
E	5	5	5	5	5	1	1	5	5	0	5	0	17	37	5	5	5	5

NOTE: In this table, W is defined as the number of pairwise wins, M is the median of the ranks, L is the number of ranks that are larger than the median, P is the sum of the best $[(m + 1)/2]$ (majority half) ranks and S is the total sum of the ranks. Also, "RS" stands for rank sum rule, "MR" for median rank rule, "OBO" for one-by-one rule and "NR" for the new rule.

manipulate. Thus, we tried to combine all the criteria of the OBO and the MR rules. Moreover, rank sum of the majority should be harder to manipulate than the rank sum by all the judges. With some trial and error, the following rule, called the NR, is proposed as a good choice.

Our proposed NR will sort the contestants sequentially in the following order:

1 number of pairwise wins W;
2 median of the ranks M;
3 number of ranks that are larger than the median L;
4 the sum of the best $[(m + 1)/2]$ ranks P, where $[x]$ means the largest integer less than or equal to x;
5 the total sum of the ranks S; and
6 if there are two-way ties at the 5th stage, the pairwise winner is ranked higher.

For Step 1, the winner is the one with a larger value, and for Steps 2–5, the smaller the value, the better.

The new Step 6 is obvious, because when there are two candidates and an odd number of judges, pairwise comparison should be the tie breaker. Step 4 is used to reduce the influence of the outliers produced by the manipulators. Table 8.2 illustrates the intermediate steps in implementing the rank sum (RS) rule, the current ISU rule (OBO), the median rank rule (MR), and the new rule (NR). The rule RS is included for its simplicity and popularity. It is probably one of the most often used rules in panel decision, and it is the most robust rule against manipulation among all weighted rank sum rules (Saari 1995, p. 283).

Comparisons of the new and current rules

In this article, we concentrate on small numbers of candidates and judges. In particular, we consider the number of judges to be an odd number, no more than 9, and we concentrate on the special case of $n = 3$ contestants for tractability. Certain properties can be verified by an exhaustive search in this case. Also, from a practical viewpoint, the competition typically focuses on the top few candidates. There is no advantage in manipulating the scores if the candidate has little chance to get into the top three or receive the gold medal.

When there are only three contestants, a judge can choose only one out of the six following three-tuples for ranking the three candidates in the order of $[A, B, C]'$,

$$\begin{bmatrix}1\\2\\3\end{bmatrix}, \begin{bmatrix}1\\3\\2\end{bmatrix}, \begin{bmatrix}2\\1\\3\end{bmatrix}, \begin{bmatrix}2\\3\\1\end{bmatrix}, \begin{bmatrix}3\\1\\2\end{bmatrix}, \begin{bmatrix}3\\2\\1\end{bmatrix}. \tag{8.1}$$

There are $t_m = 6^m$ possible sets of rankings for m judges, for example, $t_5 = 6^5 = 7776$ for five judges. If we let k_i be the number of judges who choose the ith three-tuple in the order of (8.1), then the rankings of the m judges can be represented by $\boldsymbol{k} = (k_1, k_2, \dots, k_6)$ with k_i being nonnegative integers subject to $k_1 + k_2 + \cdots + k_6 = m$. Both representations were used in our exhaustive proofs. By exhaustive listing, we found that for three judges, the new rule and the other rules lead to similar rankings. Differences in the rules occur for the cases of 5, 7, and 9 judges.

Condorcet Efficiency

Both OBO and our new rule always pick the Condorcet winner by their construction. Not being able to pick the Condorcet winner can be considered a weakness of a consensus rule. The probability that a rule chooses the Condorcet winner is called the Condorcet efficiency (Berg and Lepelley 1994). The Condorcet efficiency depends on how the judges vote. A special case is the impartial culture (IC) model (see Gehrlein 1977) which means that the three players have equal ability, that is, the judges have an equal chance to pick any of the six-tuples in (8.1). Under the IC model, when $m = 9$, the Condorcet efficiencies are 90.48% and 92.18% for the MR and RS rules, respectively.

Violation of Monotonicity

The results of a rule can be questionable when certain monotonicity properties are violated. One situation is when a candidate's score is raised, but her overall placement becomes worse. We define this as the violation of Type I monotonicity (already stated as P1). Another example is defined by the following scenario. Suppose candidate A is already determined to be the winner. Suddenly the judging panel found that one of the other contestant's score was worse than recorded. When the score is corrected, A loses her first place standing. We called this behavior a violation of Type II monotonicity.

Not all reasonable rules satisfy Type I monotonicity, but this property is satisfied by all the four rules in this article.

Theorem 8.2. *The new rule satisfies the property P1 for all values of m and n.*

Proof *Suppose the jth judge lowers the rank value r_{ij} of candidate i (lower rank means better). Then if i is better than i' by pairwise comparison before the change, i is still better than i' after the change. Hence the order by W will be no worse for skater i. The change by M, L, and S will be no worse from the proof in the MR. It can be seen that P will also not be worse after the change.* ■

Table 8.3 ***Percentage of the violation of type II monotonicity (n = 3).***

Number of judges	Number of possible sets of rankings	Type II violation (in %)			
		RS	MR	OBO	NR
5	7776	9.26	5.92	5.53	4.75
7	279, 936	9.60	10.30	5.47	3.89
9	10, 077, 696	8.96	10.37	4.91	3.97

Table 8.3 shows that for $n = 3$ the NR has fewer instances of the Type II monotonicity violation than the other three rules.

Robustness against manipulation

There is a fundamental dilemma when we try to reduce the chance of manipulators to achieve their goal. In doing so, we unavoidably reduce the chance of including delicate differences from honest judges. Because we cannot tell a manipulator (otherwise we could suspend his score) from a honest judge, we are trading sensitivity for the robustness of a decision rule.

To defeat manipulation, we need to understand the manipulator's intent and how he votes. The first obvious intent is that he wants a particular candidate, say B, to win. Another intent would be that he also wants another candidate, say A, to lose. We feel it is unlikely that a manipulator only wants someone to lose without any intent to make another candidate win. Thus, we will study only the two manipulation strategies. Consider Strategy I, where the manipulator's intent is to let B win. In this case, he always ranks B first and the other stronger contender last. That is, if his true ratings for (A, B, C) are $(1, 2, 3)$, $(1, 3, 2)$, or $(2, 1, 3)$, he will change it to $(3, 1, 2)$; otherwise he changes $(2, 3, 1)$, $(3, 1, 2)$, and $(3, 2, 1)$ to $(2, 1, 3)$. Strategy II is to let B win and A lose. In this case he will fix his rank at $(3, 1, 2)$.

The success of manipulation depends on other judges' rankings, or equivalently the difference between the contestants. For example, if the performers have big differences, then one or two judges cannot change the results. Changes in placement due to manipulation are possible only when the performances are similar. The extreme case is the IC model. Because this model requires no additional assumption of the performers' abilities, it is the most studied case in the literature (Saari 1995; Nurmi 1999). However, it is not a suitable model for manipulation study because when the candidates are of the same ability, the choice of winner is not important. Only in the case of unequal performance will people detect the unfairness due to manipulation. Nevertheless, the IC model is an extreme case of unequal performance and it is automatically included in this study.

SCORE DISTRIBUTION UNDER NORMALITY ASSUMPTION

We model the performance in the following manner. Let there be n contestants and m^* qualified judges or spectators which is assumed to be a very large number. After the performance, the score of the m^* judges forms a distribution. Let the scores of the ith contestant form a distribution $F_i(x)$, with mean (or median) μ_i, $i = 1, 2, \ldots, n$ satisfying

$$\mu_1 \geq \mu_2 \geq \cdots \geq \mu_n. \tag{8.2}$$

Let the m judges in the game be a random sample from the qualified judges. Thus, the judges' scores for contestant i are independent and identically distributed random variables from $F_i(x)$. Without loss of generality, we assume the higher the mean, the better.

We first assume that $F_i(x)$'s are normal distributions with a common variance. This approach has been widely used in rating. A review of examples in paired comparison modeling, which include the Bradley–Terry model and the Thurstone–Mosteller model, was given by David (1988). Glickman (1999) and Glickman and Jensen (in press) applied this method to rank players in chess and tennis games. In addition, the above normal model was implemented in the AccelRat software (Matthews 2003), and it is now used by the American Go Association (AGA) ratings system. Moreover, Bassett and Persky (1994) also modeled ranking as a measurement error problem with the error following a normal distribution. Though the finite range of the score system is violated with the assumption of normality, we consider that this violation is not serious because the normal distribution has a very short tail. Moreover, only the ranks are used in the rules. Let the three means of the top three candidates A, B, and C be denoted by $a + d_1 \geq a \geq a - d_2$, respectively. When we let d_1 and d_2 vary, the standard deviation (σ) can be any value without loss of generality. Moreover, the result is invariant to the location parameter a. We let $a = 3$ and $\sigma = 0.5$ in our computation.

The sensitivity is defined as the probability that the correct winner A is ranked the best, that is, P(A win). Ideally, P(A win) should be 1.0 when d_1 and d_2 are positive. The robustness is evaluated by $P(B| \sim B) \equiv P(B$ wins after manipulation $|B$ would not win without manipulation$)$ for manipulation Strategy I and the same $P(B| \sim B)$ for Strategy II plus $P(\sim A|A) \equiv P(A$ loses after manipulation $|A$ would win without manipulation$)$. A good rule should have a high value of sensitivity P(A win) and small values of $P(B| \sim B)$ and $P(\sim A|A)$. These probabilities can be computed first by using the normal distribution to compute the likelihood of each of the t_m cases defined after (8.1). Then for each fixed normal model, the three probabilities P(A win), $P(B| \sim B)$, and $P(\sim A|A)$ can be easily computed. In all the rules, there is the possibility of ties. For example, if A is changed from the sole winner to a shared winner between two candidates, do we consider this as a success of the manipulation? In our study, we count this as a half success, that is, we put 1/2 for the manipulation success in this case. Similarly, if A is changed from a shared winner to a loser, it is also 1/2 success for the manipulation.

The following three sets of means, $(3 + d, 3, 3 - d)$, $(3 + 2d, 3, 3 - d)$, and $(3 + d, 3, 3 - 2d)$, were used to represent the performance differences. Note that when $d = 0$, it is the equal ability case. Figure 8.1 shows the sensitivity P(A win) without manipulation and $P(B| \sim B)$ with one and two manipulators for $m = 9$ judges. In this figure, the extreme left column shows that the sensitivities of the four rules are almost identical. The differences in sensitivities are at the third decimal places. This surprising result simplifies the comparison of robustness. The three graphics in the middle column show one manipulator's success rate; the NR and OBO rules are the most robust and the RS is the least. For two manipulators, the MR is better when the three candidates are close, but the NR and OBO are better when the candidates' differences are larger. The RS is again the least robust one among the rules considered.

Figure 8.2 shows the robustness against manipulation with Strategy II. Note that the sensitivity is the same as Figure 8.1 because when the manipulation strategies are not used, the choice of strategy makes no difference. We see the same patterns for $P(B \mid \sim B)$, but our method is better for $P(\sim A|A)$. Similar results for the two types of susceptibility to manipulation were found where number of judges were $m = 5$ and 7.

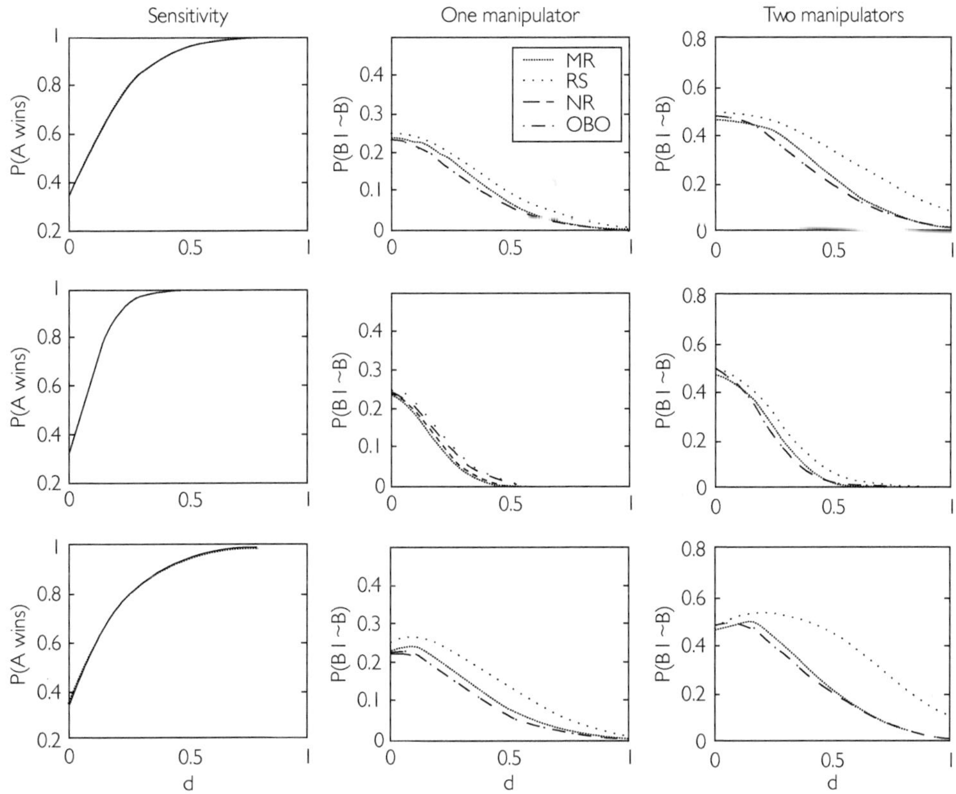

Figure 8.1 Sensitivity of the four rules and their robustness against manipulation Strategy I (to make B win) with $m = 9$ judges and $n = 3$ contestants. The three rows are for three mean structures: the top row corresponds to normal models ($\sigma = 0.5$) with means (3 + d, 3, 3 − d) for candidates [A, B, C]; the middle row with means (3 + 2d, 3, 3 − d) and the bottom row with means (3 + d, 3, 3 − 2d). In each plot, the solid line for MR, the dotted line is for RS, the dashed line for NR and the dashdot line for OBO. In this figure, the OBO rule performs nearly the same as the NR, but always inferior (at the third decimal place and indistinguishable in the graph). The number of manipulators is given at the top of each column. For the right two columns, the smaller the value the better.

OTHER SCORE DISTRIBUTIONS

Let $F_i(x)$, $i = 1, 2, 3$, defined in the section on score distribution, be any three score distributions. Then each judge's ranking preference will be one of the three-tuple listed in (8.1) with corresponding probabilities $\boldsymbol{p} = (p_1, p_2, \ldots, p_6)$ determined by $F_i(x)$s. In other words, the preferences for m judges follows a multinomial distribution with m trials and occurrence probabilities $\boldsymbol{p}$. For any given $\boldsymbol{p}$, we can exhaustively calculate the probabilities of all the 6^m possible sets of ranking by the judges and consequently we can evaluate the robustness of each decision rule. There is no sensitivity because the winners are rule dependent, unlike the normal model where we know who are the "true winners." However, we can still study the robustness by the definition of winner from its own rule. This is one standard procedure in studying manipulation (see Saari 1995).

We chose to randomly pick 10,000 $\boldsymbol{p}$'s (or equivalently $F_i(x)'$s) from the simplex

$$p_1 + p_2 + \cdots + p_6 = 1. \tag{8.3}$$

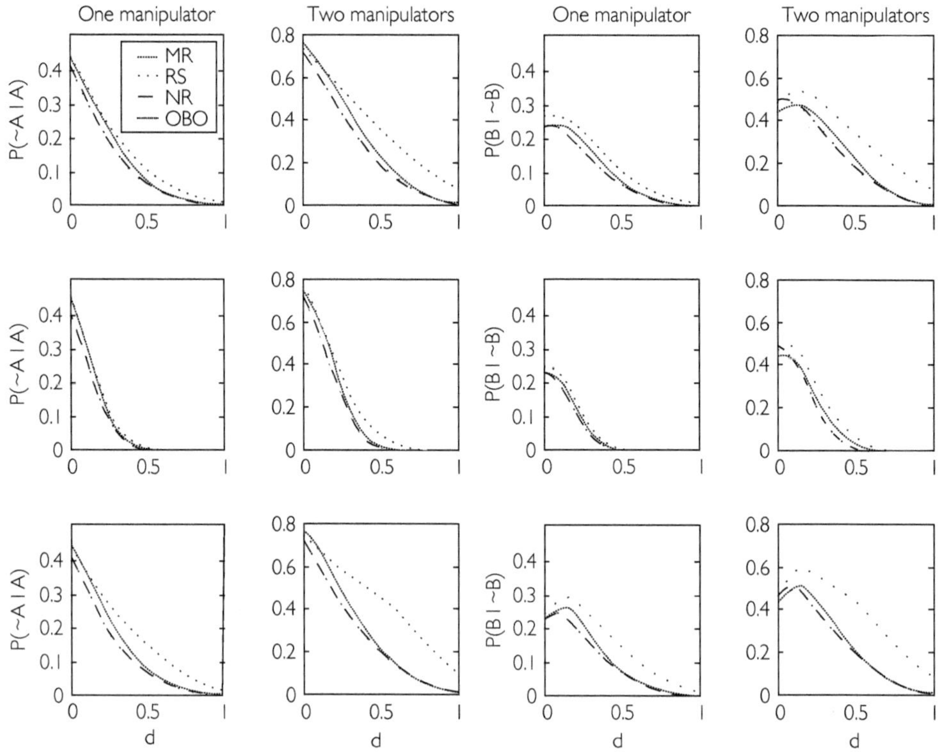

Figure 8.2 Robustness against manipulation Strategy II (to make B win and A lose) with $m = 9$ judges and $n = 3$ contestants. In addition to P(B|∼B), P(∼A|A) is added to compare the rules' robustness and the smaller the P(∼A|A), the more robust. The line styles and score distributions are the same as Figure 8.1. Again, The OBO rule is nearly the same as the NR and they are indistinguishable from the curves.

For each of them, $P(B|\sim B)$ and $P(\sim A|A)$ were evaluated. Table 8.4 shows the average success rates from these 10,000 cases under $m = 9$ with one or two manipulators. The results are similar to what we have seen in Figures 8.1 and 8.2, except that the NR is better in $P(B|\sim B)$ with two manipulators.

Table 8.4 ***The success rate of manipulation averaging over 10,000 randomly picked ps from the simplex $p_1 + p_2 + \cdots + p_6 = 1$ with m = 9 judges and n = 3 contestants. The smaller the number the better.***

Manipulators	Strategy I				Strategy II							
	P(B|∼B)				P(B|∼B)				P(∼ A|A)			
	RS	MR	OBO	NR	RS	MR	OBO	NR	RS	MR	OBO	NR
1	0.25	0.21	0.198	0.198	0.22	0.20	0.193	0.192	0.45	0.39	0.395	0.394
2	0.44	0.40	0.382	0.377	0.40	0.39	0.383	0.378	0.70	0.66	0.639	0.639

Number of Ties

As mentioned before, the number of ties produced from a rule affects its properties. Consider the extreme case where a rule makes everyone the first place winner. This rule is obviously robust against any manipulation. Hence, it is important to know whether the robustness of any method is due to the number of ties. Table 8.5 states the percentage of ties for all rules. Note that the new rule gives fewer ties than the other rules. For the same sensitivity, the rule producing fewer ties should be considered the better rule.

More than three candidates

Under our new rule, when there are only three candidates, the winner is always a Condorcet winner because of the number of pairwise wins. For larger number of contestants, however, the situation will be more complicated. Problems such as those discussed in Tables 8.1 and 8.2 can also occur under the new rule. One reason is that many winning cycles can be formed such as the situation in Figure 8.3, where *E* beats *A*, *C*, and *D* and becomes the winner by the new rule, but she can be ranked worse than *B*. A concrete example for this figure is that two judges

Table 8.5 ***Number and Percentage of cases that have two-way and three-way ties when n = 3.***

Type of ties	Number of judges	Exact number				Percentage			
		RS	MR	OBO	NR	RS	MR	OBO	NR
Two-way	5	1560	1080	0	0	20.06%	13.89%	0	0
	7	55,440	30,240	0	0	19.80%	10.80%	0	0
	9	1,900,080	589,680	15,120	0	18.85%	5.85%	0.15%	0
Three-way	5	360	0	360	0	4.63%	0	4.63%	0
	7	10,080	0	10,080	0	3.60%	0	3.60%	0
	9	290,640	94,080	290,640	94,080	2.88%	0.93%	2.88%	0.93%

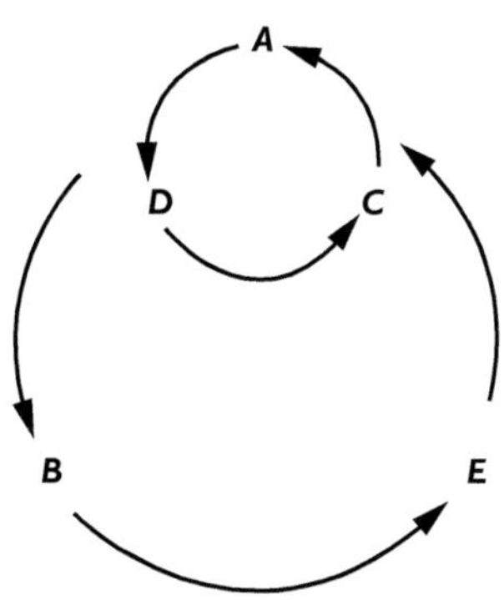

Figure 8.3 An example of the nontransitivity of ranking of a large number of contestants, where A → D means A is ranked better than D in pairwise comparison. Here, E is better than A, C, and D, and A, C, and D are better than B, but B is better than E, while the pairwise comparison of A, C, and D is also circular. Such a scenario occurs when two judges rank with preference order of ABEDC, two others with DCBEA, and one with ECADB. In this case, B is ranked better than E in four out of the five judges.

rank with preference order of *ABEDC*, two others with *DCBEA*, and one with *ECADB*. In this case, *B* is ranked better than *E* in four out of the five judges.

However, counter-intuitive cases occur in every rule due to the impossibility theorem. One objective way to compare them is to check the sensitivity with known true winner. We still use the normal model discussed in the earlier section on "Score distribution under normality assumption." The exhaustive method is no longer feasible due to the huge number of rank preferences ($m^{n!}$). We use simulation to evaluate their performance. The two criteria for measuring manipulation success are the same as those for only three candidates. The manipulation strategies are simple extensions of the previous ones, that is, (8.1) to let *B* win, the judge will make *B* number 1 and reverse the order of the other ranks, and (8.2) to let *B* win and *A* lose, he will rank *B* first, *A* last, and reverse the other ranks. The results are very similar to those for the three candidates. Only 5 and 10 candidates are presented in Figures 8.4 and 8.5. We can see that now the improvement of NR over OBO is more pronounced. In addition, the RS rule becomes more susceptible to manipulation as n or d increases. The reason that RS is doing so poorly can be seen from Table 8.1. We have simulated a more subtle manipulation strategy by lowering *A*'s rank only by 2 and putting *B* as the first while keeping the order of the other contestants unchanged (Strategy III). The results also show that the new rule is more resistant to manipulation.

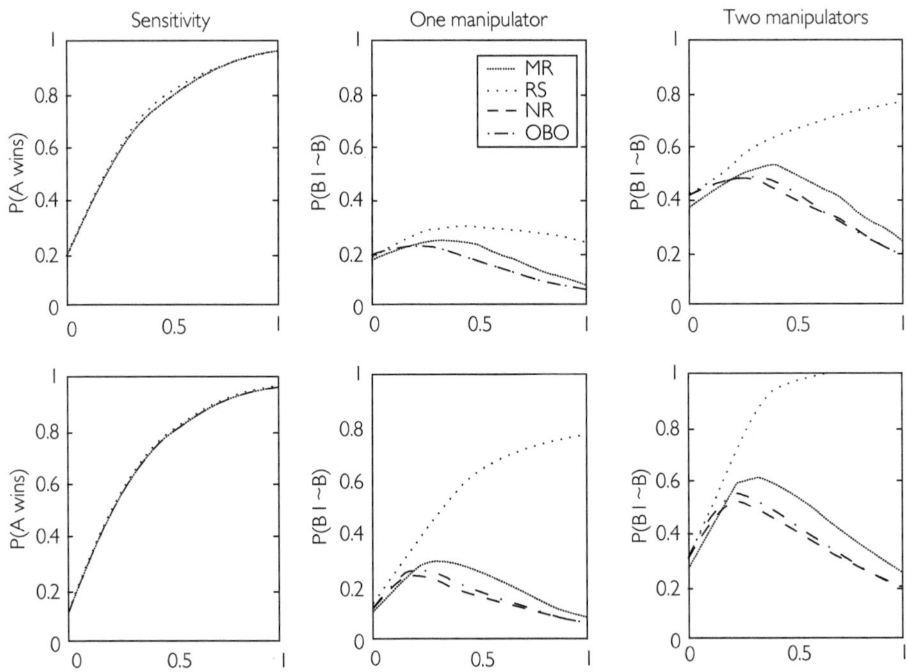

Figure 8.4 Sensitivity of the four rules and their robustness against manipulation Strategy I with $m = 9$ judges and five or ten contestants. The top row presents results for $n = 5$ contestants and the bottom row is for $n = 10$ contestants. The scores are simulated from normal distributions with mean $6 - id$, $i = 1, \ldots n$ and standard deviation $\sigma = 0.5$. The line styles and manipulation measures are the same as Figure 8.1.

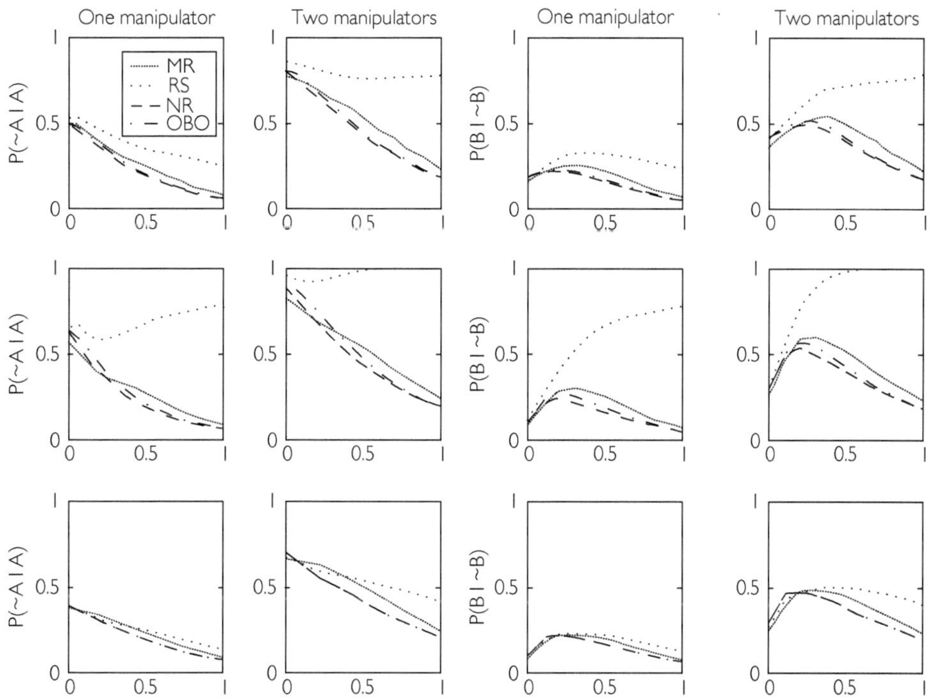

Figure 8.5 Robustness against manipulation Strategy II, which lowers A to the worst and raises B to the best, and Strategy III, which lowers A by two places then moves B to the first. The top two rows show results for Strategy II, with 5 (top row) and 10 (middle row) contestants. The bottom row shows results for Strategy III with 10 players, all with 9 judges. The score distributions and the line styles are the same as Figure 8.4. Note that in the bottom row, the OBO and the NR are indistinguishable from the curves.

Summary and discussion

When the outcome of a sport is determined by subjective evaluation by judges, it is vulnerable to manipulation. A new rule is constructed to reduce the success rate of a manipulator without losing some of the important requirements of a good rule such as high Condorcet efficiency, small chance of ties and fewer instances of monotonicity violation. The new rule contains more steps than the old, but each step is either used before or simple to comprehend. Our new rule is still relatively easy to implement.

The new Olympic decision rule for skaters adapted on June 7, 2002, in Kyoto made several changes that included a stricter ethics code and the use of scores from a random selection of nine judges from the total of 14 judges. One obvious reason to change back to the original scores instead of ranks is its simplicity. Now everybody can understand how the final placement is determined. However, as we mentioned at the beginning of this article, and apparently known to many previous rule founders, the use of the original scores can be disastrous. The judge who assigns with a larger range has more weight in the final decision than the other judges. We do not see any way of amending this, unless the scoring is defined so rigidly that every judge has to use same range. Our feeling is that the rank rule will return either when the ISU finalizes its new rule, or when a controversial decision occurs in the future. The random

Table 8.6 ***Summary of comparisons on a few important properties for with m = 9 judges and n = 3 contestants (the larger the values, the more desirable).***

Category		RS	MR	OBO	NR
Type I monotonicity		100%	100%	100%	100%
Type II monotonicity		91.04%	89.63%	95.09%	96.03%
Condorcet efficiency	IC model	92.18%	90.48%	100%	100%
Percentage of ties	two-way	worst	3rd	2nd	best
	three-way	2nd	best	2nd	best
Robustness in $P(B\|\sim B)$	1 manipulator	worst	better	best	best
	2 manipulators	worse	better	better	better
Robustness in $P(\sim A\|A)$	1 or 2 manipulators	worst	better	best	best

sample rule is advantageous only when the number of bad judges is a constant. If the proportion of bad judges remains unchanged, the random sample rule has no advantage, but it may produce a controversy if the unused scores are somehow revealed.

It is interesting to ask what would have happened if the new rule had been applied to the infamous pairs competition in the 2002 Winter Olympic figure skating. Unfortunately, the new rule could not have changed the initial results. The reason is that the top two pairs were so good that they dominated the ranks 1 and 2. The French judge's vote changed the 1 to 2 rank ratios from 4:5 to 5:4. In this case, one manipulator had the deterministic power and no reasonable rule can defeat this sort of manipulation.

We have also studied a bottom-up elimination rating method that eliminates the worst candidates first then the second worst, etc. In our exhaustive search, we have found that this method is very robust against manipulation. However, like most elimination rules (Nanson 1882; Nurmi 1999), it violates the Type I monotonicity. Due to the conflict between sensitivity and robustness, it is unlikely that any good rule could dominate other good rules in all aspects. Table 8.6 summarizes the comparison of the four studied rules with regard to important properties. It shows that the 2002 ISU OBO rule is better than the 1998 MR rule, and it is fair to say that the NR is a viable alternative to the current rules.

REFERENCES

Arrow, K. J. (1963), *Social Choice and Individual Values* (2nd ed.), New York: John Wiley.

Bassett, G. W., and Persky, J. (1994), "Rating skating," *Journal of the American Statistical Association*, 89, 1075–1079.

Berg, S., and Lepelley, D. (1994), "On probability models in voting theory," *Statistica Neerlandia*, 48, 133–146.

Borda, J. de (1781), *Memoire sur les elections au scrutin*, Histoire de l'Academie Royale des Sciences. (Translated in Sommerlad and McLean (1989), 122–129.)

Berry, S. M. (2002), "Gold medals, protests, and trading ratings: The 2002 Winter Olympics figure skating," *Chance*, 15, 14–18.
Condorcet, M. de (1785), *Essai sur l'application de l'analyse à la probabilité des décisions rendues à la pluralité des voix*, Paris: Imprimerie Royale. (Translated in Sommerlad and McLean (1989), 69–80.)
David, H. A. (1988), *The Method of Paired Comparisons* (2nd ed.), London: Chapman and Hall.
Gehrlein, W. V. (1997), "Condorcet's paradox and the Condorcet efficiency of voting rules," *Mathematica Japonica*, 45, 173–199.
Gibbard, A. F. (1973), "Manipulation of voting scheme: a general result," *Econometrica*, 41, 587–601.
Glickman, M. E. (1999), "Parameter estimation in large dynamic paired comparison experiments," *Applied Statistics*, 48, 377–394.
Glickman, M. E., and Jensen, S. T. (in press), "Adaptive paired comparison design," *Journal of Statistical Planning and Inference*.
Matthews, P. (2003), "AccelRat Software for Go Player Ratings and Tournament Pairings," available at http://ourworld.compuserve.com/homepages/accelrat/; accessed on October 5, 2003.
Nanson, E. J. (1882), "Methods of Election," *Transactions and Proceedings of the Royal Society of Victoria XIX*, 197–240.
Nurmi, H. (1999), *Voting Paradoxes and How to Deal with Them*, Berlin: Springer-Verlag.
Saari, D. G. (1995), *Basic Geometry of Voting*, Berlin: Springer-Verlag.
Satterthwaite, M. A. (1975), "Strategy proofness and arrow's conditions: Existence and correspondence theorems for voting procedures and social welfare functions," *Journal of Economic Theory*, 10, 187–217.
Sommerlad, F., and McLean, I. (1989), *The Political Theory of Condorcet*, University of Oxford Working Paper.
U. S. Figure Skating Association (1998), *The Official Book of Figure Skating*, New York: Simon & Schuster.

CHAPTER 9

An analysis of decathlon data

TREVOR F. COX AND RYAN T. DUNN

Abstract

Data in the form of times, distances and scores for athletes competing in the decathlon at the meetings of the International Amateur Athletic Federation from 1991 to 1999 were collected and then subjected to statistical analysis. The first part of the analysis concentrated on the multivariate description of the data by using cluster analysis and classical scaling. The second part of the analysis questioned whether the decathlon could be considered fair or unfair to particular athletes, e.g. the track specialist or the field specialist, since the spirit of the decathlon is that all 10 events are equally regarded. Statistical techniques used were based on Bartlett's test of homogeneity of variance, Kendall's coefficient of concordance, Spearman's rank correlation and cumulative sum charts. The tentative conclusion is that the decathlon favours those athletes who do well at the field events.

Keywords: *athletics, classical scaling, cumulative sum charts, sports data*

Introduction

Various authors of articles on statistics in sport have analysed athletics data. For example Morton (1997) looked at the effects of lane allocation on times in running races. Grubb (1998) devised models for comparing athletic performances. Dawkins et al. (1994) analysed Olympic heptathlon data by using cluster analysis and correspondence analysis.

This paper looks at data for the decathlon collected at the biennial International Amateur Athletic Federation (IAAF) World Athletics Championships held in 1991 (Tokyo), 1993 (Stuttgart), 1995 (Gothenberg), 1997 (Athens) and 1999 (Seville). Data were taken from the IAAF Web site. The decathlon takes place over 2 days and consists of the following track and field events: (day 1) 100 m race, long jump (LJ), shot-put (SP), high jump (HJ), 400 m race; (day 2) 110 m hurdles (110 mh), discus (Dis), pole-vault (PV), javelin (Jav) and 1500 m race. The athletes achieve times in minutes and seconds for track events and distances or heights in metres for field events. Point scores are awarded for each event on the basis of these times and distances, using IAAF formulae which are not readily available.

The data taken from the Web site consisted of times, distances and points awarded for each athlete for each event. However, there were some missing points scores. A plot of points awarded against times or distances indicated a very nearly perfect linear relationship. Hence, using simple linear regression, it was possible to estimate the missing scores which were then used as replacement scores. Only a few scores were actually missing and so statistical analyses were unlikely to be much affected anyway.

The analysis given here is in two parts. Firstly, dimension reduction multivariate methods are used to describe the data. The second part discusses how the decathlon can be considered to be fair or unfair to particular athletes. For example is the order of the 10 events fair to those athletes whose best events are the track events? Does the decathlon favour athletes who are good at the 100 m race or perhaps the javelin? This would go against the spirit of the decathlon, which is that all 10 disciplines are equally regarded.

Multivariate description

Although the events in the decathlon do not change over the years, the individual athletes obviously do. It is more difficult to analyse the data in terms of the athletes than it is for the events and the latter are concentrated on here.

First a cluster analysis was carried out on 'distances' between events, defined by $\delta_{rs} = 1 - \rho_{rs}$, where ρ_{rs} is the correlation between the results for the rth and sth events, but noting that negative times have been used to put times and distances on an equal footing (athletes try to minimize times but to maximize distances). Alternatively the scores for the events could have been used with identical results because of the linear relationship between scores and times or distances. The single linkage, the complete linkage and Ward's linkage cluster methods (see, for example, Everitt et al. (2001)) were tried. Complete linkage and Ward's linkage gave similar results, which were also consistent over the five meetings. For brevity, Figure 9.1 shows the dendrograms using Ward's linkage for the five meetings and also for the combined data.

Table 9.1 shows possible clusters gleaned from the dendrograms for the five meetings. It can be seen that the discus and the shot-put always occur together in the same cluster. The 110 m hurdles and the 400 m races usually fall in the same cluster as the 100 m race, as does the long jump. The high jump, pole-vault, javelin and 1500 m race were not so consistent with cluster membership.

Next the data were subjected to classical scaling (see, for example, Cox and Cox (2000)) using $\{\delta_{rs}\}$ as before as dissimilarities. For each meeting, approximately 50% of the total variation could be explained by a two-dimensional configuration. For ease of presentation these two-dimensional configurations are given in Figure 9.2, rather than three- or four-dimensional configurations which could have accounted for much more of the variation. The configurations for the last four meetings were matched to that of the first by using Procrustes analysis. The main feature in the plots is that the track events are separated out from the field events with the track events to the left of the configurations and the field events to the right. For most of the configurations the javelin is not close to any other event.

Fairness of the decathlon

There are several ways in which we can question whether the decathlon is 'fair'. The decathlon tests athletes' all-round ability. To win they must be good at many events, both track and field. It could be argued that the decathlon should not be biased towards any one event or any group of events such as the track events. For example, the athlete who excels at the 100 m race should

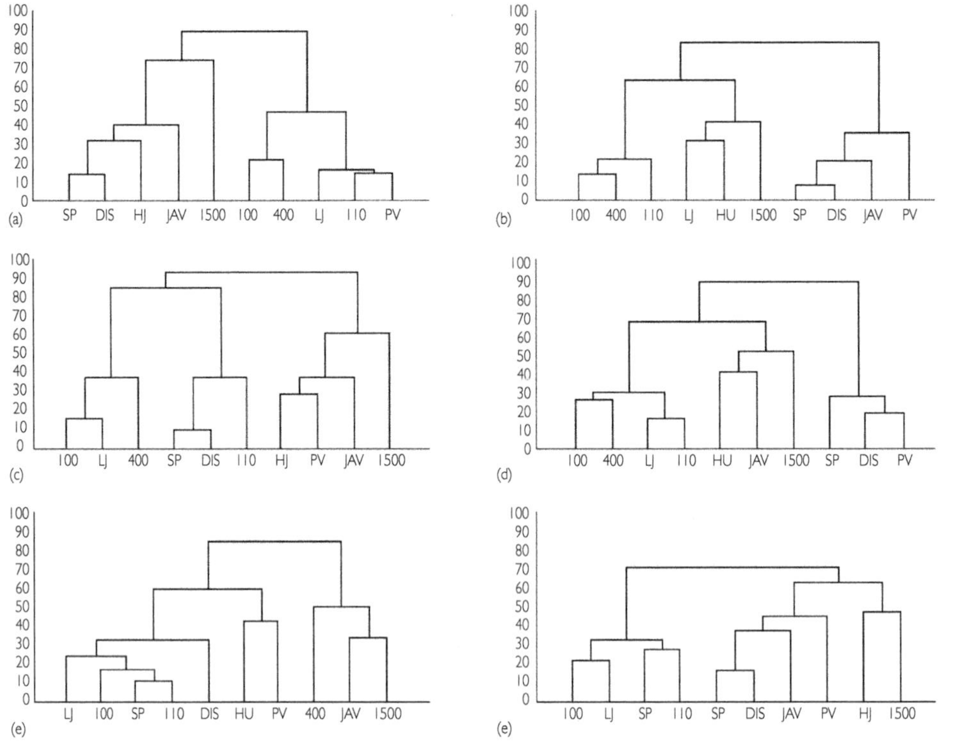

Figure 9.1 Dendrograms for the decathlon events: (a) 1991; (b) 1993; (c) 1995; (d) 1997; (e) 1999; (f) all years combined.

Table 9.1 ***Cluster analysis of the decathlon events.***

Meeting	Clusters
1991	(100 m, 400 m, LJ, 110 mh, PV) (SP, Dis, HJ, Jav) (1500 m)
1993	(100 m, 400 m, 110 mh) (SP, Dis, Jav, PV) (LJ, HJ, 1500 m)
1995	(100 m, LJ, 400 m) (SP, Dis, 110 mh) (HJ, PV, Jav, 1500 m)
1997	(100 m, 400 m, LJ, 110 mh) (HJ, Jav, 1500 m) (SP, Dis, PV)
1999	(LJ, 100 m, SP, 110 mh, Dis) (HJ, PV) (400 m, Jav, 1500 m)
Combined	(100 m, 400 m, LJ, 110 mh) (SP, Dis, Jav, PV) (HJ, 1500 m)

not be more likely to win the decathlon than an athlete who excels at the discus, i.e. the points advantage for a top athlete over the other competitors should not favour any particular event or group of events.

Are the formulae used for converting times and distances into points fair? Should the formulae take into account the world records for the particular events? (We carried out a small exercise in which points for each event were awarded, not on a linear scale, but using a quadratic scale for

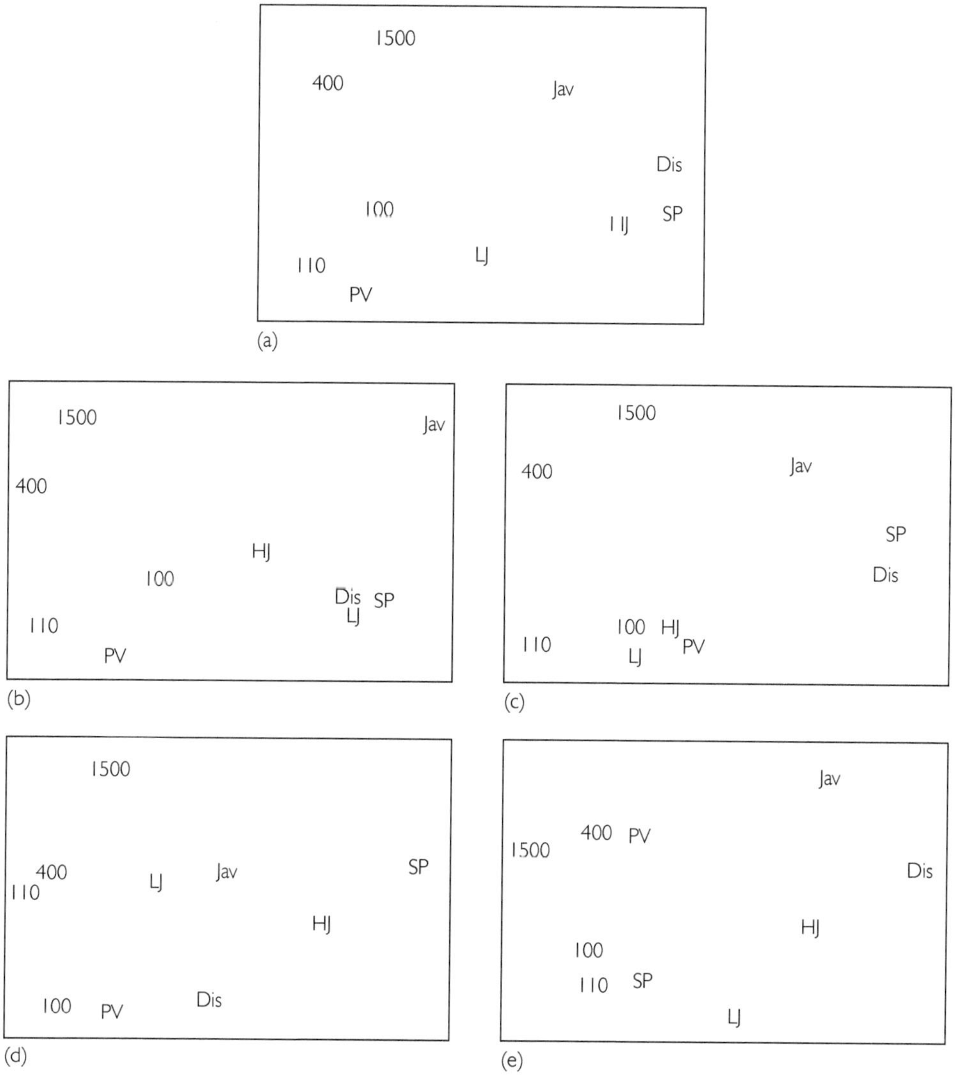

Figure 9.2 Classical scaling of the decathlon events: (a) 1991; (b) 1993; (c) 1995; (d) 1997; (e) 1999.

(a) boosting scores for the best athletes and
(b) downweighting the scores for the best athletes.

The main problem encountered was of calibration—what constitutes a poor or good performance and how many points should be awarded for such?) It is easy to see why the linear formulae are used for simplicity, especially as they easily allow for the world record total number of points for the decathlon to have meaning and to perpetuate. However, performance is not linear: the extra effort needed in running 100 m in 8 s compared with 9 s will be at least an order of magnitude greater than the extra effort needed in improving the time from 10 s to 9 s.

Is the order in which the events take place fair? We can see the reasoning in keeping the track events spaced out in time to help athletes to recover from exhaustion, especially by putting the

400 m race at the end of day 1 and the 1500 m race as the very last event. But is this fair on the 1500 m specialist, who would perhaps like to have his best event scheduled first so that he can attain some boost to morale early on? For many of the competitors the decathlon is virtually over for them by the time that it comes to the 1500 m event. Again, for world records to perpetuate, the order of events must stay the same. The above points of fairness are raised and some analyses are given below to investigate further.

Now because of the linear nature of the scoring method for the decathlon, in one sense it does not matter whether the population mean scores for the 10 events are equal. If the 400 m race gives all competitors 100 more points on average than the 100 m race, then this is of no consequence to the positions of the athletes in the final ranking. However, the population standard deviations (or some other measure of variation) ought to be equal for the 10 events. If for instance the standard deviation for the 100 m race was much larger than that for the 400 m race, then the athlete who excels at the former would gain a better score advantage over the other competitors than would the athlete who excelled at the latter. To investigate this the results for the five meetings were pooled and then Bartlett's test for homogeneity of variance was carried out to compare the variances for the 10 events. For each event there were 91 observations. (Strictly speaking we do not have independent observations because some athletes competed at more than one meeting and because of the correlation of an individual's results for the 10 events.) Table 9.2 gives the sample standard deviations and the 95% Bonferroni confidence intervals for the population standard deviations. The value of Bartlett's statistic was 358.5 ($p = 0.000$). The 1500 m event is seen to have a very high standard deviation. The value of Bartlett's statistic (84.8) is much reduced if this event is not included; however, it is still very significant ($p = 0.000$).

The order of the 10 sample standard deviations is

400 m 100 m 110 mh Dis SP LJ HJ Jav PV 1500 m.

Note that the three smallest standard deviations are for track events. The field events have larger standard deviations. This suggests that the decathlon could favour those athletes who are good in the field events.

Table 9.2 ***Test for homogeneity of variances.***

Event	Sample standard deviation	Confidence interval
100 m	56.2	[46.3 70.6]
LJ	74.3	[61.3 93.4]
SP	69.0	[56.9 86.8]
HJ	77.0	[63.5 96.8]
400 m	51.9	[42.8 65.3]
110 mh	58.6	[48.3 73.7]
Dis	66.0	[54.5 83.0]
PV	106.4	[87.8 133.9]
Jav	94.8	[78.2 119.3]
1500 m	206.2	[170.1 259.4]

Table 9.3 ***Kruskal–Wallis test for equal medians.***

Event	Median score
100 m	863
LJ	876
SP	749
HJ	803
400 m	868
110 mh	910
Dis	733
PV	849
Jav	750
1500 m	716

Out of interest, the Kruskal–Wallis test was carried out to test whether the population median scores were equal for the 10 events. Table 9.3 shows the medians for the 10 events. The value of the test statistic was 422.1 ($p = 0.000$) and, if the value for the 1500 m event is removed, 370.1 ($p = 0.000$). Thus the hypothesis of equal median scores is rejected. The events in order of median score, from smallest to largest, are

1500 m Dis SP Jav HJ PV 100 m 400 m LJ 110 mh.

The next method used to investigate the fairness of events was Kendall's coefficient of concordance, W (see Kendall and Gibbons (1990)). This treats each event as a 'judge' of the athletes, ranking them on their position in the event. Kendall's coefficient of concordance measures how well the 10 events agree on the ranking of the athletes. Table 9.4 gives the value of W for the five meetings.

The values of W are all significant at 1% apart from that for the 1995 meeting which is not significant. So, for four of the meetings, the events 'agree' with each other on the ranking of the athletes. The 1995 meeting appears discordant to the other meetings. Looking at the data, for this meeting athletes who did well overall tended to have several events with 'disappointing'

Table 9.4 ***Kendall's coefficient of concordance for the events.***

Meeting	W
1991	0.307
1993	0.363
1995	0.151
1997	0.254
1999	0.264

results, whereas, for the other meetings, athletes who did well overall tended to do well for all events. For example, in 1995, Smith of Canada was fourth overall but was placed 16th for the shot-put, 18th for the 110 m hurdles, 17th for the discus and 15th for the 1500 m race. Similarly, Spada of Switzerland who was 17th overall managed good placings of fourth in the shot-put, fourth in the discus and fourth in the 1500 m race.

Spearman's ρ, a coefficient of rank correlation, was calculated for each event with the overall ranking of athletes. Results are in Table 9.5.

The events placed in order of Spearman's rank correlation are

1500 m HJ 400 m Jav 100 mh PV 100 m SP Dis LJ.

The correlations for the 1500 m race offer no surprises although two are negative and one is zero. The 1500 m race is the last event and by this stage most athletes will usually know their final position. Disappointed athletes might not put too much effort into this final strenuous event, knowing that it will not make very much difference to their overall performance. Note that field events have four of the five largest correlations.

If athletes who finish high in the rankings gain advantage from a particular event, then for these athletes the proportion of their overall score earned from this event will be high. This can be tested by using the following. Let the athletes be labelled from 1 to n according to their overall position in the decathlon, i.e. first, second etc. Let x_{ij} be the score for athlete j for event i. Let y_{ij} be the proportion of the jth athlete's score produced by event i, i.e.

$$y_{ij} = x_{ij} / \sum_i x_{ij}.$$

Let

$$\bar{y}_i = n^{-1} \sum_j y_{ij}$$

be the mean proportion for the ith event.

Table 9.5 ***Spearman's rank correlation for the events with the overall ranking of the athletes.***

Meeting	Spearman's ρ for the following events:									
	100 m	LJ	SP	HJ	400 m	110 mh	Dis	PV	Jav	1500 m
1991	0.65	0.88	0.80	0.53	0.41	0.67	0.58	0.69	0.51	−0.01
1993	0.75	0.59	0.70	0.29	0.64	0.49	0.86	0.60	0.65	0.53
1995	0.45	0.65	0.28	0.58	0.31	0.46	0.60	0.63	0.30	−0.35
1997	0.45	0.61	0.58	0.24	0.56	0.47	0.56	0.57	0.57	0.00
1999	0.60	0.77	0.71	0.42	0.26	0.76	0.66	0.40	0.24	0.40
Mean	0.58	0.70	0.61	0.41	0.44	0.57	0.65	0.58	0.45	0.11

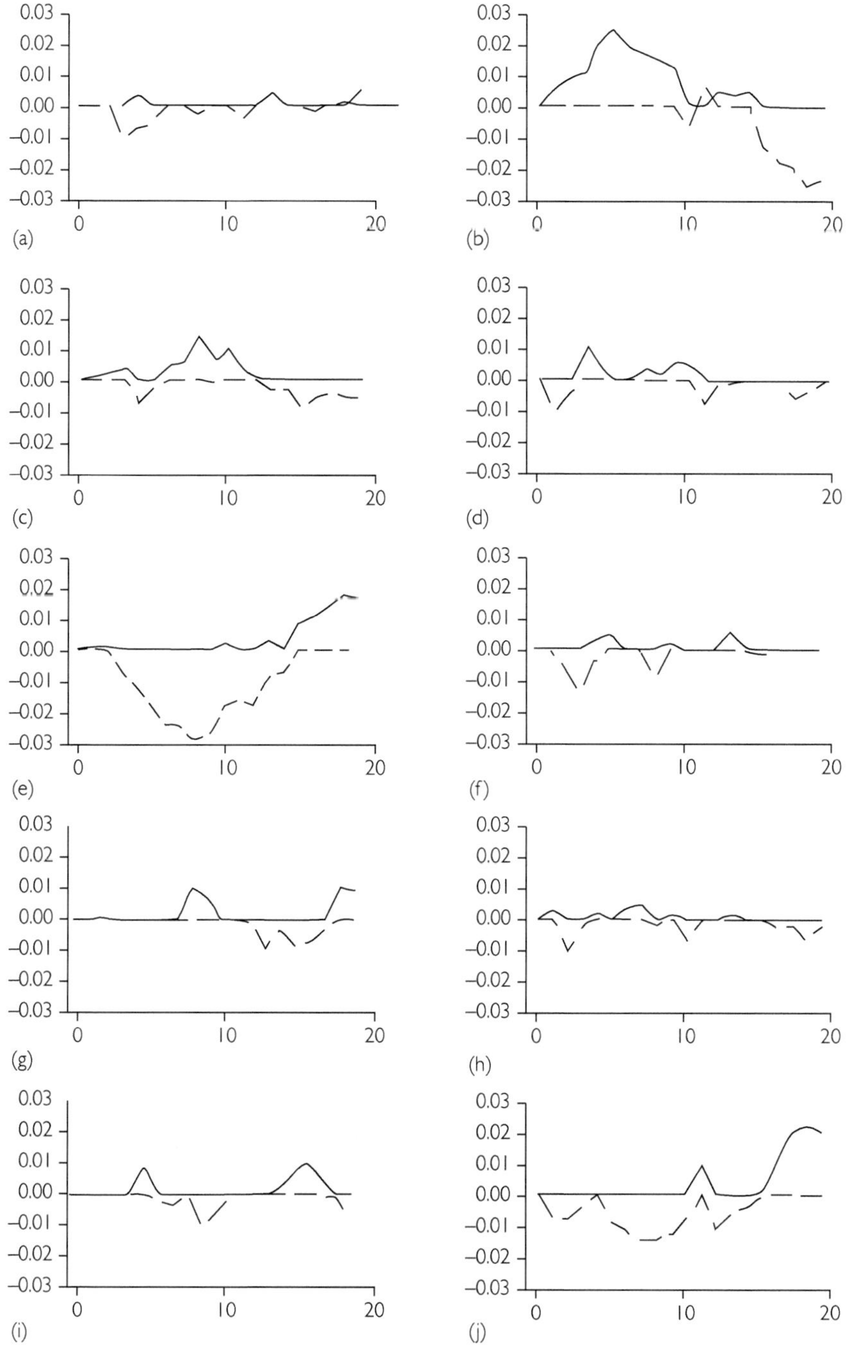

Figure 9.3 CUSUM charts for the decathlon events: (a) 100 m; (b) long jump; (c) shot-put; (d) high jump; (e) 400 m; (f) 110 m hurdles; (g) discus; (h) pole-vault; (i) javelin; (j) 1500 m.

Table 9.6 ***Events with ∧ and ∨ CUSUMs.***

Meeting	Events with the following CUSUM shapes:	
	∧	∨
1991	LJ	400 m
1993	SP	110 mh
1995	HJ	110 mh
1997	SP	100 m
1999	PV	400 m

Consider the cumulative sum (CUSUM)

$$Y_{iJ} = \sum_{j=1}^{J} (y_{ij} - \bar{y}_i)$$

for the ith event. If the events are fair, then the plotted CUSUMs (Y_{iJ} plotted against J) should show no particular pattern and be rather flat. This is because the CUSUM is accumulating the proportion of overall scores, corrected for the mean, and this should not depend on the final positions of the athletes. If athletes who do well overall tend to benefit from a particular event, then the CUSUM for that event should rise steeply and then drop away to 0. Denote the shape of this type of CUSUM by ∧. If there is an event that the best athletes tend not to do well in, then the CUSUM will drop steeply and then rise to 0. Denote the shape of this type of CUSUM by ∨. Only the CUSUMs for the 1991 meeting are shown here. These are in Figure 9.3. It can be seen that the CUSUM for the long jump has the ∧ shape and that for the 400 m event the ∨ shape. If this pattern were to be repeated for the other meetings, it could be said that the best athletes benefit too much from the 400 m event. For each meeting there appeared to be one event with a ∧ CUSUM shape and one event with a ∨ CUSUM shape. Table 9.6 lists these events.

It can be seen that there is no one event that consistently has a ∧ CUSUM shape, nor a ∨ CUSUM shape. However, it is interesting that all the ∧ CUSUM shapes are for field events, whereas the ∨ CUSUM shapes are for track events.

Overall, and not surprisingly, the analyses show that the decathlon events broadly fall into two groups: the track events and the field events. The javelin is a technically difficult event and stands out from the other field events. The shot-put and discus tend to form a pair. The CUSUM results show that no one event tends to benefit the more successful athletes. However, the tentative conclusion from the above analyses is that the decathlon favours those athletes who can do well at the field events.

REFERENCES

Cox, T. F. and Cox, M. A. A. (2000) *Multidimensional Scaling*, 2nd edn. Boca Raton: Chapman and Hall–CRC.

Dawkins, B. P., Andrea, P. M. and O'Connor, P. M. (1994) Analysis of Olympic heptathlon data. *J. Am. Statist. Ass.*, **89**, 1100–1106.
Everitt, B. S., Landau, S. and Leese, M. (2001) *Cluster Analysis*, 4th edn. London: Arnold.
Grubb, H. J. (1998) Models for comparing athletic performances. *Statistician*, **47**, 509–521.
Kendall, M. G. and Gibbons, J. D. (1990) *Rank Correlation Methods*, 5th edn. London: Arnold.
Morton, R. H. (1997) Statistical effects of lane allocation on times in running races. *Statistician*, **46**, 101–104.

PART III

American Football

Introduction

SHARIF MELOUK

American football is a sport played globally, but is most popular in the United States, Canada, Europe, and Japan. In the United States, the organized forms of the sport begin at youth levels and continue up through collegiate and professional levels. At each level, teams are grouped into leagues or divisions. Teams then compete with each other in head-to-head games during a period of time termed the regular season. The goal of every team is to win as many games as possible during the regular season to qualify for the immediately following playoff season, which determines the team that wins the championship of its league. An example is college football in the United States, where a season includes a 12-game regular season that is followed by a playoff season. A complicated evaluation process determines the teams to play for a championship.

American football is played by two teams that exchange possession of a ball (i.e., the football) to score points. The team with the highest total of points at the end of the game wins. At the collegiate and professional levels, a game lasts 60 minutes and is divided into two halves of 30 minutes with each half further divided into quarters of 15 minutes. A halftime period separates the two halves. Due to play and time stoppages, an entire game typically lasts about three hours. In the event of a tie game at the end of 60 minutes, one or more overtime periods may be played to determine the winner.

The U.S. version of the game is played on a rectangular field measuring 120 yards in length and about 53 yards in width. Each end of the field has an area termed an end zone measuring 10 yards in length and a U-shaped goal post. Figure P III.1 illustrates the field of play used in American football.

The numerical values on the field help with tracking and positioning of the football during game play. Vertical line markers on the field are separated by 10 yards. It should be noted that the play field in Canadian football has different dimensions with a rectangular field measuring 150 yards in length and 65 yards in width. The field has goal posts at each end, but the end zones on a Canadian football field are 20 yards in length. All subsequent discussion of the field relates to the U.S. version of football.

The end zones and goal posts play a primary role in scoring points. Each team fields an offensive unit and a defensive unit with each unit consisting of 11 players, each of whom plays

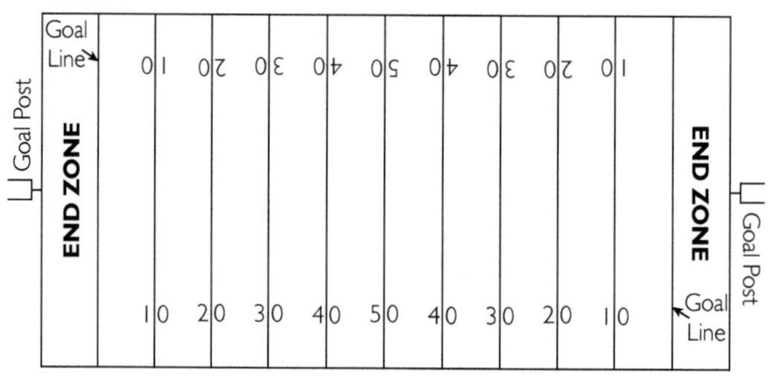

Figure P III.1 American football field.

a specific position and performs a specialized role during game action. The teams alternate possession of the football (a prolate spheroid approximately 11 inches long and 22 inches in circumference at the center, similar to but slightly more elongated than the ball used in rugby) in an attempt to score points. In the Canadian version of the sport, each unit consists of 12 players, and the ball used is slightly smaller than football used in the U.S. In both U.S. and Canadian football, a team typically scores points via its offense, although it is possible for the defense to score points for its team as well.

Routine game play takes place on the 100 yards of field between the end zones and involves the offense attempting to move the football down the field against the defense of the opposing team via running and/or passing plays. In general, the offense is given up to four plays (or downs) to advance the football at least ten yards against the defense. If successful, the offense receives a new set of downs to advance at least ten yards. This process continues until the offense scores points, is unable to advance the football, or commits a turnover (a fumble or an interception). An offense scores points by advancing the football into the end zone (i.e., a touchdown) or by kicking the football through the goal post (i.e., a field goal). If there is no scoring opportunity and it is unlikely after three plays in a set of downs that an offense will be able to advance the ball at least ten yards, a team possessing the ball typically kicks (a.k.a., punts) the football to the opposing team, which then attempts to advance the football and score points starting from the newly established field position. The game is very strategic in nature with specific plays and player formations dictating play on the field.

Given the popularity of American football worldwide and its complexity, much research effort has been devoted to examining various aspects of the game. Statistical research plays an important part in the investigation of the sport from the perspective of the fans, players, and management. In the era of Big Data, massive amounts of data are now electronically available regarding every facet of the sport, leading to many research efforts incorporating statistical modeling and analysis.

Many statisticians have focused on developing ranking methods to determine which teams are best. These types of models usually produce a rank ordering of teams from best to worst so as to determine the teams most deserving of competing for championships at the end of a regular season. All leagues have methods to determine the teams that participate in the playoffs, and some of these methods employ statistical modeling and analysis as part of the process.

Research on developing prediction models is quite prevalent in the literature. Being able to successfully predict the outcome of games certainly has benefits. However, as with other sports, prognostication is challenging. With the massive amount of data available, a seemingly unlimited number of factors exist for inclusion in prediction models.

Much research exists on modeling the actual play of the game. These models have a strategic focus with the objective of aiding in-game decision makers with their play calling. Models that capture the situation of a game (e.g., time remaining in the game, current score, current field position of the ball) as well as historical tendencies of teams, players, and coaches may provide an advantage to decision makers and influence the outcome of a game.

A cliché, yet likely truthful, phrase in sports is that "players win games," simply meaning that the players on the field, not other external factors, determine the outcomes of games. So, a critical aspect of any team sport, including American football, is team composition. Several works attempt to construct player evaluation models to aid teams in personnel decisions. Team construction is a complex task with many constraints complicating the process.

Finally, a newly emerging area of statistics-based research deals with the violent aspect of American football. Due to its physical nature, injuries are commonplace, some of which are serious and life altering; for example, awareness of and concern for brain injuries, including chronic traumatic encephalopathy, have increased sharply over the past decade. Thus, researchers are examining data to help determine which players are at the greatest risk for severe injury and develop diagnostic models for decisions and actions to improve player safety.

The featured articles on American football in this anthology provide more specific examples of statistical research and its usefulness in addressing various issues in the sport. Three articles relate to determining a champion in college football in the United States. A formula is used to determine the top two teams at the end of the regular season. These teams then play a single game to determine the champion for that year. Part of the formula is based on results from computer algorithms that employ statistical methods. Stern et al. (2004) describe the history of the college football championship process and offer ways that statisticians could assist in the process. The authors suggest that statisticians are well-equipped to address any mathematical component of the process. Berry (2003) studies part of this ranking process by investigating the computer algorithms and suggests that margin of victory should be included in team rankings determination. Mease (2003) also studies college football team ranking using a penalized maximum likelihood approach. The model only considers the win/loss record of teams. Results show the rankings produced are more consistent with viewpoints of college football experts and pundits, as opposed to the current evaluation method.

Two articles focus on in-game strategic decision making. Willoughby (2001) examines a strategic decision common to the version of American football played in Canada. The decision involves the potential placement of a player in the end zone during a field goal attempt. This decision impacts the starting field position of the opposing team's next possession. They use linear regression to capture the relationship between starting field position and the expected number of points scored based on that field position. Berry (2004) analyzes the strategic effectiveness of calling a timeout just before a field goal attempt near the end of a close game. This strategy is termed "icing" the field goal kicker, with hopes of increasing the pressure on the kicker by making him think about the upcoming kick. Based on historical data and using a logistic modeling approach, results show the strategy does influence the outcome of the field goal attempt in that the percentage of successful kicks drops in these pressure situations.

White and Berry (2002) delve into player evaluation research and use a tiered logistic modeling approach to evaluate the effectiveness of quarterbacks in the National Football League, the professional league in the United States. The quarterback position in American football is the most influential position in the sport because this offensive position player is in physical possession of the football most of the time and must make critical in-game decisions quickly. Thus, the quarterback position is a focal point in team personnel decisions. The authors compare their results to the rating system currently in use by many professional teams.

REFERENCES

Berry, S.M. (2003). College Football Rankings: The BCS and the CLT. *Chance*, 16(2), 46–49.
Berry, S.M. (2004). The Cold-Foot Effect. *Chance*, 17(4), 47–51.

Mease, D. (2003). A Penalized Maximum Likelihood Approach for the Ranking of College Football Teams Independent of Victory Margins. *The American Statistician*, 57(4), 241–248.

Stern, H.S, Massey, K., Harville, D.A., Billingsley, T., Colleyand, W. and Mease, D. (2004). Statistics and the College Football Championship. *The American Statistician*, 58(3), 179–195.

White, C. and Berry, S. (2002). Tiered Polychotomous Regression: Ranking NFL Quarterbacks. *The American Statistician*, 56(1), 10–21.

Willoughby, K.A. (2001). The Return of a Missed Field Goal in Canadian Football. *Chance*, 14(3), 29–33.

CHAPTER 10

College football rankings: The BCS and the CLT

SCOTT M. BERRY

Computer. It's such a simple word and a simple machine that does only as it's told. Despite this, the word evokes very strange responses. Invariably, when I am explaining an application of statistics in sports, people are very interested. They ask many questions, and some even ask, "How did you do it?" Depending on the audience, I explain various aspects of my model: what's included, what assumptions are made, and the data available, trying to present the model without sounding like a nerd. After a few sentences they ask, "Did you use a *computer*?" When I begrudgingly say, "yes," there are two parts to the reaction. The first is instant credibility—it must be right if I used a *computer*! The second is dismay—after all, what does a computer know about sports?

Because of this reaction I was amazed when a collection of college football bowls, called the Bowl Championship Series, formed an alliance to select eight teams to play in the four bowls of the BCS. They designed a methodology to select the top two teams in the country to play for the "national championship." The BCS standings are determined by four categories, one of which is the *computer ratings*. Seven different computer rating techniques are used to rank all college football teams. The participants in the national championship game in college football are in part determined by statistical ranking algorithms.

The BCS standings started in 1999. Changes have been made each year to the algorithm, but the general framework has remained the same. All Division I (highest level) college football teams are scored in four categories and their scores are summed to form their BCS score (lower better). The four scores are: the average ranking in two opinion polls (coaches and football writers), a trimmed-mean (excluding highest and lowest) of their computer rankings, a rank of strength of schedule divided by 25, and the number of losses. The BCS score is

$$\begin{aligned}\text{BCS Score} &= (0.25)[\text{Average rank in polls}]\\ &+ (0.25)[\text{Trimmed-mean rank in computer polls}]\\ &+ (0.25)[\text{Strength of Schedule rank/25}]\\ &+ (0.25)[\text{Number of losses}].\end{aligned}$$

When BCS standings occur that are deemed undesirable the algorithm is tweaked to prevent them in the future. There is a "Nebraska Rule," adopted when the University of Nebraska was ranked very high despite more losses than a number of teams behind them in the standings. The number of losses was added as a category to prevent such occurrences. A "quality wins" category was added in 2001. Altered slightly in 2002, bonus points (subtracted from the BCS score) are given for a win over a top 10-ranked team. For a win over the first-place team 1.0 points are subtracted, down to 0.1 for a win over the 10th place team (these are based on final season rankings). The third alteration was to force all computer rankings to ignore the margin of victory. The computer ranking algorithms were thus restricted to only considering who won each game. A win by 1 or by 40 must be deemed informationally equivalent.

In this article I investigate computer algorithms, specifically addressing how much the margin of victory helps (if at all) in rating college football teams. I investigate algorithms with the focus of predicting future games. Before discussing the algorithms, I understand that there are multiple reasons to rank college football teams, not just predicting who would win between competitors. The BCS standings are constructed to reward teams for their *accomplishments*. From an informational point of view, there are problems with the BCS standings. First, there is a good deal of redundancy in the algorithm. The first two aspects of the algorithm, the polls and the computer rankings, both incorporate the strength of schedule and the number of losses for each team. These are the last two aspects of the algorithm. The computer rating systems and the polls certainly factor in the strength of schedule. When a team plays well against a quality opponent they will be ranked higher in the polls and their computer ranking will increase. By including strength of schedule and number of losses as a separate aspect of the algorithm, they are weighed very high.

Investigating margin of victory

Hal Stern has been collecting the results of college games for at least the last 10 years. He generously provided the data to me. While the data is extremely impressive, I was over-whelmed to find a web site with the results of every college game for the last 133 years! Jay Howell's web site (www.cae.wisc.edu/~dwilson/rsfc/history/howell) has a text-based file with the results from each year. For each game, the date, the two competitors, the home team (or neutral site), and the final score are reported. I restrict my analysis to the years 1970–2001. The 2002 games are not yet included on the web site. These 32 years and 281 teams provide 22,661 games, including games played between Division I and non-Division I teams. I do have access to the 2002 scores from another web site which I analyse separately in a later section (www.bol.ulca.edu/~prwolfe/cfootball/scores.htm).

For game i, let h_i be the home team and v_i the visiting team. Likewise, let H_i be the home team's score and V_i the visiting team's score for game i. For two teams, i and j, I create a scalar measure of their abilities, labeled θ_i and θ_j, respectively. I capture the importance of being the home team with the parameter γ. The larger the value of θ, the better the team. For a given game the strength of the home team is $\theta_h + \gamma$ and the strength of the visiting team is θ_v. The algorithm is simple—if $\theta_h + \gamma$ is larger than θ_v then the home team is predicted to win the game. I use the difference, $\theta_h + \gamma - \theta_v$ as a measure of how close the game will be. In order to select the values of θ, I institute a function to measure how well the θs match the scores. Based on n observed games, the *loss* function is

$$\sum_{i=1}^{n} |(\theta_{h_i} + \gamma - \theta_{v_i}) - (H_i - V_i)|^p.$$

The case where $p = 1$ is referred to as absolute error loss. The computer rating selects the value of the θs for each of the teams, and the value of γ, which minimizes the loss function. This creates the rating for each team, θ, and a home field advantage, γ, which most closely matches the game results. One additional complication remains. Because the difference between θs is all that matters, I could add or subtract any constant from all the θs and get the same "rankings." To keep things simple I restrict the mean of the θs to be 0.

By changing the value of p I create a class of rankings, which weigh the margin of victory differently. Suppose team A plays three games, all on neutral fields, where each opponent has *known* values of $\theta = 0$. They win the first two games by one and two points and lose the third by 12. Let θ_A be the rating for team A. The loss function is

$$|\theta_A - 1|^p + |\theta_A - 2|^p + |\theta_A + 12|^p.$$

If $p = 1$ then the value of θ_A, which minimizes the loss function, is $\theta_A = 1$ (in this case the median differential will minimize the loss function). The value of the loss function, at this minimum, is 14. Team A would then be favored to beat a team with $\theta = 0$. For a value of $p = 2$, the deviations get squared. Thus, bigger losses or wins are weighted very heavily. In this case the value of $\theta_A = -3$ minimizes the loss function (in this case the mean differential will be the minimum). With a $\theta_A = -3$ team A would be an underdog against a team with $\theta = 0$. For larger values of p the margin of victory is more important.

Finding the values of θ to minimize the loss function is obviously a more challenging problem than the one described above. There are 281 teams, so there are 282 parameters to estimate: 281 team ability parameters plus one home field advantage parameter. This is where the computer comes in!

Simulating ranking systems

Using the 32 years of data, I simulate the use of the ranking system described for different values of p. I use the first four weeks of a season to simultaneously estimate the values of θ and the home field advantage for every team. I then use these rankings to predict the winners of all the week five games. The games in week five are then included in the set of known games and the rating for each team is recalculated to predict the winners of the week six games. This continues for the remainder of the season (I truncate at 16 weeks and include all bowl games in week 16). This is repeated for each of the 32 years of data. In all, 15,728 games are predicted. This simulation of the rating system is repeated for different values of p. I evaluate the different values of p by the proportion of the games predicted correctly. I also look at the average absolute deviations and standard deviations of the predictions from the estimated differences, $(\theta_h + \gamma—\theta_v)—(H_i—V_i)$. Table 10.1 presents the results for various values of p. For the value $p = 0$, each game differential is assumed to be one—thus ignoring the margin

Table 10.1 ***Proportion of games that are predicted correctly using a prospective algorithm with the given loss function parameter p.***

						p						
Week	0	0.5	1.0	1.5	1.6	1.7	1.8	1.9	2.0	2.1	2.2	2.5
5	0.618	0.632	0.671	0.677	0.677	0.675	0.677	0.679	0.677	0.676	0.676	0.676
6	0.608	0.640	0.686	0.706	0.702	0.704	0.703	0.704	0.704	0.704	0.702	0.699
7	0.627	0.618	0.692	0.721	0.722	0.720	0.721	0.719	0.718	0.719	0.717	0.714
8	0.635	0.632	0.684	0.711	0.711	0.715	0.716	0.716	0.714	0.713	0.715	0.716
9	0.639	0.668	0.719	0.745	0.751	0.748	0.750	0.749	0.750	0.750	0.749	0.746
10	0.651	0.668	0.702	0.726	0.728	0.729	0.728	0.733	0.734	0.732	0.733	0.732
11	0.654	0.670	0.732	0.750	0.749	0.752	0.752	0.755	0.754	0.751	0.751	0.751
12	0.632	0.648	0.698	0.704	0.703	0.705	0.704	0.705	0.704	0.705	0.703	0.703
13	0.673	0.667	0.724	0.728	0.735	0.736	0.733	0.738	0.733	0.732	0.728	0.722
14	0.629	0.709	0.723	0.756	0.751	0.756	0.765	0.761	0.761	0.756	0.756	0.751
15	0.422	0.600	0.600	0.622	0.622	0.644	0.644	0.667	0.667	0.667	0.667	0.667
16	0.516	0.587	0.618	0.636	0.638	0.638	0.632	0.632	0.632	0.630	0.624	0.626
T	0.631	0.647	0.697	0.716	0.717	0.717	0.718	0.719	0.718	0.717	0.716	0.715

of victory. Figure 10.1 reports the proportion correct for each value of p when various weeks are combined. Weeks 5–7, 8–10, 11–13, and 14–16 are combined. The cumulative total is also plotted.

The proportion of correct games is very close for most of the weeks. Looking at the row of totals across weeks, one sees that value $p = 1.9$ gives the highest proportion of correct games at 0.719. The values of p in the range of 1.5 to 2.5 result in similar prediction success. Notice the worst algorithm is the one which ignores the margin of victory. The success proportion is reduced from 0.719 to 0.631 when the margin of victory is ignored. This reflects poorly on the decision to force the computer algorithms to ignore the margin of victory. Clearly, there is valuable information in the score—winning by one and 21 are very different, and these results demonstrate this fact.

When I planned this analysis I expected $p = 2$ to perform the best. I was not all that surprised when $p = 1.9$ came out the best. Why would I expect $p = 2$ to perform the best? In my training as a statistician the most astonishing mathematical idea I learned was the central limit theorem. The CLT states that the average (or sum) of independent random variables will have approximately a normal distribution. This approximation gets better as the number of variables being averaged (or summed) increases. To me the amazing aspect is that this is true regardless of the distribution (discrete, continuous, or a mixture) of the underlying random variables (as long as a mean and standard deviation exist!). A score in a football game is the sum of points scored on many possessions. The distribution of points scored on one possession is discrete, and may not look anything like a normal distribution. Despite this, when these variables are added up over the course of a game the points will be approximately normally distributed. The possessions are not perfectly independent, but they don't have to be for the CLT to work.

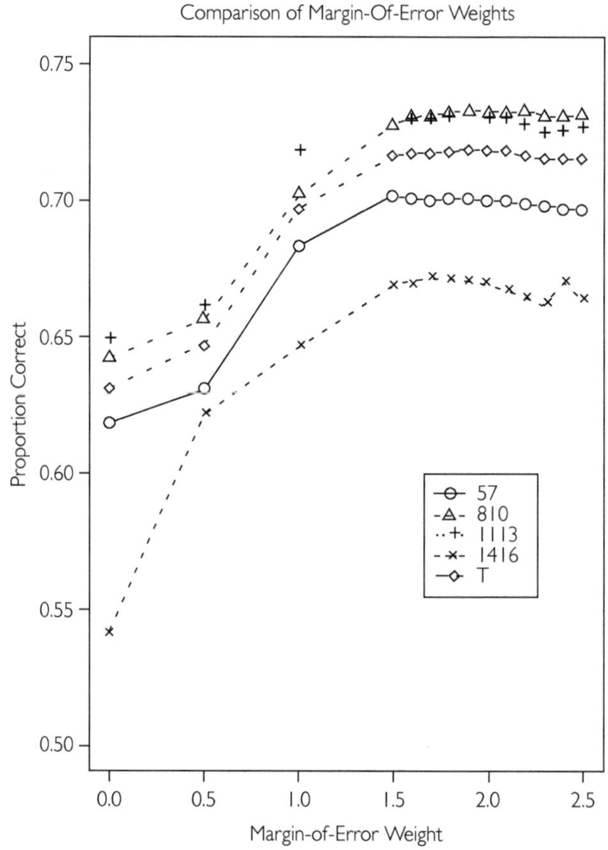

Figure 10.1 The proportion of games predicted correctly for various margin-of-error weights, p.

Given that the score, or the difference in scores, in a foot-ball game is normally distributed, then the best estimator of this normal distribution is based on minimizing mean squared error—the loss function described above with $p = 2$. While $p = 2$ didn't come out the best predictor overall, it might be evidence that discounting margin of victory slightly is warranted. I am not convinced that $p = 1.9$ is truly better than $p = 2$.

Adopting the normal distribution assumption, which, based on the empirical study, is quite reasonable, we can also attach a measure of variation for a game. Using the $p = 2$ algorithm, the standard deviation of the games, averaged over all the weeks, is 18.11. The predictive standard deviation certainly depended on the week. In week five, the standard deviation was estimated to be 22.46, decreasing down to 16.38 in week 16. This is based on estimates of the θs based on more information. For simplicity I assume that the standard deviation in a college football game is 17. I think this standard deviation is much larger than a (nonstatistical) college football fan would think. Table 10.2 presents the probability of a team winning based on the difference in their ratings and the normal distribution. The difference between the two teams is analogous to a "point spread." I think these probabilities of winning are smaller than most fans would think. Generally, it is thought that a 10-point favorite is a near sure-thing.

Table 10.2 ***Probability that the favorite beats the underdog based on the normal distribution assumption.***

Difference (Point Spread)	Probability Favorite Wins
0	0.500
1	0.523
3	0.570
7	0.660
10	0.722
17	0.841
20	0.880
25	0.929
34	0.977
50	0.998

2002 revisited

From a statistical, or computer rating, point of view, the 2002 college football season was very interesting. The Ohio State University and Miami University (of Florida) ended the regular season unbeaten. The two would play in the Fiesta Bowl for the national championship. Miami was the two-time defending champion and a prohibitive favorite over the Ohio State team, which had many very close calls during the season.

I used the algorithm with $p = 2$ to rate all teams during the 2002 season (the home field advantage, γ, was estimated to be 2.8). Interestingly, neither of the teams playing in the national championship game were ranked in the top two in this ranking. Table 10.3 presents the rating of the top 20 teams, both including and excluding the bowl games. The data set I have for the 2002 season included all college games and thus the ratings may seem high, when the mean is 0, but this is relative to the 704 teams in all divisions. I was amazed to see Kansas State come out on top. They are predicted by this algorithm to be favored to beat any other college football team. Kansas State was not even selected as one of the eight BCS teams, though I heard many complaints that they were not selected. Southern California was an interesting team that many felt was the best team at the end of the season. Much of the aura surrounding Miami was due to their two previous national titles and 34 consecutive wins, including the last two years. Their 2002 season was not as impressive. They played many close games and did not dominate the way they did in the previous two seasons. I expected that Ohio State would come out lower than the top couple of spots. They had relatively few convincing wins and they played a number of mediocre teams very closely.

Using the ratings before the bowl games I predict the results of the four BCS games using the $p = 2$ algorithm. In the Rose Bowl, Oklahoma (71.6) faced Washington State (63.5). Oklahoma was favored to win by 8.1 points, with a 0.683 probability of winning. Oklahoma won 34–14, for a differential of 20, which is 11.9 points higher than predicted. In the Sugar Bowl, Georgia (66.6) beat Florida State (63.0) 26–13. The 3.6 point advantage for Georgia gave them a predicted probability of 0.584 of winning the game. In the Orange Bowl, Southern California

Table 10.3 ***The ratings and ranks of the top 20 teams after and before the 2002 bowl season.***

Team	After Rating	Rank	Before Rating	Rank
Kansas State	78.9	1	80.6	1
Southern Cal	76.2	2	76.9	2
Oklahoma	72.3	3	71.6	3
Miami	69.9	4	71.3	4
Texas	68.0	5	67.7	5
Georgia	66.4	6	66.6	6
Penn State	65.1	7	65.2	7
Iowa	64.8	8	65.1	8
Ohio State	64.5	9	62.9	13
Alabama	63.6	10	64.2	9
Florida State	62.3	11	63.0	12
Washington State	61.8	12	63.5	11
Auburn	61.8	13	61.8	14
Boise State	61.6	14	61.2	15
Maryland	61.1	15	59.5	19
North Carolina State	60.3	16	58.3	23
Oregon State	60.2	17	63.6	10
Michigan	59.8	18	59.1	20
Colorado	59.6	19	60.7	16
California	58.8	20	59.8	17

(76.9) faced Iowa (65.1). With an 11.8-point spread, Southern California had a 0.756 probability of winning—and won 38–17. In the national championship game, the Fiesta Bowl, Miami (71.3) was predicted to win by 8.4 over Ohio State (62.9) for a 0.689 probability of Miami winning. When Ohio State won 31–24 in double overtime many people labeled it one of the greatest upsets in college football history. These ratings show that while Miami was favored, a 0.311 probability of Ohio State winning does not seem that extreme.

Discussion

There are different reasons to rate college football teams. In this paper I focus on a rating system which best predicts future games. There are no components of "accomplishment" in this model. I have no problem with The Ohio State University being named national champion. After all, they won every game, which *is* all that matters in sports. This is not true in predicting future results. I would not suggest that the BCS standings use a computer rating system exclusively—after all, they are interested in accomplishment as much as raw skill. I do think that if they are going to incorporate a computer rating algorithm, they should allow margin of victory to be included. This article demonstrates that using the margin of victory results in a clear advantage in predicting future games.

College football ratings are neat, but the demonstration of the central limit theorem is much more exciting. I know it is only a mathematical theorem—maybe it would be appreciated more if a *computer* had dictated it.

CHAPTER 11

General tiered polychotomous regression: Ranking NFL quarterbacks

CHRIS WHITE AND SCOTT BERRY

Abstract

Multinomial response modeling is still in debate, with numerous parametric and nonparametric methods being popular. We investigate a tiered logistic regression technique that can handle complex functional forms while still maintaining a parametric framework. Using our model, we find the value of different plays in the NFL. We apply these results to ranking NFL quarterbacks and compare our rankings to the rankings found using the NFL quarterback rating. The main application of this article is to a sports topic, but our model could be used in any polychotomous regression setting.

Keywords: *BIC, logistic, multinomial*

Introduction

In the National Football League, the official measure of a quarterback's performance is his "quarterback rating." This rating is used to rank quarterbacks and is so widely accepted that it is used in some contracts as an incentive. Quarterback Donovan McNabb has an incentive clause in his contract with the Philadelphia Eagles that pays him a bonus if his quarterback rating is 100 or above. Akili Smith, Tony Banks, and many others have also signed contracts that include bonuses if their quarterback ratings are above a certain threshold. It is by far the most widespread measure used to rank and differentiate quarterbacks.

The NFL quarterback rating is a linear combination of four categories: completion percentage, (average) yards per pass, (average) touchdowns per pass, and (average) interceptions per pass. It is the sum of the following four parts, multiplied by $\frac{100}{6}$:

1. $\dfrac{\text{Completion percentage} - 0.3}{0.2}$

2. $\dfrac{\text{Yards per pass attempt} - 3}{4}$

3. $\dfrac{\text{Touchdowns per pass attempt}}{0.05}$

4. $\dfrac{0.095 - \text{Interceptions per pass attempt}}{0.04}$

Each of the above parts is truncated to be between 0 and 2.375. Thus, any negative values are set to 0 and any values greater than 2.375 are set to 2.375.

As an example, consider Randall Cunningham's 1998 season. He had 259 completions out of 425 attempts for 3,704 yards. He had 34 touchdown passes and 10 interceptions. For his season:

1. $\dfrac{\frac{259}{425} - 0.3}{0.2} = 1.5471$

2. $\dfrac{\frac{3704}{425} - 3}{4} = 1.4288$

3. $\dfrac{\frac{34}{425}}{0.05} = 1.6000$

4. $\dfrac{0.095 - \frac{10}{425}}{0.04} = 1.7868.$

So Randall Cunningham's 1998–1999 NFL rating is $(1.5471 + 1.4288 + 1.6000 + 1.7868) * \frac{100}{6} = 106.045$. In this rating, all incompletions are equivalent, and all interceptions count equally against a quarterback. Similarly, all 20-yard touchdown passes are considered equal, regardless of the circumstance, and the above categories ignore sacks and fumbles.

To illustrate some of the drawbacks in more detail, consider the following examples from the 1998 NFL season. In week 3, Jeff George of the Oakland Raiders faced an intense pass rush on first down and 10 from the Denver Bronco 10-yard line and was sacked for an 11-yard loss. In week 10, Trent Dilfer of the Tampa Bay Buccaneers faced a first down and 10 from the Tennessee Oiler 10-yard line. He sensed a strong pass rush and threw the ball away for an incompletion, which is a much better alternative than being sacked for an 11-yard loss. The NFL quarterback rating penalizes quarterbacks for incompletions such as Dilfer's but does not penalize quarterbacks for sacks. Dilfer's team was better off because he threw the ball away for an incompletion, but the NFL rating penalizes him for this while George's rating is unaffected.

It would also be advantageous to fumble rather than throw an incompletion. In the 1998 week 15 San Diego Chargers versus Seattle Seahawks game, Ryan Leaf faced a second down and 14 and fumbled as he was being sacked. The fumble was returned by the defense for a touchdown, but the NFL rating does not penalize him. Leaf's team would have been much better off if he had simply thrown the ball away for an incompletion, but again the NFL rating would penalize him for this.

In week 10, the New Orleans Saints played the Minnesota Vikings and Brad Johnson of the Vikings faced a third down and three from the New Orleans five-yard line. He threw an interception that was returned 95 yards for a touchdown. In contrast, John Elway threw an interception on third down and 21 from his own 35-yard line in the week six Denver Broncos versus Seattle Seahawks game. The pass traveled 47 yards with no return on the interception. Both of these plays would count equally against each quarterback in the official quarterback rating, despite the fact that Brad Johnson severely hurt his team, while Elway's interception was better than most punts.

Finally, in the week four Minnesota Vikings versus Chicago Bears game, Erik Kramer—facing first down and 10—threw a 23-yard touchdown pass. In the week two Arizona Cardinals versus Seattle Seahawks game, Jake Plummer threw a 23-yard touchdown pass facing fourth down and 16. Fourth down and 16 is a much more difficult situation, given that you must get 16 yards in a single play, along with the fact that the defense is expecting a pass. Thus, should both of these plays count equally in measuring a quarterback or should Plummer be rewarded more?

We propose a method for rating quarterbacks (U.S. Patent Pending) that better measures the contribution to the team. We assign a value to each play and use these values to summarize a quarterback's effectiveness. A quarterback's ultimate goal is to win the football game, so one option would be to determine how much each play contributed toward or against the probability of winning. Because of multiple types of scores and a continuous clock it is very difficult to measure the probability of winning. The method we propose rewards or penalizes the quarterback according to the "expected points" gained or lost on each play. A football team is not necessarily trying to maximize the number of points scored on the given play, nor is a team attempting to maximize the number of points scored on the drive. If so, no team would ever punt. Teams are trying to maximize the number of points scored eventually, which is why teams punt. They hope to pin the other team back, not allow them to advance, and then get the ball back in good field position and be the next team to score. Since this is what teams are maximizing, we model eventual points scored as our response variable. This article uses the term expected points to refer to expected eventual points. Stern (1998) modeled eventual points on first down and 10 from various field positions using a least squares approach.

Assume that teams score an average of 3.9 eventual points on third down and 5 from the other team's 10-yard line. Then if the quarterback throws a touchdown pass, he basically "gained" 3.1 points for his team (7 points – 3.9 points). If he throws an interception that is run back for a touchdown, then he "lost" 10.9 points for his team (lost 3.9 points in addition to the 7-point touchdown). If he throws an incomplete pass, then his team faces a fourth down and 5 from the 10-yard line, where teams score an average of, say, 2.5 eventual points. Thus, an incompletion would cost his team 1.4 points (3.9 points – 2.5 points). Using this approach, quarterbacks are rewarded or penalized according to the value of the play, instead of having all 10-yard touchdown passes count equally, all interceptions count equally, and so on.

Numerous game situations affect expected points; the down, the yards to first down, the yard line, the time remaining, the score, the caliber of the defense, and so on. We focus on three explanatory variables: down (denoted "Down"), yards to go for a first down (denoted "ToGo"), and yards to the goal line (denoted "ToGoal"). To measure the differential points on each play, we create a model for the expected points as a combination of Down, ToGo, and ToGoal. Taking the difference between the expected points before and after a play measures the "value" of the play. We use the average value of all plays for each quarterback to rank them.

We fit a polychotomous regression model to estimate the expected number of points for each game situation. We use play-by-play data from the 1998 NFL season. The next section describes the data. This is followed by a section on the expected points model describing the tree-based polychotomous regression model. The section about using the model to compare previously mentioned plays describes how the model measures the impact of a play by reviewing each of the scenarios presented in the introduction. This is followed by a section on using the model to rate quarterbacks, presenting quarterback rankings for 1998 and comparing them to the NFL rating system. Concluding remarks are presented in the final section.

The data

The original play-by-play data was in the format presented in Figure 11.1, using the first three possessions of the 1998–1999 Dallas Cowboys versus Arizona Cardinals game in week 11 of

```
J.Nedney kicks 69 from AC30 to DC1, C.Warren ret. DC24 for 23 (M.Maddox).

Dallas Cowboys at 15:00
1-10-DC24 T.Aikman pass incomplete to D. Johnston.
2-10-DC24 T.Aikman pass to B.Davis to DC43 for 19 yards (T.Knight). P1
1-10-DC43 E.Smith right end to DC44 for 1 yard (J.Miller).
Dallas Cowboys time out at 15:00. First time out.
2-9-D C44 E.Smith up middle to DC46 for 2 yards (R.McKinnon, T.Bennett).
3-7-DC46 T.Aikman (shotgun) pass incomplete (T. Bennett). Aikman hit when
throwing.
Arizona Cardinals time out at 15:00. First time out.
4-7-DC46 T.Gowin punts 48 yards, out of bounds at AC6, Center-D.Hellestrae.

Arizona Cardinals at 12:33 Official time out at 12:33.
  1-10-AC6 J.Plummer pass to J.McWilliams to AC 14 for 8 (R.Godfrey, O.Stoutmire).
  2-2-AC14 A.Murrell left end to AC15 for 1 yard (C.Hennings, D.Coakley).
  3-1-AC15 J.Plummer pass to F.Sanders, P-OOB at AC29 for 14 yards
 (D.Woodson).
  AC-J.Dexter PENALIZED 5 yards for Illegal Formation. No Play
  3-6-AC10 J.Plummer pass to E.Metcalf to AC20 for 10 yards (G. Teague). P1
  1-10-AC20 J.Plummer pass incomplete to R.Moore.
  Arizona Cardinals time out at 12:33. Second time out.
  2-10-AC20 J.Plummer pass incompleate to J.McWilliams.
  3-10-AC20 12 men in huddle
  AC-F.Brock PENALIZED 5 yards for Illegal Substitution.No Play
  3-15-AC15 J.Plummer pass to L.Centers to AC10 for -5 yards (O.Stoutmire).
  4-20-AC10 S.Player punts 48 yards to DC42, Center-T.Junkin. D.|Sanders ret.
  to DC43 for 1 yard (K.Lassiter).

Dallas Cowboys at 9:27
Official time out at 9:27.
1-10-DC43 T.Aikman pass to E.Mills to AC42 for 15 (T.Knight, R.McKinnon). P2
1-10-AC42 E.Smith right end to AC35 for 7 yards (J.Miller).
2-3-AC35 E.Smith right tackle to AC17 for 18 (A.Wadsworth, R.Swinger). R3
1-10-AC17 E.Smith right end, pushed out of bounds at AC3 for 14 yards
(T.Bennett). Aikman handed toMills who pitched to Smith. R4
• 1-3-AC3 C.Warren left guard for 3 and TOUHDOWN. R5
• R.Cunningham extra point is GOOD. Center-D.Hellestrae.
Holder-E.Bjornson.
========== DC 7 AC 0, 5 plays, 57 yards, 2:42 drive, 8:15 elapsed ========
```

Figure 11.1 Original play-by-play data.

Down	ToGo	Yards Gained	ToGoal	Player	Play Type	Eventual points
1	10	0	76	T.Aikman	pass	7
2	10	19	76	T.Aikman	pass	7
1	10	1	57	E.Smith	run	7
2	9	2	56	E.Smith	run	7
3	7	0	54	T.Aikman	pass	7
4	7	48	54	T.Gowin	punt	7
1	10	8	94	J.Plummer	pass	–7
2	2	1	86	A.Murrell	run	–7
3	1	–5	85	J.Plummer	penalty	–7
3	6	10	90	J.Plummer	pass	–7
1	10	0	80	J.Plummer	pass	–7
2	10	0	80	J.Plummer	pass	–7
3	10	–5	80	Team	penalty	–7
3	15	–5	85	J.Plummer	pass	–7
4	20	47	90	S.Player	punt	–7
1	10	15	57	T.Aikman	pass	7
1	10	7	42	E.Smith	run	7
2	3	18	35	E.Smith	run	7
1	10	14	17	E.Smith	run	7
1	3	3	3	C.Warren	run	7

Figure 11.2 Structured data.

the season as an example. We wrote a program to create a structured dataset containing our variables of interest from this prose-based file. Our full dataset consisted of each play from the 1998–1999 season, which was more than 35,000 plays. The play-by-play data follow a common structure for each play which allows us to restructure the information into a dataset that includes only the relevant information. The Figure 11.1 data were restructured into the Figure 11.2 data. In the Figure 11.2 output, eventual points are in the last column. To find the eventual points for a given play, determine which team scored next. If the same team scored next, then the eventual points for the current play would be equal to how many points were scored. If the other team scored, eventual points would be equal to the negative of the amount that the other team scored. For example, in the Figure 11.1 output Dallas had the ball first and did not score on the drive. Arizona did not score on the subsequent drive. On the next drive, Dallas scored 7 points on a touchdown. Thus, each play on the scoring drive receives 7 eventual points. Each play on the Arizona drive receives –7 eventual points and each play on the first Dallas drive receives 7 eventual points.

We pooled eventual points that were –6 or –8 to be –7 and pooled points that were 6 or 8 to be 7. We want all touchdowns to count equally, without regard to the extra point. If neither team scored before the end of the first half or the end of the game, then eventual points for plays after the last score were recorded as zero.

Expected points model

The three explanatory variables are Down, ToGo, and ToGoal. The possibilities for down are {1, 2, 3, 4}. We do not want to force the difference between first down and second down to be the same as the difference between third down and fourth down. This would be unrealistic, since moving from third down to fourth down is usually more costly in terms of expected points than moving from first down to second down. Thus, we use dummy variables for each down. Since down has four possibilities, we use three dummy variables. Thus, D2, D3, and D4 denote indicator variables for second, third, and fourth down, respectively. To model expected points from D2, D3, D4, ToGo, and ToGoal, we develop a model for the probability of each of the seven possible outcomes. From each situation, a team could eventually score a touchdown (+7), kick a field goal (+3), record a safety (+2), allow a safety (−2), allow a field goal (−3), allow a touchdown (−7), or no points could be scored. To estimate the expected points we average over the probabilities of each outcome to find the expected eventual points. Estimating the probability of $k > 2$ outcomes from explanatory variables is referred to as polychotomous regression.

The response variable expected points has the seven outcomes $\{-7, -3, -2, 0, 2, 3, 7\}$. McCullagh and Nelder (1989) discussed various models for polychotomous regression, including an example of tiered logistic regression, which they referred to as nested or hierarchical response modeling. Hosmer and Lemeshow (1989) described multiple logistic regression (multiple independent variables) and also briefly described the polychotomous regression setting. Kooperberg, Bose, and Stone (1997) described model selection procedures and introduced a polychotomous regression algorithm involving linear splines. They discussed maximum likelihood estimation, stepwise selection in model fitting, the Akaike information criterion, cross-validation, and the use of an independent test set in the model selection stage. The authors applied their procedure to a phoneme recognition dataset. Albert and Chib (1993) presented a Bayesian approach to polychotomous response data using the idea of data augmentation.

This article uses a tiered logistic regression model. The response variable in logistic regression has two possible outcomes, while in our problem the response variable, eventual points, has seven outcomes. To fit a tiered logistic regression model, we split the outcomes into $k - 1$ tiers, where k is the number of outcomes. Here we have seven outcomes, so we split the outcomes into six pairs, structured as in Figure 11.3. There are numerous choices for how to split the outcomes so we chose one that was nicely organized. For a discussion of the effect of different tier structures see White (2000). At each dichotomous split we fit a logistic regression model. In Figure 11.3, each of the terminal nodes represents an outcome. To obtain the probability of a particular outcome, multiply the appropriate probabilities for the tiers leading to the terminal node of interest. For example, the outcome "no points" involves only the first tier. To find the probability of "no points," find the probability of 0 (i.e., no success) in the first-tier logistic regression. To find the probability of "negative safety," use tiers 1, 2, and 4. In the first tier and the fourth tier, find the probability of a success, and in the second tier find the probability of no success. Multiplying the three probabilities together gives us the probability of the next score being a safety for the other team.

We present the mathematical details for this model. Let Y_i be the outcome for play i, for $i = 1, \ldots, n$, and let the vector $\mathbf{X}_i$ be the independent variables for play i; thus, $\mathbf{X}_i = (\text{D2}_i, \text{D3}_i, \text{D4}_i, \text{ToGo}_i, \text{ToGoal}_i)$. Now consider a tier indicator for the outcome of play i, denoted $\Phi_i = (\phi_{i1}, \phi_{i2}, \phi_{i3}, \phi_{i4}, \phi_{i5}, \phi_{i6})$, where

$$\phi_{it} = \begin{cases} 1 & \text{if tier } t \text{ is used in the outcome for play } i \\ 0 & \text{otherwise.} \end{cases}$$

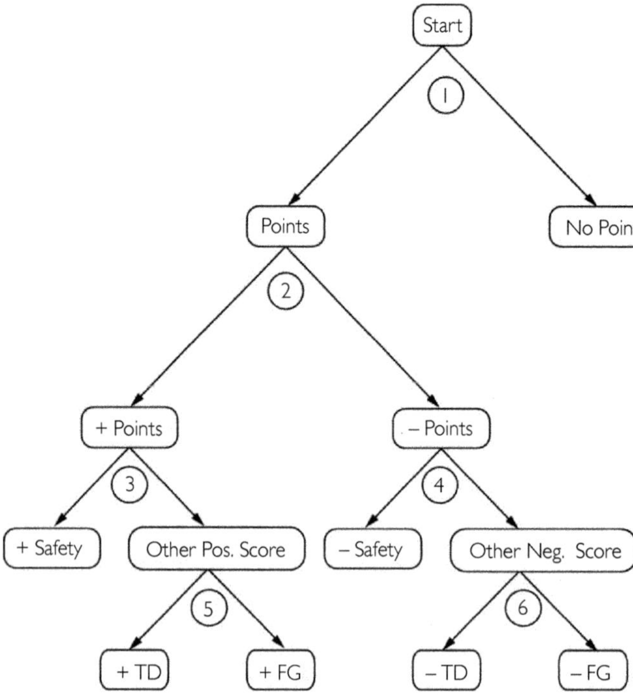

Figure 11.3 Tiers for the outcomes, with each tier numbered.

For example, considering the "no points" outcome in Figure 11.3, the only tier involved is tier 1. Thus for all "no points" plays, $\Phi_i = (1,0,0,0,0,0)$. For the seven outcomes,

$$\Phi_i = \begin{cases} (1,1,0,1,0,1) & \text{if} \quad Y_i = -7 \\ (1,1,0,1,0,1) & \text{if} \quad Y_i = -3 \\ (1,1,0,1,0,0) & \text{if} \quad Y_i = -2 \\ (1,0,0,0,0,0) & \text{if} \quad Y_i = 0 \\ (1,1,1,0,0,0) & \text{if} \quad Y_i = +2 \\ (1,1,1,0,1,0) & \text{if} \quad Y_i = +3 \\ (1,1,1,0,1,0) & \text{if} \quad Y_i = +7. \end{cases}$$

For each play i, let the vector $\Lambda_i = (\lambda_{i1}, \lambda_{i2}, \lambda_{i3}, \lambda_{i4}, \lambda_{i5}, \lambda_{i6})$ be the *left indicator*, where

$$\lambda_{ij} = \begin{cases} 1 & \text{if the outcome for play } i \text{ requires a left at tier } t \\ 0 & \text{otherwise.} \end{cases}$$

For the outcome "positive touchdown" in Figure 11.3 we need to go left at tiers 1, 2, and 5. Thus, for any "positive touchdown" play, $\Lambda_i = (1,1,0,0,1,0)$. Similarly, the seven outcomes are

$$\Lambda_i = \begin{cases} (1,0,0,0,0,1) & \text{if} \quad Y_i = -7 \\ (1,0,0,0,0,0) & \text{if} \quad Y_i = -3 \\ (1,0,0,1,0,0) & \text{if} \quad Y_i = -2 \\ (0,0,0,0,0,0) & \text{if} \quad Y_i = 0 \\ (1,1,1,0,0,0) & \text{if} \quad Y_i = +2 \\ (1,1,0,0,0,0) & \text{if} \quad Y_i = +3 \\ (1,1,0,0,1,0) & \text{if} \quad Y_i = +7. \end{cases}$$

This notation creates a convenient characterization of each outcome. Finally, let n_t denote the number of plays at each tier t, as delineated in Figure 11.3. For play i with independent variables $\mathbf{X}_i$, the probability of outcome Y_i is

$$P(Y_i \mid \mathbf{X}_i, \mathrm{L}_1, \ldots, \mathrm{L}_6) = \prod_{t=1}^{6} \left[\frac{\exp(\mathrm{L}_t)^{\lambda_{it}}}{1 + \exp(\mathrm{L}_t)} \cdot \phi_{it} + (1 - \phi_{it}) \right],$$

where L_t is the logit function at tier t. The likelihood function is

$$\begin{aligned} & f(\mathbf{Y} \mid \mathbf{X}_1, \ldots, \mathbf{X}_n, \mathrm{L}_1, \ldots, \mathrm{L}_6) \\ &= \prod_{i=1}^{n} P(Y_i|\mathbf{X}_i) \\ &= \prod_{i=1}^{n} \prod_{t=1}^{6} \left[\frac{\exp(\mathrm{L}_t)^{\lambda_{it}}}{1 + \exp(\mathrm{L}_t)} \cdot \phi_{it} + (1 - \phi_{it}) \right] \\ &= \prod_{t=1}^{6} \prod_{\{i:\phi_{it}=1\}} \left[\frac{\exp(\mathrm{L}_t)^{\lambda_{it}}}{1 + \exp(\mathrm{L}_t)} \right]. \end{aligned}$$

The problem is now reduced to fitting $k - 1 = 6$ logistic regressions. One option would be to fit the usual linear logistic regression model at each tier. Another choice would be to fit a more complicated model at each tier, such as one that uses smoothed versions of the variables. We decided to determine the best model at each tier and then use the combination of the six best models as our overall model. We used S-Plus to fit models with various explanatory variables at each tier, shown in Table 11.1. In Table 11.1, sm(·) represents a nonlinear smoothed version of the variable in parentheses (estimated using cubic B-splines) and ToGo*ToGoal is the interaction term ToGo multiplied by ToGoal.

At each tier, we calculate the Bayesian information criterion (BIC) for each model and chose the model with the minimum BIC as the best model. The BIC, due to Schwarz (1978), was discussed by Kass and Raftery (1995). It is beneficial in selecting between models because it takes into account the likelihood but also realizes that models with more terms will have higher likelihoods. Because of this, a "penalty" term is included in the BIC that penalizes models with more degrees of freedom. In our case, the BIC at a particular tier t is

$$\mathrm{BIC} = -2 \cdot \left[\log \left(\prod_{i:\phi_{it}=1} \frac{\exp(\mathrm{L}_t)^{\lambda_{it}}}{1 + \exp(\mathrm{L}_t)} \right) - \frac{1}{2} \mathrm{d}_t \log(n_t) \right],$$

Table 11.1 ***Explanatory variables considered.***

Fit #	Explanatory variables
1	D2, D3, D4, ToGo, ToGoal
2	D2, D3, D4, sm(ToGo), ToGoal
3	D2, D3, D4, ToGo, sm(ToGoal)
4	D2, D3, D4, ToGo, ToGoal, ToGo*ToGoal
5	D2, D3, D4, sm(ToGo), sm(ToGoal)
6	D2, D3, D4, sm(ToGo), sm(ToGoal), ToGo*ToGoal
7	D2, D3, D4, sm(ToGo), sm(ToGoal), sm(ToGo*ToGoal)
8	sm(ToGo), sm(ToGoal), separately for each down
9	sm(ToGo), sm(ToGoal), sm(ToGo*ToGoal), separately for each down

where

$$d_t = \text{model degrees of freedom for } \mathrm{L}_t.$$

By using the BIC form as stated, one problem arises. Finding the minimum BIC at each tier does not necessarily assure us of ending up with the overall combination of six models that has overall minimum BIC. To remedy this issue, we minimize an alternative form of the BIC at each tier that uses $\log(n)$ instead of $\log(n_t)$. This alternative form is written as

$$\mathrm{BIC}^* = -2 \cdot \left[\log \left(\prod_{i:\phi_{it}=1} \frac{\exp(\mathrm{L}_t)^{\lambda_{it}}}{1+\exp(\mathrm{L}_t)} \right) - \frac{1}{2} \mathrm{d}_t \log(n) \right],$$

where the only difference is the last term. BIC* is a form of the BIC that penalizes even more for additional terms in the model. Using this form of the BIC at each tier, the overall BIC can simply be written $\mathrm{BIC} = \mathrm{BIC}_1^* + \mathrm{BIC}_2^* + \cdots + \mathrm{BIC}_6^*$. In this case, finding the minimum BIC* at each tier *does* ensure finding the model with overall minimum BIC. The proof of this result is presented in Appendix A.

The explanatory variables that had the minimum BIC at each tier are presented in Table 11.2. At the first tier (points vs. no points) we have the prediction equation

$$\begin{aligned}
&\mathrm{logit}(\hat{P}_{\text{points}}) \\
&= 2.315 - 0.013\ \mathrm{D2} - 0.046\ \mathrm{D3} \\
&\quad -0.189\mathrm{D4} - 0.017\ \mathrm{ToGo} + \mathrm{sm(ToGoal)} \\
&= \hat{\mathrm{L}}_1
\end{aligned}$$

which we can convert from logit scale to probability scale using

$$\hat{P}_{\text{points}} = \frac{\exp(\hat{\mathrm{L}}_1)}{1+\exp(\hat{\mathrm{L}}_1)}.$$

Table 11.2 ***Minimum BIC models at each tier.***

Tier	Model with Minimum BIC
1: Points vs. No Points	D2, D3, D4, ToGo, sm(ToGoal)
2: Pos. Points vs. Neg. Points	D2, D3, D4, ToGo, sm(ToGoal)
3: Pos. Safety vs. Other Pos. Score	D2, D3, D4, ToGo, ToGoal
4: Neg. Safety vs. Other Neg. Score	D2, D3, D4, ToGo, sm(ToGoal)
5: Pos. TD vs. Pos. FG	sm(ToGo), sm(ToGoal), separately for each down
6: Neg. TD vs. Neg. FG	D2, D3, D4, ToGo, ToGoal

We estimated the logit equation using S-Plus, with the command

$$\text{logistic 1} < \text{-gam}(Y1 \sim D2 + D3 + D4 \\ + \text{ToGo} + s(\text{ToGoal}), \text{family} = \text{binomial}).$$

At the second tier (+ points vs. – points) we have

$$\begin{aligned}\text{logit}(\hat{P}_{+\text{points}}) &= 3.074 - 0.177\ \text{D2} \\ &\quad - 0.439\ \text{D3} - 0.831\ \text{D4} \\ &\quad - 0.009\ \text{ToGo} + \text{sm(ToGoal)} \\ &= \hat{\text{L}}_2.\end{aligned}$$

At the third tier (+ safety vs. other positive score) we have

$$\begin{aligned}\text{logit}(\hat{P}_{+\text{safety}}) &= -5.558 - 0.118\ \text{D2} \\ &\quad + 0.465\ \text{D3} - 0.667\ \text{D4} - 0.016\ \text{ToGo} \\ &\quad + 0.011\ \text{ToGoal} \\ &= \hat{\text{L}}_3.\end{aligned}$$

At the fourth tier (– safety vs. other negative score) we have

$$\begin{aligned}\text{logit}(\hat{P}_{-\text{safety}}) &= -6.908 + 0.145\ \text{D2} + 0.215\ \text{D3} \\ &\quad - 0.032\ \text{D4} - 0.008\ \text{ToGo} + \text{sm(ToGoal)} \\ &= \hat{\text{L}}_4.\end{aligned}$$

At the fifth tier (+ touchdown vs. + field goal) we have

$$\begin{aligned}\text{logit}(\hat{P}_{+\text{TD}}) &= (\text{sm}_1(\text{ToGo}) + \text{sm}_2(\text{ToGoal})) \cdot \text{D1} \\ &\quad + (\text{sm}_3(\text{ToGo}) + \text{sm}_4(\text{ToGoal})) \cdot \text{D2} \\ &\quad + (\text{sm}_5(\text{ToGo}) + \text{sm}_6(\text{ToGoal})) \cdot \text{D3} \\ &\quad + (\text{sm}_7(\text{ToGo}) + \text{sm}_8(\text{ToGoal})) \cdot \text{D4} \\ &= \hat{\text{L}}_5,\end{aligned}$$

where D1 is defined in the same manner as D2, D3, and D4, that is,

$$D1 = \begin{cases} 1 & \text{if down} = 1 \\ 0 & \text{if down} \neq 1. \end{cases}$$

The subscripts on the smoothed independent variables (e.g., sm_1 (ToGo)) were used to make it clear to the reader that separate smoothing functions were fit for each down. At the final tier (– touchdown vs. – field goal) we have

$$\begin{aligned} \text{logit}(\hat{P}_{-TD}) &= 0.476 - 0.026 * D2 \\ &\quad - 0.031 * D3 - 0.040 * D4 \\ &\quad + 0.00063 * \text{ToGo} - 0.00016 * \text{ToGoal} \\ &= \hat{L}_6. \end{aligned}$$

As mentioned earlier, we determine the predicted probability of a particular outcome by multiplying the applicable probabilities from each tier. For example, to find the predicted probability of a touchdown while facing first down and 10 with 10 yards to the goal, we use the probabilities given in Figure 11.4.

The predicted probability of positive touchdown (vs. positive field goal) in the left bottom tier is $\hat{P}_{+TD} = 0.637$. Continuing up the diagram, we find that the predicted probability of an

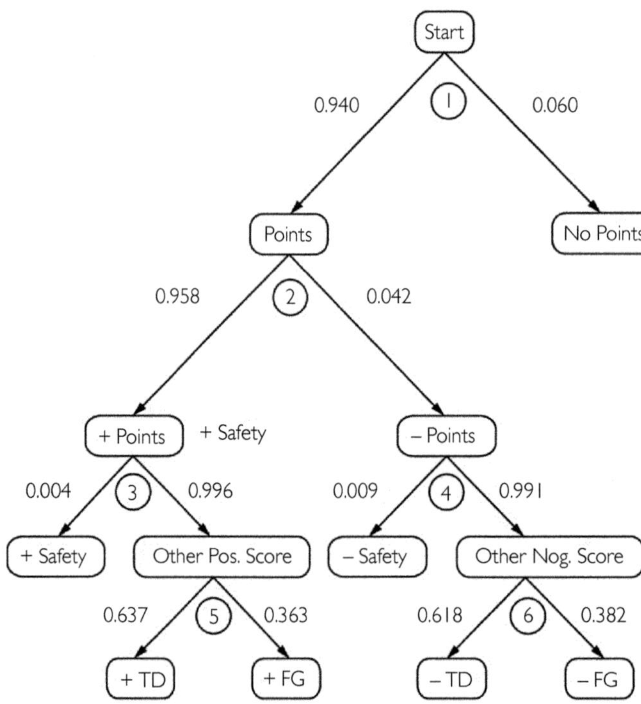

Figure 11.4 Logistic regression probabilities for first down and 10, 10 yards to goal.

"other positive score" (vs. positive safety) is $\hat{P}_{\text{other pos. score}} = 0.996$. The predicted probability of positive points (vs. negative points) is $\hat{P}_{\text{+points}} = 0.958$, and the predicted probability of points scored (vs. no points scored) is $\hat{P}_{\text{points}} = 0.940$. Multiplying these together, we get the predicted probability of a positive touchdown to be (0.637)(0.996)(0.958)(0.940) = 0.571. The predicted probabilities of the other six outcomes can be found in a similar manner, where the predicted probability of a positive field goal is 0.326, a positive safety is 0.00332, a negative safety is 0.000374, a negative field goal is 0.0150, a negative touchdown is 0.0243, and no points is 0.0598. Note that the probabilities do sum to 1.

To find the predicted expected points, we simply weight each outcome by its probability, which results in 4.77 predicted expected points. Figure 11.5 shows the expected points for downs 1 through 4 with 10 yards to go. Figure 11.6 presents three-dimensional graphs of the seven outcome probabilities for first down. Each of the graphs in Figure 11.6 is consistent with conventional wisdom. The probability of 0 eventual points decreases as a team gets closer to the opponent's goal line. Similarly, the probability of the opponent scoring a touchdown or field goal also decreases as a team gets closer to the opponent's goal line. The probability of the opponent scoring a safety is relatively high when you are close to your own goal line, and the probability of a touchdown gets higher as a team gets closer to the opponent's goal line and as yards to first down decreases. It is worth mentioning again that we are displaying the probability of *eventual* outcomes—not outcomes on the next play. The probability of a field goal on the next play when 99 yards from the goal line is zero. In our graph, this probability is not zero. This is because we found the probability of three eventual points, meaning the probability that the *next* score in the game will be a field goal for the current offensive team.

We present outcome probability graphs for two fourth down outcomes in Figure 11.7. The probability of an eventual field goal increases steadily as a team crosses midfield until curving down slightly as a team gets within five yards of a touchdown. The probability of a touchdown increases as a team approaches midfield and then starts to decrease (because now a field goal is more likely) and then increases again when a team gets very close to the opponent's goal line.

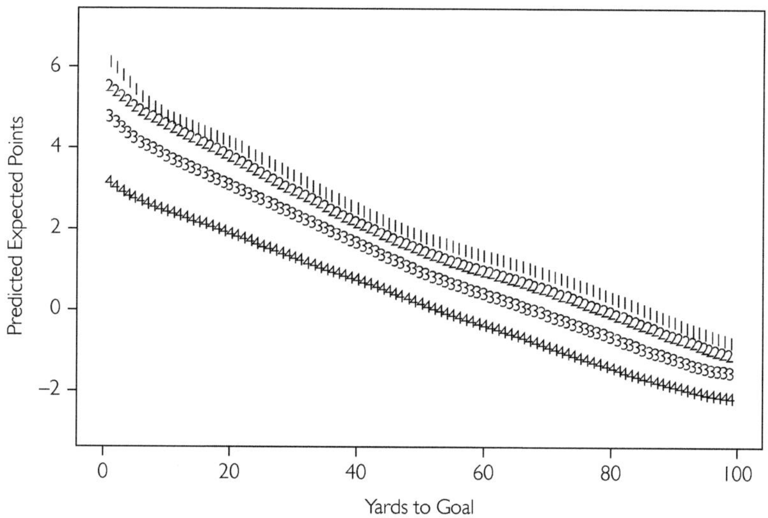

Figure 11.5 Downs one through four with 10 yards to go.

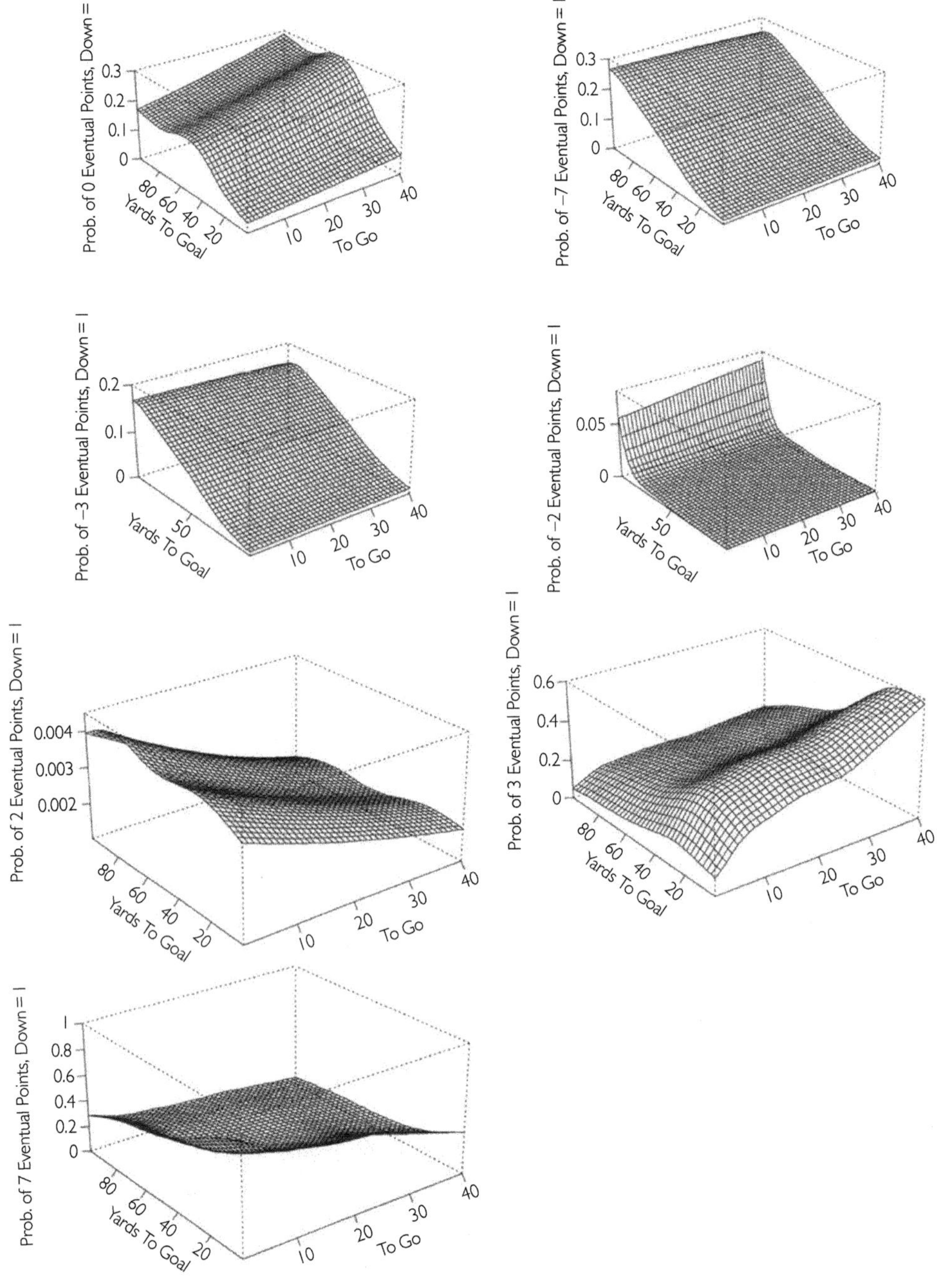

Figure 11.6 Predicted probabilities for each outcome, first down.

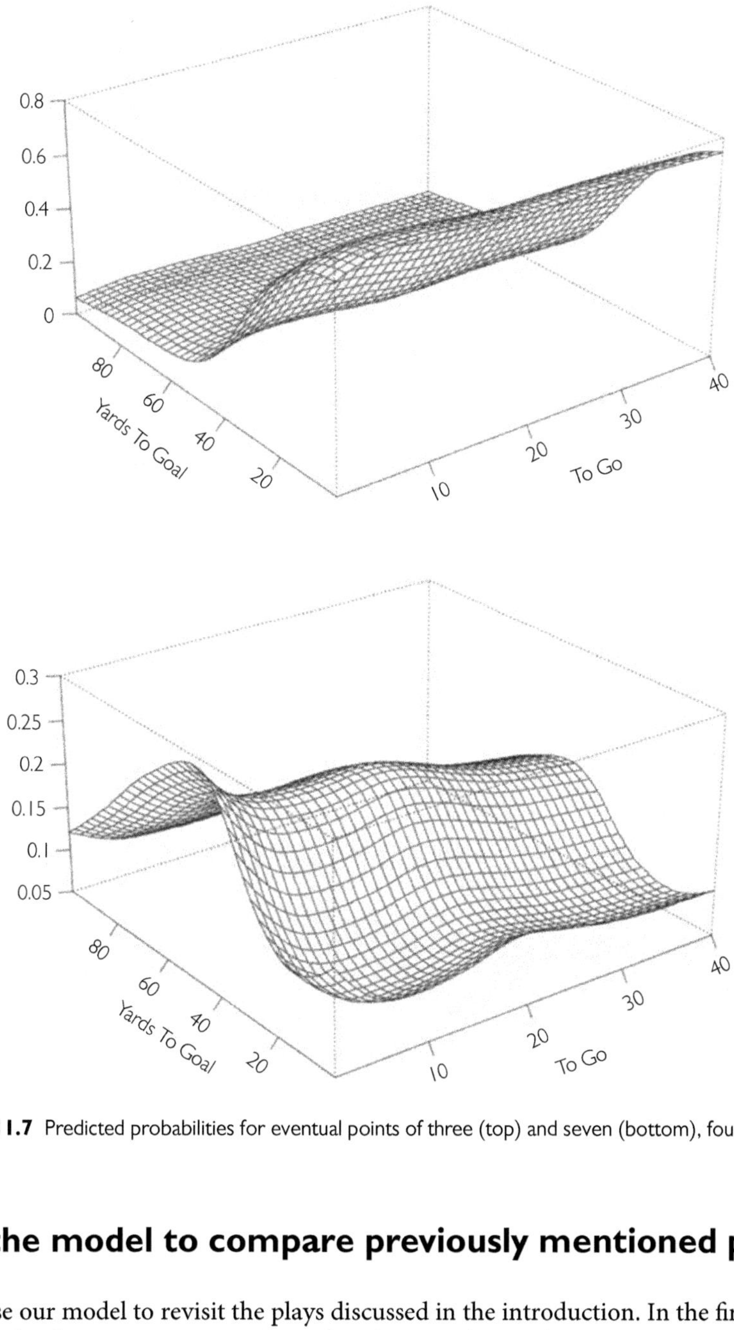

Figure 11.7 Predicted probabilities for eventual points of three (top) and seven (bottom), fourth down.

Using the model to compare previously mentioned plays

We now use our model to revisit the plays discussed in the introduction. In the first scenario, Jeff George faced an intense pass rush on first down and 10 at the Denver 10-yard line which gives an expected points of 4.77. After being sacked for an 11-yard loss, Oakland faced a second down and 21 from the Denver 21-yard line, which drops expected points to 3.68. Hence George (and the Oakland offense) lost 1.09 expected points on the play. In contrast, Trent Dilfer threw an incompletion from the same situation, resulting in second down and 10 from the 10. This puts his team in a position of 4.55 expected points, thus losing only 0.22 expected points.

When Ryan Leaf faced a second down and 14 with 57 yards to the goal line, the expected points were 0.95. By fumbling and then allowing Seattle to return the fumble for a touchdown, Leaf cost his team 7.95 points (0.95 + 7 points for touchdown). If he had simply thrown the ball away for an incompletion he would have only lost 0.53 expected points, but the NFL quarterback rating would penalize him for this incompletion while ignoring his fumble.

Erik Kramer faced a first down and 10 and threw a 23-yard touchdown pass, going from 3.85 expected points to 7, thus gaining 3.15 expected points. Similarly, Jake Plummer threw a 23-yard touchdown pass on fourth down and 16, going from 1.64 expected points to 7. He gained 5.36 expected points. In the last two examples, the outcomes were the same but the downs were different. This illustrates the fact that the down has a big impact on the importance and the difficulty of the play.

Finally, Brad Johnson threw an interception that was returned for a touchdown on third down and 3 from the 5-yard line. He lost 11.32 expected points, going from 4.32 to −7. John Elway faced a third down and 21 from his own 35-yard line and threw a 47-yard interception with no return, going from −0.20 expected points to −0.19, thus gaining 0.01 expected points. This case illustrates that all interceptions are not the same—some can even be beneficial!

Using the model to rate quarterbacks

To rate quarterbacks, we decided to exclude plays in the last two minutes of each half. We did this for various reasons, one of the reasons being that the winning team will often play a "prevent defense" which gives up small gains in hopes of preventing a score, and we did not want this to confound our results. We also included defensive pass interference plays as completed passes since they are positive plays for the passing team. Finally, we counted intentional grounding penalties as sacks, since there is a loss of yardage and loss of down.

Although we refer to these as quarterback ratings they are certainly confounded with the ability of the quarterback's offense. A great quarterback without good blocking or without good receivers will have a low rating that might not be reflective of his true individual ability. Similarly, a quarterback who plays with a great running back is at an advantage since the defenses cannot concentrate on passes. Thus, it would be more accurate to say that we are ranking the offense's passing ability with the particular quarterback in control. When comparing quarterbacks on the same team, such as Flutie vs. Johnson or Elway vs. Brister, we can better control for the strength of the offense. The NFL rating also does not account for this limitation. We also do not account for the strength of the defense. One quarterback could have had a more difficult schedule and thus faced more talented defenses. The NFL rating similarly does not account for this.

To obtain our rankings, we use the expected point values previously discussed. For each play, we took the difference between the end-result point value and the initial point value to get the value of the play. Once we determine the point value of each play, we find the average point value over all plays for each quarterback. Each average, along with the corresponding rank of each average, is presented in Table 11.3. All quarterbacks from 1998–1999 who participated in a significant number of plays are included. We also include the NFL rating and corresponding rank for each quarterback in Table 11.3. Any differences between our ranking and the NFL ranking that are 10 or greater are denoted with a ⇑ (for a quarterback who moved up 10 or more positions) or a ⇓ (for a quarterback who moved down 10 or more positions).

Table 11.3 ***Quarterback rankings.***

Player	NFL rating		Our ranking All plays		Our ranking No runs		Our ranking No sacks, runs	
R. Cunningham	106.0	1	0.46	1	0.45	1	0.53	1
R. Johnson	102.9	2	0.11	⇓ 22	0.07	24	0.43	4
V. Testaverde	101.6	3	0.38	2	0.36	2	0.47	3
S. Young	101.1	4	0.25	9	0.20	10	0.36	7
C. Chandler	100.9	5	0.30	5	0.28	6	0.52	2
B. Brister	99.0	6	0.27	7	0.19	11	0.25	17
J. Elway	93.0	7	0.31	4	0.30	5	0.39	6
N. O'Donnell	90.2	8	0.00	⇓ 31	0.00	34	0.17	24
M. Brunell	89.9	9	0.23	10	0.24	8	0.35	10
B. Johnson	89.0	10	0.30	6	0.32	3	0.42	5
T. Aikman	88.5	11	0.26	8	0.28	7	0.31	11
S. Beuerlein	88.2	12	0.15	15	0.19	13	0.35	8
B. Favre	87.8	13	0.21	11	0.20	9	0.29	12
D. Flutie	87.4	14	0.32	⇑ 3	0.31	4	0.35	9
J. Garrett	84.5	15	0.15	16	0.19	12	0.28	13
C. Batch	83.5	16	0.12	21	0.09	22	0.26	15
B.J. Tolliver	83.1	17	0.12	20	0.15	16	0.21	21
E. Kramer	83.1	18	0.17	14	0.16	14	0.26	16
E. Zeier	82.0	19	−0.01	⇓ 32	−0.01	32	0.16	29
T. Green	81.8	20	0.05	26	0.04	25	0.20	23
D. Bledsoe	80.9	21	0.13	19	0.12	17	0.27	14
R. Gannon	80.1	22	0.14	17	0.10	20	0.21	20
S. McNair	80.1	23	0.17	⇑ 13	0.12	18	0.22	19
D. Marino	80.0	24	0.08	24	0.08	23	0.16	26
W. Moon	76.6	25	0.02	29	0.04	27	0.16	28
J. Plummer	75.0	26	0.07	25	0.04	26	0.17	25
T. Dilfer	74.0	27	0.05	27	0.02	29	0.13	30.5
J. Harbaugh	72.9	28	0.04	28	0.02	28	0.11	32
J. Kitna	72.3	29	0.17	⇑ 12	0.15	15	0.24	18
P. Manning	71.2	30	0.09	23	0.10	21	0.16	27
K. Graham	70.8	31	0.14	⇑ 18	0.11	19	0.21	21
K. Detmer	67.7	32	0.00	30	0.01	30	0.04	36
T. Banks	68.6	33	−0.04	35	−0.06	35	0.06	34
D. Kanell	67.3	34	−0.13	41	−0.14	41	−0.02	41
D. Wuerffel	66.3	35	−0.10	39	−0.13	40	0.09	33
G. Foley	64.9	36	−0.12	40	−0.12	39	0.00	39
R. Peete	64.7	37	−0.02	33	−0.03	33	0.13	30.5
K. Stewart	62.9	38	−0.05	36	−0.08	36	0.01	37
K. Collins	62.0	39	−0.09	38	−0.10	37	0.05	35
D. Hollas	60.6	40	−0.20	42	−0.17	42	−0.01	40
E. Grbac	53.1	41	−0.04	34	−0.05	34	0.01	38
C. Whelihan	48.0	42	−0.09	37	−0.11	38	−0.04	42
B. Hoying	45.6	43	−0.26	44	−0.29	44	−0.16	44
R. Leaf	39.0	44	−0.22	43	−0.23	43	−0.11	43

We present three different rankings depending on which plays are included. We use the first ranking presented (our ranking, all plays) as our main ranking. It includes passes, sacks, interceptions, and runs. Our second ranking excludes runs, while our third ranking includes only passes and interceptions. These two supplemental rankings are presented to see if there are any unusual differences. For example, Rob Johnson is ranked 22nd using our main rating but jumps to 4th if sacks and runs are excluded. This implies that Johnson was frequently sacked (he had 30 sacks relative to only 107 pass attempts!) and also explains how we ranked him 22nd, since we account for sacks, but the NFL rating ranked him 2nd. He is one of the best examples why the NFL rating is inadequate. Sacks are very costly to a team and Johnson was sacked much more than the average quarterback. As a comparison, fellow Buffalo Bills quarterback Doug Flutie had many more pass attempts (354), yet was sacked only 12 times. A quarterback rating should account for such costly plays.

Overall, the NFL ranking seems to place the highly regarded quarterbacks at the top and the lesser quarterbacks at the bottom. The rank correlation between the NFL ranking and our main ranking was 0.84. But there are definitely some deviations. First of all, the NFL rating ranks Bubby Brister of the Denver Broncos ahead of John Elway of the Denver Broncos. Any knowledgeable football enthusiast would wholeheartedly choose Elway over Brister. Our ranking does have Elway ahead of Brister. Another perplexing result is the fact that the NFL rating ranks Rob Johnson of the Buffalo Bills higher than Doug Flutie of the Buffalo Bills. Flutie had a terrific year, being named the starting quarterback ahead of Johnson and being selected to the Pro Bowl. Our ranking has Flutie ahead of Johnson.

Other rankings that differ significantly are for quarterbacks Neil O'Donnell and Eric Zeier. Neil O'Donnell is ranked 8th in the NFL rating but only 31st in our rating. This can be explained partially by his relatively high number of sacks (30 sacks relative to 343 pass attempts). O'Donnell also fumbled on three of the sacks, all of which were returned by the defense for touchdowns. Zeier's discrepancy can also be partially attributed to a relatively large number of sacks (18 sacks relative to 181 pass attempts).

Excluding runs had little effect on the ratings. Most of the quarterbacks that have a history of running did go down slightly (as expected) when runs were excluded, such as Steve Young, Steve McNair, and Kordell Stewart.

The play with the largest positive expected points gained belonged to Charlie Batch of the Detroit Lions. On third down and 10 with 98 yards to the goal line, Batch threw a touchdown pass. He went from −1.56 expected points to 7 points, thus gaining 8.56 expected points. The quarterback play with the largest negative expected points gained was mentioned earlier and belonged to Brad Johnson of the Minnesota Vikings. His interception on third down and 3 from the five-yard line was returned 95 yards for a touchdown, going from 4.32 expected points to −7 points, thus the expected points "gained" were −11.32.

We can also determine who had the most big plays. If we define a big play to be more than 3 gained expected points, then Randall Cunningham of the Minnesota Vikings had the most with 31 big plays (again we exclude plays in the last two minutes of each half). If we define a big play to be 5 gained expected points, then again Cunningham had the most with 11. The top quarterbacks with plays gaining more than 3 and 5 expected points are presented in Table 11.4. We can similarly determine who had the most bad plays. The quarterbacks with the highest number of plays gaining less than −3 and −5 expected points are presented in Table 11.5.

Another interesting result is that many of the quarterbacks have negative average expected points. Thus, on average they contribute negatively with respect to expected points.

Table 11.4 ***Number of plays with > 3 and 5 expected points gained.***

# Plays > 3		# Plays > 5	
R. Cunningham	31	R. Cunningham	11
C. Chandler	21	S. Young	8
D. Bledsoe	20	B. Favre	8
S. Young	19	C. Chandler	7
J. Elway	19	T. Dilfer	6
S. Beuerlein	18	V. Testaverde	5
B. Favre	18	T. Green	4
V. Testaverde	16	D. Flutie	4
D. Marino	15	11 tied at	3
M. Brunell	14		
P. Manning	14		
J. Plummer	13		
T. Green	12		
R. Gannon	11		

Table 11.5 ***Number of plays with < − 3 and −5 expected points gained.***

# Plays < −3		# Plays < −5	
J. Plummer	19	B. Favre	6
B. Favre	18	V. Testaverde	5
T. Dilfer	16	K. Stewart	5
P. Manning	15	J. Plummer	5
S. Young	14	D. Hollas	5
D. Hollas	14	R. Leaf	4
T. Green	14	P. Manning	4
K. Stewart	13	D. Kanell	4
K. Collins	13	C. Chandler	4
R. Leaf	12	T. Dilfer	3
D. Bledsoe	11	S. Young	3
D. Marino	11	N. O'Donnell	3
T. Banks	10	G. Foley	3
V. Testaverde	9	E. Grbac	3
C. Whelihan	9	D. Marino	3
S. Beuerlein	9	D. Bledsoe	3
C. Chandler	9	11 tied at	2

Concluding remarks

Modeling a polychotomous response as a function of explanatory variables is a challenging statistical problem. This is especially true in the NFL example in which there are large deviations from monotonicity and lots of data. Our tiered logistic regression model adapts well to this scenario and the conditional structure provides extra insight into the data. Our model could be used in any multinomial response case and is especially useful in capturing complex functional forms.

The biggest hurdle that a ranking like ours must overcome is the complexity that is involved. The general public is used to simple formulas, so a ranking like ours can be intimidating. While acknowledging this limitation, we feel like the value added more than offsets the added complexity. The method we use is a major improvement through the fact that it does not force all plays with the same end-result to have the same value. Our expected point method of rating players could be extended to rating running backs, punters, offenses, defenses, and so on.

Appendix

The overall BIC can be expressed as

$$\begin{aligned}
\text{BIC} &= -2 \cdot \left[\log \left(\prod_{i=1}^{n} \prod_{t=1}^{T} \left[\frac{\exp(\text{L}_t)^{\lambda_{it}}}{1+\exp(\text{L}_t)} \cdot \phi_{it} + (1-\phi_{it}) \right] \right) \right. \\
&\quad \left. - \frac{1}{2}(\text{model degrees of freedom}) \cdot \log(n) \right] \\
&= -2 \cdot \left[\log \left(\prod_{t=1}^{T} \prod_{i:\phi_{it}=1} \left[\frac{\exp(\text{L}_t)^{\lambda_{it}}}{1+\exp(\text{L}_t)} \right] \right) - \frac{1}{2}(\text{d}_1 + \cdots + \text{d}_T) \cdot \log(n) \right] \\
&= -2 \cdot \log \left(\prod_{t=1}^{T} \prod_{i:\phi_{it}=1} \left[\frac{\exp(\text{L}_t)^{\lambda_{it}}}{1+\exp(\text{L}_t)} \right] \right) + (\text{d}_1 + \cdots + \text{d}_T) \cdot \log(n) \\
&= -2 \sum_{t=1}^{T} \sum_{i:\phi_{it}=1} \log \left(\frac{\exp(\text{L}_t)^{\lambda_{it}}}{1+\exp(\text{L}_t)} \right) + (\text{d}_1 + \cdots + \text{d}_T) \cdot \log(n) \\
&= -2 \left[\sum_{i:\phi_{i1}=1} \log \left(\frac{\exp(\text{L}_t)^{\lambda_{i1}}}{1+\exp(\text{L}_1)} \right) + \cdots + \sum_{i:\phi_{iT}=1} \log \left(\frac{\exp(\text{L}_T)^{\lambda_{iT}}}{1+\exp(\text{L}_T)} \right) \right] \\
&\quad + (\text{d}_1 + \cdots + \text{d}_T) \cdot \log(n) \\
&= -2 \left[\sum_{i:\phi_{i1}=1} \log \left(\frac{\exp(\text{L}_1)^{\lambda_{i1}}}{1+\exp(\text{L}_1)} \right) + \text{d}_1 \log(n_1) + \cdots \right. \\
&\quad \left. + \sum_{i:\phi_{iT}=1} \log \left(\frac{\exp(\text{L}_T)^{\lambda_{iT}}}{1+\exp(L_T)} \right) + \text{d}_T \log(n_T) \right] + \text{d}_1 \log(n) + \cdots + \text{d}_T \log(n) - \text{d}_1 \log(n_1) \\
&\quad - \cdots - \text{d}_T \log(n_T)
\end{aligned}$$

$$= -2\left[\sum_{i:\phi_{i1}=1}\log\left(\frac{\exp(L_1)^{\lambda_{i1}}}{1+\exp(L_1)}\right) + d_1\log(n_1) + \cdots\right.$$

$$\left. + \sum_{i:\phi_{iT}=1}\log\left(\frac{\exp(L_T)^{\lambda_{iT}}}{1+\exp(L_T)}\right) + d_T\log(n_T)\right] + d_1\cdot\log\left(\frac{n}{n_1}\right) + \cdots + d_T\cdot\log\left(\frac{n}{n_T}\right)$$

$$= \mathrm{BIC}_1 + \cdots + \mathrm{BIC}_T + d_1\cdot\log\left(\frac{n}{n_1}\right) + \cdots + d_T\cdot\log\left(\frac{n}{n_T}\right).$$

Thus, minimizing the BIC at each tier does *not* ensure that the overall model will have overall minimum BIC. However, using a similar form of the BIC would ensure an overall minimum. This form uses $\log(n)$ instead of $\log(n_t)$ and is expressed as

$$\mathrm{BIC}_t^* = -2\sum_{i:\phi_{it}=1}\log\left(\frac{\exp(L_t)^{\lambda_{it}}}{1+\exp(L_t)}\right) + d_t\log(n).$$

REFERENCES

Albert, J. H., and Chib, S. (1993), "Bayesian analysis of binary and polychotomous response data," *Journal of the American Statistical Association*, 88, 669–679.

Hosmer, D., and Lemeshow, S. (1989), *Applied Logistic Regression*, New York: Wiley.

Kass, R., and Raftery, A. (1995), "Bayes factors," *Journal of the American Statistical Association*, 90, 773–795.

Kooperberg, C., Bose, S., and Stone, C. (1997), "Polychotomous regression," *Journal of the American Statistical Association*, 92, 117–127.

McCullagh, P., and Nelder, J. A. (1989), *Generalized Linear Models*, New York: Chapman and Hall.

Schwarz, G. (1978), "Estimating the dimension of a model," *The Annals of Statistics*, 6, 461–464.

Stern, H. (1998), "American football," in *Statistics in Sport*, ed. J. Bennett, London: Edward Arnold, pp. 3–23.

White, C. (2000), "Polychotomous Regression Applied to Futility Analysis and Expected Points Modeling," unpublished Ph.D. dissertation, Texas A&M University, Dept. of Statistics.

CHAPTER 12

The cold-foot effect

SCOTT BERRY AND CRAIG WOOD

The 2004 Super Bowl featured the favored New England Patriots playing the underdog Carolina Panthers. With the Panthers trailing 29–22 and 1:08 remaining in the game they scored on a 12-yard touchdown pass. The successful extra point tied the game at 29. New England took over at their own 40-yard-line after the kickoff went out of bounds. Four Tom Brady completions led the Patriots to the Panthers' 23-yard-line, Patriots kicker Adam Vinatieri set up to kick a game-winning 40-yard field goal with nine seconds remaining. The Panthers called a time-out. Vinatieri had to wait two minutes before making his attempt. The Panthers called the time-out to "ice" Vinatieri, hoping that, because he had to wait two minutes, thinking about the importance of his upcoming kick, he would be more likely to miss. Vinatieri made his kick and the Patriots won the Super Bowl.

Icing a kicker is not an uncommon strategy. Throughout the regular season it is standard practice when facing an important kick. Icing is used in other sports as well. When a basketball player faces an important free throw the opposing team will try to ice him by calling a time-out. While icing may be standard practice, does it work? Does making the kicker think about his field goal attempt for an extra two minutes alter his probability of success? It is a strange concept—giving the opponent more time to think about their upcoming kick. In many situations more time is deemed a positive thing. A golfer likes to take his time reading a putt and gathering his thoughts. Basketball coaches call time-out to regroup the team and baseball hitters are taking more and more time between pitches. Does having to wait before you attempt a very specialized task create "bad thoughts" or "nervous feelings?" We investigate whether icing the kicker is a successful strategy.

The data

The distance of a field goal is typically 17 yards beyond the current yard-line. For example, a team on the opponents' 10-yard-line will "hike" the ball seven yards back to the holder.

New England Patriots kicker Adam Vinatieri (4) kicks a 41-yard field goal in the fourth quarter of Super Bowl XXXVIII to beat the Carolina Panthers 32–29 in Houston, Sunday, Feb. 1, 2004.

The goal posts are placed 10 yards behind the goal line. Thus, the distance for this attempt would be 27 yards. The shortest possible field goal attempt is 18 yards. The record for the longest kick is 63 yards.

We collected the result and information about each field goal attempt during the 2002 and 2003 National Football League (NFL) regular season and playoffs. We recorded the kicker, the length of the kick, the score of the game, the time left in the game, and whether a time-out was called by the defense before the kick. Other factors possibly affecting the kick were collected. The type of turf (grass or artificial turf) and weather conditions were recorded. Conditions were recorded as sun, clouds, rain, snow, or indoors. The average wind speed (we do not know the wind direction relative to the direction of the kick) and the temperature for outdoor games were recorded.

There were 52 different field goal kickers, combining for a total of 2,003 attempts in these two seasons. Of these 2,003 kicks, 1,565 were successful (78.1%). We classified a pressure kick as one that would create a lead or tie for the team kicking and three minutes or less remaining in the game (or overtime). This encompasses the situations in which the kicker knows this is probably the last shot for his team and his kick would be critical to the outcome of the game. There were 139 (7%) pressure kicks over the two years, of which 101 were successful (a success rate of 67%). There were 38 times in which the defense called a time-out prior to the pressure kick and 24 of these were successful (67%). The percentage of successful kicks is less in pressure situations, but this may be due to more difficult kicks—in more difficult physical conditions. Teams in desperate situations are more likely to attempt more difficult kicks. In the following section we create a model for the probability of a successful kick, accounting for the various physical effects to better isolate a possible effect of pressure and icing.

Modeling the probability of a successful kick

There have been other papers modeling the probability of a successful kick. In an interesting geometry model Berry and Berry (1985) decomposed a kick attempt with a distance and directional component. Berry (1999) advanced this model using hierarchical modeling and knowledge of each missed kick being short or wide. Morrison and Kalwani (1993) and Bilder and Loughlin (1998) also model the success of NFL field goal kickers. Each of these papers use a logistic-based model for the probability of success. We also use a logistic-based model, but use a nonparametric function for the effect of distance on the kick.

We model the probability, P, of a successful kick, as a function of distance, using the following logistic model:

$$\log\left(\frac{P}{1-P}\right) = f(d)$$

The function $f(d)$ is a piecewise linear function, which is defined using "knot" values at each five-yard interval:

$$f(d) = \begin{cases} \beta_{18} & d = 18 \\ \beta_{18} + \beta_{20} & d = 20 \\ \beta_{18} + \beta_{20} + \beta_{25} & d = 25 \\ \beta_{18} + \beta_{20} + \beta_{25} + \beta_{30} & d = 30 \\ \beta_{18} + \beta_{20} + \ldots + \beta_{35} & d = 35 \\ \beta_{18} + \beta_{20} + \ldots + \beta_{40} & d = 40 \\ \beta_{18} + \beta_{20} + \ldots + \beta_{45} & d = 45 \\ \beta_{18} + \beta_{20} + \ldots + \beta_{50} & d = 50 \\ \beta_{18} + \beta_{20} + \ldots + \beta_{55} & d = 55 \\ \beta_{18} + \beta_{20} + \ldots + \beta_{60} & d = 60 \end{cases}$$

For any points between these five-yard intervals we interpolate. For example, f(33) $= \beta_{18} + \beta_{20} + \beta_{25} + \beta_{30} + \frac{3}{5}\beta_{35}$. The log-odds ratio between a 30-yard field goal and a 35-yard field goal is $\exp(\beta_{35})$. We know that the farther the field goal attempt the less likely it will be a success, so we restrict each β (other than β_{18}, which is an intercept) to be less than 0. We place a truncated normal prior distribution on each β, with a mean of -0.25 and a standard deviation of 1.

Using the full dataset and ignoring the difference in kickers, Figure 12.1 shows the posterior mean of the probability of a success for yardages between 18 and 60 yards. The standard deviation for the probability of success is also shown on Figure 12.1.

To account for the different kickers we add a parameter for each. Let γ_i be the efficacy for kicker i. We include this in the model as

$$\log\left(\frac{P}{1-P}\right) = f(d) + \gamma_i.$$

The odds of a success, $\left(\frac{P}{1-P}\right)$, can be written as

$$\left(\frac{P}{1-P}\right) = \exp(f(d))\exp(\gamma_i).$$

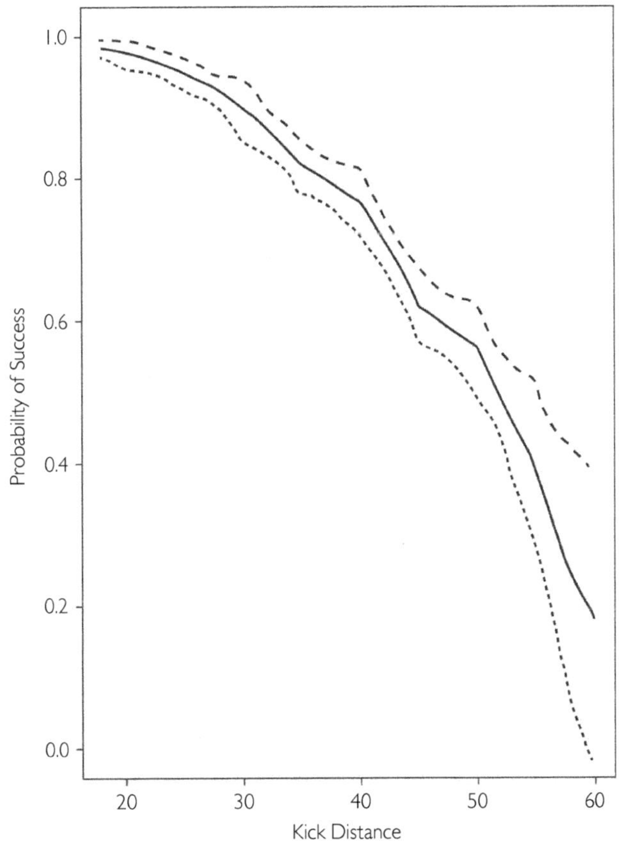

Figure 12.1 The posterior mean and 95% interval for the probability of success from each distance assuming there is no difference in kickers.

The value of exp (γ_i) has a multiplicative effect on the odds of success. Thus, a value of $\gamma_i < 0$ ($\exp(\gamma_i) < 1$), reduces the probability of success, while a value of $\gamma_i > 0$ ($\exp(\gamma_i) > 1$) increases the probability of success. We create a normal distribution of kickers with a mean of 0 and a standard deviation of σ (σ has a non-important prior distribution). The relative efficacies of the kickers are constant over the distances. If a kicker is better from one distance he is better from all distances. This is probably not exactly correct, as some kickers are more accurate and other have better distance, but these subtle differences in kickers is minor.

Table 12.1 summarizes the results of the model with kicker parameters added. Figure 12.2 presents the probability of a successful kick for several different kickers. Notice the proportional nature of the kickers. Jason Hanson (Detroit Lions) is considered the best kicker and has the same positive proportional effect to the odds of a successful kick, compared to the other kickers.

Using a multiplicative effect to the odds of success we can now incorporate other aspects of the kick, including the investigation of pressure and icing on a kicker. For the presence of a physical or psychological variable, say Z = 1 corresponds to "cold weather" (temp < 40), and Z = 0 otherwise, we use a model

$$\log\left(\frac{p}{1-p}\right) = f(d) + \gamma_i + \theta Z.$$

Table 12.1 ***A summary of the top, bottom, and middle kickers. This model includes a distance variable and a variable for each kicker. The posterior mean and standard deviation of each parameter (γ) and the relative risk ($\exp(\gamma)$) is reported.***

	Kicker	γ		Relative Risk	
		Mean	SD	Mean	SD
1	Hanson	0.492	0.310	1.718	0.567
2	Vanderjagt	0.454	0.283	1.640	0.477
3	Stover	0.389	0.267	1.528	0.414
4	Janikowski	0.388	0.266	1.526	0.411
5	Brien	0.292	0.331	1.417	0.503
...					
26	Hall	0.009	0.233	1.037	0.242
...					
50	Tuthill	−0.441	0.392	0.695	0.281
51	Peterson	−0.540	0.283	0.606	0.173
52	Marler	−0.575	0.266	0.583	0.155

Therefore θ captures the effects of cold weather. If $\theta < 0$ then cold weather decreases the probability of success. We include cold weather, as well as other interesting aspects of field goals in the next section.

What affects a field goal try?

In this section we investigate the many factors of a field goal try. We investigate physical and psychological effects in the model described above. This has the benefit of accounting for the distance of the kick and the kicker. Any factors that have an effect on a kick, do so beyond the effects of the kicker and distance. The variables we investigate include:

Weather: Each game is classified as being sunny (829), clouds (717), snow (39), rain (132), or inside (385). We use the following variables in the model: *Snow/Rain*, *clouds*, and *inside*. The temperature of the game at kickoff is also recorded. We characterize games as being *cold* when the start temperature is 40° F or lower. We characterize the game as *hot* when the start temperature is 80° F or higher. We also record when the kick takes place on *grass*.

A kick is a *pressure* kick when it takes place within the last three minutes or overtime and could create a tie or put the kicking team ahead. A kicker is considered *iced* when the opponent calls a time-out before a pressure kick.

We investigate each of these factors by placing the variable in the model (as described in the previous section) with the kicker effects and distance. Table 12.2 reports a summary of the results. The results match the prevailing thought on the difficulty of a kick. A kick inside a dome is more likely to be successful; there is an estimated mean increase of 37% in the odds of a successful kick. For an average kicker, the estimated probability of a successful 40-yard

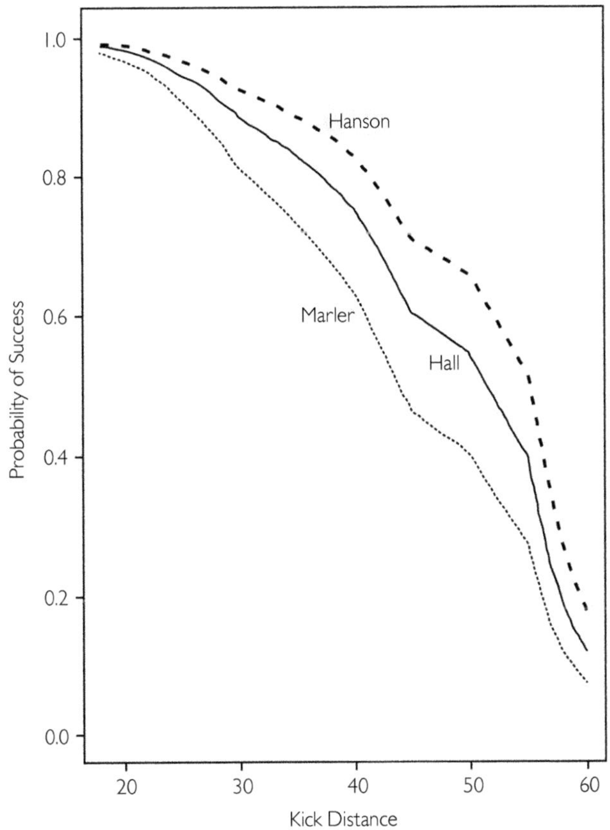

Figure 12.2 The posterior mean probability of success for three different kickers. Hanson is the highest rated kicker, Hall is the mean kicker, and Marler is the lowest rated kicker. This model includes the distance of the kick and the different kickers.

Table 12.2 ***A summary of the physical and psychological variables for each kick. These results are for the nine separate models with the above variables, including distance and the kicker. The posterior mean and standard deviation of each parameter (θ) and the relative risk ($\exp(\theta)$) is reported.***

Variable	θ		Relative Risk		Pr(> 0)	N
	Mean	SD	Mean	SD		
Inside	0.304	0.160	1.372	0.222	0.979	385
Clouds	0.177	0.115	1.202	0.138	0.940	717
Snow/Rain	−0.429	0.194	0.664	0.132	0.011	171
Wind 15	−0.281	0.162	0.765	0.125	0.031	264
Cold	−0.062	0.154	0.952	0.149	0.339	641
Hot	0.076	0.207	1.102	0.233	0.652	198
Grass	−0.379	0.123	0.690	0.086	0.004	1308
Pressure	−0.004	0.213	1.018	0.222	0.466	139
Ice	−0.553	0.357	0.614	0.229	0.074	38

kick (in sunny weather) is 0.759 (odds of 3.15). For a 40-yard kick, for an average kicker, in an indoor stadium this mean probability increases to 0.812 (odds of 4.31). The probability that being inside has a beneficial effect on kicks is 0.979.

Clouds also have a small beneficial effect on kicks, with an estimated 20.2% increase in the probability of success on each kick. This results in a mean success probability of 0.791 on a 40-yard kick for an average kicker on a cloudy day. There is a 98.9% chance that snow or rain will reduce the probability of success on a kick. There is an estimated 33.6% reduction in the odds of a successful kick. For an average kicker, from 40 yards, the mean probability of a successful field goal in the snow or rain is 0.676. High winds also are likely (0.969) to reduce the probability of a successful kick. The effect is less than that of snow or rain, but still a meaningful 23.5% reduction in the odds of success. From 40 yards, an average kicker in high winds would have an estimated 0.707 probability of success.

Cold and hot weather have the expected direction of effect, but in neither case is the effect statistically strong. In cold weather there is an estimated 4.8% decrease in the odds of success, while in hot weather there is a 10.2% estimated increase in the odds of success. This results in mean probabilities of success, for an average kicker from 40 yards, of 0.750 and 0.776. Kicking off grass is highly likely (0.996) to be more difficult than kicking off of artificial turf, with a reduction of 31.0% in the odds of success. This results in a mean probability of success of 0.685 off grass, from 40 yards for an average kicker.

The first of the two "psychological" variables, pressure, has little effect on the probability of success. There is a mean decrease of 1.8% to the odds of success when in a pressure situation. This effect is smaller than for any of the physical variables. This is interesting in part because the success rate for pressure kicks is approximately 10% less than for nonpressure kicks. While pressure has seemingly no effect, icing the kicker has a pretty strong negative effect. There is a 92.6% chance that icing the kicker actually reduces the probability of success. The mean reduction in the odds of success is 38.6%. For an average kicker, from 40 yards, the estimated probability of a successful kick after being iced is 0.659. The small sample size (38) of icing

Table 12.3 ***A summary of the physical and psychological variables for each kick included in one model, with distance and kicker included. The posterior mean and standard deviation of each parameter (θ) and the relative risk ($\exp(\theta)$) is reported.***

Variable	θ		Relative Risk		Pr(> 0)	N
	Mean	SD	Mean	SD		
Inside	0.283	0.262	1.376	0.387	0.877	385
Clouds	0.229	0.151	1.272	0.191	0.940	717
SnowRain	−0.169	0.214	0.864	0.189	0.216	171
Wind15	−0.266	0.186	0.780	0.149	0.070	264
Cold	−0.243	0.179	0.797	0.144	0.088	641
Hot	0.053	0.200	1.075	0.212	0.606	198
Grass	−0.396	0.155	0.681	0.109	0.012	1308
Pressure	0.226	0.247	1.292	0.318	0.829	139
Ice	−0.696	0.370	0.535	0.212	0.030	38

attempts makes it hard to conclude it highly likely that icing works (0.926), but the estimated effect size appears large. Reducing the probability of a successful kick from 0.759 to 0.659 is a very important difference.

When adding icing to the distance and identity of the kicker, it has an important negative effect. Some of the physical variables have an effect on kicks and may account for some of the negative aspects of the iced kicks. It may be that they were all on grass and in cold weather. We build a model using all of the variables in concert. Table 12.3 presents a summary of this model. The effects of the variables are similar to the model with just the kicker and distance effects. In this model a pressure kick, with the kicker not iced, is more likely to be successful. The data is not strong on this, with a 0.829 probability of pressure having a positive effect. This full model has a higher probability (0.970) that icing has a negative effect on kickers than the smaller model. The 46.5% reduction in risk is also a larger decrease than the smaller model. An average kicker, facing a 40-yard attempt, in a pressure situation (with sunny conditions, no

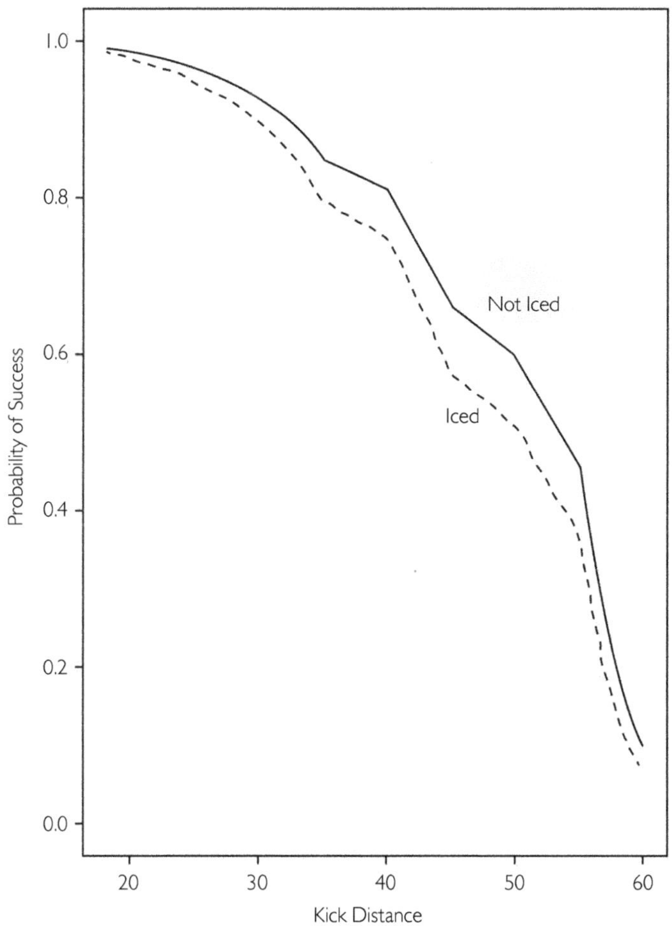

Figure 12.3 The posterior mean probability for a successful kick for an average kicker, in a pressure situation. The two curves represent the kicker not iced (solid) and iced (dashed).

Table 12.4 ***The results for the 101 pressure kicks and the 38 icing kicks. The number of successes/trials and the proportion of successes are reported.***

	≤ 30	31–40	41–50	51+
Pressure	5/6 (0.833)	22/26 (0.846)	25/34 (0.735)	7/16 (0.438)
Ice	4/5 (0.800)	10/16 (0.625)	8/13 (0.615)	2/4 (0.500)

precipitation, low wind, warm weather, and artificial turf) has a 0.847 probability of success. If a time-out is taken, icing the kicker, the probability of success is estimated to be reduced to 0.747. Figure 12.3 shows the posterior mean probability of success for an average kicker in a pressure situation when the kicker is and is not iced.

The conclusion from the model is that icing the kicker works—it is likely icing the kicker reduces the probability of a successful kick. We are surprised by the strength of this conclusion and hence investigate the 38 examples of icings and 101 pressure kicks without icings in more detail. Table 12.4 shows the results of icing kicks and nonicing kicks, broken into distance intervals. The middle distance intervals are those with the most attempts, and within these intervals there is a substantial difference between the pressure kicks and the icing kicks. Clearly the evidence is not overwhelming, but it is compelling. Before the data presentation we doubted the icing effect, but we now believe there is a compelling case.

Discussion

A common topic of research within sports lately has been to investigate "psychological effects." Whether this is looking for a hot hand, clutch hitting, or a pressure element in kicking field goals. Here we present a compelling case that icing a field goal kicker on a pressure kick may decrease the probability of success. This implies that a psychological effect of pressure exists, and is compounded by more time to dwell on the kick. The hot-hand effect describes the possibility that a player, such as a basketball shooter, has streaks in which he is a better shooter. The existence of the hot-hand effect is a controversial topic. As an alternative to this hot-hand effect we have presented evidence of a "cold-foot" effect!

FURTHER READING

Berry, D.A. and Berry, T.D. 1985. The probability of a field goal: rating kickers. *American Statistician*, 39, 152–155.

Berry, S.M. 1999. A geometry model for NFL field goal kickers. *Chance*, 12:3.

Bilder, C.R. and Loughlin, T.M. 1998. "It's Good!" An analysis of the probability of success for placekicks. *Chance*, 11:2, 20–24.

Morrison, D.G. and Kalwani, M.U. 1993. The best NFL field goal kickers: are they lucky or good? *Chance*, 6:3, 30–37.

CHAPTER 13

A penalized maximum likelihood approach for the ranking of college football teams independent of victory margins

DAVID MEASE

Abstract
This article introduces a penalized maximum likelihood approach for ranking all NCAA Division 1-A college football teams. The model does not consider margin of victory and is based solely on win/loss data. Despite the simplicity, the model leads to rankings which exhibit greater agreement with expert opinion (i.e., AP and Coaches polls) than the models historically and currently used by the Bowl Championship Series (BCS). The model can be implemented using standard statistical software packages.

Keywords: *Bayesian, paired comparisons, rankings*

Introduction

Prior to 1998, the national championship of U.S. Division 1-A college football was based solely on two opinion polls, commonly referred to as the Coaches and the AP (Associated Press) polls. The ranking of the teams in these two polls is determined by coaches and sports writers, respectively, who vote weekly for the top 25 teams based on the teams' performances in all games played. Up until 1998, the team that finished the year in the number one position in both polls was deemed to be the national champion. In cases in which the two polls did not agree on the choice of the top team, the two teams involved shared the title of national champion, which occurred most recently in 1997 with Michigan and Nebraska.

With the 1998 season came the inception of a new system under which the top two teams at the end of the season would play one final game for the championship, thereby eliminating the possibility of a shared title. This new system was called the Bowl Championship Series, or BCS. The BCS system employed rankings produced by a number of computer models in addition to the rankings of the AP and Coaches polls in order to determine the top two teams. The purpose of these computer models was to lessen dependence on the AP and Coaches polls which, although historically trusted as representing expert opinion, have often been criticized on the following two accounts. First, the human pollsters are not objective observers and may have biases toward certain schools based on regional loyalty, historical perception, and so on. Second, it is impossible for a human pollster to recall all outcomes of all games involving the 117 Division 1-A teams over the course of an 11- to 14-week season, even if he or she had witnessed or read about every game.

Although the computer models employed by the BCS do not have any such bias and are able to consider the outcomes of all games played, they have also proven to be extremely controversial as a result of many instances in which they produced nonintuitive rankings which differed significantly from the AP and Coaches polls. For instance, in 2001 the University of Oregon, which finished second in both the AP and Coaches polls, finished eighth in one of the eight BCS computer models, and finished seventh in two of the others. The low ranking of Oregon in these three computer polls was thought to be attributable to their many victories by narrow point margins, since the four BCS computer models that did not use margin of victory ranked Oregon no lower than third. This was not the first time that controversy resulted from the computers polls weighing margin of victory much more heavily than public opinion, and as a result it was mandated that all computer polls either ignore margin of victory or be excluded from the BCS system beginning in 2002. The idea was that by forcing computers to ignore margin of victory, the resulting rankings would be more consistent with the public's opinion, which tends to be more a function of a team's winning percentage and quality of opposition than a function of the point margins. Furthermore, this would remove any incentive for a team to "run up the score" in a game that is a foregone conclusion, which is universally considered to be bad sportsmanship.

It should be noted that the belief that the BCS computer models are more influenced by margin of victory than public opinion is not shared by everyone. Some people, including the creators of some of these computer models, would argue the human pollsters themselves can be highly influenced by large victory margins, citing examples in which a team that wins by a large margin climbs higher in the polls than a team that wins by a small margin. Although it is in fact possible for human pollsters to be influenced in such a way, a number of instances similar to the one involving Oregon described above were enough to convince the BCS that the computer polls tended to weigh margin of victory too heavily (even after a restriction was made limiting the maximum margin to 21 points). Further evidence that human pollsters tend to put relatively small weight on margin of victory was given by Harville (2003), who noted that the version of his model which ignores margin of victory agreed more strongly with the human pollsters for the football season he considered. Sentiment for eliminating margin of victory from the computer polls also arose as a result of the fact that a computer poll cannot discern between a large margin of victory resulting from one team dominating another for an entire game versus a large victory margin as a result of a large amount of scoring occurring after the game has already effectively been decided. A human pollster, on the other hand, can differentiate these two situations.

In this article we examine the problem of constructing a computer model to rank the (currently) 117 Division 1-A football teams without using victory margins. A new computer model is proposed that is shown to produce rankings which on average agree more strongly with the AP and Coaches polls than the models used by the BCS as well as other competing models. The proposed model uses a penalized likelihood approach which results in a ranking process that attempts to mimic the thought processes of the human pollsters of the AP and Coaches polls. The article is organized as follows. The next section gives a simple example illustrating the complexities involved in ranking teams without using victory margins. The section following that describes some of the models used in the statistics literature for ranking football teams. The subsequent section discusses the models that are or have been used by the BCS. The two sections after that present the proposed model and discuss its implementation using statistical software. The section entitled "Results and comparisons with competing models" compares the rankings produced by the proposed model to those of some competing models, including the BCS computer models. The final section describes some possible modifications to the model.

A simple example

To illustrate the complexity in fairly ranking teams based solely on win/loss data, consider the following season in which five teams (A, B, C, D, and E) played a total of eight games with the following outcomes:

- Game 1: Team A defeated Team C
- Game 2: Team A defeated Team E
- Game 3: Team B defeated Team A
- Game 4: Team B defeated Team E
- Game 5: Team C defeated Team D
- Game 6: Team C defeated Team E
- Game 7: Team D defeated Team E
- Game 8: Team D defeated Team E

Team B finishes the season with a win/loss record of 2 wins and zero losses (denoted 2–0), Team E finishes with a record of 0–5, and Teams A, C, and D all finish with records of 2–1. Now suppose no scores are available and we seek to fairly rank the five teams based only on the above outcomes. We can judge the strength of each team based on its win/loss record as well as on the strength of its opponents, who in turn are judged based on their win/loss records and the strength of their opponents, and so on. An argument for ranking these five teams may go as follows.

Team B should be ranked as the first-place team, having no losses and having defeated Team A, who is a strong opponent since if not for being defeated by Team B, would have no losses. Team E should be ranked as the worst team, having no wins and having loss to Team D twice, who is a weak opponent since if not for defeating Team E (twice) would have no wins. Teams A, C, and D remain to be ranked from second to fourth. Although all three have the same record of 2–1, Team A should be ranked second since their only loss came to Team B, who is a strong opponent since they are the number one team. Finally, Team C should be ranked above Team D since Team C defeated Team D while Team D defeated only Team E, who is a weak opponent since they are ranked as the worst team. Thus, the ranking of the teams from first to last should be B, A, C, D, and E.

Although such an argument is tractable for the five teams playing eight games in the example, for the 117 Division 1-A teams who each play 11 to 14 games on average the problem becomes extremely difficult for a human pollster to handle. The voters of the AP and Coaches polls attempt to reason in a similar fashion as above, but it is impossible for them to simultaneously consider all games played. This is the motivation for developing a computer model that can simultaneously judge the strength of all 117 teams based on winning percentage and strength of opposition.

Football ranking models in the statistical literature

A number of articles exist in the statistics literature dealing with the ranking of college football teams. Many of these use a linear model approach in which margin of victory is modeled as a function of the two competitors in each game as well as a number of covariates. For instance, Harville (1977) proposed a linear model which considered as covariates the location of the game (i.e., which team was the home team), the date of the game, and the division membership of the two teams involved in each game. In order to downplay victories by large margins he also considered two variations on this model, one in which the margin of victory was truncated at 15 points and one in which the margin of victory was truncated at 1 point. Note that latter model in essence ignores margin of victory and as such offers a possible solution to the problem of interest in this article. Results from Harville's models will be considered in the section on "Results and comparisons with competing models." In addition to truncating margin of victory at a certain value as Harville did, other approaches to downplaying the effect of large victory margins in linear models were offered by Stern (1992) and Bassett (1997). In Stern's (1992) model a smaller weight was given to games in which the victory margin was large, while Bassett (1997) proposed using absolute error as opposed to least squares.

In addition to linear model approaches, likelihood-based approaches can also be found in the statistics literature. Keener (1993) considered a number of models for ranking college teams, including a version of what is often called the Bradley–Terry model after an article dealing with paired comparisons for experimental design (Bradley and Terry 1952). In this model, each Team i is assigned a parameter θ_i that can be thought of as representing its strength. The probability that team X defeats team Y is then given by $\theta_X/(\theta_X + \theta_Y)$, and the estimates for the θ_i values can then be obtained by maximizing the likelihood over all the game outcomes for the entire season. Ordering the MLE estimates for the θ_i values then determines a ranking of the teams. Although Keener's model used a weighting factor to weigh large margins of victory more than close victories, it is not necessary.

It is important to note that a problem with the Bradley–Terry model is that if team i has no losses, the likelihood becomes larger as θ_i becomes larger. It follows that the maximum likelihood estimator for the θ_i value of any undefeated team is infinity. As a consequence, the model will produce a tie for first among all undefeated teams. This is problematic due to the substantial variability in the strength of opponents among the different Division 1-A teams. It is undesirable to restrict all teams with at least one loss to be ranked below all undefeated teams. Moreover, a two- or three-way tie for first is a rather unsatisfactory answer to the question of who is number one. Keener applies the model to a season in which all teams have at least one loss so that this is not a problem; however, for general use the model would need to be modified since it is not uncommon to have one or more undefeated team in Division 1-A football.

A different likelihood-based approach was proposed by Thompson (1975) for ranking the professional teams of the National Football League (NFL). Similar to the model above, each team i was assigned a strength parameter θ_i, but in this model the probability that team X defeated team Y was given by $\Phi(k(\theta_X - \theta_Y))$, where θ_X and θ_Y are the θ_i parameters for teams X and Y, respectively, and Φ is the standard normal cumulative distribution function. The reason for the inclusion of the parameter k in the model is that Thompson restricted the values for the θ_i's to be the integers $1, \ldots, n$, where n is the number of teams. Although this restriction eliminates the problems associated with undefeated teams described in the above paragraph, it also makes maximization of the likelihood extremely difficult, requiring a search over $n!$ possibilities. Thompson avoided this problem by further restricting that no team can be ranked below a team having a lower winning percentage. While such a restriction may be reasonable for the teams of the NFL, which play schedules that are relatively comparable in difficulty, such a restriction is undesirable when ranking college football teams due to the variability in the schedule strengths mentioned above. Note that the use of the normal CDF for paired comparisons can be traced back to Thurstone (1927).

Competing models used by the BCS

In the five years of existence up to the time of the writing of this article, ten different models have been used by the BCS for ranking the Division 1-A college football teams. Moreover, some of these ten models have been modified during this period. Due to the large number of models and the fact that in some cases the technical details regarding the models are not public, we will not give descriptions of each specific model. Instead, we will evaluate these models based on the rankings they produced for the specific football seasons during which each model was included in the BCS. This will be done in the section on "Results and comparisons with competing models." Although we do not provide a description of each BCS model, it should be noted that the BCS models cover a wide range of complexity. Some models take into account only win/loss data while others consider such factors as location, date, rankings from previous seasons, and margin of victory (prior to 2002).

The proposed model

To begin, let us suppose the intrinsic performance level of each team i varies from day to day according to a normal distribution with mean θ_i and variance of 1/2 (for simplicity). Treating the performance level as random is consistent with the fact that even very good teams can be "upset" by weaker teams. The use of the normal distribution was used largely because it results in a model that produces reasonable rankings, but it can also be motivated by supposing that the performance level is a sum of a large number of independent factors and invoking the central limit theorem. Further, suppose that this intrinsic performance level of each team is independent of its opponent's level and that every game is won by the team that has the greater performance level on that day. Under these assumptions the probability that team X defeats team Y is $\Phi(\theta_X - \theta_Y)$, similar to the Thompson model described earlier. Note that the same expression also arises from the model of Harville (2003) by supposing that the difference in the score when team X plays team Y is normal with mean $\theta_X—\theta_Y$ and variance 1 and that only win/loss data is available. However, we prefer the interpretation of the θ_i parameter as a mean performance level rather than a mean score for modeling win/loss data. This is the basic idea behind our likelihood, which is given in its entirety for the n teams by

$$l(\theta) = \underbrace{\prod_{(i,j)\in S} [\Phi(\theta_i - \theta_j)]^{n_{ij}}}_{\text{Part 1}} \times \overbrace{\prod_{i=1}^{n} \Phi(\theta_i)\Phi(-\theta_i)}^{\text{Part 2}} \times \underbrace{\Phi(\theta_{n+1})\Phi(-\theta_{n+1}) \times \prod_{(i,j)\in S^*} [\Phi(\theta_i - \theta_j)]^{n_{ij}}}_{\text{Part 3}}$$

and discussed in detail in the following paragraphs. This likelihood function $l(\theta)$ is maximized as a function of $\theta = (\theta_1, \ldots, \theta_{n+1})$, where θ_{n+1} is a nuisance parameter. We will rank the teams according to the MLEs of their respective θ_i values such that the team with the largest value of θ_i is ranked as the top team.

Part 1 of the likelihood captures the motivation from the first paragraph in this section where we proposed that the probability that team X defeats team Y is $\Phi(\theta_X—\theta_Y)$. For Part 1 we define the set S to consist of all ordered pairs (i, j) for which team i defeated team j and both team i and team j are members of Division 1-A. The value n_{ij} is the number of times team i defeated team j. Tie games are counted as half a win and half a loss.

In the maximization of Part 1, a team earns a large value of θ_i by defeating other teams who themselves have earned relatively large θ_i values. In this way, maximization of Part 1 simultaneously judges the strength of each team based on its win/loss record and based on the strength of its opponents, which is in turn determined based on their win/loss records and the strength of their opponents. Thus, the process of maximizing Part 1 alone mimics the thought processes a human would attempt to use to rank the teams as described in the section "A simple example." However, there are a couple of problems with using Part 1 alone.

The first problem is the difficulty mentioned earlier in conjunction with the Bradley–Terry model. That is, if a team i has no losses, Part 1 increases without bound as θ_i becomes larger so that the MLE for any undefeated team is infinite. To solve this problem we penalize the likelihood by multiplying by $\prod_{i=1}^{n} \Phi(\theta_i)\Phi(-\theta_i)$ as given by Part 2. For more discussion on estimation based on penalized likelihood the reader is directed to Green (1998) and Silverman (1985).

Although the main reason for using the penalty term above is that it restricts the MLE estimates of the θ_i's to be finite, it will also be shown that it leads to rankings which consistently agree with human polls (i.e., AP and Coaches polls). Furthermore, the use of this penalty term is easily justified. Two justifications for the penalty term are given in the following two paragraphs.

One way to view the penalty term is as Bayesian prior distribution. To illustrate this we will associate with a team X the parameter ξ_X defined by $\xi_X = \Phi(\theta_X)$. Note that these ξ parameters can be thought of as the probability of defeating a team with a θ_i value of zero. Now if we treat the unpenalized likelihood described before as the conditional distribution of the game outcomes given the ξ's and assign independent Beta distributions with parameters α and β as the prior for the ξ of each team, the posterior of the ξ's is proportional to the unpenalized likelihood and

$$\prod_{i=1}^{n} \xi_i^{\alpha-1}(1 - \xi_i)^{\beta-1} = \prod_{i=1}^{n} \Phi(\theta_i)^{\alpha-1}(1 - \Phi(\theta_i))^{\beta-1}.$$

Taking $\alpha = \beta = 2$ gives the likelihood proposed. The motivation for taking $\alpha = \beta$ is to give an equal prior probability of winning and losing for each team, while the choice of the value 2 for these two parameters was selected after testing the model on actual football seasons.

A second way to view the penalty term is by considering one additional "virtual" team with a θ_i value of zero. Giving every team exactly one win and one loss to this virtual team also results in a likelihood with the penalty term proposed. (Note that in general one could give α—1 wins and β—1 losses to this virtual team.) Viewing the penalty term in this way makes it clear how the problems presented by undefeated teams are solved, since when the virtual team is considered each team actually has at least one loss. Furthermore, this view of the penalty term also suggests why it is simple to fit the model using existing binary response regression procedures to be described in the section on "Implementation of the model using statistical software."

Finally, maximization of the product of Parts 1 and 2 will produce an adequate ranking of the teams, but without using Part 3 we would be ignoring all games in which one opponent was a member of Division 1-A but the other was not (recall the set S does not include such pairings). Because such games are not extremely common, it is not unreasonable to ignore such games. In fact, the BCS computer model of Wes Colley does exactly that. However, it is foreseeable that this could lead to considerable controversy if, say, a certain team's only loss was to a non-Division 1-A opponent. In such a case, ignoring this loss would make this team undefeated and ranked as one of the best teams, despite having a loss to an extremely weak opponent. In place of ignoring these games, another solution would be to expand the set S to include all games of any team who has played a Division 1-A team. However, these extra games would involve yet another set of teams for which we would be ignoring games against opponents not in the set. Thus, unless we are willing to rank almost all college football teams in the nation, a different solution must be found.

The solution that was chosen is to combine all teams not in Division 1-A who play a Division 1-A opponent into one generic n + 1st team. The set S^* is then taken to be all ordered pairs (i, j) for which team i defeated team j and one of i or j is equal to $n + 1$ and the other is in $\{1, \ldots, n\}$. The term $\prod_{(i,j)\in S*} [\Phi(\theta_i - \theta_j)]^{n_{ij}}$ in Part 3 of the likelihood then accounts for all games involving Division 1-A teams that were not included in the set S. As before, the n_{ij} are the number of times team i defeated team j. Note that most of the games in the set S^* are losses for the generic non-Division 1-A team, and as such θ_{n+1} invariably has a very small MLE value. Thus, the reward for a Division 1-A team beating a team not in Division 1-A is very small while a loss to a non-Division 1-A has a strong negative effect on the ranking, as should be the case. Finally, the term $\Phi(\theta_{n+1})\Phi(-\theta_{n+1})$ is included in Part 3 to penalize the likelihood of this generic n + 1st team just as Part 2 did for the other n teams.

Implementation of the model using statistical software

Standard statistical software packages can be used to maximize the likelihood function for the model by fitting a binary regression model employing the probit link function. The procedure for doing this is similar to the method used by Fienberg and Larntz (1976) for fitting the Bradley–Terry model. The data matrix should consist of one column for each team in Division 1-A as well as one additional column to represent the generic non-Division 1-A team. For each game played involving at least one Division 1-A team, a row is included in the data matrix in which the value 1 is placed in the column for the winner of the game and a −1 is placed in the column for the loser of the game while the values for the other columns are all set to zero. In addition to these rows, the data matrix should also contain two additional rows for each team. For any given team these two rows are as follows. The first row should contain a 1 in the column for that team with the rest of the entries being zero, while the second row should contain a −1 for that team with the rest of the values being zero. These two additional rows provide the penalty term for the likelihood. Finally, the response vector should be set to a vector of all 1's equal in length to the number of rows in the data matrix.

If a probit regression model with no intercept is fit to the data matrix and response vector constructed in this way it can be verified that the likelihood being maximized is exactly the likelihood for the proposed model. As such, the parameter estimates output from the software corresponding to each of the different columns of the data matrix will be the maximum likelihood estimates for the θ_is in the proposed model. Thus, the single parameter estimate for the generic non-Division 1-A team can be discarded and the remaining values can be used to rank all the Division 1-A teams. Code for implementing this procedure in SAS for the example presented in the section "A simple example" can be found at http://members.accesstoledo.com/measefam/SAScode.html.

Note that if tie games exist in the data the above procedure must be modified slightly since not all n_{ij} will be integers. In these cases, each row in the data matrix described above that does not correspond to a tie game should be entered twice, while rows corresponding to tie games should still be entered only once. In this way the likelihood being maximized is proportional to the square of the desired likelihood, and the ranking of the teams produced will be correct.

Results and comparisons with competing models

This section examines the rankings resulting from the competing models introduced earlier. We will compare these rankings to those of the proposed model in terms of similarity to the AP and Coaches Polls. This similarity will be quantified in terms of the average absolute difference between the rankings of each model and the average ranking of the AP and Coaches Polls, although one may argue that a measure which gives large differences smaller weight (or larger weight) than does the absolute value function may be more appropriate.

Harville (1977) gave his rankings for the 1975 college football season. These rankings are included in Table 13.1 along with the rankings for the model proposed in this article ("Mease"). The teams listed are the top 15 teams in terms of average rankings of the AP and Coaches Polls ("Poll Avg."). The three columns listed for Harville's model correspond to the original model ("no cap") as well as the two models in which the margin of victory is limited to 15 points and 1 point. The final row in the table gives the average absolute difference between the rankings for each model and the AP and Coaches average rankings for the 15 teams. From these values it can be seen that for the 1975 season the proposed model agrees more closely with the AP and Coaches Polls than the "15 point" and "no cap" Harville models, while the "1 point" Harville performs better than the proposed model with regard to average absolute difference from the AP and Coaches Polls.

Keener (1993) gave rankings based on his model for the 1989 college football season. Table 13.2 gives these rankings for the top 15 teams as determined by the average of the AP and Coaches Polls for 1989 ("Poll Avg."). From this table it can be seen that the model proposed in

Table 13.1 ***Comparison to Harville's Models (1975 Season).***

Poll avg.	Team	Record (wins-losses-ties)	Mease	Harville 1 point	Harville 15 point	Harville no cap
1	Oklahoma	11–1–0	2	1	6	3
2	Arizona State	12–0–0	1	2	9	14
3	Alabama	11–1–0	5	3	4	1
4	Ohio State	11–1–0	3	4	2	2
5	UCLA	9–2–1	16	16	16	15
6.5	Texas	10–2–0	7	7	3	5
6.5	Arkansas	10–2–0	9	10	5	6
8	Michigan	8–2–2	13	19	14	8
9	Nebraska	10–2–0	6	5	7	4
10	Penn State	9–3–0	12	9	10	11
11.5	Texas A&M	10–2–0	10	6	12	12
12	Maryland	9–2–1	27	23	15	17
14	Miami(Ohio)	11–1–0	8	8	25	26
14	Pitt	8–4–0	21	13	13	7
14.5	California	8–3–0	15	18	22	22
0	Avg. Abs. Diff.		3.93	3.67	4.13	4.53

Table 13.2 ***Comparison to Keener's Model (1989 Season).***

Poll avg.	Team	Record	Mease	Keener
1	Miami(Florida)	11–1–0	2	1
2.5	Notre Dame	12–1–0	1	3
2.5	Florida State	10–2–0	5	5
4	Colorado	11–1–0	3	7
5	Tennessee	11–1–0	4	11
6	Auburn	10–2–0	9	8
7.5	Michigan	10–2–0	8	9
8	Alabama	10–2–0	6	12
8.5	Southern California	9–2–1	10	4
10	Illinois	10–2–0	7	14
11.5	Nebraska	10–2–0	14	13
11.5	Clemson	10–2–0	11	6
13	Arkansas	10–2–0	13	17
15	Penn State	8–3–1	16	18
16	Michigan State	8–4–0	19	15
0	Avg. Abs. Diff.		1.60	2.87

this article has an average absolute ranking difference of 1.60 from the AP and Coaches average rankings. By comparison, the Keener model has an average difference of 2.87 and as such did not perform as well in terms of similarity to the AP and Coaches Polls as the model proposed in this article. It is also interesting to note that Houston, the second ranked team in the Keener model, is not included in the table since it was not ranked by the Coaches Poll, which ranked only 20 teams at that time. The model proposed in this article ranked Houston as 15th, which is similar to their AP ranking of 14th.

The final tables listed give comparisons with the BCS models. Tables 13.3–13.7 give the rankings for the BCS models for the five college football seasons beginning in the years 1998, 1999, 2000, 2001, and 2002, respectively, along with the rankings for the model proposed in this article ("Mease"). The rankings for all models as well as the average of the AP and Coaches Polls ("Poll Avg") exclude the bowl games played at the end of the season. This was done since the primary function of the BCS is to select teams to participate in these bowl games. The teams included in Tables 13.3–13.7 are the top 15 teams as computed by the overall BCS ranking system prior to the bowl games and are listed in the order of these overall BCS rankings.

From Table 13.3 it can be seen that the model proposed in this article outperformed two of the three models used by the BCS in 1998 in terms of average absolute difference from the AP and Coaches average ranking. In Tables 13.4–13.6 the proposed model outperformed three, seven, and seven of the eight models used in the years 1999, 2000, and 2001, respectively, and Table 13.7 shows the proposed model outperformed four of the seven models used in 2002. Table 13.8 summarizes the average absolute ranking differences from the AP and Coaches

Table 13.3 ***Comparison to 1998 BCS Models***

Poll avg.	Team	Record (wins-losses)	Mease	Jeff Sagarin	New York Times	Anderson/ Hester
1	Tennessee	12–0	1	2	2	1
2	Florida State	11–1	2	3	1	2
4	Kansas State	11–1	4	1	5	4
3	Ohio State	10–1	8	6	3	7
5.5	UCLA	10–1	3	4	6	3
8.5	Texas A&M	11–2	5	5	4	6
5.5	Arizona	11–1	6	9	9	5
7	Florida	9–2	10	8	11	10
8.5	Wisconsin	10–1	9	10	10	9
10	Tulane	11–0	7	14	23	8
15	Nebraska	9–3	11	7	15	11
12.5	Virginia	9–2	15	18	17	13
11	Arkansas	9–2	14	12	22	17
13	Georgia Tech	9–2	16	20	12	16
17.5	Syracuse	8–3	24	16	7	24
0	Avg. Abs. Diff.		2.47	3.07	3.80	2.33

Table 13.4 ***Comparison to 1999 BCS Models.***

Poll avg.	Team	Record	Mease	Jeff Sagarin	New York Times	Anderson/ Hester	Richard Billingsley	Dunkel	Kenneth Massey	Herman Mathews	David Rothman
1	Florida State	11–0	1	1	1	1	1	1	1	1	1
2	Virginia Tech	11–0	2	2	2	3	2	2	2	2	2
3	Nebraska	11–1	3	3	4	2	3	3	3	3	3
5.5	Alabama	10–2	4	6	3	4	5	7	6	4	4
5.5	Tennessee	9–2	9	5	5	8	7	6	5	5	6
7	Kansas State	10–1	6	4	6	5	4	5	4	6	5
4	Wisconsin	9–2	13	7	8	12	8	4	7	11	9
8	Michigan	9–2	8	9	7	6	10	9	8	7	10
9	Michigan State	9–2	7	8	10	7	6	8	9	8	8
10	Florida	9–3	10	11	16	9	9	12	12	9	7
15	Penn State	9–3	11	10	20	11	11	10	10	10	11
11	Marshall	12–0	5	13	11	15	33	31	11	22	12
12	Minnesota	8–3	24	15	21	21	14	19	17	15	15
15.5	Texas A&M	8–3	16	17	15	14	13	16	15	19	16
16	Texas	9–4	14	14	21	13	17	13	16	14	13
0	Avg. Abs. Diff.		2.77	1.57	2.43	2.83	3.10	2.90	1.30	2.50	1.77

Table 13.5 ***Comparison to 2000 BCS Models.***

Poll avg.	Team	Record	Mease	Jeff Sagarin	New York Times	Anderson/ Hester	Richard Billingsley	Dunkel	Kenneth Massey	Herman Mathews	David Rothman
1	Oklahoma	12–0	1	3	3	1	1	3	2	2	1
3	Florida State	11–1	2	1	1	3	2	1	1	1	2
2	Miami(Florida)	10–1	3	2	2	4	3	2	3	3	3
4	Washington	10–1	4	8	5	2	10	11	5	4	4
5.5	Virginia Tech	10–1	6	5	4	6	5	5	4	7	7
5.5	Oregon State	10–1	5	7	8	5	7	9	8	5	5
7	Florida	10–2	7	6	6	7	4	4	7	6	9
8.5	Nebraska	9–2	9	4	10	9	6	13	6	8	6
10	Kansas State	10–3	11	9	12	12	8	12	11	11	8
9.5	Oregon	9–2	8	14	15	8	12	17	14	9	11
10	Notre Dame	9–2	10	16	8	10	14	15	15	10	12
12	Texas	9–2	15	10	11	15	11	6	9	12	10
16	Georgia Tech	9–2	12	11	7	11	9	8	10	13	14
14.5	TCU	10–1	14	12	20	20	16	7	12	14	15
14.5	Clemson	9–2	13	15	19	13	13	21	13	15	13
0	Avg. Abs. Diff.		1.00	2.47	2.73	1.60	2.33	4.33	2.33	0.87	1.33

Table 13.6 ***Comparison to 2001 BCS Models.***

Poll avg.	Team	Record	Mease	Anderson/ Hester	Wes Colley	Richard Billingsley	Kenneth Massey	David Rothman	Jeff Sagarin	Herman Mathews	Peter Wolfe
1	Miami (Florida)	11–0	1	1	1	1	1	1	1	1	1
4	Nebraska	11–1	2	2	2	2	3	2	3	2	2
3	Colorado	10–2	4	4	5	4	4	5	5	5	3
2	Oregon	10–1	3	3	3	3	2	8	7	6	7
5	Florida	9–2	7	9	8	7	8	4	2	3	5
8	Tennessee	10–2	5	5	4	8	6	7	8	7	4
9	Texas	10–2	8	8	9	10	9	3	4	4	6
7	Illinois	10–1	6	7	6	6	12	13	12	10	12
11	Stanford	9–2	10	6	7	11	5	9	9	8	8
6	Maryland	10–1	9	14	10	5	10	11	11	14	11
10	Oklahoma	10–2	11	10	11	9	13	6	6	9	9
13	Washington State	9–2	12	12	12	12	7	10	10	11	10
12	LSU	9–3	15	11	13	14	14	12	18	13	14
14	South Carolina	8–3	18	20	19	19	17	17	23	23	17
20.5	Washington	8–3	16	13	15	15	11	16	25	17	13
0	Avg. Abs. Diff.		1.90	2.70	2.30	1.57	3.03	3.03	3.63	3.10	2.90

Table 13.7 ***Comparison to 2002 BCS Models*.***

Poll avg.	Team	Record	Mease	Anderson/ Hester	Wes Colley	Richard Billingsley	Kenneth Massey	New York Times	Jeff Sagarin	Peter Wolfe
1	Miami (Florida)	12–0	2	2	1	1	1	1	1	2
2	Ohio State	13–0	1	1	2	2	2	3	2	1
4	Georgia	12–1	3	3	3	3	4	4	3	3
5	USC	10–2	4	5	4	6	3	2	4	4
3	Iowa	11–1	5	4	5	5	8	5	5	5
7	Washington State	10–2	8	8	8	9	5	9	6	6
8	Oklahoma	11–2	6	7	7	4	7	6	8	7
6	Kansas State	10–2	11	14	12	11	10	7	11	10
11.5	Notre Dame	10–2	7	6	6	8	6	13.5	7	8
9	Texas	10–2	9	10	9	7	11	11	9	11
11.5	Michigan	9–3	10	9	10	16	9	8	10	9
10	Penn State	9–3	13	11	13	14	14	10	15	13
13.5	Colorado	9–4	15	13	15	22	13	16	13	15
15.5	Florida State	9–4	14	12	11	23	12	18	12	12
13.5	West Virginia	9–3	18	18	16	15	18	15	18	17
0	Avg. Abs. Diff.		2.03	2.17	2.03	3.10	2.43	1.67	1.97	2.10

*NOTE: Rankings of all models were adjusted to exclude Alabama, which was on probation in 2002 and consequently not ranked by the Coaches Poll or the BCS.

Table 13.8 ***Summary of average absolute differences for all BCS Models.***

Model	1998	1999	2000	2001	2002	Average
Proposed Model (Mease)	2.47	2.77	1.00	1.90	2.03	2.03
David Rothman		1.77	1.33	3.03		2.04
Herman Mathews		2.50	0.87	3.10		2.16
Wes Colley				2.30	2.03	2.17
Kenneth Massey		1.30	2.33	3.03	2.43	2.27
Anderson/Hester	2.33	2.83	1.60	2.70	2.17	2.33
Peter Wolfe				2.90	2.10	2.50
Richard Billingsley		3.10	2.33	1.57	3.10	2.53
Jeff Sagarin	3.07	1.57	2.47	3.63	1.97	2.54
New York Times	3.80	2.43	2.73		1.67	2.66
Dunkel		2.90	4.33			3.62

average rankings over the five seasons in which the BCS existed. From this comparison it can be seen that the model proposed in this article in fact had the smallest average difference from the AP and Coaches average over the five year period among all BCS models.

Possible modifications to the model

Although the purpose of this article is to present a model that performs well using only win/loss data, the model in fact can be easily adapted to incorporate other available information such as the location of the game, the date of the game, and even the margin of victory if so desired.

The location of each game played is almost always on the campus of one of the two competing teams. This team is called the "home team" and is thought to generally have an extra advantage due to crowd support and a number of other factors. This "home-field" advantage can be incorporated into the model by replacing θ_i—θ_j by θ_i—θ_j + λ if team i is the home team or by θ_i—θ_j—λ if team j is the home team. The likelihood can then be maximized over this single home-field advantage parameter λ along with θ. The teams can be ranked based on their respective values of θ_i as before. The model considered by Harville (1977, 2003) uses an analogous single parameter to capture the home-field advantage. Alternatively, one can consider a separate home-field advantage parameter for each team as was done by Harville and Smith (1994) for college basketball teams. Although this model is more flexible and may in fact be more realistic, it is not as useful for ranking teams since it will generally result in two different ranks for each team depending on whether the teams are ranked with their home-field advantage included or not.

In addition to the location of the games, it is sometimes suggested that the date of the games should be considered when ranking the teams. Specifically, some feel that games that occur later in the season should be weighed more heavily in the final rankings than games that occur earlier in the season. One method for doing this in the proposed model would be to incorporate these weights into the values of n_{ij}. As a simple example, if the games in the second half of the season were thought to be twice as important as the games in the first half, one could double the n_{ij} values for all games occurring in the second half of the seasons. Note that in general if one scales the weights for the games such that they are integers, the model can still easily be fit using standard software as described in the section on "Implementation of the model using statistical software." The only necessary change would be to create replicate lines for some entries in the input matrix to reflect these weights, analogous to the method for dealing with tie games described earlier.

Similar to weighting games based on date, games could also be weighted based on some measure of difference in the final score. That is, the n_{ij} values could be used to account for margin of victory. Again, by scaling the weights to be integers, standard software can still be used to fit the model. Careful selection of the n_{ij} may produce a model which agrees even more strongly with the polls by giving more weight to large victories (as some suggest that human pollsters do) but not to the large extent that previous models have done.

A final possible modification to the model is to replace the normal CDF function Φ by some other function. The logit CDF function is one possible alternative that still allows for the model to be fit using standard software. Although the logit CDF and normal CDF have quite different tail behavior, the use of the logit CDF function did not affect the resulting rankings substantially for the football seasons considered.

Summary

We have described a penalized maximum likelihood model for ranking college football teams independent of victory margins. By analysing resulting rankings from actual college football seasons we have shown that the model on average agrees more strongly with the AP and Coaches Polls than many of the more complex models discussed in the statistics literature and used by the BCS. We have shown that the model can be implemented using binary response regression procedures available in most standard statistics software packages.

Although no one model for ranking teams is necessarily "better" than any other model, the proposed model has many attractive features. Its strong agreement with the AP and Coaches polls suggests that it is consistent with popular/expert opinion, while at the same is free from the problems of personal bias and limited memory inherent in human polls. Second, the model ignores all factors other than wins and losses and is simple in form, suggesting a parsimonious solution to the problem of ranking college football teams. Finally, the model is motivated by statistical theory and can be fit using standard statistical software.

REFERENCES

Bassett, G. W. (1997), "Robust sports ratings based on least absolute errors," *The American Statistician*, 51, 99–105.

Bradley, R. A., and Terry, M. E. (1952), "The rank analysis of incomplete block diagrams. I. The method of paired comparisons," *Biometrika*, 39, 324–345.

Fienberg, S. E., and Larntz, K. (1976), "Log linear representation for paired and multiple comparisons models," *Biometrika*, 63, 245–254.

Green, P. (1998), "Penalized likelihood," in *Encyclopaedia of Statistical Sciences*, updated Vol. 2, New York: Wiley.

Harville, D. (1977), "The use of linear-model methodology to rate high school or college football teams," *Journal of the American Statistical Association*, 72, 278–289.

—— (2003), "The selection or seeding of college basketball or football teams for postseason competition," *Journal of the American Statistical Association*, 98, 17–27.

Harville, D., and Smith, M. (1994), "The home-court advantage: how large is it, and does it vary from team to team?," *The American Statistician*, 48, 22–28.

Keener, J. P. (1993), "The Perron–Frobenius theorem and the rating of football teams," *SIAM Review*, 35, 80–93.

Silverman, B. W. (1985) "Penalized maximum likelihood estimation," *Encyclopaedia of Statistical Sciences*, 6, 664–667.

Stern, H. (1992), "Who's Number One?—Rating football teams," in *Proceedings of the Section on Statistics in Sports* 1992, pp. 1–6.

Thompson, M. (1975), "On any given Sunday: Fair competitor orderings with maximum likelihood methods," *Journal of the American Statistical Association*, 70, 536–541.

Thurstone, L. L. (1927), "The method of paired comparisons for social values," *Journal of Abnormal and Social Psychology*, 21, 384–400.

CHAPTER 14

Determining optimal strategy requires finding the point value of field position

The return of a missed field goal in Canadian football

KEITH A. WILLOUGHBY

Late in the first quarter of the 1987 Grey Cup, the championship game of the Canadian Football League (CFL), Toronto Argonauts' kicker Lance Chomye attempted a 46-yard field goal. His kick went wide right. Edmonton Eskimo Henry "Gizmo" Williams caught the ball five yards deep in the end zone. He began to return the ball upfield, evading Toronto linebacker Don Moen then obtaining a key block from teammate Steve Benjamin. Williams finally stopped in the Argonauts' end zone with a 115-yard (that's right, 115-yard!) touchdown return. His spectacular effort was Edmonton's first touchdown of the game; they went on to earn a 38–36 victory over the Argonauts.

The return of the missed field goal is an important strategic decision encountered in Canadian football. Since the latter stages of the 19th century, professional teams have played a brand of football unique to Canada. In many vital aspects, this sport differs from the (more well-known) version of football played in the United States. Canadian teams have 12 players on the field at a time, while American football features squads of 11. Canadian teams are allowed three downs to advance the ball 10 yards. Four downs are used in American football. The respective fields are of different dimensions. Canadian football is played on surfaces 110 yards long (plus end zones 20 yards deep), while American football uses 100 yard fields (with end zones 10 yards in length). Figures 14.1 and 14.2 illustrate the dimensions of the football fields.

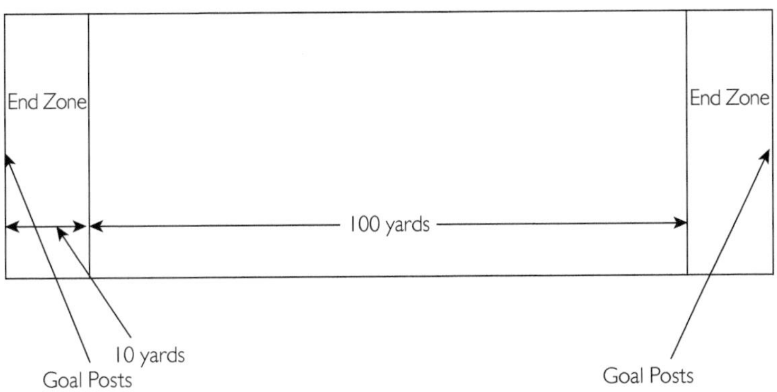

Figure 14.1 American football field—dimensions.

The missed field goal in Canadian football

Perhaps the most striking difference between the two brands of football occurs when a team attempts a field goal. This play involves one team member attempting to kick the ball between

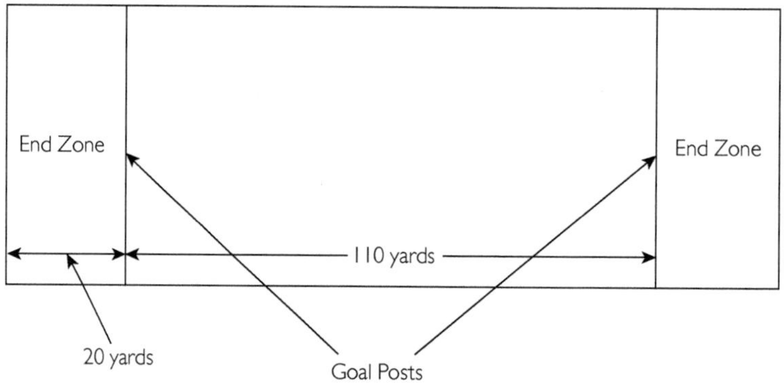

Figure 14.2 Canadian football field—dimensions.

two goalposts. A field-goal attempt is ruled unsuccessful when the ball fails to pass between the goalposts. In American football, the goalposts are positioned at the back of the 10-yard end zones (refer to Fig. 14.1). Consequently, when a field-goal attempt is unsuccessful, the ball usually travels out of play (beyond the boundary of the end zone). The referee blows his whistle and play is stopped.

In Canadian football, a missed field-goal attempt sets in motion a crucial decision. Inasmuch as the goalposts are positioned at the *front* (not back) of the end zone, and since the end zones are 20 yards deep, the ball may still be in the field of play after an unsuccessful field-goal attempt (refer to Fig. 14.2). For this reason, Team A will position one player (e.g., Edmonton's "Gizmo" Williams in the aforementioned 1987 Grey Cup) in the end zone behind the goalposts. In the event that team B misses the field goal, Team A's player must decide whether or not to run the ball out of the end zone. Doing so is referred to as "returning the ball."

Returning the football out of the end zone, or failing to do so, has some scoring and strategic implications. When the player on Team A decides to return the missed field goal out of the end zone, Team A will initiate possession of the football at the yard line on which its player is eventually tackled. Obviously, it is to Team A's benefit to return the unsuccessful field goal as far up the field as possible. Returning the ball to only, say, Team A's one- or two-yard line would not be advantageous. Team A would then suffer from poor field position. It would be forced to move the ball a great distance before reaching Team B's end zone. However, returning the ball a substantial distance up the field would represent a beneficial outcome for Team A. It would then have better field position, closer to Team B's end zone. Touch-downs resulting from the return of unsuccessful field goals represent Team A's best possible outcome for this decision.

However, Team A's player may decide it is not worthwhile to return the missed field goal. Such an action is called "conceding." The usual procedure involved in conceding is for the player, on retrieving the ball, to simply go down on one knee. This results in a stoppage of play. Team B receives one point for this unsuccessful field goal. Team A then begins possession of the football at either of the two following places (whichever is further from Team A's end zone):

- their own 35-yard line
- the yard line at which Team B had possession of the football before attempting the field goal

To clarify, suppose Team B had possession at Team A's 20-yard line. If Team B missed the field goal and Team A conceded a single point, Team A would begin possession of the ball at their own 35-yard line (the 35-yard line is further from Team A's end zone than the 20-yard line). However, if Team B was at Team A's 40-yard line before unsuccessfully kicking the field goal (and a single point was conceded), Team A would start a drive from their own 40-yard line (the 40-yard line is further from Team A's end zone than the 35-yard line).

For the sake of completeness, we ought to address the event of Team A deciding to return the missed field goal but the player being subsequently tackled within Team A's end zone. In other words, Team A's player did not make it out of his own end zone. This situation is viewed identically to the "conceding" case. Team B obtains a single point, and Team A begins possession of the football at either its own 35 or the point of last possession by Team B (again, whichever is further from Team A's end zone).

A missed field goal is by no means a rare occurrence in Canadian football. For example, during the 1998 CFL season a total of 424 field goals were attempted. Of these attempts, 306 resulted in successful field goals (a success rate of 72.2%). A total of 55 of the 118 unsuccessful field goals were returned out of the end zone, with the other 63 being conceded. (Data are from the CFL Web site www.cfl.ca.)

The decision

Some key factors affect the return/concede decision. The current score differential in the game is likely to have an effect on the decision to return the missed field goal. For example, if the score was tied (particularly late in the game), then Team A would most likely return the ball. Conceding a single point in this case would give the lead to Team B. Further, if Team A held a one-point advantage in the final few seconds of a game, it would surely decide to return the missed field goal. Getting the ball out of the end zone would result in victory for Team A. However, consider a scenario in which the current score and time remaining in the game are of marginal consequence. (For example, the concede/return decision could occur during the initial stages of the game.) Then, the decision is left largely up to the player preparing to return the missed field goal. Essentially, the concede/return decision becomes one of *points* versus *field position*. Conceding gives the opposition a single point but provides a team with possession of the football at its own 35-yard line (or beyond). Returning the unsuccessful field goal saves Team A one point. However, it could begin a drive with relatively poor field position (if the player on Team A were only able to return the ball a short distance beyond his end zone).

The crucial issue, then, becomes one of determining at what yard line a team is indifferent between returning the ball or conceding a single point. In other words, how far up the field does the player on Team A need to get? Getting to his own one yard line does not seem like a good outcome. However, returning the ball out to, say, midfield would offer tremendous advantages. Team A saves one point *and* obtains great field position.

The value of field position

The literature on the quantitative value of field position in professional football is sparse. Former quarterback Virgil Carter and operations researcher Robert Machol examined 56 games during the initial seven weeks of the 1971 National Football League (NFL) season. Their goal was to determine the expected value (in terms of points scored) of beginning a drive at various yard lines on the field of play. As one might expect, starting a drive at a yard line closer to the opponent's end zone had a larger expected point value than initiating possession of the football near one's own end zone. Beginning a drive relatively close to the opponent's end zone ensures that a team will stand a reasonably good chance of scoring, say, a field goal, even if they can only move the football a short distance.

To determine how far up the field a particular team needs to return the ball, we will follow, to a certain extent, the approach of Carter and Machol.

Specifically we develop a linear model, relating the field position at the start of the drive to the ultimate outcome of the drive. The field position refers to the distance from Team A's end zone at which the drive began. For example, a value of 3 for the explanatory variable (starting field position) would mean that Team A initiated possession of the football at its own three yard line—very poor field position, indeed! If the value of the explanatory variable were 55, then Team A began a drive at midfield (recall that Canadian football fields feature a distance of 110 yards between end zones). We point out that, if the explanatory variable's value was 80, then Team A started its drive from *Team B's* 30 yard line.

Initially, it is useful to think of the response variable as being the score (if any) obtained by Team A during a particular drive. Generally, there are three potential scoring opportunities when a team has possession of the football—a touchdown (usually worth seven points), a field goal (three points), or a single point from a missed field goal. However, we also wanted to include the effect of Team A *not* scoring on a drive and subsequently turning possession of the ball over to Team B. Thus, we need to offer an important clarification to our definition of the response variable:

- If Team A scores on a drive, then the value of the response variable is simply Team A's particular score (seven points, three points, or one point).
- If Team A *does not* score on a drive, then the value of the response variable is the score obtained by *Team B* on its subsequent drive. These possible scores would be seven points, three points, one point, or zero points (the last case resulting if Team B failed to score during its possession of the football).

This image was originally posted to Flickr by Joe Bielawa at http://flickr.com/photos/88596146@N00/7970687326. Reproduced under a Creative Commons licence

For example, suppose Team A began a drive on its 40-yard line and eventually kicked a field goal. The resulting (explanatory variable, response variable) pair would be (40, 3). Suppose Team A initiated possession of the football at its own five-yard line and was unable to score on this drive. Suppose further that Team B scored a touchdown on its subsequent drive. Then, the resulting pair would be (5, −7). The negative number is used for the value of the response variable since Team B, not Team A, scored the touchdown. As a final example, assume Team A started a drive at Team B's 50-yard line and was unable to score on this drive. If Team B also failed to score on its subsequent drive, then the resulting pair would be (60, 0).

Why do we modify our definition of the response variable? Quite simply, we want to explicitly consider the dangerous effect of poor field position. If Team A begins a drive deep in its own end and fails to move the ball far up the field (or, even worse, fumbles the ball to the opposition), the likelihood is fairly high that Team B will obtain possession of the football with great field position. Team B then stands a reasonably good chance of scoring (seven points, three points, or one point) on this drive. Generally, it is not advantageous for Team A to begin possession of the football with poor field position. In the event that Team A scored, we do *not* consider Team B's subsequent possession of the football. After Team A scores, it initiates a kick-off to Team B. In this case, Team B's starting field position is independent of where Team A began its scoring drive. Carter and Machol had a slightly different method for accounting for the effect of bad field position. They tried to account for *all* possible outcomes. Thus the outcome was either a score (seven or three) or turning the ball over to the opponent at the X yard line. This led them to a system of equations that could be used to find the value of different field position. We did not have enough data to carry out that analysis.

Results

The data for fitting our linear model were obtained from the Saskatchewan Roughriders, a CFL franchise. We used each of their games during the 18-game 1998 regular season. In all, we examined 272 drives, with 80 of these (29.4%) resulting in scores. For each of these drives, we recorded the beginning field position, the points scored by the Roughriders, and the points scored by the opposition (if the Roughriders failed to score any points on their drive). As an aside, we note that the minimum value for the explanatory variable was 2 (meaning that the Roughriders began a possession at their own two-yard line), while the maximum value was 95 (the Roughriders were able to start a drive at the opposition's 15-yard line). The average starting position was about the 36-yard line. The data are shown in Figure 14.3. We fit a line to

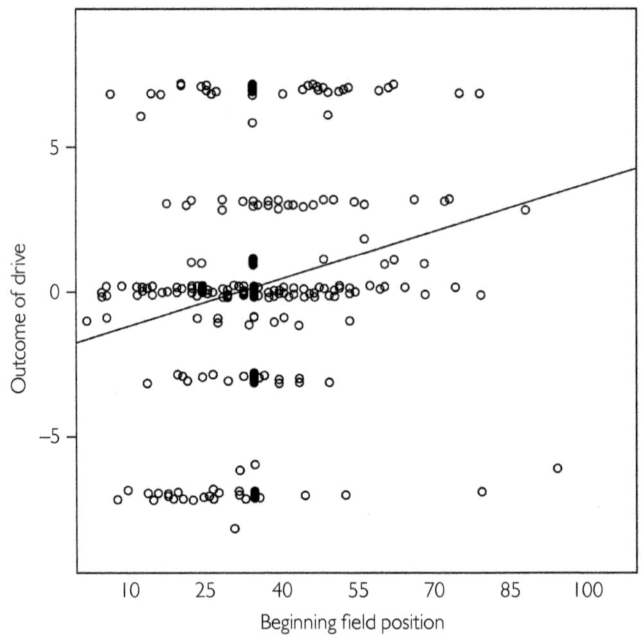

Figure 14.3 Scatterplot showing relationship between the outcome of a drive (as defined in the text) and starting field position. Points have been jittered in the vertical direction to display overlapping points by adding a small random deviation. The best fitting regression line is also shown.

these data and obtained the following results: Expected points scored = −1.738 + 0.054 (field position). The slope coefficient gives the increase in expected points scored per yard of field position.

There are some key insights that can be obtained from this linear regression equation. Initiating possession of the football deep in one's end is indeed dangerous! For example, if the explanatory variable is 5, then the expected value of points scored is

$$-1.738 + .054(5) = -1.468.$$

Starting a drive near one's own end zone and not being able to move the ball very far up the field could provide the opposition with excellent field position. The opposition then stands an exceptional chance of scoring on their subsequent drive.

For the CFL franchise currently under study, beginning a drive close to the opposition's goal line does not necessarily result in a touchdown (seven points). For instance, if the field position is equal to 94 (Team A is starting a drive on Team B's 16-yard line), then the expected value of points scored is $-1.738 + 0.054(94) = 3.338$. Team A expects to score *some* points, but it may not *always* be a touchdown. This last result seems low, the result might be different for a team with more offensive success. It is noteworthy that the slope here agrees reasonably well with the Carter and Machol (1971) result. They found a slope of 0.07.

Making the decision

Recall that the return/concede decision involves points versus field position. Teams that concede give up a single point, but gain some valuable field position. Teams that return the missed field goal save the point, but may not end up with reasonably good field position. The value of the slope coefficient in our linear regression equation suggests that each additional yard of field position is "worth" 0.054 points (for this particular team). As a result, this team would equate about 18.52 yards with *one point* (let's round this up to 19, since football teams deal with yardage in discrete amounts). Thus, 19 yards of field position and a single point have roughly the same value. In other words, this team would be indifferent between giving up one point and gaining 19 yards of field position.

Consider first the case in which Team B attempts a field goal from at or within Team A's 35-yard line. If Team A concedes the single point, then it will obtain possession of the football at their own 35-yard line. Since Team A is indifferent between losing one point and gaining 19 yards of field position, it will only be attractive to return the football if the player on Team A is confident that he can get out to *at least* the 35 − 19 = 16 yard line.

The appropriate decision is different if Team B attempts a field goal from beyond Team A's 35-yard line. Now if Team A concedes the single point, then it will start a drive from where Team B had possession of the football. We shall use the symbol N to refer to the yard line of Team B's possession before attempting the field goal (obviously, $N > 35$). Again, since Team A is indifferent between losing one point and gaining 19 yards of field position, the player on team A ought to return the football only if he is confident that he can get out to *at least* the $(N - 19)$-yard line. Being unable to return the ball that far means that they have sacrificed too much field position, even though they did not give up one point.

In the 1987 Grey Cup story that started this article, N was 39. The decision rule obtained by our statistical analysis suggests that the returner needed to be confident that he could get to the 20-yard line. Apparently, he was confident and correct.

ACKNOWLEDGEMENTS

Without implicating them, we wish to thank the following employees of the Saskatchewan Roughriders for their valuable assistance in this project: Mr. Cal Murphy (Former Head Football Coach) and Ms. Tennille Grimeau (Marketing and Sales Assistant).

REFERENCES AND FURTHER READING

Carter, V., and Machol, R.E. (1971), "Operations research on football," *Operations Research*, 19, 541–545.

LaRoi, H. (1987), "The Giz just had to have a Grey Cup ring," *Edmonton Journal*, November 30, 1987, D10.

CHAPTER 15

Statistics and the college football championship

HAL S. STERN

Editor's Note: Presently, there is a controversy associated with the identification of a champion in United States college football. Teams cannot play a complete round robin and the best team is decided on the basis of a complicated Bowl Championship Series (BCS) rating system and the results of postseason "bowl" games. Because the rating problem is essentially a statistical problem, *The American Statistician* invited Hal Stern to describe the history of the college football championship and explain the contributions that statistical thinking might make. To respond to Stern's article, we invited several authors of the BCS computer ratings systems and several statisticians familiar with ranking systems.

—James Albert, Editor, *The American Statistician*

Abstract

The U.S. college football champion is determined each winter by the Bowl Championship Series (BCS), a set of four college football games and an associated ranking system that helps to determine the participants in the four games. One game each winter (the specific game rotates among the four participating games) hosts the national championship game between the top two teams in the BCS ranking. The BCS has been a controversial system since its implementation prior to the 1998 season with the most recent 2003 season producing a disputed championship, the very thing the BCS system was developed to avoid. This current article reviews the history of the college football national championship, the rise of the BCS, the BCS ranking system, and the contributions that statistical thinking can make toward improving the BCS. Though the problem of optimally ranking sports teams is a difficult one there is clearly room for improvement in the present system!

Keywords: *quantitative reasoning, rankings, ratings, sports*

On sports and championships

Professional sports leagues in the United States tend to be organized as collections of small (four to eight teams) divisions that play unbalanced schedules (more games against opponents within the division than against other teams). This promotes strong local rivalries but implies that one cannot just compare records at the end of the season to determine a champion. Instead the regular season is used mainly to identify participants for a post-season championship tournament. It is the post-season tournament's champion (e.g., the World Series winner in base-ball and the Super Bowl winner in football) that is universally recognized as that season's champion. Division or conference titles earned during the year are celebrated but not with the same level of enthusiasm as the championship.

This is of course not the only model for professional sports. For example, football (soccer) leagues in Europe are generally organized as 18- to 20-team leagues that play a balanced regular season schedule composed of home-and-away games with each team in the league. The regular season championship in such leagues is highly valued. There are also tournaments and competitions, both intranational (e.g., the FA Cup in England) and international (e.g., Europe's Champions League), considered as separate events. The tournaments resemble the American post-season tournaments. Tournament titles are valued but not necessarily more than the regular season title.

American college sports (except football) have evolved a kind of compromise of the two models. University teams are organized in athletic conferences. The conference regular season title is usually based on a balanced schedule, though this is not always true now that conferences have grown in size. Regular season conference champions along with some of the best nonchampions are invited to a post-season national championship tournament. As in professional sports the post-season championship tournaments have become the most sought-after title, the one that is highly recognized. Unlike the professionals it seems that regular season conference championships remain highly valued.

College football's most competitive level (Division I-A) is unique among American sports. There is no post-season championship tournament. (There are championship tournaments in the football divisions for smaller schools). Instead for many years the college football national champion has been crowned based on the final votes of two polls (the Associated Press poll of sportswriters and the *USA Today*/ESPN poll of coaches). In recognition of the fact that the championship is settled by polls rather than on the field, the college football title has long been known as the "mythical" national championship.

The remainder of this article considers the way in which college football currently selects its national champion and how ideas from the field of statistics might improve the process.

American college football—past and present

It is natural to start with a brief history of the search for a college football national champion.

The early days

A first key point is that there has long been interest in knowing which college football team is the nation's best. This is not a modern phenomenon. The earliest recognized rating system designed to address the topic was developed by Frank Dickinson of the University of Illinois in 1924. Teams were awarded points for winning games, with the number of points awarded determined by the quality of the opponent. Dickinson stopped producing his ratings in 1940, citing his age, a declining interest in college football, and the large number of alternative systems that had been developed (Dickinson 1941). Thus, by 1940 there is evidence of great national interest in identifying the best college football team in the land.

The rise of the polls

In 1936 the Associate Press (AP) news agency began a national poll of sportswriters to identify the top teams. In 1950 United Press International (UPI), a competing news agency, developed a poll of coaches (currently run by *USA Today* newspaper and the ESPN cable television network). The two polls quickly became the most highly regarded source of information regarding the top teams. The top-ranked poll team (assuming the two polls agreed) was recognized by all as the national champion. The two polls agreed most of the time during these early years (with disagreements in 1954, 1957, and 1965).

During this era a number of college football "bowl" games began to develop. (The first bowl game, the Rose Bowl, actually dates back to the early 1900s but all the others began in 1935 or later). The bowl games are played after the regular season in the second half of December and the early part of January. Top teams are invited to play one additional game against another top team, usually from another region of the country. The games are major tourist attractions for the host cities. There are currently 28 bowl games each year. Initially the bowl games were viewed as separate from the business of identifying a national champion. But first the AP (in 1968) and then the UPI (in 1974) decided to select their champion after the bowl games were played. This raised the importance of the bowl games.

The two polls remain the final arbiters of the "mythical" college football championship to this day. In response to a couple of controversial split championships around 1990, conferences, teams, and bowls began to formulate alliances and coalitions designed to ensure that the top two teams would meet in one of the bowls allowing for an unambiguous champion. There were some successes during this period but still a number of "failures" where two different champions were crowned (due to the absence of a particular conference, team, or bowl from the arrangement).

The retirement of Tom Osborne in 1997 and the start of the BCS

In 1997 there were two undefeated teams at the end of the regular season, Michigan and Nebraska. Prior to the bowl games Michigan was ranked ahead of Nebraska in both polls. Michigan was committed to playing against a West Coast team in the Rose Bowl and thus not available to meet Nebraska in a championship game. Two things happened during the bowl season that year: (1) Tom Osborne, Nebraska's long time coach, announced prior to the bowl games his intention to retire at the end of the season; and (2) Nebraska and Michigan both won their bowl games. When the polls were announced after the bowl games the writers had maintained Michigan as the top team but the coaches selected Nebraska, jumping them over Michigan despite the fact that both teams won. The resulting controversy over the split championship was one of the factors that led to the creation of the Bowl Championship Series or BCS. This agreement involved all of the major conferences and bowls and finally "guaranteed" (or so we thought) that the two best teams would meet each year to determine a unique national champion.

Why is there no tournament?

Though not the main point of the article it is worth a short paragraph to consider the question of why there is no tournament for the top tier of college football. Among the reasons given by opponents of a tournament are: (1) the danger of injury to players in the additional games; (2) the pressure that additional games would put on the student-athletes' academic programs; (3) a desire to avoid having the college football season run into a second semester; (4) a sense that the occasional split championship controversy is good for the game; and (5) recognition that the bowls have been good allies to college football. Though the issues raised on this list are valid concerns, none would seem to pose an insurmountable obstacle. In particular, the decision of many college football teams and conferences to add preliminary games in August and conference championships in December, as well as the experience of other college football divisions, seem to suggest that limiting the scope of the season is not in reality a major concern. It is similarly hard to argue that the presence of tournaments in other sports has decreased their popularity or reduced the joy fans have in arguing about who the true best team really was. Finally, it should be noted that every tournament plan that has been mentioned would try to make use of some of the existing bowl games while leaving the others no worse off than they are now. What then can be the reason for the absence of a tournament? One possibility is that having a championship tournament sponsored by college football's lead organization (the NCAA) would require the member schools of the football-dominant BCS conferences to split the many of millions of dollars of revenue generated by such an event more equally with their fellow institutions of higher learning than is required by the current system.

The Bowl Championship Series (BCS)

The BCS resulted from negotiations among six major athletic conferences (plus the University of Notre Dame), four major bowl games, and the television networks. The eight slots in the four participating bowl games would be earned by the champions of the six major conferences plus two at-large teams. A BCS rating system was developed to assist in the selection of the teams. The actual selection of teams is guided by a number of rules, including special rules that govern Notre Dame's participation. The most important rule for our purposes is the rule which stipulates that the two most highly ranked teams in the BCS rating system compete in the bowl game designated as hosting that year's national championship game.

Design of the BCS rating system

The BCS rating system consisted initially of four elements: polls, computer ratings, a measure of schedule strength, and team record. A team's rating is the sum of four numbers, one corresponding to each element. The average rank of the team in the coaches' and writers' polls is the first number used in the BCS rating. Because the writers were reluctant to play too large a role in determining the two teams to compete for the championship, the BCS organizers also included a number of computer ratings. A couple of points about terminology are needed here. The reference to these as computer ratings is a bit misleading; the ratings in question are determined by mathematically based algorithms for rating football teams. Because of the large number of teams involved the systems are implemented on computers, but that is not really the key feature. There is also some ambiguity in the terms ranking and ratings. Typically the term ratings is used when there is information contained in the numerical scores assigned to the teams and the term rankings is used when the primary goal is only to put the teams in order. We try to follow this distinction but it is difficult given the variation in usage throughout the literature. The number of quantitative rating systems included has varied from three to eight with the 2003 BCS including seven ratings: those of Anderson & Hester, Billingsley, Colley, Massey, *New York Times*, Sagarin, and Wolfe. These are discussed further below. The average rank of each team in the seven computer ratings (actually the worst ranking is dropped for each team) is the second number used in the BCS ranking system. The teams' schedules are ranked for difficulty (using a measure that combines the records of each team's opponents and the records of their opponents' opponents) and that ranking divided by 25 is the third number in the BCS sum. Finally teams receive one point for each game that they lose. The system has been modified over time to correct perceived inequities. We return to this point below, for now we note only that the current version of the BCS includes one additional factor besides the four described thus far, a quality win deduction that removes points from a team's BCS score for wins over other highly ranked BCS teams. The top ten teams in the final 2003 regular season BCS standings along with the different elements of the system are presented in Table 15.1.

A system without a guiding principle

From its beginning the BCS ranking system has been hurt by the fact that its main objective is poorly defined. It is designed to select the "top two teams" but this term lacks a formal definition. The relative weights of the included elements were determined using trial and error to ensure that the BCS would have selected the top teams (as defined by an informal consensus) in earlier years. This is not a terrible approach, but it is worth pointing out that there are likely many combinations of the BCS ranking system elements that would pass such a test. How to choose among them? This is where it would be helpful to have in mind a particular set of principles or goals. For example, one might elect to identify the two teams judged most likely to defeat all the others at the end of the season. Other possible goals are described later. Without a specific goal it is all too easy to overreact to short-term fluctuations and random events. The events that have unfolded since the formation of the BCS in 1998 have exposed this as a major weakness in the BCS system.

There is a joke told in which an insurance company is described as an automobile traveling on the highway at high speed. The chief executive officer has the steering wheel, the marketing department is stepping on the gas, the chief financial officer is applying the brakes, and the actuary is in the backseat screaming directions using a map made by looking out the rear window. BCS developments since 1998 lead one to believe that this insurance company joke could easily be retold about the BCS. There have been frequent adjustments to the BCS in order to correct for perceived failures most visible in the rear window. Specifically:

- 2000: Undefeated Oklahoma was an obvious choice for the championship game. There were several once-beaten teams that were candidates to play Oklahoma. The BCS system selected once-beaten Florida State even though Florida State had lost to once-beaten Miami during the season. This bothered many fans (who were not quite so bothered that once-beaten Miami had lost to once-beaten Washington during the same season). It is at this point that a quality win bonus was added to the BCS. After the

Table 15.1 ***Final (pre-bowl) 2003 Bowl Championship Series (BCS) Rankings. Only the top ten teams shown. Details for computing the various components are provided in the accompanying text. SOS stands for "strength of schedule."***

BCS Rank	Team	Polls (AP & ESPN/USAT)	Avg Poll	Computer ratings (A&H, RB, CM, KM NYT, JS, PW)	Avg Comp. Rtg.	SOS Rank	SOS Points	Games Lost	Sub Total	Quality Win Ded.	Total
1	Oklahoma	(3,3)	3	(1,1,1,2,5,1,1)	1.17	11	0.44	1	5.61	–0.5	5.11
2	LSU	(2,2)	2	(2,2,2,1,2,2,2)	1.83	29	1.16	1	5.99		5.99
3	USC	(1,1)	1	(3,3,3,3,1,4,3)	2.67	37	1.48	1	6.15		6.15
4	Michigan	(4,4)	4	(7,4,6,5,3,5,5)	4.67	14	0.56	2	11.23	–0.6	10.63
5	Ohio St	(7,6)	6.5	(6,6,4,4,8,6,7)	5.50	7	0.28	2	14.28		14.28
6	Texas	(5,5)	5	(5,9,8,7,4,8,10)	6.83	20	0.80	2	14.63	–0.1	14.53
7	Florida St	(9,8)	8.5	(8,8,5,8,7,7,6)	6.83	15	0.60	2	17.93		17.93
8	Tennessee	(6,7)	6.5	(10,7,10,11,9,10,11)	9.50	46	1.84	2	19.84	–0.2	19.64
9	Miami(Fl)	(10,9)	9.5	(9,5,7,9,10,11,9)	8.17	13	0.52	2	20.19	–0.4	19.79
10	Kansas St	(8,10)	9	(16,12,12,12,6,13,13)	11.33	10	0.40	3	23.73	–1.0	22.73

initial BCS rankings are developed each year an adjustment is made for wins against other top BCS teams. Had the quality win bonus been in effect for 2000, Miami would have been chosen. This would turn out to be a theme for the BCS—the current year's system would definitely have gotten last year's decision correct!

- 2001: The following year it was undefeated Miami that was the obvious choice. This time the BCS selected once-beaten Nebraska over once-beaten Oregon despite the fact that Nebraska had lost badly in their last regular season game. Popular perception this time was that the computer ratings paid too much attention to the large margin of victory in Nebraska's early season triumphs while not putting enough value on Oregon's steady but unspectacular performances. It was decided that the computer ratings should not use the scores of the games.
- 2003: Controversy of a different type arose during this most recent season. There were no undefeated teams, but there were three once-beaten teams from strong conferences. How to choose among them? Let's examine the data in Table 15.1. The human polls agreed that USC and LSU were the top two teams. The computers were equally insistent that Oklahoma and LSU were the top two teams. Oklahoma and LSU were selected to play in the national championship game which was won by LSU. The final coaches poll, as per the BCS agreement, elevated LSU to the top ranking. The writers' poll, not obligated by the BCS agreement, maintained USC as its top ranked team. The BCS, a system expressly created to identify a single champion, had failed to do so. What next? It seems clear that another "tweak" of the rules is in order. The plans for the 2004 season have not yet been announced though a couple of possibilities include: (1) only using the full BCS formula if the polls do not agree on the top two teams; (2) only considering teams that won their conference championship (this would have eliminated Oklahoma in 2003 and Nebraska in 2001). When asked for his opinion about the outcome in 2003, BCS coordinator Mike Tranghese offered that "People know I'm an anti-computer person" hinting that the weight of the computer ratings would be reduced.

The role of statistical thinking

The problem facing college football is to find a method for selecting the top two teams at the end of the football season. Each team plays between 11 and 13 games against a wide range of competition. The wide range of competition means that just looking at the records of the various teams is not sufficient. A team with one loss in a powerful conference may well be stronger than a team with no losses that plays in a weaker conference. Choosing the top two teams in such an environment is clearly a difficult task. It is basically impossible unless one can define what one means by the "top two teams." As mentioned earlier, one possible goal would be to identify the two teams that would be predicted to defeat all others in a post-season tournament. This is not, however, the only possible goal. Morris (1978) proposed that it would be ideal to try and predict what would have happened if a complete round-robin schedule (each team playing every other team) were to be played. Games that were played could perhaps be combined with statistical predictions for the unplayed parts of the round-robin. Another plausible goal is to try, as Dickinson did, to recognize the two teams that have "achieved" the most during the season. This of course requires a formal approach to measuring achievement!

It is clear that different goals might lead us to identify different teams at the end of the year. A team that lost two games early in the year without its star player but won every subsequent game by 40 points might be the team most likely to defeat all the others at the end of the season but clearly did not achieve as much as an undefeated team in a reasonably strong conference.

The situation facing college football is not all that different from the one that faces a data analyst trying to identify a one-number summary of a distribution. The mean, median, and mode all have clear and compelling arguments for being the best summary. The best choice in a particular context depends on the objectives of the person who wants the summary. To help decide we often ask about the loss that will be incurred by choosing a wrong summary. Though formal specification of a loss function is unlikely in the college football world, a full discussion of the ramification of different rating systems would clearly seem valuable. Some of the problems faced by the BCS (what to do if the polls and computers disagree, what game information should be used in the computer ratings, what to do if team A is ranked above team B but team B has defeated team A during the season) could have and should have been anticipated in advance. The following section provides a short discussion of the computer rating piece of the BCS; this discussion is intended to serve as an illustration of one area in which statistical expertise would prove valuable to BCS organizers.

Computer ratings and the BCS

It is disappointing, if not surprising, that most of the blame for 2003's BCS fiasco is being placed on the computer ratings. The distinction between human rankings and computer-generated mathematically based ratings is somewhat artificial. The computers are only systematically and emotionlessly applying a series of human-generated rules to determine the ratings of the teams. There is a great deal of research demonstrating that human opinions tend to be clouded by a number of consistent biases. There is a bias in favor of recent events, a bias in favor of familiar people/teams, and many other biases. In choosing the top teams it would seem desirable to have rules applied consistently instead of based on which games a particular poll participant happened to watch or remember hearing about. Despite this there is clear evidence that many reporters and college football officials are much more comfortable with the human polls than with the mathematically based ratings. For the moment we put this preference for human polls aside and consider the computer ratings part of the BCS in more detail.

A ratings primer

An entire article on just rating methods would be possible (and even quite interesting). Such an article would not in my opinion address the true difficulties that plague the BCS. Instead, we settle here for a short primer on statistical methods for rating teams to illustrate some of the issues that BCS organizers need to address. In particular, a range of rating methods that are based on different amounts of information are considered. Several references to the team rating literature appear below; interested readers are directed to Mease (2003) and Massey (2004) and the references therein for more information.

One of the simplest rating methods and a useful starting point for any discussion of college sports is the ratings percentage index (or RPI). This is a device used by the NCAA (college sports' governing body) to rank teams in a variety of sports. The RPI is a weighted average of a team's winning percentage (with weight 0.25), the average winning percentage of its opponents (with weight 0.50), and the average of its opponents' opponents winning percentage (with weight 0.25). The RPI trades off the two key aspects of a team's performance: the achievement of the team (winning percentage) and the quality of the opposition. (The last two components of the RPI are the schedule strength measure used by the BCS.) The RPI is simple and easy to interpret. Of course, the RPI does have some weaknesses; for one thing it is possible to win a game and have your rating go down! Another problem is that the measure of opponent's strength (their record) is clearly not the ideal measure. The RPI also does not take the site of the game into account.

At a next level of sophistication, but still ignoring the scores of games, there are a variety of paired-comparison methods. This includes, for example, the Bradley–Terry model (Bradley and Terry 1952), which postulates the existence of team strength parameters π_i such that the probability that team i defeats team j is $p_{ij} = \pi_i/(\pi_i + \pi_j)$. There are a number of methods for determining optimal values of π_i. One common approach is to maximize the likelihood of the observed game outcomes. There is a small problem with undefeated or winless teams but this can be easily handled in a variety of ways (see, e.g., Mease 2003). This approach improves upon the RPI by measuring opponents' strength implicitly rather than with a simple measure like winning percentage. There are a wide variety of related methods available because the method of paired comparisons (comparing a large number of treatments two at a time) is used in a variety of fields of application. David (1988) is a comprehensive reference on paired comparisons. The majority of the BCS methods currently in use are of this type.

It is natural to ask what can be done if we allow ourselves to make use of game scores. One answer can be found in the long history of rating teams using least-squares methods (see, e.g., Leake 1976; Stefani 1977; Harville 1977; Stern 1995). The basic idea is to choose ratings R_i for each team to minimize the difference between the observed outcome Y_{ij} when team i plays team j and the outcome predicted by the ratings, that is, to minimize $\sum_{\text{games}} (Y_{ij} - (R_i - R_j))^2$. Such methods can also be viewed as maximum likelihood ratings under a normal model for game outcomes. This approach can be easily modified to incorporate a home field advantage and to downweight the influence of large values of Y_{ij} (see, e.g., Harville 2003). Both modifications are important. The least squares idea can also be used without the scores by defining the game outcome Y_{ij} as 1 or -1 depending on which team wins the game.

It is possible to increase further the amount of information used in our rating system by including results from previous seasons. This is done by Harville (1980) and Glickman and Stern (1998). Such methods begin each season by recognizing that the final ratings of the previous year contain some information about the quality of the teams for the coming year. Though this is undeniably true one can easily imagine that this would be an undesirable feature for a system whose goal is to unbiasedly select the top teams for the current year.

The value of information

One obvious way that statistics and statisticians might be of assistance to the BCS process is by evaluating the many rating methods and identifying their strengths and weaknesses. The discussion of the previous section begins this process. To illustrate how statistics can inform our choice of rating methods, let us suppose for the moment that we agree the goal of a rating system is to accurately predict future games. We repeat here that this is only one possible goal and not even the goal that most people would use to describe the BCS's mission. Given that goal, however, it is possible to apply statistical thinking to assess the various methods.

Table 15.2 gives the proportion of game outcomes predicted correctly in a variety of sports by a sequence of increasingly sophisticated methods. The first column gives the proportion of games correctly predicted knowing only the site of the game; that is, this method always predicts the home team will win. This is clearly a crude system because it does not use any information about the relative strengths of the competing teams. The fact that the proportion of correct predictions is always greater than 0.5 reinforces the notion that the site of the game is an important piece of information. (Interestingly, this important piece of information is excluded from most of the BCS computer ratings.) The second column is a form of compromise between least-squares and paired comparisons methods. It uses the least-squares idea, but replaces the score of each game by +1 or −1 depending upon whether a team won or lost its game. This column also uses the site of the game. Note that using the results of past games improves the accuracy of our predictions by a fair amount as would be expected. In these experiments no least squares predictions were attempted until one-third of the season was completed; this initial portion of each season was used to initialize the ratings. Then each week's games were used to update the ratings before predicting the next week's games. The third column indicates the value of adding the scores to our least squares rating method. In each sport there is some benefit to incorporating the score of the game if we would like to predict future outcomes. The biggest increase is in college football! The final column of Table 15.2 may be thought of as an upper bound; it reflects the frequency with which professional oddsmakers' predictions are found to be correct. Least squares ratings do quite well, near but not quite at our upper bound. The RPI is not included in Table 15.2. Separate experiments in college basketball and college hockey (not shown) suggest that the RPI (69% correct predictions) is less effective than least squares ratings.

Table 15.2 ***Comparison of the proportion of correct predictions using four different prediction methods. The column labeled "Home" gives results for a method that always predicts the home team will win; "LS(1/–1)" makes predictions based on least squares ratings with 1/–1 game outcomes; "LS(scores)" makes predictions based on least squares ratings with actual game outcomes; "Oddsmaker" reflects the predictions of professional oddsmakers. Least squares ratings are used to predict the final two-thirds of each season with the first one-third of the season used to initialize the rankings.***

Sport	Years	Home	LS (1/–1)	LS (scores)	Oddsmaker
Pro football	1988–1993	0.58 (1342)	0.63 (862)	0.65 (862)	0.67 (1303)
Pro basketball	1985–1986	0.66 (1886)	0.69 (1255)	0.70 (1255)	0.71 (1827)
Pro baseball	1986	0.53 (3884)	0.56 (643)	0.56 (643)	0.55 (938)
College football	1992–1996	0.62 (3038)	0.69 (2104)	0.73 (2104)	0.75 (1551)
College basketball	1995	0.68 (2258)	0.72 (1724)	0.72 (1724)	0.75 (2068)
College hockey	1991–1992	0.62 (1405)	0.72 (998)	0.73 (998)	—

NOTE: Numbers in parentheses are number of games predicted.

Table 15.3 ***Comparison of the proportion of correct predictions for methods using information from previous seasons. The parameter σ_s^2 determines how quickly old information is discounted; large values of σ_s^2 correspond to ratings being developed primarily on the current season's results while small values of σ_s^2 correspond to ratings that change slowly over time. Results are for prediction of 1027 professional football games (the second halves of nine seasons). Predictive accuracy is highest for moderate values of σ_s^2.***

Prediction method	Percent correct
Oddsmaker	66.3
State-space ($\sigma_s^2 = \infty$) (same as least squares)	63.6
State-space ($\sigma_s^2 = 100$)	63.9
State-space ($\sigma_s^2 = 9$)	64.9
State-space ($\sigma_s^2 = 6.25$)	64.9
State-space ($\sigma_s^2 = 1$)	62.0

Table 15.3 provides information about the value of information from previous seasons. Again, the aim here is not to argue that such information should be used but only that statistical thinking allows one to provide objective data about the value of such information. The results in Table 15.3 are based on a crude version of the state-space model for rating teams used by Glickman and Stern (1998). In the crude version of the model applied here the data for each year are fit by a normal-theory least squares model and the parameters/ratings for the different years are modeled as a random walk with between-season variance σ_s^2. The value of σ_s^2 is fixed at a number of different values for the prediction experiments in the table. The resulting models are applied to forecast games in the second-half of nine professional football seasons in the late 1980s and early 1990s. When σ_s^2 is small, the team ratings are assumed to vary only a little from year-to-year. When $\sigma_s^2 = \infty$, the ratings from one year have no effect on the predictions the next year. The data show that some sharing of information from one year to the next (i.e., values of σ_s^2 in between the two extremes) improves predictive performance. A complete data analysis in Glickman and Stern (1998) finds that the best estimate for σ_s^2 is in the same range (6 to 9) that provides optimal predictions in Table 15.3. It should be emphasized again that there is no claim that prediction is an appropriate goal for the BCS. Tables 15.2 and 15.3 merely demonstrate how statistical thinking can be used to support good decision making in the context of a specified goal.

Computer ratings in the BCS

The previous section provides a quick (but hopefully clear) discussion of a number of quantitative rating methods. It is interesting to briefly review what we know about the BCS components in the context of that discussion. The information provided here was obtained by visiting the relevant Web sites (pointers to the various rating sites can be found at the Massey or Billingsley Web sites; www.mratings.com and www.cfrc.com, respectively).

Sagarin's BCS rating is based on the chess rating system developed by Elo (1978). Billingsley is a long-time observer of college football who has developed a series of rules for updating a team's rating based on its most recent result. Interestingly, Billingsley's method does use information from the previous seasons to initialize the ratings whereas most (all?) of the other methods do not! Wolfe's ratings are described as being based on the Bradley–Terry method. Colley's method is described in great detail on his Web site (www.colleyrankings.com); it turns out to be closely related to least-squares ratings with 1/—1 outcomes though that is not the motivation presented. Colley's method also includes a bit of what statisticians generally refer to as shrinkage estimation. Massey's ratings are well-documented; they represent a variation on the Bradley–Terry paired-comparison idea. (Incidentally, both Massey and Sagarin maintain ratings with and without score information on their Web sites). The Anderson & Hester and the *New York Times* ratings provide very few details. In the case of the *New York Times* it has been reported that even the people who produce the ratings each week don't know the algorithm. If true, this is disturbing especially given that many observers find the week-to-week changes of the *New York Times*' ratings to be quite unusual. The week-to-week changes in the *New York Times*' results suggest that it may be giving extremely large influence to recent results compared with the other ratings.

Given that scores are not allowed to be used in the BCS computer ratings, each of the systems now in use basically represents a formal quantitative combination of team performance and schedule strength based only on win/loss information. The variation we see among the seven included ratings is due to choices made about: whether to include/exclude site information, whether schedule strength should be measured by examining the records/ranking of teams when the game was played or revised to account for later performances, whether recent games are weighed more than early games, and a variety of other ad hoc choices (constants, shrinkage). It is interesting to note that the pre-BCS system using only polls was often criticized for not considering schedule strength. The BCS ranking includes schedule strength implicitly within each computer rating and then again explicitly as an additional term. Could the schedule strength pendulum have swung too far in the opposite direction? Is it possible that the BCS now over corrects for schedule strength? The argument here is that it would be beneficial if BCS organizers were aware of the issues involved so that an informed choice could be made.

Summary: What can statisticians do?

I take the goal of this article as being to review the BCS and to ask how statistics might be used to improve the situation. The latter question is interesting to think about.

The BCS is not the first time that statisticians have been excluded from the discussion of a policy question of great public interest, and it probably won't be the last time this happens either. Also the BCS is surely not the most important policy question currently lacking statistical input. Why worry about it at all? One reason is that the BCS raises familiar questions about how statistics and statisticians can get themselves invited to the table when policies based on quantitative data are discussed. Moreover, the BCS raises those questions in front of a relatively large audience of sports fans. As a discipline, statistics focuses on the collection, analysis, interpretation, and reporting of data. These are the key elements facing the BCS organizers and thus statistical input is essential.

One natural reaction is to suggest that statisticians should be put in charge of developing the mathematically based rating systems that should be included in the BCS ranking system. However, I think it would be a mistake for statisticians to claim a unique ability to develop the ideal rating system. I've argued that the best rating system must depend on the goal for which the system is being developed. It is unlikely that we could agree as a discipline on the correct goal and besides that's not necessarily our decision anyway.

Instead perhaps we can better frame the public debate by reinforcing some of the very basic ideas from our field. We should be asking the questions that drive the public discussion such as:

- What is the objective of the BCS rankings? The discussion of ratings systems in the previous section focuses on prediction as the objective. A recent article by Mease (2003) in this same journal focuses on how closely a computer rating system mimics the human polls. This seems to match what the BCS folks would have liked ... at least for this year! The bottom line is that for the BCS to succeed a goal or objective must be identified.
- What sources of information should be included in the BCS rankings? The discussion of mathematically based rating systems in the section on "Computer ratings and the BCS" identifies a number of key issues in this vein. These issues clearly need to be discussed in the context of "computer" ratings but they should also be discussed in the context of the human polls. Who votes in these polls and what information is provided to the voters?
- How sensitive is the BCS method to slight changes or variations in the data? At the least answering this question would require that the details of the existing rating systems be revealed to a group of experts able to judge their sensibility.

If we can get sportswriters and fans asking the above questions, then statistics will have contributed mightily to reforming the BCS.

REFERENCES AND FURTHER READING

Bradley, R. A., and Terry, M. E. (1952), "The rank analysis of incomplete block designs: I, The method of paired comparisons," *Biometrika*, 39, 324–345.

Billingsley, R. (2004), College Football Research Center [on-line]. Available at www.cfrc.com.

David, H. A. (1988), *The Method of Paired Comparisons* (2nd ed.), Oxford: Oxford University Press.

Dickinson, F. G. (1941), "My football ratings—From Grange to Harmon," *What's What in Football Weekly Magazine*, Omaha, NE: What's What Publishing Co.

Elo, A. E. (1978), *The Rating of Chessplayers, Past and Present*, New York: Arco.

Glickman, M. E., and Stern, H. S. (1998), "A state-space model for National Football League scores," *Journal of the American Statistical Association*, 93, 25–35.

Harville, D. (1977), "The use of linear model methodology to rate high school or college football teams," *Journal of the American Statistical Association*, 72, 278–289.

—— (1980), "Predictions for National Football League games via linear-model methodology," *Journal of the American Statistical Association*, 75, 516–524.

—— (2003), "The selection or seeding of college basketball or football teams for postseason competition," *Journal of the American Statistical Association*, 98, 17–27.

Leake, R. J. (1976), "A method for ranking teams with an application to college football," in *Management Science in Sports*, eds. R. E. Machol, S. P. Ladany, and D. G. Morrison, Amsterdam: North-Holland, pp. 27–46.

Massey, K. (2004), Massey Ratings [on-line]. Available at www.mratings.com.

Mease, D. (2003), "A penalized maximum likelihood approach for the ranking of college football teams independent of victory margin," *The American Statistician*, 57, 241–248.
Morris, C. (1978), Discussion of "football ratings and predictions via linear models," in *American Statistical Association Proceedings of the Social Statistics Section*, Alexandria, VA: ASA, pp. 87–88.
Stefani, R. T. (1977), "Football and basketball predictions using least squares," *IEEE Transactions on Systems, Man, and Cybernetics*, 7, 117–120.
Stern, H. (1995), "Who's number one in college football? ... And how might we decide?" *Chance*, 8, 7–14.

DISCUSSION

KENNETH MASSEY

The BCS challenge

The Bowl Championship Series has granted me the singular experience of being an academic suddenly thrust into the world of jocks, executives, and media. I produce the Massey Ratings, a computer-implemented mathematical algorithm for ranking college football teams that has been a component of the BCS formula since 1999. As a college football fan, I am honored to be part of the system, but as a mathematician this role has more often been a source of frustration. In his article, Hal Stern discussed the instabilities and weakly defined objectives of the current BCS formula. Reactionary patches have been successful only in hindsight, and recent debacles have many fans justifiably upset with the BCS process.

The Pigeonhole Principle states that if the number of pigeons exceeds the number of holes, then some pigeon must be left out. USC was the unfortunate pigeon last season, as LSU and Oklahoma qualified for a two-team playoff, otherwise known as the BCS championship game. Given that the system cannot accommodate three or more deserving teams, what predefined criteria should be employed to narrow the field to exactly two?

The BCS formula's task, while easy enough to state, is challenging to implement. Those who believe there exists a perfect system, impervious to controversy, are mistaken. Often the separation between teams is so slight that no obvious consensus is possible, and the decision comes down to splitting hairs. This unenviable task is further complicated by conflicting or double standards. Consider these examples:

- A strong schedule is often touted as essential to a worthy champion's resume, yet no team should be penalized for what they cannot control. Two of USC's opponents, Notre Dame and Hawaii, lost during the last week of the 2003 season. The effect was enough to propel LSU into the title game instead of USC.
- At the end of the 2003 regular season, five of seven computers had Oklahoma ranked Number 1 and LSU ranked Number 2. Fortunately for LSU, the Massey Ratings had these two reversed, otherwise USC would have beaten out LSU by the slim margin of 6.15 to 6.16 in the BCS rating.
- Miami Ohio finished the 2003 season with one loss, on the road to a highly ranked Iowa team early in the year. Suppose they had won that game. As the nation's only unbeaten team, albeit from a mid-major conference, would they have been more deserving than one-loss LSU, who faced a grueling SEC schedule?
- BCS computers are not permitted to incorporate margin of victory in their calculations, yet surely Oklahoma would have been a less controversial qualifier last season

had they not been humiliated by 28 points in the Big 12 championship game, but rather lost by only a field goal.

- Head-to-head competition is traditionally a significant factor in *who's better* arguments. In 2001, Washington beat Miami, who in turn beat Florida State. Each ended the season with one loss. FSU played for the championship, public sentiment was for Miami, while nobody lobbied for Washington.

In each scenario, the intermediate value theorem guarantees there is a line to cross, but unfortunately we cannot all agree where it is! Still, recent years have shown that there must be some majority opinion, perhaps media driven, that is eager to quarrel with the BCS selections. In response, the already awkward BCS formula has been modified to correct perceived inadequacies.

Stern correctly points out the lack of clear guidelines with regard to the objectives and implementation of the BCS formula. As I write, the commissioners are once again planning a change, this time promising a streamlined system fully endorsed by a team of mathematical and statistical consultants.

One thing I've learned from the process is the importance of public relations. Comparable to the U.S. tax code, the current BCS formula is a strange concoction of polls, computers, strength of schedule, quality wins, and losses. A gray box clouds the inner workings, so that the average fan feels the whole thing is at the mercy of a few computer nerds stuck away in some top-secret laboratory.

With Murphy's law and a critical public at work, even if all contingencies could be anticipated, there is no way to guarantee a happy outcome every year. There is, however, room for considerable improvements to the current system, which are outlined in the next section.

An improved BCS

We have already shown that the choice of two most deserving teams for a national championship game is not always clear-cut. In fact, there are no formal guidelines for what actually constitutes *deserving*. Certainly win/loss record and vague notions of schedule strength and impressive play are the primary factors. But who should measure these things, what metric should be used, and how should they be balanced?

The public is most comfortable with the human polls (AP and Coaches), despite their flaws. Unchecked by any public voting record, each reporter or coach is susceptible to bias, often unconsciously. Ballots are often filled out hurriedly, based on the highlight reels of the previous day's games. No human can monitor the hundreds of games played during a season for subtle schedule connections that yield valuable comparative information. Short memories abound, and teams are invariably penalized more for late season or particularly impressionable losses. Yet the polls retain credibility because by the end of the season two teams have usually emerged at the top, and deservedly so in the minds of most fans.

The original BCS formula was spearheaded by former SEC commissioner Roy Kramer. Realizing the dangers associated with relying solely on the fickle polls, the decision was made to incorporate an objective mathematical component, comprised of several *computer rankings*. These enrich the information content of the formula, and serve to keep the traditional polls honest.

Shrouded in mystery, computers have garnered disfavor from fans and a constant barrage of undeserved criticism from the media. One culprit in the BCS controversy has been a poorly defined objective. Established computer models were chosen, but no guidelines set for how to deal with margin of victory, home field, or the passage of time. After the 2000 season, margin of victory is no longer factored into our equations.

Most of the blame should be placed on the redundant components: strength of schedule, and the more recently introduced quality wins and losses. Polls and computer algorithms already rely heavily on these factors. The extra components are dubious at best, and induce significant bias, especially with regard to strength of schedule. In 2003, a pseudo-ranking comprised of these components had USC ranked Number 7, and generated an effect far exceeding the relatively minimal difference between polls and computers.

Having isolated the fundamental flaw of the BCS, let us now suggest an additional improvement. Each week newspapers publish the top 25 polls, listing also the Borda voting point total obtained by assigning 25 points for a Number 1 vote, 24 for Number 2, and so on. Often ignored, this provides very valuable information regarding closeness of the vote. In the past, the BCS has used only ordinal rankings. I, along with David Rothman, Jeff Sagarin, and Peter Wolfe, have suggested a more fluid poll measure based on the Borda count. Specifically, if a team receives p points from n voters then we define

$$r = 26 - \frac{p}{n}.$$

Consider the following hypothetical poll, with $n = 65$ voters.

Team	Rank	Points	r
Virginia Tech	1	1600	1.38
Harvard	2	1520	2.62
M.I.T.	3	1519	2.63
Princeton	4	1450	3.69

Unlike the ordinals, transformed Borda counts accurately reflect the extremely close call between Harvard and M.I.T., and likewise the relatively vast gap in front of Number 4 Princeton.

A streamlined BCS formula would consist of three viable components: AP, coaches, and computers. Weighting each equally, the BCS rating for each team could be defined as the arithmetic mean of the respective rankings. Interestingly, the geometric mean would make it slightly less likely that the computers could overturn a consensus poll matchup. The public's palate for third roots, however, is suspect.

In retrospect, the BCS formula outlined here, using the current set of computer ratings, Borda adjusted polls, and void of extraneous components, would have yielded the *correct* championship matchup every year of the BCS's existence. Furthermore the proposed system would be more easily grasped by the casual fan.

Computer ratings

Computer ratings have been embraced by sports ranging from chess to World Cup soccer. Slowly, college football fans are becoming aware of the merits of this alternative to the traditional polls. Unfortunately, there are aspects to the ranking problem that render the methodology too complex for a satisfactory layman's description.

In particular, each season is short and conference oriented, and must be analyzed in isolation from historical bias. This creates disparate, weakly connected schedule graphs. Insufficient, intransitive, and inherently noisy data make it difficult to obtain accurate assessments of team strengths. It is not unlikely that the best team from a given year could lose two games, while an inferior team, riding emotion and luck, goes undefeated.

Stern mentioned several alternative rating methods for fitting the season's game data. Optimality is therefore an intrinsic property of the mathematical model, and objectivity applies only to the implementation. Linear (least squares) and logistic (Bradley's pairwise comparison) regression form the basis for many well-documented rating systems. Instead of tidy calculator formulas, most computer ratings require an iterative solution to a system of equations. Strength of schedule is implicit in the coupling of parameters, and the solution is a vector of ratings satisfying some defining model objective. Statisticians are familiar with the probabilistic models upon which ratings systems are based. Surprisingly, certain ideal model properties have yet to be realized simultaneously in a rigorous manner. That, however, is a subject for another time.

Conclusion

I remain optimistic that by the time you read this, the BCS will have announced an improved formula. For the first time, they have hired trained consultants to test proposals and give sound statistical advice. With the BCS contract extended through 2009, we will have ample opportunity to compare the reformed system with its predecessors. I am confident the result will be another feather in the statistician's cap.

DISCUSSION

DAVID A. HARVILLE

Hal Stern has provided the readers of *The American Statistician* with a very nice overview of the development and evolution of the BCS, with emphasis on the BCS ranking system and on the potential for improvement. In response, I offer the following comments.

1 *Political Nature of the BCS.* It is informative to view the BCS in general and the BCS ranking system in particular, as the result of a political process involving diverse and influential constituencies. Prominent among those constituencies are the networks that televise the bowl games, the "bowls," the coaches, teams, and university administrators, the media, and the fans. Different constituencies tend to have somewhat different interests and objectives. In developing and periodically modifying the ranking system, the BCS decision makers have attempted to please or at least placate the various constituencies. Their approach has been reactive and has taken the form of "including something for everyone."

   A significant theme in Hal Stern's article is that "the BCS ranking system has been hurt by the fact that its main objective is poorly defined." The diversity of the various BCS constituencies may make agreement on a main objective difficult if not impossible. In any case, what would seem to be desirable is a more systematic (and less reactive) approach in which objectives are articulated and prioritized and in which possibly difficult choices are confronted and made.

2 *Polls Versus Computer Rankings.* For reasons that are in large part economic, it is in the best interests of various BCS constituencies "to give the public what it wants." The polls tend to reflect public opinion and are much better at doing so than computer ranking systems. As Hal Stern points out, the factors that resonate with the public can change from year to year in ways that are difficult to anticipate.

   On those occasions where the computer rankings differ significantly from the results of the polls, the differences are regarded by many as reflecting poorly on the computer ranking systems. Mease (2003) considered agreement with the polls to be an important criterion in the design and evaluation of computer ranking systems. A possible reaction is that if computer ranking systems are to be judged primarily on their ability to duplicate the results of the polls, then why bother to include them in the BCS?

   How does the predictive accuracy of computer ranking systems compare with that of the polls? The frequency with which the least squares algorithm identifies the winning team provides some information about what is achievable with a computer ranking system. And, the teams favored by the betting line could be regarded as having been chosen by a process that bears some resemblance to a poll. The accuracy figures given in Hal Stern's Table 15.2 appear to include predictions for games played over a considerable part of the season. Intuition and the available data suggest that the predictive accuracy of least squares relative to that of the betting line tends to improve as the season progresses. Moreover, from the perspective of the BCS, the only accuracy of much relevance is that at the end of the regular season. In any case, when it comes to predictive accuracy, there does not seem to be much evidence that least squares is either significantly better or significantly worse than the polls.

If there is a case to be made for the inclusion in the BCS of computer ranking systems, it would seemingly need to be made on the basis of what they have to offer in the way of "objectivity" and "transparency" and on the capability they provide for the exclusion of "inappropriate" information.

3 *Accounting for the Home-Field Advantage.* The home-field advantage (the advantage of playing at home versus playing at a neutral site) is estimated to be nearly four points (e.g., Harville 2003). The BCS ranking system takes little account of the home-field advantage. Any effects on the polls of information about the site of a game tend to be rather transient. Some (a majority?) of the computer ranking systems employed in the BCS do not account for the home-field advantage.

Because of the nature of college football schedules, the failure of the BCS ranking system to adequately account for the home-field advantage introduces systematic biases that favor teams from certain conferences at the expense of teams from certain other conferences. For example, it works in favor of teams from the Big 10 Conference. During the 2003 season, each of the 11 Big 10 teams played 8 games against other Big 10 teams, 4 at home and 4 "on the road." In addition, each played 4 games against out-of-conference opponents. Of the 44 out-of-conference games played by Big 10 teams, 32 were home games. One Big 10 team (Ohio State) played all 4 of its out-of-conference games on its home field.

This phenomenon is driven by economic considerations. Big 10 teams tend to have large stadiums and numerous fans. Many of the opponents in their out-of-conference home games are from conferences like the Mid-American Conference, composed of teams with smaller stadiums and less numerous fans. The (14) Mid-American teams played a total of 56 out-of-conference games, 27 at home and 29 on the road.

The impact of the difference between the two conferences (in out-of-conference home games versus out-of-conference road games) is exacerbated by the effects of a related factor. The majority (17) of the out-of-conference home games played by Mid-American teams were against non-Division I-A opponents, while only 2 of those played by Big 10 teams were against such opponents. When victories against non-Division I-A opponents are ignored as they are in compiling the SOS (strength of schedule) component of the BCS ranking system, the imbalance in the ratio of home games to away games becomes much greater. Only 10 of the 39 games played by Mid-American teams against out-of-conference Division I-A opponents were played on the home field of the Mid-American team; 30 of the 42 games played by Big 10 teams against such opponents were played on the home field of the Big 10 team.

The predominance of home games among the out-of-conference games of teams from conferences like the Big 10 tends to result in more victories for those teams than would otherwise be the case. In the case of a ranking system that fails to account for the home-field advantage, the ranking of any such team is biased upward. Not only is its winning percentage likely to be inflated, but so are those of its within-conference opponents and those of its within-conference opponents' within-conference opponents, within-conference opponents. (For similar reasons, the ranking of a team from a conference like the Mid-American Conference is biased downward.)

Consider, for example, the SOS columns in Hal Stern's Table 15.1, which indicate that Ohio State played a more difficult schedule than any of the other top teams in the BCS rankings. It seems very unlikely that the difficulty of Ohio State's schedule would have been ranked that highly had the home-field advantage been taken into account.

4 *Margin of Victory*. Ranking systems can be classified as to whether their use of margin of victory is: (1) unrestricted; (2) partially restricted (as in a margin of victory that exceeds a certain limit is treated exactly the same as one that equals the limit); or (3) completely restricted (i.e., restricted to the use of win-loss-tie information). With the complete or partial restriction of the use of margin of victory, the creation of any incentive to "run up the score" is avoided (at least to a considerable extent).

As discussed by Hal Stern, it was decided following the 2001 season to impose a complete restriction on the use of margin of victory by the computer ranking systems employed in the BCS. That change does not seem to have been motivated by any concerns about running up the score—it could not have been expected to have much effect on that practice because the polls would continue to be influenced by margin of victory. Rather, the change seems to have been made in response to pressure from the Pac 10 Conference, which felt that Oregon (a conference member) should have fared better in the 2001 computer rankings and would have fared better had there been less emphasis on margin of victory. It is ironic that two years later (in 2003) this change seems to have worked to the detriment of another member of the Pac 10 Conference (USC). Had the change not been made, the computer rankings would have been influenced somewhat by margin of victory (as were the results of the polls), with the presumed consequence that USC would have fared better in the computer rankings and would have been among the top two teams in the overall BCS rankings. The 2003 computer rankings were unfairly targeted for criticism because of their divergence from the polls. That divergence came about because of the change instituted by the BCS decision makers, and any criticism should have been directed at them.

Neither the unrestricted nor the completely restricted use of margin of victory seems appropriate. The unrestricted use of margin of victory encourages the undesirable practice of running up the score and may do so even if the larger margins are "downweighted" (Harville 2003, sec. 3.3). Moreover, the college football schedule is highly unbalanced and is such that each team plays only about 12 games. Consequently, when the use of margin of victory is completely restricted, it becomes virtually impossible to accurately rank certain teams, for example, an undefeated team with a weak schedule (Harville 2003, sec. 7). The partially restricted use of margin of victory is seemingly the best option. However, a satisfactory implementation of that option requires something more than "simple truncation" (Harville 2003, sec. 3.3).

5 *Shrinkage.* In his section entitled "A ratings primer," Hal Stern reviews the use of least squares and maximum likelihood methods to obtain ratings for team effects (the R_i's). Variations on those methods are obtained by regarding the team effects as random and by adopting a Bayesian approach (Glickman and Stern 1998) or an empirical BLUP (best linear unbiased prediction) or empirical Bayes approach (Harville 1977, 1980, 2003). Typically, the team effects are taken to be independently and normally distributed with mean 0 and with common variance, say σ_R^2. In relatively simple settings, the Bayes, empirical Bayes, and empirical BLUP approaches are characterized by a phenomenon known as "shrinkage." Similar ratings can be obtained via a penalized maximum likelihood approach (Mease 2003).

At first glance, these approaches would seem to have considerable appeal. However, unless the value of σ_R^2 is assumed to be relatively large (much larger than warranted by the inherent variability among the teams), their adoption can have undesirable (and possibly unforeseen) consequences. Teams that are very successful but that play very weak schedules may be ranked much too highly—refer to Rothman (2004) for some historical examples. A possible explanation is that, conditional on the Division I-A schedule, it is quite unrealistic to assume that the team effects are distributed independently. The team effects of teams that play each other tend to be much more alike than those of teams that do not play each other.

More realistic assumptions about the team effects could be obtained via a hierarchical approach in which (at the first stage) different means are introduced for different conferences, and it is the deviations of the team effects from the conference means that are assumed to be independently and identically distributed. However, those assumptions are subject to the criticism that their adoption would allow a team to "borrow strength (or weakness)" from its fellow conference members and would thereby tend to favor the weaker teams in the stronger conferences at the expense of the stronger teams in the weaker conferences.

6 *Role of Statistics/Statisticians.* As Hal Stern points out in his Summary, "the BCS is not the first time that statisticians have been excluded from the discussion of a policy question of great public interest." In fact, it is not the first time even within the confines of college sports. The BCS was preceded by the RPI (ratings percentage index), which was devised by the NCAA (National Collegiate Athletic Association) staff with little if any external input—the "inspiration" for the SOS component of the BCS ranking system appears to have come from the RPI. As a statistical instrument, the RPI is fundamentally flawed (e.g., Harville 2003), and its flaws manifest themselves in a variety of undesirable ways that have been described by some of the more discerning members of the press (e.g., Wolff 2001, 2003). Nevertheless, the NCAA persists in its use.

It seems clear that the current BCS ranking system has some serious deficiencies and that significant improvement requires something more than the kind of reactive "tweaking" that has been the forte of the BCS decision makers. As discussed earlier (and as intimated by Hal Stern), what would seem to be needed is a more systematic approach that includes the establishment and prioritization of criteria that reflect the interests of the various BCS constituencies. Depending on the nature of the criteria, a considerable amount of statistical expertise may be needed to devise an appropriate ranking system. The involvement of statisticians could prove helpful in "shaping" discussions about the establishment and prioritization of the criteria as well as in devising an appropriate system.

The membership of the NCAA and the BCS consists of institutions of higher learning. The faculties of those institutions include some of the most distinguished statisticians in the world. That the NCAA and the BCS employ statistical instruments of such dubious merits as the RPI and the current BCS ranking system should be a source of some embarrassment to those university officials having oversight responsibilities for those organizations.

ADDITIONAL REFERENCES

Rothman, D. (2004), Comment on "A Penalized Maximum Likelihood Approach for the Ranking of College Football Teams Independent of Victory Margins," by D. Mease (letter to the editor), *The American Statistician*, 58, 176.

Wolff, A. (2001), "Dirty pool," *Sports Illustrated*, March 26, 40–48.

—— (2003), "A madness to the method?," *Sports Illustrated*, March 3, 48–55.

DISCUSSION

RICHARD BILLINGSLEY

The first impression I have after reading Hal Stern's article on the college football championship is that it is one of the best—if not the best—article I have ever seen published on the subject. I believe that Stern did an excellent job of researching and articulating the information. There are a few points that I would like to expand on, and perhaps shed a little different light on, although in general I don't believe I would change or dispute anything Stern had to say.

I totally agree that the first mistake, if one could call it that, of the BCS, is that the first commissioner, Roy Kramer, did not publicly present a more specific "mission statement." I'm just not sure that Kramer felt it was necessary. The formula used by the BCS was designed by Charles Bloom of the Southeastern Conference and I believe Kramer, after reviewing the formula, thought the goal was very clear. The goal was that two teams would be chosen to play for the national championship based on a public formula that emphasized strength of schedule over wins, losses, and known human prejudices. It's a common misconception that the BCS was created to avoid split championships because from the very onset everyone knew—the commissioners, the media, and the fans—that split championships were still possible. We all just hoped it would never take place. In 2003 it did, and somehow all of a sudden the media proclaimed the BCS failed.

The truth is that the Bowl Championship did not fail in 2003, nor has it failed in its entire existence. Instead 2003 was a perfect example of how the BCS succeeded. The sole purpose of the BCS is, in times of uncertainty about the two best teams in the nation, to allow the formula to be the deciding factor. It's that plain and simple. The only reason for any dissention is that in several seasons both human elements, the AP/Coaches polls, were in unison but happened to be at odds with the computer element. In all cases the strength of schedule (SOS) component was the deciding factor, and it was strong enough to override what I believe was a human error in judgment. In my opinion it's not the first time the human element made an error in judgment and, unfortunately, I'm sure it won't be the last. Overall I think the BCS is light years better than what we had previously. I do believe, however, it can still be improved, but that is another article in itself. I understand that radical changes are coming to the BCS formula, and I will reserve further comment until there is an official announcement.

Another point that I would like to expand on is about the lack of a college football playoff. This is an incredibly complex issue, but the core of the problem is that college football is a unique sport, totally unlike any other. There are currently 117 teams in Division I-A, far greater than the playoff-manageable NFL. Additionally, college football simply cannot be compared to basketball because the games cannot physically be played at the same pace. College basketball dwindles more than 60 teams down to 4 teams in two weeks. This is not humanly possible in football. People tend to forget those two simplistic yet unavoidable obstacles when thinking about a college football playoff. The media wants it, the average fan clamors for it, yet very few people actually stop to think about the logistical limitations in creating a full-fledged playoff. The major issues beyond the physical limitations are those of administration. The NCAA is governed by college presidents and all legislation must come through those presidents. Until the governing body changes its attitude about extending the season into the second semester, and agree to some financial agreement with member institutions, a full-fledged playoff will never happen.

One more point I will briefly touch on is the credit Hal Stern gives to the retirement of Tom Osborne and the creation of the BCS. Although it's an interesting observation, it's highly likely that it is nothing more than a mere coincidence. The vision for the BCS was created by Roy Kramer in 1992 when the Bowl Coalition was established. The only reason the Bowl Championship Series did not come into full existence at that time was the resistance to participate on the part of the Rose Bowl, the Pac 10, and the Big 10. Negotiations began in 1992 to bring all three of those groups into the fold. To say that Tom Osborne's retirement spearheaded the formation of the BCS would be an overstatement. Those negotiations were already well underway before the end of the 1997 regular season when Osborne announced his retirement. The Big 10 and Pac 10 were already leaning in the direction of BCS membership before the 1997 season. The Rose Bowl Committee, which is made up of gentlemen who are classically "old school," simply took longer to be convinced that participating in the BCS was right for the granddaddy of all bowls. I doubt very seriously that Tom Osborne's retirement swayed their position as much as the persistent, persuasive voice of Roy Kramer who worked tirelessly for six years to bring the BCS into existence.

On a parting note I want to make this statement. It may surprise you to know that I personally do not believe that a computer rating should be the sole source for determining the two participants in a championship game. The reason is that I don't believe a computer rating, regardless of how good it is (and I believe I have one of the best), is as good as a combination of computers and a knowledgeable, responsible, human element. I admire math, and would be the first to recognize the modern marvels attributed to its existence. We could not have put a man on the moon without it. At one point, however, I recall telling the BCS commissioners that "I am attempting to put two teams in the Rose Bowl, not send them to the moon." Therefore I totally agree with Hal Stern that statisticians and mathematicians may not hold the ultimate key in determining college football championships. Blending a responsible human element which is open to understanding and using computer data seems to be the best that can exist at this time. Unfortunately, very few of the AP writers and coaches are taking their responsibility seriously, and none seem to be open to the remote possibility that a good, reliable, proven, computer rating can generate some very valuable data.

DISCUSSION

WES COLLEY

This article gives a brief treatment of the statistical methods of the BCS computer rankings, which is certainly a timely and topical endeavor, given the controversy that has surrounded the BCS in recent years.

I'm very pleased to see that the author identifies as a critical issue for the BCS is to identify the goal of the computer rankings. In fact, to me this is the outstanding issue, in terms of both the understanding on the part of the BCS conference commissioners and of fan/pundit confusion.

There are a couple of issues with the historical summary that seem important to address:

1 On bowl games: Using bowl games as a tournament where, say, the Peach Bowl is Southeast Number 4 versus Southeast Number 1, and the Holiday Bowl is Southwest Number 3 versus Southwest Number 2, and so on, does have an important disadvantage, which is that the names of the bowls would likely become meaningless, as the pride in being "Peach Bowl Champion" would go away, because it would only mean being "first-round regional winner." A large advantage of the bowl games is that 14 teams are "champions" of something, with trophies to show for it. Second, bowl games are great for fans, in that they can plan a single trip to a bowl game during the holidays to see their team play its final game for a championship. Would fans be equally able (and eager) to go to the Peach, then the Music City, then the Orange, then the Sugar? Very few have the economic status to accommodate that schedule. The solution would be to play regional games at one team's home field, but that totally removes any semblance of the bowls.

2 On revenue splitting with other conferences: One common misconception is that non-BCS conference teams cannot enter a BCS bowl game. That is absolutely false. A non-BCS team in the top 10 is eligible for a BCS bowl. To date there has been only one non-BCS team close to the top 10 at the end of the season (Tulane, whose SOS was so miserable that there was little outcry), which is unfortunate. Furthermore, only twice has there been a non-top-10 BCS conference team in a BCS bowl. Should lightning finally strike, and a non-BCS team play in a BCS bowl, the payoff would be identical for both teams, so there is *not* an undeserved financial advantage for BCS conferences. In short, if there were no BCS, only twice would teams other than those picked likely have been picked in the previous bowl system, and in both of those cases, two other BCS teams would almost certainly have taken their places (anyway).

In the historical summary of BCS modifications, the 2001 section isn't quite right. Nebraska lost to Colorado in the final game of the regular season. Since both

Nebraska and Colorado had finished the regular Big XII season with one conference loss, by a conference tie-breaker rule (head-to-head), Colorado advanced to the championship game, where it defeated highly rated Texas. As such, Colorado was regarded as "obviously the second best team in the country," though with the early loss to Fresno St., it did not deserve a shot to play Miami, so "a quality Oregon team," the highest ranked one-loss team should go. To everyone's surprise, Oregon drilled Colorado in the Rose Bowl, so the argument quickly became that Oregon, not Colorado, was "obviously" the second best team in the nation, and had "obviously" deserved to play Miami all along.

3 In the 2003 section, it should be mentioned that (now) what's likely to happen for the fall of 2004 is the removal of the "tweak columns" at the end of the BCS formula, and have a simple (AP + Coaches + Computer)/3 formula, where the computer part is the average of the computer rankings. This is an important difference in that it is a "de-tweak," more than it is a tweak. Billingsley noted that if you removed the tweak columns, LSU and USC would have played each other as desired.

The only other comment that might be made in the article is (when known) what the motivation for each of the computer rankings is, because I think this is the most salient issue at hand. Sagarin's motivation for his Elo method is simply that he'd rather use his proprietary method, the motivation for which is maximizing predictiveness. Colley's method, to the contrary, is based on "deservedness," trying to measure which team has done the most to "earn" the right to play for the title.

Finally, just a note to help the author. He alludes to a fictitious example of a team that lost early but then won by 40 in its last several games; in fact, a very good real example of that exists. Florida State lost to Clemson and Southern Mississippi early in a season in the late 1980s, but proceeded to demolish all comers afterward, including an Orange Bowl romp over Nebraska. Almost everyone agreed FSU was the "best" team, but because of the two early losses, was not the "most deserving" team. As such, FSU was not voted to be national champion.

Ultimately, I think the most likely adjustment, (AP + Coaches + Computers)/3, is right-headed. I would go just a small step further and allow the computer rankers to use the system of their choice, rather than restricting us to win-loss only (that would affect three of the seven, I think). Statistical and intellectual diversity in the systems is important in establishing some sense of the parameter space. If Sagarin's non-Elo method and mine produce remotely similar rankings, one has to believe that we've cornered the fox pretty well. If you restrict Sagarin, that's less clear, since the restriction makes his method more similar to mine. Despite a fairly wide array of methods, each of the authors has fairly strong convictions about his system and its motivations. Using a diverse set of methods, with a diverse set of motivations is surely the best path toward "truth."

Beyond the math itself, probably the largest problem faced by the BCS has been confusion on the part of fans, media, players, and coaches, in terms of what the computer rankings are. It has been nearly impossible to explain that when the "tweak columns" obscure the issue nearly beyond hope to most people. The common fan thinks that the computer rankings still have subjective input, and that the tweak columns are somehow part of the computer ranking component. I constantly get e-mails that say, "Why do you *think* LSU is better than Oklahoma?" and the like. I have to explain that what I *think* is irrelevant, that the system is purely mathematical, with zero subjective input. Most fans are quite surprised by this, though it is precisely this feature that is the whole point of the computer rankings in the first place. The BCS has done a poor job of explaining that to the public.

This is exacerbated by the relative "sequestering" of the computer rankers by the BCS and media. For example, to "humanize" myself to I began posting on my Web site pictures from games I had attended, only to receive an irate e-mail three days later from one of the commissioners to remove the pictures at once or be removed from the BCS. But the media are bigger culprits—during hours and hours of discussion of the BCS on ESPN in December 2003, not a single one of us was consulted. Media "experts" on the BCS were brought forth, as if we, the actual rankers, were simply unreachable, which is absolutely wrong. All of us are more than happy to discuss our systems with the media and, in fact, think it's important to do so.

This "sequestering"—when layered on top of the perceived arcanity of the BCS system—creates an "Ivory Tower" problem. Fans think of us as unyielding academics, wearing gowns, blithering away, and scratching on wax tablets in our ivory towers, only occasionally blessing the plebs below with our biddings. More likely, we're sitting in the stands next to them in Athens, Gainesville, or Norman, eating hot dogs and yelling our heads off.

So what we, the computer "geeks," most would like is for the BCS to adopt a simpler "non-tweaked" system (which apparently is likely), allow each ranker to use the system of his choice, and adopt a policy of expressing faith in the computer systems, and making their authors more available to the media and fans.

DISCUSSION

DAVID MEASE

The problem the BCS faces each year of selecting the "best" two teams in college football is extremely important but quite difficult. The importance of the problem is well appreciated by any fan of college football such as myself, while even those who have never seen a single game can at the very least relate to the huge financial ramifications of the selection. Hal Stern gives an excellent description of the difficulty of the problem and correctly points out that the "main objective is poorly defined." This stems from the fact that there is no clear notion of what merits one team being ranked higher than another.

Current proposed changes to the system may lessen the importance of the BCS rankings in determining which teams play for the national championship. However, some ranking by which teams can be selected will necessarily be part of any system, since even a postseason tournament would need to select, say, the top 4, 8, or 16 teams from among the currently 117 Division I-A teams.

As Hal mentions, many of the difficulties that arise and often lead to controversy in the rankings can and should be anticipated. Indeed, many common themes repeat themselves year to year, and even week to week. I personally follow the weekly rankings of the AP and Coaches Polls closely, in addition to keeping my own "computer" model updated and tracking the computer models of others. What follows is a list of challenges faced by any ranking system, whether human or computer, subjective or algorithmic. Understanding these issues may help improve the system.

1 *Weekly Invariance.* With respect to the top 25 teams ranked by the AP and Coaches Polls there is an unwritten rule that once Team A is ranked ahead of Team B, the relative ordering of the two teams should remain unchanged provided neither team loses. People have come to expect this consistency from week to week. Many people have difficulty with the BCS computer models when this does not hold, although it is perfectly logical that the order should change if Team B has defeated a more difficult opponent than Team A. The controversy described by Hal surrounding Tom Osborne's retirement was largely a result of a violation of this principle.
2 *Prior Information.* One complaint that has been made about the AP and Coaches Polls is that the voting tends to favor traditionally powerful teams; that is, teams are perceived as being rewarded in the current year's rankings for success in past seasons. One attractive feature of objective computer rankings is that they can be made free of this bias by using a flat prior on all teams at the beginning of the season. The drawback to this of course is that the prior information that is thrown out is actually valuable, at least from a predictive point of view, as Hal discussed in the article. Furthermore, not using prior information means that rankings are extremely

volatile early in the season, causing issues with the first point previously. This is one reason the BCS does not publish its rankings until midway through the season.

3 *Prediction versus Accomplishment/Reward.* As Hal mentioned in the article, two possible ideas of what merits Team A being ranked above Team B are (1) whether Team A would be predicted to beat Team B in an accurate predictive model or (2) whether Team A "deserves" to be ranked above Team B by virtue of having accomplished more. The AP and Coaches rankings tend to strike an interesting balance between the two. If a certain Team A is perceived as having very little chance of beating a Team B (say because of being undersized or very weak in some aspect), then they will be extremely reluctant to rank Team A ahead of Team B. This speaks toward prediction. On the other hand, predictive models and odds-makers would have no problem predicting a team with three losses might beat a team with one loss; whereas, the AP and Coaches polls rarely would rank a three-loss team over a one-loss team provided both teams come from major conferences. The sentiment seems to be that three losses disqualify that team from deserving the better ranking, even if it is generally believed they would be more likely to win.

   Furthermore, a ranking procedure based solely on a predictive model has problems in and of itself. Many covariates that are highly predictive of winning are *consequences* of winning more than *causes* of winning. For instance, a large number of running plays (as opposed to passing plays) is generally very predictive of winning. This is largely due to the fact that teams which have a lead tend to choose more running plays. The reasoning is that running plays use more time off the game clock, thereby leaving less time for the opponent to catch up. If a predictive model were used to rank teams, a team may be able to improve their ranking simply by executing a greater proportion of running plays.

4 *Quantifying Large Victories.* In addition to the number of running plays being predictive of victory, the margin of victory, or the magnitude of the difference between the two teams' final scores, is also quite predictive of future victories. Hal Stern gives evidence of this in his Table 15.2. However, similar to the situation described with running plays, a ranking model that includes margin of victory can allow teams to artificially increase their rankings by "running up the score," that is, making an effort to score more points at the end of a game in which they have a large lead. Such behavior is generally considered poor sportsmanship. In fact, this initially caused so much controversy that the BCS eventually made the decision that none of their computer models would be permitted to include margin of victory. Hal alluded to one particular instance. Of course, the drawback to not including margin of victory in the models is that there is no way to reward or appreciate a victory in which one team clearly outperformed another. Humans can appreciate this and are able to separate such victories from artificial score inflation provided they watch or at least read about the game. However, even the voters in the AP and Coaches polls cannot keep track of all the games played.

5 *Conference Champions.* As Hal described, the teams are organized into a number of small groups called conferences. Each conference necessarily has a conference champion, an honor a team earns either by virtue of having the best winning percentage within the conference, or in some cases by winning a designated conference

championship game. There has been much debate especially of late as to what extent, if any, teams should be given an additional reward in the rankings for winning their conference championship. Currently the BCS does not give any additional credit for this distinction in its rankings, excepting that which is reflected in the subjective AP and Coaches polls.

6 *Order of Games*. One criticism of the AP and Coaches polls over the years has been their short memory; more recent games are perceived to carry more weight than games earlier in the season. A benefit of the computer polls was supposed to be that they could give every game equal weight. In fact, the BCS indicated that this was their preference. However, there has been some debate as to whether completely equal weight might be going too far in the other direction, such as the case with Oklahoma playing in the most recent BCS national championship game despite having lost their final game of the regular season. Of course, in this case, it was also significant that this game was their conference championship game and that it was perceived as being a lopsided loss. These are two factors that are not included in the computer polls as already discussed.

7 *Weak Conferences*. The organization of teams into conferences has an important history and serves a valuable function, while at the same time making the ranking problem more difficult. Because these conferences vary substantially in quality and there are a limited number of games between teams from different conferences (most teams play the majority of their games against opponents within their own conference), the standard errors for cross-conference comparisons are much larger than they would be for the same number of games with independent pairings of the teams. The problem of how to rank a team with an outstanding winning percentage from a weak conference arises almost every year, at least once. The argument on one side is a predictive argument: such teams would be predicted to have a relatively small chance of winning a game against even a mediocre team from a strong conference. The other side is the accomplishment argument: this team has lost none or maybe only one game but to a strong opponent, and deserves to be ranked as high as any team having two or three losses. Many of these situations resolve themselves when the team in question eventually loses. This was the case with the Horned Frogs of Texas Christian University in the most recent season who, until losing late in the year, were in position in the BCS rankings to earn an automatic invitation to a BCS Bowl game. The history of college football is cluttered with other such teams that have not lost and created more controversy. Most notable is Brigham Young University, which actually earned a national championship in 1984. Human pollsters usually discredit these teams due to the historically weak reputation of their conferences. Computer polls, on the other hand, can vary immensely with their treatment of such teams, depending on "strength of schedule" penalties as well as the performance of the other teams in that conference against out-of-conference opponents during that particular season.

8 *Home Field*. As Hal explained, the home-field advantage is substantial in college football. Additionally, human pollsters do seem to be more impressed by a win that is in an opponent's stadium in front of a hostile crowd. Despite this, home-field advantage is not included in many computer polls. Although it is simple to include one additional parameter to account for this in most models, there do exist arguments against its inclusion. First, teams generally play fairly balanced schedules with about one half of the games in their own stadiums and one half on the road, so it tends to not affect rankings substantially in most cases. Further, some early BCS models which included margin of victory actually subtracted an estimated home field advantage of three or four points from the final score of the home team. Thus, a team that thrilled their fans by kicking a game-winning field goal (worth three points) might have actually been better off attempting to score a touchdown (worth six points but harder to get), as the former may not have truly given them a win in such a model. Finally, it is also possible that some teams may have a larger home field advantage than others so that a single parameter for every team would not necessarily be fair. Note, however, that in the case of college basketball Harville and Smith (1994) found that variation in home advantage from team to team was relatively small.

9 *Comparing Teams with Equal Losses.* The conclusion of this last season presented an interesting controversy that seems to suggest one major difference between the way a human poll and a computer poll evaluate a team's performance over a season. The situation was described by Hal and involved USC, LSU, and Oklahoma each of which had only one loss at the end of the season (prior to the bowl games). Many computer polls ranked these teams with Oklahoma first followed by LSU and then USC, the exact opposite order of the AP and Coaches Polls. These computer rankings were criticized by many as being illogical, although interestingly this ranking is the exact inverse order of the ranks of the opponents to whom these teams lost (Kansas State, Florida, and California). It would seem that while humans were evaluating the teams based on a number of different factors, the computers were evaluating these teams largely based on the quality of the opponents to whom they lost. This is arguably quite logical. Furthermore, one needs to keep in mind that the computers did not consider that Oklahoma lost in a lopsided game that was the conference championship and occurred very recently but that USC's loss was in triple overtime on the road early in the season.

10 *Interaction Effects.* It is a tradition among the best teams in the AP and Coaches Polls that if there are two teams A and B each having only one loss total and Team A has defeated Team B (accounting for Team B's one loss) then Team A will be ranked ahead of Team B. While this can lead to circular arguments if carried further, at least at the top level it has come to be expected. However, this is often the opposite of what many analytic algorithms or predictive models would give. The loss of Team A was presumably to a weaker team while Team B's only loss came at the hands of Team A, which is a very good team. Most predictive models and computer rankings do not include what we would call interaction effects; it seems that it is generally believed there are simply too many possible pairwise interactions and that these interactions are not effective for prediction in sports such as football. (Another researcher and I are currently studying this.) However, people perceive interactions, or at least feel that if two teams have the same number of losses but one team has defeated the other, then the victor deserves the higher ranking. It is a compelling argument but not necessarily a good predictive rule.

In conclusion, the problem of ranking college football teams is quite challenging and not necessarily well-posed. There are often no right answers, but many answers are historically more accepted than others. Whatever the fate of computer models in determining the college football national championship, if nothing else we have learned from the controversy to think more analytically about the problem and issues involved.

REFERENCE

Harville, D., and Smith, M. (1994), "The home-court advantage: How large is it, and does it vary from team to team?," *The American Statistician*, 48, 22–28.

REPLY

TAS Editor Jim Albert deserves a great deal of thanks for his efforts in putting together this discussion. Readers are fortunate to have the views of Bowl Championship Series (BCS) insiders Billingsley, Colley, and Massey as well as those of knowledgeable outsiders Harville and Mease. There is relatively little to argue with in their discussion. Each discussant contributes valuable insights about the way the current system works and the issues that must be addressed in order to improve it. The lack of controversy here likely reflects the fact that both the BCS insiders and the outsiders are strong proponents of the use of quantitative methods in sports. In fact one limitation of this group is that it does not include representatives of those in the media or in the BCS hierarchy who question the value of such tools.

The importance of nontechnical issues

Several of the discussants note, correctly in my opinion, that nontechnical issues are at least as important as technical issues in developing a system to identify the top two teams (assuming we accept that as an initial statement of our goal). Massey cites "the importance of public relations" as a key lesson learned from his involvement in the BCS. Colley notes that the current BCS formula—with its quality win bonus and obscure schedule strength calculation—leads to confusion as to what the computer ratings are about. In fact the creators of the BCS computer ratings have often been ridiculed by the media with little actual attention given to what these researchers are trying to achieve. Colley's observation that many hours of media attention were showered on so-called "BCS experts" during the December 2003 BCS controversy (regarding Oklahoma, LSU, USC) with essentially no attention given to the BCS computer rating participants is especially striking. Public mistrust of the computer ratings is not a phenomenon unique to college football. Recently in California and in other states there has been evidence of public mistrust about college admissions. As with the football ratings such mistrust seems to have roots in a lack of information about the underlying process. There are no quick fixes for such situations, only education can improve things.

Harville emphasizes the political nature of the BCS task with interested parties including television networks, bowl game organizers, university administrations, teams, coaches, media, and fans. One way to describe the current system is that BCS folks are trying to give the public what it wants *within* the political constraints imposed by the various participants. In fact I'd argue that it is the attempt to satisfy the public that has yielded the frequent modifications of the BCS rules over the last several years. When it comes to satisfying the public the human polls have a significant advantage over the computer ratings. The polls respond quite quickly to the voice of the public—unpopular ratings one week generate intense public discussion which leads to more popular ratings the next week. By their very nature the computer ratings ignore such public pressure. (One friend of mine was distraught when his final prebowl ratings rated small school Miami University of Ohio as the top team ahead of USC, LSU, and Oklahoma. Once the algorithm is programmed it is out of the programmer's control!)

What happens next to the BCS?

The concerns of the previous section reinforce the argument that BCS organizers need a more extensive discussion of priorities and goals. Since the end of the season some discussion has occurred, primarily between the BCS conferences and their colleagues from less powerful football conferences (with the occasional threat of congressional investigation thrown in). This still falls short of the extensive multi-party discussions that might lead to a stable resolution.

Our editor had hoped that the BCS rules for 2004 would be in place by the time this material went to press. We do know that starting with the 2006 season (when new contracts for televising the games begin) a fifth game will be added to the program and played the week after the current set of games. Unfortunately, the new 2004 formula for determining which teams play for the national championship has not been officially unveiled as this rejoinder is written. The administrator of the BCS has offered several hints though. These hints are consistent with the speculation contained in the discussion. The new formula will apparently use only the two human polls and a composite computer rating (with each of the three receiving equal weight). This eliminates the strength of schedule and quality win bonus components of the current system. It also appears that Massey's suggestion to use the vote totals in the human polls rather than just the rankings is a good possibility. Massey indicates that this would have led to correct decisions in the past. Of course we know by now that, as in the stock market, good BCS performance in the past is no guarantee of good future performance. One does wonder why the same logic would not apply to the computer ratings—a case in which there is a clear number one in a particular computer rating should be treated differently than a case in which two teams have nearly the same rating.

The BCS administrator has been much less specific about changes that might occur within the computer component of the formula. The *New York Times* has decided it will no longer participate in the BCS. The *Times* cited a conflict of interest in influencing the events that they cover. Possible modifications of the remaining computer ratings include requiring the use of home-field advantage, allowing some use of margin of victory, and modifying the way in which the ratings are averaged. BCS officials also confirm Massey's information that a mathematician will be involved in the review of the process.

All of the proposals in the preceding paragraph seem like steps in the right direction. Thus, it does seem that the BCS administration has recognized that statistical thinking has contributions to make.

Why not a playoff?

I find that any discussion of the BCS, including this one, leads me back to a college football playoff. The discussants hit on some of the familiar arguments against a playoff: there would still be a team selection issue (Mease), university administrators do not want the season to extend into the second semester (Billingsley), there are too many teams and football is too physical a sport to allow a tournament like the NCAA basketball tournament (Billingsley), it would negatively impact the bowls (Colley). In response I'd offer the following quick notes. It is true that some teams would be left out of any tournament as they are in basketball, baseball, and in other college football divisions where tournaments are held. Such controversies don't last very long though, at least when compared to the long-lasting controversies engendered by the current BCS two-team playoff. As for the academic calendar, the BCS recently announced plans for a fifth game to be held the week after the current games; this suggests that a tournament finishing up in the second semester is not out of the realm of possibility. With respect to Billingsley's practicalities, doesn't the fact that Division I-AA can hold a 16-team tournament over four weeks at the end of the season suggest that the big schools could as well. Colley is correct that the role of the bowls in any tournament would have to be negotiated; there may well be negative consequences in terms of local interest for a bowl like the Peach Bowl but these negatives would need to be traded off against the much greater national interest in a Peach Bowl that was part of an NCAA championship tournament.

An opportunity

In at least one peculiar way statisticians should be happy about the current controversy surrounding college football and the BCS. The flaws of the BCS provide a unique opportunity for us to discuss the importance of quantitative thinking in the context of a popular and very public issue. If there were a non-controversial college football champion each year, then football ratings would likely be little more than a curiosity in both the sports and academic worlds. Instead it is a good bet that many people follow the details of the BCS formula with more enthusiasm than they ever invested in their introductory statistics class. Those of us interested in statistics and sports would be wise to make the most of this "teachable moment" to educate family, friends, neighbors, and perhaps even the media about the value of quantitative thinking in solving real-world problems.

PART IV

Cricket

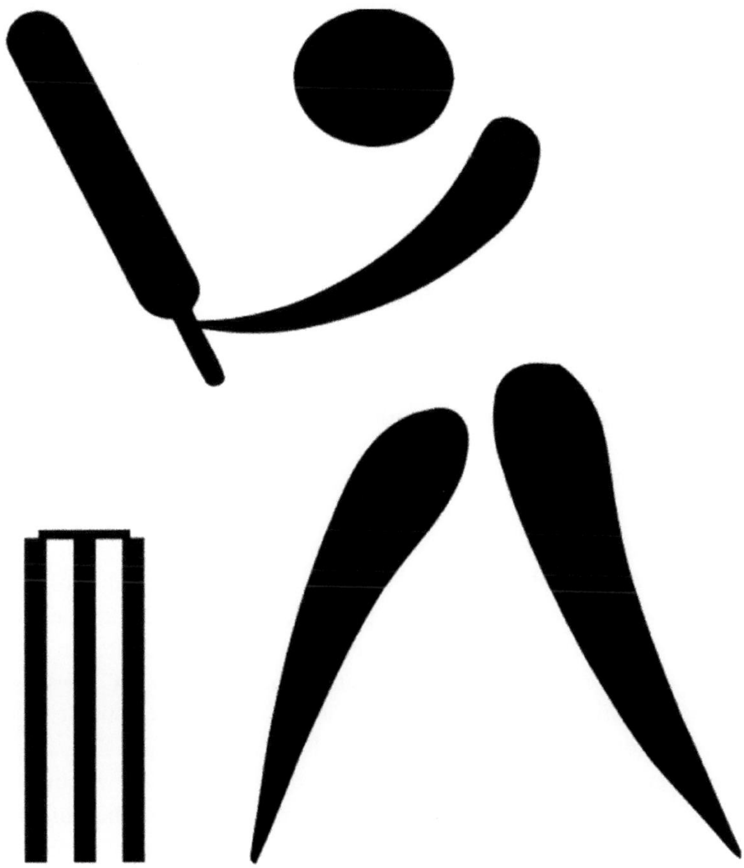

Introduction

PAUL J. VAN STADEN

This introduction provides a short overview of the sport of cricket and aims to highlight the general areas of statistical research in cricket, including the topics dealt with in the selected articles in this volume. It is not possible to give a detailed discussion on how cricket is played in this introduction. Readers unfamiliar with the sport may consult one of many books that explain the fundamentals of the sport, such as Eastaway (1992) or Knight (2013).

Cricket, a bat-and-ball game played between two teams consisting of 11 players each, is one of the oldest existing team sports in the world, with a history that can be traced back to southern England in the sixteenth century. The first set of rules, the Laws of Cricket, was compiled in 1744. Through the expansion of the British Empire the game spread from England to North America, the Caribbean, the Indian subcontinent, Australasia, and Southern Africa. The first international cricket match was between the United States and Canada in 1844 in New York, while the inaugural recognized Test match was played in 1877 between Australia and the touring England team at the Melbourne Cricket Ground. Test cricket is played exclusively by the ten Full Members of the International Cricket Council (ICC), namely Australia, Bangladesh, England, India, New Zealand, Pakistan, South Africa, Sri Lanka, the West Indies, and Zimbabwe. The ICC is the international governing body of cricket and, in addition to the above mentioned Full Members, has more than 90 Associate and Affiliate Members. Apart from Test cricket, various other types of cricket matches are played between the members of the ICC and, depending on the domestic cricket setup within a member country, by clubs, franchises, counties, provinces, states, schools, and tertiary institutions in that country. Professional cricket currently has three popular types of matches:

- First-class unlimited overs matches, with a maximum of two innings per team, are scheduled for three, four, or five days. Test cricket, considered the highest standard of first-class cricket, is the longest form of the game, scheduled as five-day matches with three two-hour sessions of play per day.
- In one-day limited overs matches, first played between English county teams in 1962, each team is given a single innings in which the maximum number of overs is restricted to between 40 and 60 overs. In particular, One-Day Internationals (ODIs) are international fixtures with the maximum number of overs limited to 50 overs per team. ODIs are contested between the ten Full Members of the ICC as well as six Associate Members who have been granted temporary ODI status by the ICC. The most important ODI tournament is the quadrennial ICC Cricket World Cup.
- The newest version of cricket, Twenty20 limited overs matches, was introduced in England in 2003. It is the shortest format in cricket, with one innings of 20 overs allocated per team. Twenty20 Internationals (T20Is) are international matches between the ten Full Members of the ICC and also eight Associate Members who have been granted temporary T20I status by the ICC.

Cricket is played on a bounded oval-shaped field with a rectangular 22-yard long pitch at its center and wooden targets, known as wickets, at the two ends of the pitch—see Fig. P IV.1. A cricket match consists of one or two innings per team, made up of consecutive overs comprising six deliveries each. During an innings, one team is batting on the pitch, two players at

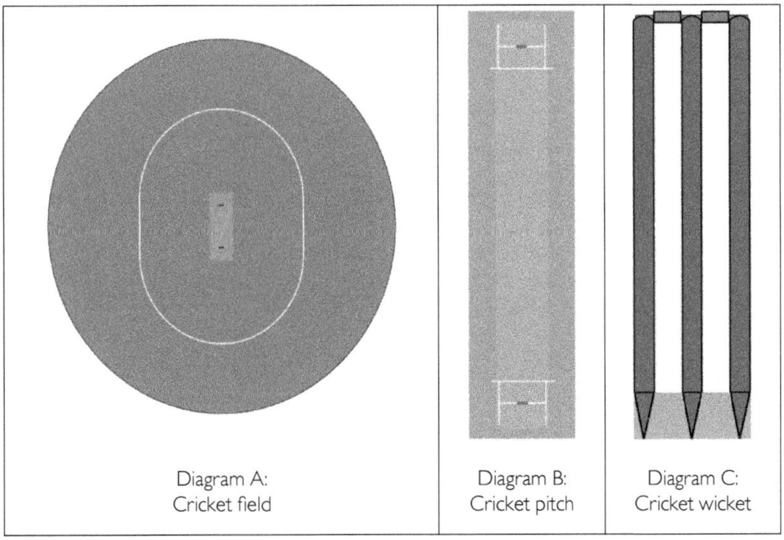

Figure P IV.1 Diagram A depicts a cricket field with the cricket pitch in the center. The cricket pitch is illustrated in more detail in Diagram B, with the wickets placed at the end of the pitch. As shown in Diagram C, each wicket consists of three stumps and two bails.

a time, while all 11 players of the opposing team play in the field. The batsmen of the batting team try to hit a cricket ball with their cricket bats to score runs. The cricket ball is delivered (in cricket terms referred to as bowled) to the batsmen by bowlers from the fielding team. The fielding team attempts to dismiss the batsmen (also referred to as taking their wickets or getting them out) in several different ways. An innings ends:

- when 10 of the 11 batsmen have been dismissed (leaving the last remaining batsman not out);
- when the batting team's captain declares the innings closed (not possible in limited overs cricket);
- when the batting team scores the required number of runs to win the match; or
- when the allocated time or overs for the match finishes.

In limited overs cricket, the team with the higher number of runs at the end of the match is the winner. Winning an unlimited overs match is not as straightforward, since, apart from scoring the most runs, the objective is to take all the opposing team's wickets. Consequently, the result of an unlimited overs match is often a draw, even though one team scored more runs than the other.

The development of limited overs cricket was motivated by the high percentage of draws in unlimited overs matches and the need for a more spectator-friendly format with decisive results. But, because cricket cannot be played in wet weather, allowance has to be made for rain interruptions in limited overs cricket. Initially very basic rain-rule methods were utilized for adjusting the number of runs required for victory (i.e., the target score). However, the farcical result in the 1992 Cricket World Cup semi-final between England and South Africa prompted the ICC to search for a more suitable rain-rule system.

The Duckworth–Lewis method, developed by Frank Duckworth and Tony Lewis, was the first proposed system published in peer-review literature (Duckworth and Lewis, 1998) and is an excellent example of how statistical methodology can be applied for problem solving in sport. In 1999, the ICC adopted it as the standard mechanism for resetting target scores in limited overs cricket. In the article by Duckworth (2004), the history behind the method is chronicled. The *Professional Edition* of the Duckworth–Lewis method, introduced by Duckworth and Lewis (2004), was used in all ODIs and T20Is from 2004, while the Duckworth–Lewis–Stern method, an updated version to account for modern scoring rates, is being used by the ICC since 1 October, 2014 (Stern, 2015). A number of alternative rain-rule systems has been presented in the literature, including the method by Preston and Thomas (2002). Whereas the purpose of the Duckworth–Lewis method is to maintain the difference between the teams through the margin of advantage, the method of Preston and Thomas (2002) aims to preserve the probability of victory.

Akin to baseball, another popular bat-and-ball game, the scoring system in cricket lends itself to the creation of a wealth of numerical information. Scorecards of all professional cricket matches are currently accessible via websites such as ESPNcricinfo (www.espncricinfo.com) and CricketArchive (www.cricketarchive.com). These websites also include records and statistics of past and current cricket players that can be utilized to construct basic performance ratios for the evaluation and comparison of players. Table P IV.1 presents as an example the international career records of Jacques Kallis from South Africa.

For batting, the table summarizes the number of matches played, the number of innings batted, the number of not-out innings (NO), the number of runs scored and balls faced, the highest score made (HS), and the number of centuries (100) and fifties (50) scored. The batting average (AV) is defined as the number of runs scored in all innings divided by the number of completed innings (Innings—NO). Only from the beginning of the 1990s has the number of balls faced by batsmen been captured in their batting records, enabling the calculation of the batting strike rate (SR) as the number of runs scored per 100 balls faced. Batsmen aim to maximize these two performance criteria by scoring as many runs as possible and losing their wickets as seldom as possible, relative to the number of balls faced.

The number of innings in which the player bowled, the number of balls bowled, runs conceded and wickets taken, the best bowling performance (most wickets taken in an innings for the fewest runs conceded) and the number of innings in which the bowler took five or more wickets (5), are typically tabulated for bowlers. There are three bowling performance criteria,

Table P IV.1 ***Test, ODI, and T20I career records of Jacques Kallis.***

Batting	Matches	Innings	NO	Runs	Balls	HS	100	50	AV	SR
Tests	166	280	40	13,289	28,903	224	45	58	55.37	45.97
ODIs	328	314	53	11,579	15,885	139	17	86	44.36	72.89
T20Is	25	23	4	666	558	73	0	5	35.05	119.35

Bowling	Matches	Innings	Balls	Runs	Wickets	Best	5	AV	ER	SR
Tests	166	272	20,232	9535	292	6–54	5	32.65	2.82	69.28
ODIs	328	283	10,750	8680	273	5–30	2	31.79	4.84	39.37
T20Is	25	19	276	333	12	4–15	0	27.75	7.23	23.00

namely the bowling average (AV), calculated by dividing the number of runs conceded by the number of wickets taken, the bowling economy rate (ER), defined as the number of runs conceded per over (i.e., six balls) bowled and the bowling strike rate (SR), given by the number of balls bowled divided by the number of wickets taken. Relative to the number of balls bowled, a bowler would ideally like to simultaneously maximize the number of wickets taken and minimize the number of runs conceded. Hence it follows that the better bowlers in cricket tend to have lower ratios for the three bowling measures.

Historically the batting and bowling averages have been the principal criteria used for rating and comparing batsmen and bowlers. The batting strike rate and the bowling economy rate are of importance in limited overs cricket where teams aim to score runs at a fast rate. Among the five standard performance criteria, a unique situation arises with the batting average in that it can be inflated due to the presence of a high proportion of not-out innings. Various solutions have been proposed to deal with the problem—see van Staden et al. (2010) and also the references therein for details.

Recently, van Staden (2009) suggested a third batting criterion, the number of balls faced in all innings divided by the number of completed innings, which he termed the survival rate. He furthermore extended a simple graphical representation by Kimber (1993) for evaluating cricketers' batting and bowling performances. With respect to batting, more advanced statistical analyses have been conducted using a nonparametric approach with the Kaplan–Meier estimator (Kimber and Hansford, 1993), statistical process control (Bracewell and Ruggiero, 2009), and Bayesian stochastic modeling (Koulis et al., 2014).

The previously mentioned batting and bowling criteria can be combined to create new performance measures. For instance, Sharp et al. (2011) defined batting and bowling indices which they then utilized for the selection of the optimal team in Twenty20 cricket. Team performance and rating is another important area of research in cricket. Allsopp and Clarke (2004) developed batting and bowling ratings for teams in Test and ODI cricket. By fitting a multinomial logistic regression model, they determined the factors affecting results in Test cricket, considering winning the coin toss, home-team advantage and the lead/deficit after the first innings of both teams. The optimum batting strategy in limited overs cricket has been studied by Preston and Thomas (2000) through a stochastic dynamic programming representation. Perera et al. (2014) analyzed match strategy in Test matches, developing and assessing declaration guidelines. Examples of recent articles using ball-by-ball commentary and scoring data from the ESPNcricinfo website (www.espncricinfo.com) include a study of the impact of fielding restrictions on run scoring in ODIs by Silva et al. (2015), and the development of a cricket simulator by Davis et al. (2015).

Finally, cricket data has been employed in innovative ways to demonstrate newly developed statistical methodologies. Cai et al. (2002) proposed a new estimation method for the hazard function in the context of survival analysis, which they used to estimate the hazard function of the runs scored by Steve Waugh of Australia in his Test cricket innings from December 1985 to August 1997. Eggar (2002) applied a test for the applicability of a first-order Markov chain model to the sequence of wins and draws in the first 269 Test matches between England and Australia. Avery (2002) used the same data set to illustrate how interconnected Markov chain models can be fitted to discrete state series with generalized log-linear modeling.

The articles referenced in this introduction represent a small sample of papers focusing on cricket. Nonetheless, the amount of published statistical research in cricket is small compared to other popular professional sports, especially baseball. A possible reason is that the various role-players within the cricket fraternity have not yet fully grasped the potential value of cricket analytics. The four selected articles in this volume present a diverse range of statistical methodologies and applications in cricket and will hopefully encourage further statistical studies in this fascinating sport.

REFERENCES

Allsopp, P.E. and Clarke, S.R. (2004). Rating Teams and Analysing Outcomes in One-Day and Test Cricket. *Journal of the Royal Statistical Society: Series A (Statistics in Society)*. 167(4). 657–667.

Avery, P.J. (2002). Fitting Interconnected Markov Chain Models—DNA Sequences and Test Cricket Matches. *Journal of the Royal Statistical* Society: Series D (The Statistician). 51(2). 267–278.

Bracewell, P.J. and Ruggiero, K. (2009). A Parametric Control Chart for Monitoring Individual Batting Performances in Cricket. *Journal of Quantitative Analysis in Sports*, 5(3). 1–19.

Cai, T., Hyndman, R.J., and Wand, M.P. (2002). Mixed Model-Based Hazard Estimation. *Journal of Computational and Graphical Statistics*, 11(4). 784–798.

Davis, J., Perera, H., and Swartz, T.B. (2015). A Simulator for Twenty20 Cricket. *Australian & New Zealand Journal of Statistics*, 57(1). 55–71.

Duckworth, F.C. (2004). A Fair Result in Foul Weather. *Significance*, 1(2). 94–96.

Duckworth, F.C. and Lewis, A.J. (1998). A Fair Method for Resetting the Target in Interrupted One-Day Cricket Matches. *Journal of the Operational Research Society*, 49(3). 220–227.

Duckworth, F.C. and Lewis, A.J. (2004). A Successful Operational Research Intervention in One-Day Cricket. *Journal of the Operational Research Society*, 55(7). 749–759.

Eastaway, R. (1992). *Cricket Explained*, St. Martin's Press, New York.

Eggar, M.H. (2002). Validity of Fitting a First-Order Markov Chain Model to Data. *Journal of the Royal Statistical Society: Series D (The Statistician)*. 51(2). 259–265.

Kimber, A. (1993). A Graphical Display for Comparing Bowlers in Cricket. *Teaching Statistics*, 15(3). 84–86.

Kimber, A.C. and Hansford, A.R. (1993). A Statistical Analysis of Batting in Cricket. *Journal of the Royal Statistical Society: Series A (Statistics in Society)*. 156(3). 443–455.

Knight, J. (2013). *Cricket for Dummies*, John Wiley and Sons, Ltd., Chichester.

Koulis, T., Muthukumarana, S., and Briercliffe, C.D. (2014). A Bayesian Stochastic Model for Batting Performance Evaluation in One-Day Cricket. *Journal of Quantitative Analysis in Sports*, 10(1). 1–13.

Perera, H., Gill, P.S., and Swartz, T.B. (2014). Declaration Guidelines in Test Cricket. *Journal of Quantitative Analysis in Sports*, 10(1). 15–26.

Preston, I. and Thomas, J. (2000). Batting Strategy in Limited Overs Cricket. *Journal of the Royal Statistical Society: Series D (The Statistician)*. 49(1). 95–106.

Preston, I. and Thomas, J. (2002). Rain Rules for Limited Overs Cricket and Probabilities of Victory. *Journal of the Royal Statistical Society: Series D (The Statistician)*. 51(2). 189–202.

Sharp, G.D., Brettenny, W.J., Gonsalves, J.W., Lourens, M., and Stretch, R.A. (2011). Integer Optimisation for the Selection of a Twenty20 Cricket Team. *Journal of the Operational Research Society*, 62(9). 1688–1694.

Silva, R.M., Manage, A.B.W., and Swartz, T.B. (2015). A Study of the Powerplay in One-Day Cricket. *European Journal of Operational Research*, 244(3). 931–938.

Stern, S.E. (2015). personal e-mail communication, 29 December, 2015.

van Staden, P.J. (2009). Comparison of Cricketers' Bowling and Batting Performances using Graphical Displays. *Current Science*, 96(6). 764–766.

van Staden, P.J., Meiring, A.T., Steyn, J.A., and Fabris-Rotelli, I.N. (2010). Meaningful Batting Averages in Cricket. *Proceedings of the 52nd Annual Conference of the South African Statistical Association for 2010*, 75–82.

CHAPTER 16

A fair result in foul weather

FRANK DUCKWORTH

The Duckworth–Lewis method for adjusting targets in rain-interrupted one-day cricket matches is used throughout the world, but, as **Frank Duckworth** explains, he originally came to statistics from physics via metallurgy, and only thought of applying statistical methods to sport when their power became apparent to him during his career in the nuclear energy industry.

As a schoolboy I was always good at maths, but come the end of my time in the sixth form people told me that maths at university was extremely disappointing; it was vague and didn't relate to reality; it was all horrible symbols and terribly complicated. So I drifted into physics at Liverpool University. And in 1961, having got a good Honours degree, the routine was to do a PhD.

The going rate at the time for a PhD in the Nuclear Physics Research Laboratory at Liverpool University was about five or six years, and I didn't really fancy that. But I learnt that you could move over into the metallurgy department; the boundary between solid-state physics and metallurgy was very thin and you could easily step from one to the other.

That was really a mistake because I wasn't cut out to be a metallurgist. It took me four years to scrape together some rather unreliable data from a badly conceived and hopelessly underfunded project, but I had just about enough to work with. And I think that's possibly where I first started to become a statistician, because I had to squeeze the most I could out of the data I had and work out some publishable conclusions. I got my PhD but I think I could say I got away with it! It was the worst piece of work I've ever done in my life and I like to think that the rest of my career has been making amends for it in some way.

A subject to make my own

In 1965 I took my first, and what turned out to be my only, permanent appointment, which was with the Central Electricity Generating Board (CEGB) at the Berkeley Nuclear Laboratories in Gloucestershire. I was ostensibly employed as a metallurgist, but it was known when they employed me that I was interested in mathematical methods. I found myself interpreting the data obtained from examining fuel elements that came out of our nuclear power stations using remote handling facilities. There were problems associated with these nuclear fuel elements and we had to try to find out why these were happening and how they might be solved. This involved numerical analysis of data. So during my first year at Berkeley I spent some considerable time in the library reading about least squares, regression and correlation—I had done no statistics whatsoever up to that point.

This image was originally posted to Flickr by nic_r at http://flickr.com/photos/37256969@N08/3739955294. Reproduced under a Creative Commons licence

Soon afterwards some operational problems started to arise with the fuel elements in a couple of our nuclear reactors; uncomfortably large numbers were leaking and had to be taken out prematurely. When we came to look at the manufacturing histories of those that had failed it turned out they had followed a particular pattern. So someone said: "Ah, this is a statistical problem; Frank Duckworth has done some statistics, get on to him", and from then on my career changed entirely. With about 400 graduates and PhDs in the whole of the CEGB research department, there was hardly anyone who had studied statistics, so here was a subject to make my own.

In 1966 the CEGB sent me on a six-week residential "Techniques of operational research" course at Lancaster University. Half of it was about statistics and that's where I learnt most of my standard statistics. I suppose that was where I started thinking about applying statistics to problems outside my area of application.

Statistics and sport

One field that had always interested me was sport. The first thing I did here was a statistical analysis of football league attendances and how they related to success of the home and away teams. In 1973 I sent my results, which were based on data from the 1971–1972 and 1972–1973 seasons, to Alan Hardacre, the Secretary of the Football League. But he just couldn't understand that anybody could tell him anything about his data that he didn't know himself. He marked my conclusions like a homework, without any comments! That's when I first realised the tremendous gulf that can exist between statistical analysis and the people who own the data.

I had always been interested in the problem of how to revise targets in rain-interrupted one-day cricket matches. Deciding the result according to who had the higher average run rate led to some very unfair conclusions. Each team actually starts with two resources: 50 overs and 10 wickets. The score a team can make depends on both these resources. If rain interrupts play and means that you now only have 25 overs, say, for your innings, the fact that you've still got 10 wickets means that you can afford to take greater risks and therefore score many more than half the runs you would have scored in 50 overs. It's a production process with two constraints; rain tightens one of those constraints but not the other. You have to make a fair correction, but the correction that had always been made was *pro rata* with overs and this led to some great injustices.

Around about 1988 I was in my garden listening to the commentary on an important one-day match, and there had been a rain interruption and a revised target had been set. This revised target was so easy that all the interest had gone out of the game. I distinctly remember the commentator, Christopher Martin Jenkins, saying that surely someone, somewhere, could come up with something better! Then he read out some letters he had received making various suggestions and I remember thinking that every one of these was unsatisfactory in some way or other. Nobody seemed to realise that it was a mathematical two-factor problem and it needed a mathematical solution.

Someone, somewhere

So right then I went indoors, wrote down a formula, played with it a little, turned it into a computer program and sent it off to Christopher Martin Jenkins. A few weeks later I got a letter back from the BBC, saying "Mr Martin Jenkins thanks you for your letter, which he has read with interest". So I thought that was the end of that, and I was reminded of my earlier problem in communicating my work on football attendances.

In the 1992 World Cup competition, the organisers adopted the "most productive overs method" for correcting targets. Unfortunately this method had not been properly thought through and its limitations were brutally exposed in the semi-final encounter between South Africa and England. What happened was that South Africa were responding to a good total set by England and had reached the point where they needed 22 runs to win with 13 balls to face and four wickets in hand. The general opinion was that it was a 50:50 situation. But a short shower of rain meant that two overs (12 balls) were lost, and the rule in place meant that the target was reduced by the runs scored in the cheapest two overs during England's innings. These amounted to just one run, so the South Africans had to return to the pitch with the totally impossible task of making 21 runs off just one ball! The match had been reduced from one of great excitement to total farce. The most productive overs method was virtually abandoned on the spot and the International Cricket Council sent out an appeal to the mathematicians of the world to come up with something better.

Some light relief

I had joined the Royal Statistical Society in 1975, because I had somehow become a statistician although I was not working in a statistical environment. At the first conference I attended,

I understood very little of the papers presented, but I met people who have been my friends ever since and we still enjoy our statistical exchanges. In 1992 I saw in the Society's newsletter, then called *News & Notes*, that they were organising an international conference in Sheffield, and they wanted suggestions for specialised topics from anyone prepared to organise them. I got in touch with the organisers and offered to arrange a session on "Statistics in sport", which they accepted, believing that it would provide a bit of light relief! I found people to present six papers, but two weeks before the conference one of the speakers had to go into hospital for a minor operation, and rather than leave the half-hour slot blank I decided that I would present something of my own in its place.

After the farcical World Cup result, I had written a short paper, which I called "A fair result in foul weather". But I hadn't known what to do with it, disheartened by my previous experiences which showed how difficult it was to interest sports administrators in mathematical concepts. But with the half-hour slot to fill I decided that I would present this paper. The parameters of the formula upon which my proposed method was based were plucked out of the air from my own "cricket intuition"—I had not done any analysis of real cricket data. In presenting it I suggested that a proper analysis of cricket data would make a useful research project.

And that proved to be vitally important because, a few weeks later, I received a letter from a mathematics lecturer at the University of the West of England in Bristol by the name of Tony Lewis. A colleague of his who had been at the conference had told him about my paper,

First appeared in the *Cape Times*. Used by permission.

First appeared in the *Cape Times*. Used by permission.

knowing that Tony himself was interested in the problem. Well, the project went ahead, Tony and I got together, and a couple of years later we felt that we had some proper parameters based on real cricket data to put into the formula.

In 1995 Tony wrote to Tim Lamb at the Test and County Cricket Board (TCCB; now the England and Wales Cricket Board) saying that he commended the formula which was based on an analysis of match data, and this did the trick; for in October of that year we were invited to go to Lord's Cricket Ground to make a formal presentation of our method to the TCCB.

One of the major problems was that our formula required a computer to make it work and we realised that the world of cricket was not ready for such sophistication. So in the weeks leading up to our presentation we worked together and succeeded in simplifying it so that the method could be invoked with nothing more than a single table of numbers and a pocket calculator. And by the time we came to make our presentation we had well and truly joined forces, and it was the "Duckworth–Lewis method" that we were offering.

A method is born

The TCCB were interested in the method, but what was perhaps even more important was that they had invited along David Richards, the Chief Executive of the International Cricket Council, which is the governing body of the sport worldwide. He was notably impressed and as a result invited us to make a formal presentation to the Chief Executives of all test playing nations in the pavilion at Lord's in July 1996. Following that the method took off.

It's used everywhere now, all round the world, but that doesn't mean there aren't people who think they can do better. We receive an average of four or five different methods to assess every year and have received many hundreds of e-mails suggesting ideas for improvement, every one of which we've answered—except the two abusive ones!

My part of the development of the method is not the best piece of work I've ever done—I consider that to be the privately published work I did on the operation of an island economy—but it has certainly been the most influential. Our names are known all over the world; we've given many radio and television interviews and made presentations in several countries, and as I write we are just off to Barbados to give a couple of seminars.

After South Africa, the host nation, were eliminated from last year's World Cup in the quarter-finals owing to their misreading of the Duckworth–Lewis charts, Tony and I were sent a copy of the cartoon reprinted above which had appeared in the *Cape Times*, a South African newspaper. With one ball to go, South Africa had thought that the batsman simply had to stay in, but in fact another run was needed. The batsman blocked the last ball and the team were jubilant. But the jubilation was short lived as it was quickly realised that they had only reached the score needed to tie the match, which wasn't good enough, and they went out of the tournament. It was not the fault of the Duckworth–Lewis method of course!

CHAPTER 17

Rating teams and analysing outcomes in one-day and test cricket

P. E. ALLSOPP AND STEPHEN R. CLARKE

Abstract

Multiple linear regression techniques are applied to determine the relative batting and bowling strengths and a common home advantage for teams playing both innings of international one-day cricket and the first innings of a test-match. It is established that in both forms of the game Australia and South Africa were rated substantially above the other teams. It is also shown that home teams generally enjoyed a significant advantage. Using the relative batting and bowling strengths of teams, together with parameters that are associated with common home advantage, winning the toss and the establishment of a first-innings lead, multinomial logistic regression techniques are applied to explore further how these factors critically affect outcomes of test-matches. It is established that in test cricket a team's first-innings batting and bowling strength, first-innings lead, batting order and home advantage are strong predictors of a winning match outcome. Contrary to popular opinion, it is found that the team batting second in a test enjoys a significant advantage. Notably, the relative superiority of teams during the fourth innings of a test-match, but not the third innings, is a strong predictor of a winning outcome. There is no evidence to suggest that teams generally gained a winning advantage as a result of winning the toss.

Keywords: *cricket, linear modelling, logistic regression, sports*

Introduction

Cricket has two distinct phases: the batting phase and the bowling phase, which provide an indication of a team's attacking and defensive abilities. Generally, the batting team strives to maximize its score by making as many runs as possible whereas the bowling team endeavours to restrict the score of the batting team. A match result can be construed as the combined effect of the batting and bowling abilities of the respective teams. In this paper we shall model the batting and bowling ratings for the International Cricket Council (ICC) one-day and test playing nations as well as gauge which factors associated with a team's first-innings performance in test cricket most critically affect outcomes of matches.

The analysis of the attacking and defensive abilities of sporting teams has been conducted in some sports such as rugby league and soccer but not cricket. We find this surprising given that the batting and bowling phases of cricket are clearly distinct. Lee (1999) used a bivariate negative binomial regression model to account for the attacking and defensive abilities of teams in rugby league and concluded that a team's ability to defend was more important than their offensive capabilities. Dixon and Coles (1997) also incorporated attacking and defensive parameters in a bivariate Poisson model when modelling scores in English soccer and found that the teams that are higher in the league table had the higher average attack and defence ratings.

One-day cricket is restricted to one innings per team, with each innings allocated a maximum of 50 overs. For completed matches the only outcomes are a win (or loss) or a tie. The team with the higher score is declared the winner. A tie results if both teams attain the same score after completing their innings. A drawn result only arises if weather or other interruptions do not allow a match to be completed to a stage where rain interruption rules are applied.

Test cricket is played between 10 of the ICC cricket playing nations and involves a maximum of two innings per team. A test-match can last up to 5 days. The possible outcomes are a win (or loss), a draw and very rarely a tie. A team wins by having a higher aggregate score after dismissing their opponents in the second innings. If both teams complete their second innings with the same aggregate score the match is a tie. Otherwise the match ends in a draw.

We have considered five seasons of results from 1997 to 2001, which include 248 one-day matches and 151 test-matches. The data were obtained via the Cricinfo archives at www.cricinfo.com. Bangladesh has not been included in the test cricket analysis because it has only recently been adopted as an ICC-sanctioned test playing nation and consequently has played in relatively few test-matches. Bangladesh, Kenya and Scotland have not been included in the one-day analysis because they also have played in relatively few matches. Note also that over this period a few matches were played on neutral grounds but these have been excluded from the study. Subsequently, in all matches there is a designated home team.

Modelling batting and bowling ratings in one-day and test cricket

In developing batting and bowling ratings for teams in one-day cricket we are faced with an inequity problem when the team batting second wins. The team batting second, unlike the first-innings team, does not expend its resource quota of 50 overs and 10 wickets when it wins a match. Consequently, their scores are truncated and are not a true measure of how well they have performed. To overcome this problem we apply techniques developed by Duckworth and Lewis (1998) and adapted by Allsopp and Clarke (2000) to estimate the projected score of the team batting second after it has exhausted its available resources. The Duckworth and Lewis method sets a revised target for the team batting second when overs in either innings have been lost because of a break in play. The target is revised in accordance with the available run scoring resources in the form of wickets and overs that the two teams have at their disposal at the time of the interruption. By adapting the Duckworth and Lewis method when the team batting second wins a match, we are in effect scaling up its score to allow for the number of further runs that it could make with its unused resources. Duckworth and Lewis have prepared detailed tables from which the resources that are available in the form of wickets and overs are expressed as a single resource percentage R. In addition we shall incorporate modifications that were proposed by de Silva et al. (2001) and use modified R-values expressed as $R_{mod} = (1.183-0.006R)R$. This adjustment accounts for the overestimation of predicted 50-over scores when the team batting second uses up very few resources and the underestimation of predicted scores when it uses nearly all its resources. If S_{actual} represents the truncated score at the point of victory, the subsequent projected score for the team batting second is estimated to be

$$S_{projected} = 100S_{actual}/(100 - R_{mod}).$$

In one-day cricket, if both teams played until all their resources were exhausted it would be expected that the distribution of the scores would be similar. Consequently, the distribution of the second-innings scores after the truncated scores have been replaced with the projected scores should also be similar to that of the first innings. For the first innings, the sample mean and standard deviation are 236.1 and 52.0 runs respectively. For the second innings, with the inclusion of the projected scores, the respective sample mean and standard deviation are 239.4 and 57.4 runs. A test for equal variances confirms that there is no significant evidence to suggest that the variances for the two innings are different ($F = 0.819$; p-value 0.116). Similarly, a two-sample t-test, with assumed equal variances, verifies that the two innings are not statistically different ($t = -0.69$; p-value 0.492). Consequently, there is no valid reason to reject the concept that the estimated second-innings scores are different from what would be expected if second-innings teams could continue batting until their available resources were exhausted.

To rate teams in test cricket we shall focus our attention on the first innings only, where we assume that teams attempt to optimize their performance to establish a substantial first-innings lead. This offers a reliable reflection of a team's batting and bowling abilities since a team's major focus is the establishment of a first-innings lead. Second-innings performances are not being considered because teams tend to be more measured in their approach and accordingly adapt their style of play as a strategic response to what has occurred in the first innings. Consequently, a team's second-innings batting and bowling performances are likely to lack the consistent approach that is more evident in a team's first innings.

A team's first-innings score in a test or an innings score in a one-day match that is played between the batting team i and the bowling team j on ground k is modelled as

$$s_{ijk} = A + a_i - b_j + h_{ik} + \varepsilon_{ijk} \tag{17.1}$$

where the indices $i, j, k = 1, \ldots, 9$ represent the nine ICC cricket playing nations. Clarke and Allsopp (2001) used similar modelling techniques to analyse the 1999 World Cup of cricket. However, whereas the response variable for the World Cup of cricket data was the margin of victory (or the final differential in runs between competing teams), for the analysis of a team's batting and bowling abilities the actual innings scores will be modelled. The response variable s_{ijk} signifies a team's score, the intercept A represents the expected score between average teams on a neutral ground and a_i and b_j signify the batting and bowling ratings of teams i and j respectively. When $k = i$, the home advantage parameter h_{ik} can be modelled as either a team's individual home advantage h_i or as a common home advantage h. More complicated home advantage models can also be considered. For example, since India, Pakistan and Sri Lanka play in very similar conditions it could be assumed that when these teams played each other neither team enjoyed a home advantage. ε_{ijk} is a zero-mean random error with constant variance. The error term is included because two competing teams will not necessarily repeat their first-innings performances the next time that they meet. Using a design matrix of indicator variables, a least squares regression model is fitted to the scores. For convenience, $\sum_{i=1}^{9} a_i = 900$ and $\sum_{j=1}^{9} b_j = 900$, which ensure that the average batting and bowling ratings are each 100. Accordingly, ratings above and below this figure provide evidence on how well a team has performed relative to the average performance.

Exploratory analysis of test cricket

Tables 17.1–17.4 provide a summary of the test-match results over the period of the study. Table 17.1 provides the overall percentage of wins, draws and losses. Australia and the West

Table 17.1 ***Summary of test cricket results for the period 1997–2001.***

Team	Matches played	% wins	% draws	% losses
Australia	42	69	14	17
South Africa	38	55	32	13
Sri Lanka	26	42	23	35
England	44	32	25	43
India	27	30	30	41
West Indies	40	28	15	58
Pakistan	30	27	27	47
New Zealand	30	27	37	37
Zimbabwe	25	12	32	56

Table 17.2 ***Summary of test cricket results for the period 1997–2001 after a first-innings lead had been established.***

Team	Matches where leading	% of matches leading	% wins after leading	% draws after leading	% losses after leading
Australia	38	90	76	5	19
South Africa	27	71	63	30	7
Sri Lanka	15	58	67	13	20
India	14	52	43	43	14
New Zealand	11	37	64	27	9
West Indies	13	33	62	15	23
Pakistan	13	33	54	23	23
England	14	32	43	50	7
Zimbabwe	6	24	17	33	50

Table 17.3 ***Comparison of home and away performance and batting order in test cricket for the period 1997–2001.***

	Home team wins	Home team draws	Home team loses	Total
Home team bats first	24	23	28	75
Home team bats second	46	15	15	76
Total	70	38	43	151

Table 17.4 ***Batting and bowling result summary for the first-innings of a test-match for the period 1997–2001.***

Team	Batting		Bowling		Differential (batting – bowling)
	Mean	Standard deviation	Mean	Standard deviation	
Australia	382	116	242	100	140
South Africa	374	116	259	119	115
Sri Lanka	316	125	297	122	19
New Zealand	319	120	329	158	−10
Pakistan	284	96	307	132	−23
India	297	180	333	91	−36
West Indies	237	84	294	108	−57
England	275	116	339	137	−64
Zimbabwe	234	107	391	143	−157

Indies have a much lower percentage of draws than the other teams, which may indicate a less defensive style of play. By contrast, New Zealand has a high percentage of draws. Table 17.2 provides the percentage of wins, draws and losses after a first-innings lead had been established. The consistently strong showing of Australia and South Africa has been well documented, which is clearly evident in the high proportion of matches in which they gain a first-innings lead. All teams except Zimbabwe showed a strong tendency to win after establishing a first-innings lead. However, only the top four ranked teams established a lead more than 50% of the time. The importance of a first-innings lead is demonstrated by South Africa and New Zealand, which have virtually identical records given a first-innings lead. However, South Africa established a lead almost twice as often as New Zealand. Australia's extremely low percentage of draws after leading contrasts with South Africa's and could indicate a propensity to go for wins even if it risks losing. A fast scoring rate in the first innings would also allow time for both teams to force a win. Of the other teams, Sri Lanka and the West Indies have a lower percentage of draws than losses after leading. By contrast India, New Zealand and England have a high proportion of draws compared with losses after a first-innings lead.

Table 17.3 allows a comparison of home and away performance in test cricket depending on batting order. The home team won 46% of matches and lost 28%, demonstrating a clear home advantage. However, the home side performed better when batting second. In fact the advantage in batting second was as strong as that enjoyed by the home side, and the home side lost more matches than it won when batting first. The team batting second won 46 + 28 = 74 or 49% of matches and lost only 26%. This is in stark contrast with the accepted wisdom of batting first if the captain wins the toss.

Table 17.4 provides a summary of the teams' first-innings batting and bowling results. The differentials in the last column provide a measure of the team's average first-innings lead. Table 17.4 demonstrates the dominance of Australia and South Africa, and the weakness of Zimbabwe, with nearly 100 runs separating them from the rest of the teams.

Although summary statistic tables such as these give an indication of relative team strengths, they also show effects such as home advantage, which in turn compromise their usefulness. The test calendar is not balanced—some teams play more series against stronger opponents, and in every series one team has a home advantage. To measure team strengths properly these effects need to be taken into account. We do this in the next section by modelling these effects in one-day cricket and the first innings of test cricket.

One-day cricket and test cricket first-innings results

In fitting model (17.1) to the one-day scores, which includes the replacement of the truncated second-innings scores with the projected scores (as outlined in the section on modelling batting and bowling ratings), the Anderson–Darling test for normality establishes that the normality assumption is not breached (p-value 0.177). The resultant parameter estimates for the batting and bowling ratings are provided in Table 17.5 and provide an easily interpretable measure of a team's relative strength in terms of runs. For example, Australia's batting rating of 116 suggests that it performs 16 runs better than expected against an average team, whereas Zimbabwe's bowling rating at 77 suggests that it generally performed 23 runs worse than expected when bowling, so their opponents score 23 runs more than average. Thus Australia batting against Zimbabwe would score 16 + 23 = 39 runs more than average. Similarly Zimbabwe batting against Australia's bowling would score 13 + 13 = 26 fewer runs than average. Hence we would expect Australia to defeat Zimbabwe on a neutral ground by 39 + 26 = 65 runs. This can also be obtained by subtracting the combined rating of Zimbabwe from that of Australia. The combined rating that is given in Table 17.5 is the sum of the batting and bowling rating, again adjusted to sum to average 100, and gives a measure of a team's overall strength.

The estimate for the common home advantage parameter is 14 runs (p-value 0.002), which suggests that the home team generally gained a significant advantage (in runs). This is consistent with de Silva and Swartz (1997), who showed that a significant common home advantage was evident in one-day cricket. Researchers such as Stefani and Clarke (1992), Clarke and Norman (1995) and Harville and Smith (1994) also considered an individual home advantage effect when modelling margins of victory. This seems reasonable in cricket, where the home advantage that is enjoyed by India might be expected to be different from that of (say) England. However, allowing for individual home advantages, by the addition of eight more parameters, does not provide a significant improvement over the common home advantage model. Various other models for home advantage were tried. For example, since India, Pakistan and Sri Lanka

Table 17.5 ***Team ratings for one-day cricket for the period 1997–2001.***

Team	Batting rating	Bowling rating	Combined rating
Australia	116	113	129
South Africa	106	119	125
Sri Lanka	111	106	117
Pakistan	100	101	101
India	114	86	100
England	82	112	94
New Zealand	97	88	85
West Indies	87	98	85
Zimbabwe	87	77	64

play under similar conditions, we might assume that there is no home advantage when these teams play each other. However, these models also failed to show any significant improvement on the common home advantage model.

When model (17.1) was fitted to the test cricket first-innings scores, the Anderson–Darling test for normality indicated that the residuals were not normally distributed (p-value 0.000). However, by transforming the response variable using a logarithmic transformation, the normality assumption is not breached (p-value 0.165). Model (17.2) represents the subsequent multiplicative model:

$$\ln(s_{ijk}) = A + a_i - b_j + h_{ik} + \varepsilon_{ijk} \tag{17.2}$$

where $\sum_{i=1}^{9} a_i = 0$ and $\sum_{j=1}^{9} b_j = 0$. Applying the inverse transformation to the resultant parameters then gives estimates for the batting and bowling rating whose product is 1. These parameter estimates multiplied by 100 are provided in Table 17.6 and give the expected

Table 17.6 ***Team ratings for test cricket teams for the period 1997–2001.***

Team	Batting rating	Bowling rating	Combined rating
Australia	128	114	146
South Africa	118	120	142
New Zealand	104	100	104
Sri Lanka	104	99	102
Pakistan	105	94	99
West Indies	86	111	96
England	92	99	91
India	96	89	85
Zimbabwe	77	80	62

percentage of the average score $A = 276$ when that team is batting or bowling against an average opponent. For example, Australia's batting rating of 128 suggests that during the first innings it generally performed 28% better than average, whereas Zimbabwe's bowling rating of only 80 implies that its opponents scored 100/80 = 125% of the average. Combining these results indicates that on a neutral ground against Zimbabwe we would expect Australia to score 128/80 = 160% of the average. This figure would be adjusted if batting on home soil. Similarly, Zimbabwe against Australia would score 77/114 = 67% of the average. Thus Australia would be expected to lead Zimbabwe on the first innings by a huge 160 – 67 = 97% of the average first-innings score, or 268 runs.

The common home advantage parameter was significant and estimated to be 112 (p-value 0.017), so the home team generally scored 12% more than expected on a neutral ground. This represents approximately 33 runs advantage. Again, there is no significant evidence to suggest that allowing for individual home advantages provides a significant improvement over the common home advantage model (p-value 0.967).

Although statistically not significant (p-value 0.167), when a parameter measuring the effect of batting first was included it was estimated at 0.93, which represents roughly 19 runs advantage to the team batting second.

The batting and bowling ratings can be combined multiplicatively to provide a measure of a team's overall strength, and they are also given in Table 17.6. Clearly, the results reinforce the outcomes that are summarized in Tables 17.1, 17.2 and 17.4, with Australia and South Africa rated substantially above average in both the batting and the bowling phases of test cricket. Notably, the West Indies has a significantly high bowling rating but is rated quite low with the bat. This is reflected in the fact that the West Indies won only 28% of its matches overall and could establish a first-innings lead only 33% of the time. This is an interesting result given that during the period of the study the West Indies had the game's premier batsman in Brian Lara playing for them. This reinforces the notion that the consistently successful teams tended to perform well in both batting and bowling. Since the chances of winning are increased after a first-innings lead has been established, solid first-innings performances in both batting and bowling tend to increase a team's chances of establishing a lead and subsequently increase its chances of winning. This is certainly the case with Australia and South Africa who won well over half their matches after establishing a first-innings lead. Comparisons with the one-day cricket ratings indicate that the rank order is preserved for the top two one-day teams. This suggests that the better-equipped teams tended to perform well in both forms of the game and underscores the recent dominance of Australia and South Africa in international cricket. Of the other teams, New Zealand appears to perform better at Test cricket whereas India performs poorer. As might be expected, the ratings for test cricket are more spread than those for one-day cricket, where restrictions that limit the balls bowled would even up results.

Modelling the factors affecting outcomes in test cricket

A multinomial logistic regression model is adopted to determine which factors associated with a team's first-innings performance most critically affect match outcomes. Using this approach we can examine, for example, whether the home or away team was more likely to win a test-match. A logistic regression model is employed because the response variable is categorical. The trichotomous response variable is the match outcome for the team batting first, which can be a win, draw or loss. It is also assumed that the response variable is ordinal because of the implicit order that is expressed in the match outcomes. The explanatory variables are the (signed) first-innings lead, a common home team advantage, the result of the toss and each team's batting superiority over the opposition's bowling (as measured by the team's batting rating divided by the opposition's bowling rating). If the cumulative probability of a win, draw and loss is denoted by γ_{ijw} for team i batting first in the second innings against team j, the outcome of a match is modelled as

$$\ln\left(\frac{\gamma_{ijw}}{1-\gamma_{ijw}}\right) = \beta_{0w} + \beta_1 h + \beta_2 x + \beta_3 \frac{a_i}{b_j} + \beta_4 \frac{a_j}{b_i} \qquad (17.3)$$

where $w = 0$ or $w = 1$ for the respective cumulative probability of a win and draw, $h = 1$ or $h = 0$ if the home advantage rests with the team batting first or second and x is the (signed) first-innings lead of team i. Note that the (a_i/b_j)-term is the relative superiority of the batting team over the bowling team in the third innings and a_j/b_i is the relative superiority of the batting team over the bowling team in the final innings.

In originally fitting model (17.3) to the match outcomes we included a term for winning the toss but this proved not only insignificant but also in the wrong direction—i.e. winning the toss reduced a team's chance of winning the match. For this reason we have excluded the toss parameter from the model. This is an interesting outcome given that the established orthodoxy contends that winning the toss provides a team with a significant advantage, particularly if the wicket appears to favour its strengths. Captains tend to bat first when winning the toss and thus to avoid batting last on a possibly deteriorating wicket.

Assuming that relatively superior teams endeavour to capitalize on their strengths as well as to exploit the weaknesses of their opponents we examined the combined effect of winning the toss and team strength and included an interaction term to see whether relatively weaker teams were more prone to defeat when losing the toss. We found no evidence to support this hypothesis. The inability of captains to capitalize on winning the toss in one-day cricket was previously demonstrated by de Silva and Swartz (1997) and Clarke and Allsopp (2001).

The parameter estimates for model (17.3) are provided in Table 17.7. Application of the Pearson and deviance goodness-of-fit tests produces χ^2-values of 288.2 (p-value 0.617) and 245.1 (p-value 0.986) respectively, indicating that model (17.3) provides an adequate fit of the

Table 17.7 *Parameter estimates associated with the first innings of a test-match for the period 1997–2001.*

Parameter	Term	Coefficient	p-value
β_{00}	Intercept (win)	0.362	0.869
β_{01}	Intercept (win and draw)	1.986	0.367
β_1	Home	0.9257	0.011
β_2	Lead	0.006286	0.000
β_3	a_i/b_j	0.321	0.788
β_4	a_j/b_i	−2.485	0.053

data. Clearly, the home team showed a strong tendency to win and the winning advantage that is gained by establishing a first-innings lead, though not surprising, is plainly evident. The latter point is highlighted in Table 17.2 where we find that most teams were inclined to win after establishing a first-innings lead. However, the model also produces a large advantage in batting second. A home team with bowling and batting ratings of 100, with no lead on the first innings over a similarly rated team, has about a 27% chance of winning if it bats first and 56% if it bats second. For the away team these reduce to 14% if it bats first and 35% if it bats second. This is clearly against current thinking but reinforces the trend that is shown in Table 17.3. There is also a clear indication that the probability of winning a test-match is highly influenced by the relative difference in strengths of the teams in the fourth innings but not the third innings. Although it was not surprising that the rating differential in the last innings is bordering on significant it was unexpected that the rating differential in the penultimate innings was insignificant. This suggests that teams are in the best position to force a win when they can exploit their superiority, whether that be batting or bowling, in the final innings. This may explain why the better teams are good at both bowling and batting—they are always in a position of strength in the final innings. It may also explain the low percentage of draws for the West Indies—its excellent bowling rating but weak batting rating means that either the West Indies or its opponents always have the upper hand in the final innings.

If we further characterize the overall strength of a team as the combined effect of its first-innings batting and bowling abilities we can restate model (17.3) as

$$\ln\left(\frac{\gamma_{ijw}}{1-\gamma_{ijw}}\right) = \beta_{0w} + \beta_1 h + \beta_2 x + \beta_3 \frac{c_i}{c_j} \tag{17.4}$$

where $c_i = a_i b_i$ are the combined batting and bowling ratings from Table 17.5 and c_i/c_j represents each team's combined batting and bowling superiority over that of its opposition. Table 17.8 provides the parameter estimates for the new model. Application of the Pearson and deviance goodness-of-fit tests produces χ^2-values of 284.9 (p-value 0.683) and 246.3 (p-value 0.985) respectively, indicating that model (17.4) provides an adequate fit of the data. In general this model results in conclusions and observations that are similar to those of the previous model with all parameters now significant. It is not surprising to find that the relative difference in the competing team's overall first-innings batting and bowling strength has been a contributing factor in the shaping of a win.

Table 17.8 ***Parameter estimates associated with the first innings of a test-match for the period 1997–2001.***

Parameter	Term	Coefficient	p-value
β_{00}	Intercept (win)	−3.1616	0.000
β_{01}	Intercept (win and draw)	−1.5437	0.022
β_2	Home	0.8575	0.017
β_3	Lead	0.006363	0.000
β_4	Combined rating differential	1.3051	0.019

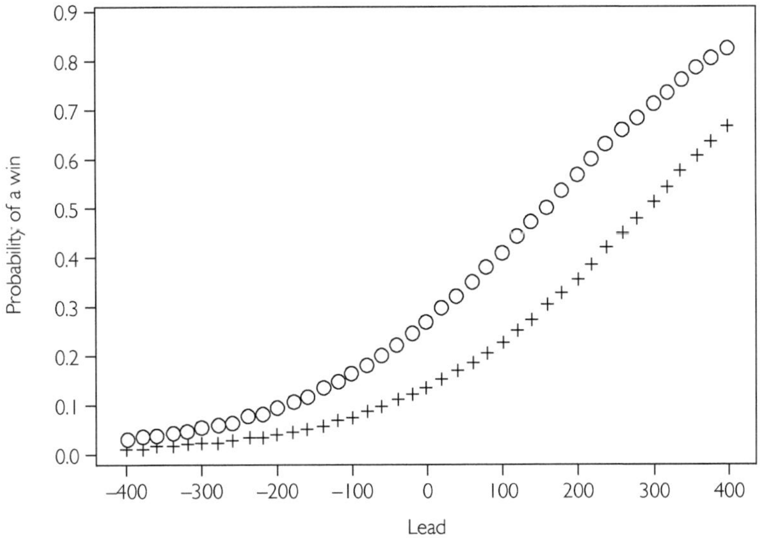

Figure 17.1 Comparison of the estimated winning probabilities of the team batting first after establishing a first-innings lead for the home (○) and away (+) teams in test cricket for the period 1997–2001.

To analyse the effect of establishing a first-innings lead, from model (17.4) we have

$$\Pr(w = 0 \mid h, x, c_i/c_j) = \frac{\exp(\beta_{00} + \beta_2 h + \beta_3 x + \beta_4 c_i/c_j)}{1 + \exp(\beta_{00} + \beta_{2h} + \beta_3 x + \beta_4 c_i/c_j)}. \tag{17.5}$$

Assuming that the team that bats first is the home team and competing teams are equally rated we have $h = 1$ and $c_i/c_j = 1$. If we set the probability of winning at 0.5 and solve equation (17.5) for the average lead that the home team needed to establish to have a better than even chance of winning we obtain 157 runs. Thus the home team needed a lead of 157 runs to have a 50% chance of winning. Conversely, if the team batting first was the away team so that $h = 0$ and $c_i/c_j = 1$, then we find that the away team needs a lead of 292 runs to have a 50% chance of winning. This confirms the clear advantage that the home team had over the away team. Figure 17.1 provides a comparison of the winning probabilities of the home and away teams for leads up to 400 runs. From Figure 17.1, if two equally rated teams ended on the same score after the completion of the first innings, so that $x = 0$, the point estimate probabilities of winning for the home and away teams were 0.27 and 0.14 respectively. However, in the section on 'One-day cricket and test cricket first-innings results' we showed that the home side establishes an expected lead of approximately 33 runs on the first innings. Such a lead produces winning chances for the home and away sides of 0.31 and 0.11.

Conclusions

In analysing each innings of one-day cricket we are confronted with a resource inequity problem when the second-innings team wins. This arises because the first-innings team has exhausted available resources in the form of overs or wickets whereas when the second-innings team wins it has unused resources available to it. Consequently, to enable informed comparisons to be made between the first and second innings a projected 50-over score can be calculated that estimates the score that the second-innings team could make if it had access to unused resources. One-day scores can then be modelled to establish a team's relative batting and bowling abilities and subsequently to establish batting and bowling ratings. Similar modelling techniques can be applied to determine the batting and bowling ratings for teams in the first innings of a test-match. However, whereas an additive model adequately fits the one-day scores, a multiplicative model is a more appropriate fit for test scores. Clearly, Australia and South Africa have been dominant in both one-day and test cricket, which is reflected in their relatively high ratings in both forms of the game. A team's home advantage has also played a part, with the home team, on average, gaining a significant advantage (in runs) in both one-day cricket and the first innings of a test cricket match.

When modelling outcomes of matches in test cricket, not surprisingly, a team's first-innings lead is a very strong predictor of a winning result. The relatively stronger teams such as Australia and South Africa were more likely to establish a first-innings lead and thus showed a stronger inclination to win. There was a significant home advantage in test cricket. This came in two parts. The home team could generally establish a significant runs advantage over its opposition in the first innings, and it has been shown that the first-innings lead is a strong predictor of a win. This has provided the home team with a distinct winning edge. However, the home team also had a higher probability of turning any given first-innings advantage into a win.

Given that the first-innings ratings provide a measure of a team's relative strength over the duration of a test-match we can use the ratings to measure the relative difference in strength between teams during the third and final innings. The resulting model confirmed the descriptive statistics that it was an advantage to bat second. This challenges the established orthodoxy of generally electing to bat first when winning the toss and is probably the main reason why winning the toss is not helpful. It was also established that a team's superiority in the final innings was much more likely to produce a winning result than if its superiority over its opposition is exposed in the third innings.

Although the results generally confirm currently held beliefs on the relative strength of teams in one-day and test cricket, in some aspects the results are surprising. The model outcomes allow confirmation and quantification of well-known effects such as home advantage and first-innings lead but also provide opportunities to challenge aspects of cricket which have become folkloric, such as the effects that are attributed to winning the toss and the preferred batting order.

REFERENCES

Allsopp, P. and Clarke, S. R. (2000) Methods for quantifying performances in one-day cricket. In *Proc. 5th Austr. Conf. Mathematics and Computers in Sport* (ed. G. Cohen and T. Langtry), pp. 28–36. Sydney: University of Technology Sydney.

Clarke, S. R. and Allsopp, P. (2001) Fair measures of performance: The World Cup of Cricket. *J. Oper. Res. Soc.*, **52**, 471–479.

Clarke, S. R. and Norman, J. M. (1995) Home ground advantage of individual clubs in English soccer. *Statistician*, **44**, 509–521.

Dixon, M. J. and Coles, S. G. (1997) Modelling association football scores and inefficiencies in the football betting market. *Appl. Statist.*, **46**, 265–280.

Duckworth, F. and Lewis, T. (1998) A fair method for resetting the target in interrupted one-day cricket matches. *J. Oper. Res. Soc.*, **49**, 220–227.

Harville, D. A. and Smith, M. H. (1994) The home-court advantage: how large is it, and does it vary from team to team? *Am. Statistn*, **48**, 22–28.

Lee, A. (1999) Modelling Rugby League data via bivariate negative binomial regression. *Austr. New Zeal. J. Statist.*, **41**, 141–152.

de Silva, B. M., Pond, G. R. and Swartz, T. B. (2001) Estimation of the magnitude of victory in one-day cricket. *Austr. New Zeal. J. Statist.*, **43**, 1369–1373.

de Silva, B. M. and Swartz, T. B. (1997) Winning the coin toss and the home team advantage in one-day international cricket matches. *New Zeal. Statist.*, **32**, 16–22.

Stefani, R. T. and Clarke, S. R. (1992) Predictions and home advantage for Australian rules football. *J. Appl. Statist.*, **19**, 251–261.

CHAPTER 18

Rain rules for limited overs cricket and probabilities of victory

IAN PRESTON AND JONATHAN THOMAS

Abstract
The paper discusses the properties of a rule for adjusting scores in limited overs cricket matches to preserve probabilities of victory across interruptions by rain. Such a rule is argued to be attractive on grounds of fairness, intelligibility and tactical neutrality. A comparison with other rules also offers a useful way of assessing the way in which the application of such rules will affect the fortunes of teams in rain-affected games. Simulations based on an estimated parameterization of hazards of dismissal and on numerical dynamic programming methods are used to compare a probability preserving rule with the now widely used Duckworth–Lewis method.

Keywords: *cricket, dynamic programming, rain rules*

Introduction

The game of cricket suffers from a requirement for clement weather of a type that is rarely guaranteed by the climates in which the game is played. Interruptions due to poor weather have proven a particular difficulty for designing rules for the limited overs version of the game, since such interruptions affect the fortunes of the two teams very differently depending on the point in the game at which they occur and, if prolonged and ignored, would destroy the contest and therefore much of the interest of the game.

Many rules have been suggested for adjusting the terms of the game in such a way as to maintain fairness between the two teams in cases of weather-affected matches. However, none have proven wholly satisfactory. For example, the 'average run rate' rule which scales down second-innings targets proportionately is widely felt to favour teams batting second whereas the 'most productive overs' rule which reduces second-innings targets only by runs scored in lowest scoring overs is equally widely seen as favouring teams batting first. The many methods, their strengths and their weaknesses are summarized by Duckworth and Lewis (1998, 1999), whose own rule has won attention and increasingly widespread adoption.

We begin with a description of the game. Limited overs cricket is played between two teams of 11 players. Each team has one 'innings' in which it 'bats' while the other team 'bowls'. At any point in the game, the team batting has two players at the centre of the pitch, known as the 'wicket', facing balls delivered to them by the bowling side. Balls are delivered in groups of 6 known as 'overs'. The batting team scores 'runs' by attempting to strike the ball and to run between the ends of the wicket while the bowling side recovers and returns the ball. In doing this the batsmen expose themselves to the risk of being dismissed ('getting out' or 'losing a wicket') in several ways. All major forms of dismissal are made more likely by a high run scoring rate and batting teams therefore need to balance the advantage of fast scoring against the risk of losing wickets.

If there is no intervention by the weather then the innings ends after a set number of overs have been bowled or all except one member of the batting team have been dismissed, whichever happens sooner. The total runs accumulated by the team batting first is then the target to be achieved by the team batting second if they are to win the match. However, the game can only be played in clement conditions. Breaks in play due to bad weather raise a problem if occurring after the game has started and lasting sufficiently long to require curtailment of the total playing time. Some fair way needs to be found of reducing the target if the overs that are available for the second innings are reduced and of inflating the target if those for the first innings are. The manner in which this is done needs to recognize that the harm done to a team's interests will differ depending on the point in the innings at which the break occurs.

The Duckworth–Lewis (DL) rule is based on the notion of correcting for a loss of 'run scoring resources'. In this it can be seen as a development of the idea underlying simpler rules based on run rates. We are sympathetic to the general notion of run scoring resources but there is more than one way to value the several dimensions to resources against each other. Duckworth and Lewis (1998, 1999) appear implicitly to interpret the value of run scoring resources as the 'average total score' given match circumstances. We explore an alternative approach in which they are valued according to their contribution to the probability of victory.

This approach has several advantages. Firstly, a rain rule which adjusts runs to preserve teams' advantage judged according to their own objectives appears to us to capture well the notion of fairness. Secondly, although the application of the rule would require complex calculations, the underlying concept is easily understood. Thirdly, rules such as this which recognize the team's own objectives are tactically neutral in the sense that they will give teams no incentive to vary their strategy in anticipation of rain. If, in all possible future circumstances, probabilities of victory after rain rule adjustment remain the same as before the interruption then teams expecting rain will have no reason to adjust their behaviour in anticipation of an application of the rule. The main disadvantage of the rule is probably the difficulty, explained below, in applying it consistently to cases where play must be abandoned.

The paper is organized as follows. The next section outlines theories of optimum batting strategy in the context of which we address the properties of rain rules in the section following that. These properties are compared with those of the DL rule in the subsequent section and numerical simulations are used to expand on these comparisons in the section 'Simulated examples' followed by a concluding section.

Optimum batting strategy

Clarke (1988, 1998) has used dynamic programming to establish optimal run scoring rates in the first and second innings of examples of limited overs games. Preston and Thomas (2000) extended this by deriving some general analytical results for problems of this type as well as estimating a model of actual strategies. We exploit their results to prove some general properties of fair rain rules.

We begin with a general formulation of the problem facing the batting team in either innings. There are T_1 balls in the first innings and T_2 in the second. In an uninterrupted game $T_1 = T_2$. Runs accumulated after the tth ball of the ith innings are denoted $R_{i,t}$. The objective of the team batting second is to maximize the probability that it accumulates the target set by the first team, $R_{2,t} > R_{1,T_1}$ for some $t \leq T_2$. Denote the value of that probability after t balls of the innings have expired, $R_{2,t}$ runs have been accumulated and k wickets have fallen by $F_{2,t}(R_{2,t}, k; R_{1,T_1}, T_2)$. (Arguments after the semicolon do not vary within innings in uninterrupted games and will be suppressed where convenient in what follows.) The objective of the team batting first is to maximize the probability that it sets a target which the second team fails to reach. Denote the value of that probability after t balls of the innings have expired, $R_{1,t}$ runs have been accumulated and k wickets have fallen by

$$F_{1,t}(R_{1,t}, k; T_1, T_2) = 1 - \mathcal{E}_{R_{1,T}} F_{2,0}(0, 0; R_{1,T_1}, T_2) \qquad (18.1)$$

where the $\mathcal{E}_{R_{1,T}}$-operator denotes expectation over the distribution of R_{1,T_1} given the chosen strategy.

We suppose that the control variable is the *expected* run rate $r_{i,t}$. The effect of run scoring rates on the probability of dismissal are captured in a dismissal hazard which depends also on the number of wickets fallen, $h_{i,t} = \eta_i(r_{i,t}, k)$. We assume that $\eta_i(\cdot)$ is positive, increasing and convex in its first argument. In the simplest case, there would be no uncertainty in run scoring other than that associated with the possibility of dismissal and actual runs achieved would equal $r_{i,t}$. More generally, we would wish to allow for the runs scored to depend uncertainly on the intended run rate regardless of dismissals so that the actual runs achieved would be a random variable with expectation $r_{i,t}$.

Let $r^*_{i,t}$ denote the optimal intended rate and $h^*_{i,t} = \eta_i(r^*_{i,t}, k)$. By Bellman's principle of optimality (see Bertsekas (1995), page 16 and following pages)

$$F_{i,t}(R_{i,t}, k) = (1 - h^*_{i,t+1})\mathcal{E}_{Ri,t+1} F_{i,t+1}(R_{i,t+1}, k) + h^*_{i,t+1} F_{i,t+1}(R_{i,t}, k + 1) \qquad (18.2)$$

where the $\mathcal{E}_{Ri,t+1}$-operator denotes expectation over the distribution of $R_{i,t+1}$ given no dismissals and expected run rate $r^*_{i,t+1}$.

Assuming for heuristic purposes that we can regard the scoring rate as a continuous variable and that first-order conditions characterize the optimum, then

$$0 = (1 - h^*_{i,t+1})\mathcal{E}_{Ri,t+1} \frac{\partial F_{i,t+1}(R_{i,t+1}, k)}{\partial R_{i,t+1}} - \frac{\partial h^*_{i,t+1}}{\partial r^*_{i,t+1}} \{\mathcal{E}_{Ri,t+1} F_{i,t+1}(R_{i,t+1}, k) - F_{i,t+1}(R_{i,t}, k + 1)\}.$$

The first term is the expected marginal gain from run accumulation whereas the second is the expected marginal loss from the increased likelihood of dismissal. Using equation (18.2) we can rewrite this as

$$0 = (1 - h^*_{i,t+1})\mathcal{E}_{Ri,t+1}\frac{\partial F_{i,t+1}(R_{i,t+1}, k)}{\partial R_{i,t+1}}$$
$$- \frac{\partial\{\ln(h^*_{i,t+1})\}}{\partial r^*_{i,t+1}}\{\mathcal{E}_{Ri,t+1}F_{i,t+1}(R_{i,t+1}, k) - F_{i,t}(R_{i,t}, k)\}.$$

Optimizing strategies can be shown to differ considerably between innings and to depend on the nature of the uncertainty in runs achieved. Preston and Thomas (2000) discussed several cases. Suppose that $k = 9$, so that all except one wicket have fallen. In the typical second-innings problem with no run scoring uncertainty, then the optimum strategy is to pursue a constant run rate equal to the required rate. For $k < 9$, there are good reasons to expect a higher run rate than this to be optimal. In contrast, in the first innings the optimum strategy is typically an increasing run rate. As the innings progresses the potential overs to be lost by early dismissal are declining and high scoring rates become more attractive. If there is uncertainty then simulations suggest that the run rate should increase modestly over the course of the second innings even for k = 9.

Probability preserving rain rules

Suppose that a rain interruption requires play to be shortened by δ balls. If this happens during the second innings, the batting team must resume with $T'_2 = T_2 - \delta$ balls available in the innings. If it happens during the first innings it is usual to reduce the number of balls available to both teams by $\delta/2$, if possible. In either case there is a need to adjust scores if one team is not to be seriously disadvantaged by the interruption. It is not regarded as acceptable to revise wickets lost and the usual practice is to revise the run target. It is clearly equivalent and expositionally more convenient to represent any adjustment as revising the total of runs accumulated to R'.

Define a probability preserving (PP) rain rule as one which adjusts run totals to maintain the probability of each team winning. Such a rule falls into the wider class of tactically neutral rain rules which are such that the probability of either team winning after an interruption is a monotonically increasing function of the probability before. It is only for rules within this class that the anticipation of rain will give batting teams no incentive to change their strategy.

Second-innings adjustments

Take firstly the case of the second innings. A PP rule decides R' by

$$\begin{aligned} F_{2,t}(R_{2,t}, k; R_{1,T_1}, T_2) &= F_{2,t}(R', k; R_{1,T_1}, T_2 - \delta) \\ &= F_{2,t+\delta}(R', k; R_{1,T_1}, T_2), \end{aligned} \tag{18.3}$$

i.e. R' is chosen so that the team has the same probability of winning after the interruption as it had before.

It is illuminating to consider adjustment in the case of the shortest possible interruption, $\delta = 1$, since this yields particularly readily interpretable expressions. If batsmen behave optimally according to the results described above, then it can be shown (see Appendix A.1) that

$$R' - R_{2,t} \simeq r^*_{2,t+1} - \frac{h^*_{2,t+1}(1 - h^*_{2,t+1})}{\partial h^*_{2,t+1}/\partial r^*_{2,t+1}}. \tag{18.4}$$

The appropriate rule therefore accords the team its optimal expected rate $r^*_{2,t+1}$ less an adjustment which is itself a function of $r^*_{2,t+1}$. Let us call this the adjusted optimal rate. The need for the adjustment arises to compensate for the certainty that the batting team survives the interruption without a loss of wickets.

First-innings adjustments

The case of the first innings is complicated slightly by the possibility that the total number of balls available may be modified in both innings. Suppose that the numbers of balls available in each innings are reduced by $\delta/2$. A PP rule decides R' by

$$\begin{aligned} F_{1,t}(R_{1,t}, k; T_1, T_2) &= F_{1,t}(R', k; T_1 - \delta/2, T_2 - \delta/2) \\ &= F_{1,t+\delta/2}(R', k; T_1, T_2 - \delta/2). \end{aligned} \tag{18.5}$$

Again R' is chosen so that the team has the same probability of winning after adjustment to the overs remaining for each team.

Suppose that the interruption lasts for two balls, $\delta = 2$, the shortest interruption leading to a reduction in both innings. Then by similar reasoning (outlined fully in Appendix A.2) it can be shown that

$$R' - R_{1,t} \simeq r^*_{1,t+1} - \frac{h^*_{1,t+1}(1 - h^*_{1,t+1})}{\partial h^*_{1,t+1}/\partial r^*_{1,t+1}} - \mathcal{E}_{R_{1,T}}\left\{ r^*_{2,1} - \frac{h^*_{2,1}(1 - h^*_{2,1})}{\partial h^*_{2,1}/\partial r^*_{2,1}} \right\}. \tag{18.6}$$

Hence the runs accorded will be close to the difference between the current adjusted optimal rate and the expected adjusted optimal rate at the start of the second innings.

Premature termination of match

All the discussion so far has assumed that the interruption allows the match to resume with overs still to be played. In such a case a PP rule is implementable with knowledge only of probability level sets in (R, t) space. However, there is a discontinuity in the application of the rule as the interruption reaches the scheduled end of the match since at this point a winner must be decided without the possibility of further play.

One natural extension of the rule to such cases would be to award victory in such a prematurely terminated match to the team which was most likely to win at the time of interruption, i.e. the team batting would be declared to have won if $F_{i,t}(R_{i,t}, k) > 0.5$. However, such a rule requires more information since it needs knowledge not only of the shape of probability level sets but also of the value of the probability. The shape of the probability level sets could in principle be independent, or close to independent, of the relative abilities of the two teams even when we would not expect this to be true of the actual value of the probabilities.

Also such an extension to the rule would not be strictly PP since the team ahead at the point of interruption would be declared the winner with certainty. As a consequence the rule would no longer be tactically neutral if there were any possibility of the match being prematurely terminated. As a referee has pointed out, for example, the second-innings strategy under anticipation of an imminent downpour terminating the match would become a matter of achieving targets over by over. Tactical neutrality could be achieved by an alternative extension randomizing the declared winner in accordance with the probabilities of victory before the interruption but such indeterminacy might not be a popular feature of a rule.

Comparison with Duckworth and Lewis rule

We start with a description of the DL rule (Duckworth and Lewis, 1998, 1999). The DL rule is not a PP rule and is not intended to be, as they made clear:

'The D/L method maintains the margin of advantage. It does not maintain the probability of winning or losing'

(Duckworth and Lewis (1999), page 27). In our opinion, each rule has things to be said in its favour. Comparing adjustments made by the DL and PP rules may none-the-less provide a worthwhile insight into how the application of the former will or will not alter probabilities of victory at different points in the match.

The DL rule can be represented as adjusting $R_{i,t}$ in line with the proportion of run scoring resources lost as a consequence of the interruption. To explain this we need to outline the notion of run scoring resources. In the interpretation offered by Duckworth and Lewis (1998), page 222, these are to be measured by the average total score in the remaining overs available given the number of wickets lost and in practice these are calculated from first-innings scores. Accordingly let $Z_{1,t}(k;\ T_1)$ denote the expected score in the remaining T_1—t balls of the first innings if k wickets have fallen.

The evolution of $Z_{1,t}(k)$ across the course of the innings is crucial to the application of the DL rule and it is therefore worth saying something about this. (In general, the behaviour depends on runs already scored so the expressions below should be thought of as involving averages across the distribution of $R_{1,t}$. We neglect this complication in the ensuing exposition.) Clearly,

$$Z_{1,t}(k) = (1 - h_{1,t+1})\{Z_{1,t+1}(k) + r_{1,t+1}\} + h_{1,t+1}Z_{1,t+1}(k+1) \tag{18.7}$$

and therefore, by rearrangement,

$$Z_{1,t}(k) - Z_{1,t+1}(k) = (1 - h_{1,t+1})r_{1,t+1} + h_{1,t+1}\{Z_{1,t+1}(k+1) - Z_{1,t+1}(k)\}.$$

The rate of decline of $Z_{1,t}(k)$ therefore depends on the typical run rate at that point in the first innings and on the likelihood and cost of dismissal.

Let us suppose that it is not too inaccurate to think of teams in the first innings seeking to maximize their expected final score. Clarke (1988), for example, is happy simply to assume this as the objective. Then

$$1 - h_{1,t+1} + \frac{\partial h_{1,t+1}}{\partial r_{1,t+1}}\{Z_{1,t+1}(k+1) + r_{1,t+1} - Z_{1,t+1}(k)\} = 0$$

and

$$Z_{1,t}(k) - Z_{1,t+1}(k) = r_{1,t+1} - \frac{h_{1,t+1}(1 - h_{1,t+1})}{\partial h_{1,t+1}/\partial r_{1,t+1}}, \tag{18.8}$$

so this decline is in fact exactly the adjusted optimal rate after t balls of the first innings and k wickets lost. This suggests that it may be possible to draw out close links between the nature of DL and PP rules.

Second innings

The DL formula for adjusting second-innings targets can be represented by

$$R' = R_{2,t} + \frac{R_{1,T_1}}{Z_{1,0}(0)} \{Z_{1,t}(k) - Z_{1,t+\delta}(k)\}. \tag{18.9}$$

If $R_{1,T_1} = Z_{1,0}(0)$, so that the target set is equal to the average score from an uninterrupted first innings, then the rule restores to the team affected the decline in resources $Z_{1,t}(k)$ across such an interruption for a first-innings team with k wickets lost. Otherwise they are accorded such a reward appropriately scaled.

We can note some differences from a PP rule. Firstly, suppose that the interruption is for one ball and take the case in which $R_{1,T_1} = Z_{1,0}(0)$. We note from the results above, equations (18.8) and (18.4), that the DL rule attributes something close to the adjusted optimal rate for a typical first-innings team whereas a PP rule would attribute the adjusted optimal rate for the team currently batting in the second innings. Since we know that optimal strategies differ in the two innings the two rules cannot make the same adjustments.

Secondly, if R_{1,T_1} and $Z_{1,0}(0)$ are not equal then the DL rule scales the adjustment by their ratio. The reason is that, at any point in the innings, the DL rule aims to revise the initial target according to the proportion of innings resources lost. The rule takes no account of runs already scored up to the point of the interruption and, as a consequence, a team could find that they had already achieved the revised target on resumption, i.e. $R' > R_{1,T_1}$ even if $\delta < T2—t$. Duckworth and Lewis (1999), page 27, recognized this explicitly:

> 'if sufficient overs are lost it may not even be necessary for them to resume their innings. This is not an inconsistency. It happens with every other method of target revision that has been used.'

This property alone makes it immediately plain that the rule cannot preserve probabilities of winning.

However, it would plainly be possible to modify the rule to avoid this property by having it revise the *remaining* target according to the proportion of *remaining* resources left:

$$R' = R_{2,t} + \frac{R_{1,T_1} - R_{2,t}}{Z_{1,t}(k)} \{Z_{1,t}(k) - Z_{1,t+\delta}(k)\}. \tag{18.10}$$

The option of providing targets by such a rule has been incorporated into the latest version of the DL computer program.

First innings

The DL rule for first-innings adjustments differs depending on which team suffers most from the interruption. We concentrate comments on the case holding 'in the great majority of instances' (Duckworth and Lewis (1998), page 223)—that in which the team batting first loses the most resources. In this case the rule stipulates

$$R' = R_{1,t} + Z_{1,t}(k) - Z_{1,t+\delta/2}(k) - Z_{1,0}(0) + Z_{1,\delta/2}(0).$$

The runs accorded are the difference between the resources lost by the batting team, $Z_{1,t}(k)$—$Z_{1,t+\delta/2}(k)$, and those that would be lost by a typical first-innings team suffering a similar interruption at the beginning of the innings, $Z_{1,\delta/2}(0)$—$Z_{1,0}(0)$.

Comparing this with equation (18.6) and noting equation (18.8) reveals differences and similarities. For a short interruption, the PP rule makes an adjustment based on the difference between the current adjusted optimal scoring rate and that expected to apply to the lost balls at the beginning of the subsequent innings. The DL rule adjustment, in contrast, reflects the difference between the same current adjusted optimal scoring rate and that applying at the beginning of the current innings.

Secondly, the averages used in the DL rule are not conditioned on the current run total $R_{1,t}$. This is probably of little significance as regards the optimal rate now and of none as regards that at the beginning of a typical first innings. However, it will matter to the optimal rate that is expected to apply at the beginning of the second innings and therefore to the PP rule. The consequences of this will depend on the state of play at the point of interruption. If the team batting first is doing well then the opposing team would need to aim at a higher run rate than otherwise and to the extent that the DL rule does not take account of this it appears that it should favour teams that are already performing well.

Simulated examples

Dynamic programming simulations allow us to investigate the magnitude and direction of differences between rain rule adjustments implied by DL and PP rules. The simulations are similar in character to those in Preston and Thomas (2000). We consider a 40-overs match and, for manageability, we adopt a specification with the over as the unit of decision. The hazard is assumed to be the same for all teams and to take the multiplicative form $\eta_i(r_{i,t}, k) = \beta r_{i,t}^{\alpha} \gamma(k)$ where $\gamma(k)$ is a step function taking the value 0.5 in the range $k < 6$, 1 in the range $5 < k < 8$ and 1.5 for $7 < k$. To allow for uncertainty in run scoring there is a probability p that the team fails to score in any over even if not dismissed. Thus each over has the possible outcomes that a wicket is lost, no wicket is lost and no runs are scored and that no wicket is lost and the intended run rate is scored. We have found the conclusions to be insensitive to the fineness of grid for run scoring rates and the simulations used here allow a choice over only integer values.

The choice of α, β and p is clearly crucial. Preston and Thomas (2000) estimated a partnership duration model using 133 matches from the 1996 Axa Equity and Law Sunday League without imposing optimal behaviour. Their estimates suggest a value for α of about 1.1. We have also explored the choice of parameter values by fitting simulated distributions of first-innings scores under optimizing behaviour to the distribution observed in the same data.

To be precise, we took a grid of 1430 combinations of values for α, β and p with $1.1 \leq \alpha \leq 1.9$, $-6.5 \leq \ln(\beta) \leq -2.25$ and $0 \leq p \leq 0.5$. For each combination we solved for the optimum strategy, simulated 10,000 games and took the fifth, 10th, 25th, 50th, 75th, 90th and 95th percentiles of the distribution of first-innings scores. A simple grid search, minimizing the equally weighted quadratic distance from the corresponding observed percentiles, gives values $\alpha = 1.6$, $\ln(\beta) = -4.5$ and $p = 0.2$. The closeness of fit is shown in Table 18.1. (More details are available on request from the authors.) We adopt these values for simulation.

The optimum batting strategies calculated for such parameter values are similar in properties to those discussed in Preston and Thomas (2000). These allow us to calculate the probability of victory given circumstances at any point in the two innings. Figure 18.1, for

Table 18.1 ***Actual and fitted percentiles for first-innings totals.***

Percentile	Actual value	Fitted value
5th	127	122
10th	139	140
25th	167	170
50th	201	202
75th	234	233
90th	265	262
95th	276	277

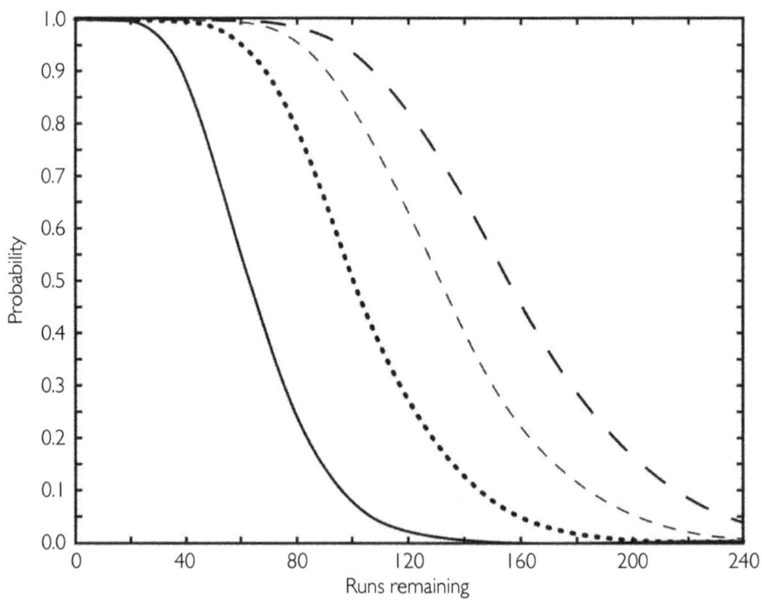

Figure 18.1 Second-innings probabilities of victory, three wickets lost:——, 10 overs left;——, 20 overs left;——, 30 overs left;——, 40 overs left.

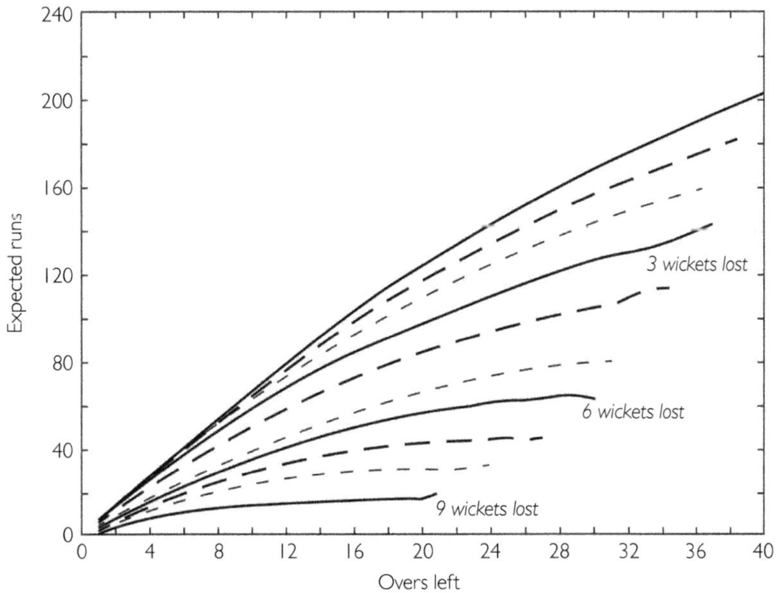

Figure 18.2 Simulated DL curves.

example, shows curves describing the second-innings probability of victory as a function of the outstanding run target, assuming three wickets lost, at four different points in the innings. These curves clearly contain all the information that is required for an application of a PP rule.

To compare with the DL rule we need to calculate expected run totals given wickets lost and balls remaining at all feasible points in the first innings. We do this by taking means across 100,000 simulated matches with teams applying the known optimizing strategies. The estimated DL curves are shown in Figure 18.2 and clearly have a sensible shape, similar to those estimated directly by Duckworth and Lewis (1998), page 222, from actual data. Note that we take the actual simulated averages without smoothing by fitting a parametric form as they did. The gaps in the curves illustrated are where the circumstances involved occurred insufficiently often (by which we mean fewer than 100 times) for accurate estimation through simulations.

These curves can then be used to calculate DL adjustments for comparison with those required by a PP rule. We have done this for many first- and second-innings situations and Figures 18.3 and 18.4 illustrate typical examples. In each case we consider an interruption after 15 overs with three wickets lost and in the second-innings case we assume that the initial target was the average set by first-innings teams. We then take three situations. In one the current scores are such that the two teams are equally likely to win whereas, in the other two, one or other team is well ahead so that the better placed team has odds of victory of 2:1.

Take for example Figure 18.3. The uppermost set of three curves corresponds to a case where 133 runs remain to be scored in 25 overs with seven wickets in hand. A team pursuing the optimum batting strategy from this point onwards would achieve the target with probability 0.33 given the parameter values used here. The three curves plot the adjustments called for under the three rules as overs left at resumption become fewer. The bottom set of curves illustrates adjustments for a case where only 105 runs needed initially to be scored and the probability of victory was correspondingly higher at 0.67.

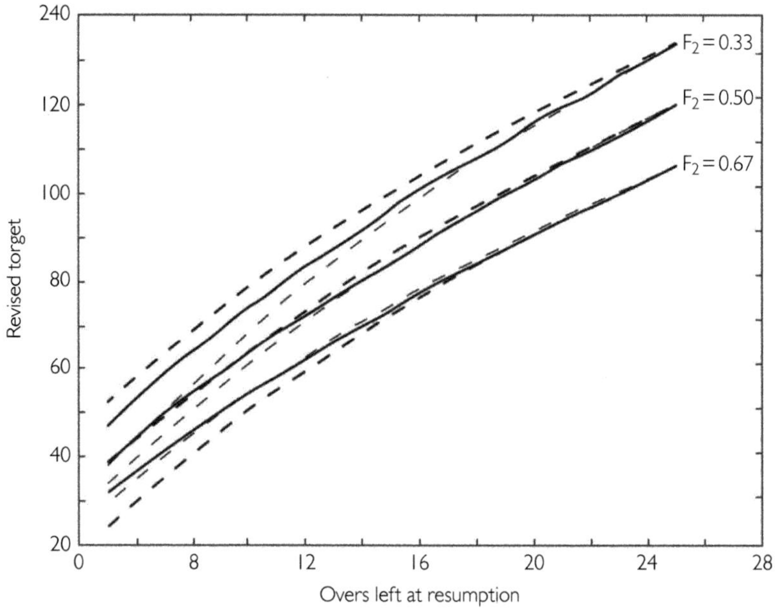

Figure 18.3 Second-innings rain rule comparisons: interruption after 15 overs and three wickets lost:——, PP rule;——, DL rule;——, modified DL rule.

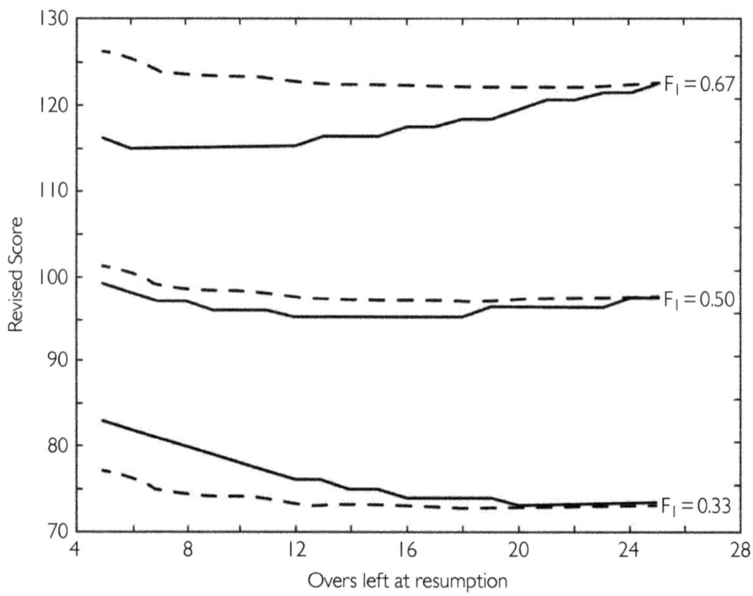

Figure 18.4 First-innings rain rule comparisons: interruption after 15 overs with three wickets lost:——, PP rule;——, DL rule.

For both first and second innings the pictures indicate compatible conclusions. In either innings the DL rule is more generous relative to a PP rule the better the batting team is performing. In each case illustrated batting teams already performing well in the innings have their probability of victory enhanced if rain intervenes and the DL rule is applied. In the first innings the DL rule benefits the bowling side only if the batting team is doing well and in the second innings only if the batting team is considerably behind.

The modified rule (18.10) applicable in the second innings is also illustrated. Interestingly the modification brings the DL rule closer to a PP rule for batting teams performing well but takes it further away for teams behind par who are treated noticeably more generously in comparison with a genuine PP rule.

An interesting question to ask is how probabilities of victory after application of the DL rule relate to probabilities before. Figure 18.5 plots these probabilities against each other for interruptions of the same length at the same point in the second innings, as runs required and wickets lost are varied. There are three distinct curves corresponding to the three different possibilities considered for the number of wickets lost. Probabilities of victory after application of the rule are not a simple function of probabilities before—for a given probability of victory before rain the probability at resumption is higher under the DL rule the fewer wickets are lost. This has interesting implications for the tactical non-neutrality of the rule. Since the rule does not preserve the ranking of situations by probability of victory, teams anticipating rain

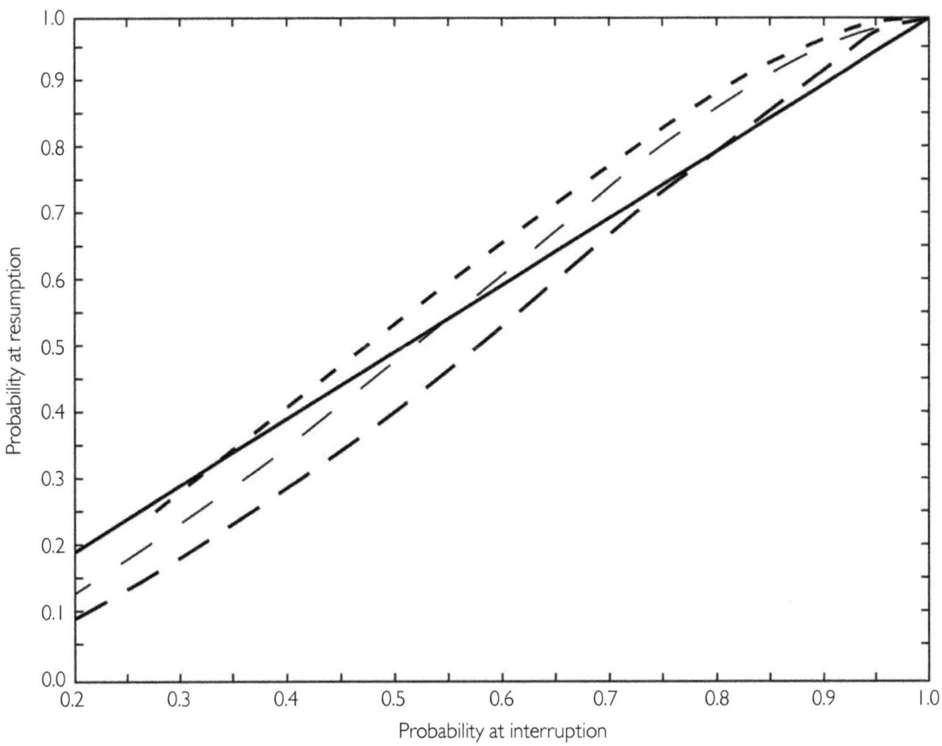

Figure 18.5 Probabilities of victory before and after interruptions: interruption of 10 overs duration after 15 overs of play:——, PP rule;——, two wickets lost;——, five wickets lost;——, eight wickets lost.

and seeking to maximize their probability of winning after anticipated application of the DL rule might behave differently from how they would if no rain were expected. The anticipation of rain can lead to changes of strategy as the team batting switches towards maximizing the probability after the anticipated application of the rule. In these simulations, for example, we can see that a team responding in this way to the DL rule would adopt a more cautious batting style than it would under a tactically neutral rule.

Practical application

Feasibility

The information that is needed to calculate the PP adjustments comprises a tabulation of the probabilities of victory $F_{1,t}(R_{1,t}, k; T_1, T_2)$ and $F_{2,t}(R_{2,t}, k; R_{1,T_1}, T_2)$. This is more extensive than the information that is required for current alternatives and could not, for example, be condensed into tables as brief as those provided by Duckworth and Lewis (1999). None-the-less the calculations could be programmed to be done quickly and simply on a computer that is capable of storing the required information and the limit on feasibility is therefore the accessibility to umpires of the computer resources required. (With advances in modern telecommunications this is likely to become decreasingly prohibitive.) If it were possible to approximate the probability functions accurately with a sufficiently parsimonious specification then computation would be rendered even simpler. That remains a topic for further research.

Examples

To facilitate a comparison with other methods and to demonstrate feasibility we provide a couple of examples. These both come from the first round of the 1992 World Cup and are both used as examples in Duckworth and Lewis (1998). Since these were 50-overs matches we repeated the above dynamic programming and simulation exercises using the same parameter values to produce estimates of the required functions for 50-overs matches.

The first example is the match between South Africa and Pakistan at Brisbane on March 8th, 1992. South Africa set a target of 211 runs in a full 50 overs. Pakistan had scored 74 runs for the loss of two wickets when rain intervened with 29 overs remaining. The match resumed with 14 overs lost and the most productive overs method in use for this tournament set a revised target of 193 which Pakistan failed to achieve. Duckworth and Lewis (1998) reported that their method would have set a lower target of 164, in line with the observation that the most productive overs method tends to penalize teams batting second. The methods of this paper provide an alternative set of estimates for the DL curves using a different set of data and a different estimation method and therefore could give an alternative DL target but in fact give exactly the same target of 164. The PP target calculated to keep the probability of victory constant across the interruption is very close at 161.

The second example is the match 4 days later between South Africa and England at Melbourne. South Africa again batted first, without interruption, and set a target of 236 runs. England had reached 63 without loss in 12 overs before rain robbed the match of nine overs. The revised target of 226 was reached by England. Duckworth and Lewis (1998) recommended a lower target of 207. The alternative DL target using our estimated parameters is quite close at 203. The PP target is 207.

Conclusion

Rain rules based on the idea of keeping probabilities of victory constant across interruptions offer an intelligible and fair alternative to other rules. Furthermore even if the case for their use is not regarded as persuasive it is still interesting to compare them with alternative rules to judge how the application of those other rules is likely to change the balance of advantage within a game. For example, simulations demonstrate that, under the widely used DL rule, probabilities of victory after an interruption are not in a one-to-one relationship with probabilities before, teams already doing well at the point of interruption are favoured and anticipation of rain may lead teams towards cautious batting.

ACKNOWLEDGEMENTS

We have benefited from discussions with or suggestions from Pelham Barton, Frank Duckworth, Jeremy Edwards, Charles Grant, Anthony Lewis, Hamish Low, Stephen Clarke, participants in seminars at University College London and the University of Birmingham and at meetings of the Royal Statistical Society and three referees. We are grateful to the Press Association and the England and Wales Cricket Board for access to data. All opinions remain our own.

Appendix A

A.1 Second-innings adjustments

From equation (18.3)

$$\begin{aligned} F_{2,t}(R_{2,t}, k; R_{1,T_1}, T_2) &= F_{2,t}(R', k; R_{1,T_1}, T_2 - \delta) \\ &= F_{2,t+\delta}(R', k; R_{1,T_1}, T_2). \end{aligned}$$

Using equation (18.2)

$$\begin{aligned} F_{2,t}(R_{2,t}, k) &= \mathcal{E}_{R_{2,t+1}} F_{2,t+1}(R_{2,t+1}, k) + h^*_{2,t+1} \left\{ F_{2,t+1}(R_{2,t}, k+1) - \mathcal{E}_{R_{2,t+1}} F_{2,t+1}(R_{2,t+1}, k) \right\} \\ &= \mathcal{E}_{R_{2,t+1}} F_{2,t+1}(R_{2,t+1}, k) - \frac{h^*_{2,t+1}(1 - h^*_{2,t+1})}{\partial h^*_{2,t+1} / \partial r^*_{2,t+1}} \mathcal{E}_{R_{2,t+1}} \frac{\partial F_{2,t+1}(R_{2,t+1}, k)}{\partial R_{2,t+1}}. \end{aligned} \qquad (18.11)$$

By averaging across Taylor series expansions around $R' = R_{2,t+1}$

$$F_{2,t+1}(R', k) \simeq \mathcal{E}_{R_{2,t+1}} F_{2,t+1}(R_{2,t+1}, k) + \mathcal{E}_{R_{2,t+1}} \frac{\partial F_{2,t+1}(R_{2,t+1}, k)}{\partial R_{2,t+1}} (R' - R_{2,t+1}). \qquad (18.12)$$

Equating the two expressions we derive equation (18.4),

$$0 \simeq \mathcal{E}_{R_{2,t+1}} \frac{\partial F_{2,t+1}(R_{2,t+1}, k)}{\partial R_{2,t+1}} \left\{ R' - R_{2,t+1} + \frac{h^*_{2,t+1}(1 - h^*_{2,t+1})}{\partial h^*_{2,t+1}/\partial r^*_{2,t+1}} \right\}$$

$$\Rightarrow R' - R_{2,t} \simeq r^*_{2,t+1} - \frac{h^*_{2,t+1}(1 - h^*_{2,t+1})}{\partial h^*_{2,t+1}/\partial r^*_{2,t+1}},$$

assuming that $\partial F_{2,t+1}(R_{2,t+1}, k)/\partial R_{2,t+1}$ is close to constant across the range of $R_{2,t+1}$ in question.

A.2 First-innings adjustments

From equation (18.5)

$$\begin{aligned} F_{1,t}(R_{1,t}, k; T_1, T_2) &= F_{1,t}(R', k; T_1 - \delta/2, T_2 - \delta/2) \\ &= F_{1,t+\delta/2}(R', k; T_1, T_2 - \delta/2). \end{aligned}$$

We expand both sides. By similar reasoning to that used in equation (18.11)

$$F_{1,t}(R_{1,t}, k) = \mathcal{E}_{R_{1,t+1}} F_{1,t+1}(R_{1,t+1}, k) - \frac{h^*_{1,t+1}(1 - h^*_{1,t+1})}{\partial h^*_{1,t+1}/\partial r^*_{1,t+1}} \mathcal{E}_{R_{1,t+1}} \frac{\partial F_{1,t+1}(R_{1,t+1}, k)}{\partial R_{1,t+1}}. \qquad (18.13)$$

On the right-hand side, we use the fact that

$$\begin{aligned} F_{1,t+1}(R', k; T_1, T_2 - 1) - F_{1,t+1}(R', k; T_1, T_2) &= \mathcal{E}_{R_{1,T}} \left\{ F_{2,0}(0, 0; R_{1,T_1}, T_2) - F_{2,1}(0, 0; R_{1,T_1}, T_2) \right\} \\ &= \mathcal{E}_{R_{1,T}} \left\{ \mathcal{E}_{R_{2,1}} F_{2,1}(R_{2,1}, 0) - F_{2,1}(0, 0) \right\} \\ &\quad + \mathcal{E}_{R_{1,T}} h^*_{2,1} \left\{ F_{2,1}(0, 1) - \mathcal{E}_{R_{2,1}} F_{2,1}(R_{2,1}, 0) \right\} \\ &\simeq \mathcal{E}_{R_{1,T}} \mathcal{E}_{R_{2,1}} \frac{\partial F_{2,1}(R_{2,1}, 0)}{\partial R_{2,1}} \left\{ r^*_{2,1} - \frac{h^*_{2,1}(1 - h^*_{2,1})}{\partial h^*_{2,1}/\partial r^*_{2,1}} \right\}, \end{aligned} \qquad (18.14)$$

where the arguments after the semicolon in the terms of the final two lines have been suppressed. The derivation of the final line from the previous line follows from a Taylor series expansion of the first expectation around $R_{2,1} = 0$ and use of the first-order condition to substitute in the second. The expectation operator $\mathcal{E}_{R_{1,T}}$ must here be read as taking expectations over the distribution of R_{1,T_1} conditional on $R_{1,t+1} = R'$ and we assume reasonably that we can ignore any difference made by the anticipated curtailing of the second-innings duration.

Combining these and using an average of Taylor series expansions of $F_{1,t+1}$ $(R', k; T_1, T_2)$ around $R' = R_{1,t+1}$ as in the derivation of equation (18.12), we derive

$$\begin{aligned} 0 \simeq \mathcal{E}_{R_{1,t+1}} & \frac{\partial F_{1,t+1}(R_{1,t+1}, k)}{\partial R_{1,t+1}} \left\{ R' - R_{1,t+1} + \frac{h^*_{1,t+1}(1 - h^*_{1,t+1})}{\partial h^*_{1,t+1}/\partial r^*_{2,t+1}} \right\} \\ + \mathcal{E}_{R_{1,T}} & \frac{\partial F_{2,1}(R_{2,1}, 0)}{\partial R_{2,1}} \left\{ r'_{2,1} - \frac{h^*_{2,1}(1 - h^*_{2,1})}{\partial h^*_{2,1}/\partial r^*_{2,1}} \right\}. \end{aligned}$$

Thus, to the extent that

$$\frac{\partial F_{1,t+1}(R_{1,t+1}, k)}{\partial R_{1,t+1}} \simeq \mathcal{E}_{R_{1,T}} \frac{\partial F_{2,1}(R_{2,1}, 0)}{\partial R_{2,1}}$$

across the relevant ranges of $R_{1,t+1}$ and $R_{2,1}$, as seems reasonable given equation (18.1), we can derive equation (18.6),

$$R' - R_{1,t} \simeq r^*_{1,t+1} - \frac{h^*_{1,t+1}(1 - h^*_{1,t+1})}{\partial h^*_{1,t+1}/\partial r^*_{1,t+1}} - \mathcal{E}_{R_{1,T}} \left\{ r^*_{2,1} - \frac{h^*_{2,1}(1 - h^*_{2,1})}{\partial h^*_{2,1}/\partial r^*_{2,1}} \right\}.$$

REFERENCES

Bertsekas, D. (1995) *Dynamic Programming and Optimal Control.* Belmont: Athena Scientific.

Clarke, S. R. (1988) Dynamic programming in one-day cricket—optimal scoring rates. *J. Ops Res. Soc.*, **39**, 331–337.

—— (1998) Test statistics. In *Statistics in Sport* (ed. J. Bennett), pp. 83–103. London: Arnold.

Duckworth, F. C. and Lewis, A. J. (1998) A fair method of resetting the target in interrupted one-day cricket matches. *J Ops Res. Soc.*, **49**, 220–227.

—— (1999) *Your Comprehensive Guide to the Duckworth/Lewis Method for Resetting Targets in One-day Cricket.* Bristol: University of the West of England.

→Preston, I. and Thomas, J. (2000) Batting strategy in limited overs cricket. *Statistician*, **49**, 95–106.

CHAPTER 19

Focus on sport

Batting strategy in limited overs cricket

IAN PRESTON AND JONATHAN THOMAS

Abstract

The paper attempts to understand batting strategies that are employed in limited overs cricket games. The question of the optimum batting strategy is posed in a simplified dynamic programming representation. We demonstrate that optimum strategies may be expected to differ fundamentally in the first and second innings, typically involving an increasing run rate when setting a target but a run rate which may decline over the course of an innings when chasing one. Data on English county level limited overs games are used to estimate a model of actual batting behaviour. The statistical framework takes the form of an interesting variant on conventional survival analysis models.

Keywords: *batting strategy, cricket, dynamic programming, survival analysis*

Introduction

Cricket is a game with many interesting strategic aspects. This paper attempts to understand batting strategies that are employed in limited overs cricket games. We investigate both those strategies that are most likely to lead to victory and those that are actually played. There are several reasons why these issues are of interest. An understanding of optimum strategies is of value in an obvious practical sense to those who play the game and want to win. Strategies played also determine the way in which probabilities of victory evolve over the course of games. An understanding of this may help in judging how teams stand at any point in a game and may therefore also be relevant to the design of rules for adjusting scores in rain-interrupted games. In principle, it could also be useful in the evaluation of the performance of players. Our results suggest that the optimum strategy depends on the way in which the likelihood of dismissal depends on the attempted scoring rate. Our empirical analysis of actual batting behaviour allows us to estimate the parameters of this relationship.

In this paper we consider the question of the optimum batting strategy in a simplified dynamic programming representation of a limited overs cricket game. Clarke (1988) has previously investigated optimum strategy in limited overs cricket by using dynamic programming methods and has simulated solutions numerically. We extend this by imposing a greater theoretical structure on the problem. This allows us to derive conclusions which can be generalized beyond his particular numerical specifications and which hopefully illuminate the source of his results. For example, his conclusion that the popular strategy of building up runs slowly has little justification, at least in the second innings, is mirrored in our findings. We suggest reasons, however, why such a strategy may have more justification in the first innings.

We also use data on English county level limited overs games to estimate a model of cricketers' behaviour. Bairam and Howells (1991a, b) and Bairam et al. (1990a, b) have investigated the empirical relationship between measures of performance, including batting and bowling inputs, and both probabilities of victory and the accumulation of points. This is a rather different exercise, however, from investigating the aspects of within-match strategy considered here. Our statistical framework takes the form of an interesting variant on conventional survival analysis models. Kimber and Hansford (1993) have previously applied survival analysis to data on individual innings. Our work goes further in its attempt to relate estimates to the effects of covariates on hazards within the context of a structural model of batting strategy.

Duckworth and Lewis (1998) have developed and estimated a target reduction rule based on ideas of neutrality between teams and balance in run scoring resources. This rule, adopted by several cricketing authorities, is undoubtedly a valuable improvement on crude proportional reduction rules and we hope that the sort of analysis in this paper can illuminate its possible foundations in the context of a model of cricketing behaviour.

The paper is structured as follows. The next section briefly outlines the main features of the game and considers optimal strategies. Several unresolved issues that are investigated through numerical dynamic programming and simulation are reported in the section 'Simulations'. An empirical specification is described in 'Empirical specification: the second innings' and the section that follows presents the results. The last section gives a summary and discusses directions for future research.

Batting strategy as a stochastic dynamic programming problem

Limited overs cricket is a game between two teams of 11 players. There are two 'innings' with each team 'batting' once and 'bowling' once. At any point in the game, the team batting has two players at the centre of the pitch, known as the 'wicket', facing balls delivered to them by the bowling side. Balls are delivered in groups of 6 known as 'overs'. The batting team scores 'runs' by attempting to strike the ball and to run between the ends of the wicket while the bowling side recovers and returns the ball. In doing this the batsmen expose themselves to the risk of being dismissed ('getting out' or 'losing a wicket') in several ways. We assume that all major forms of dismissal are made more likely by a higher run scoring rate. This seems reasonable if batsmen have any control over their scoring rate since otherwise it would be difficult to see why batsmen should show any inhibition in their rate of scoring. The innings ends after a set number of overs have been bowled or all but one of the members of the batting team have been dismissed, whichever happens sooner. The total runs accumulated by the team batting first is then the target to be achieved by the team batting second if they are to win the match. In some versions of the game, though none of the games analysed in the empirical section below, there are restrictions on the fielding side during certain overs of the innings, designed to encourage greater adventurousness by the batting side. The '15-overs rule' is an example of this. More detailed outlines of the game can be found in many books, such as Marylebone Cricket Club (1998) or Morrison (1998), and on several World Wide Web sites, such as the Cricinfo site at http://www.cricket.org.

Several questions of strategy immediately arise. How fast should a batting team aim to score its runs at any point in either innings? How should the increased probability of dismissal be balanced against the value of accumulating more runs? First we consider these issues in the context of the second innings where the batting team is chasing a well-defined given target. The more complicated case of strategy in the first innings is then addressed.

The second-innings problem

The issue can be formally addressed within the context of a stochastic dynamic programming problem and it is natural to start with the second innings. Suppose that the team batting second must score a target of $\bar{R}$ runs in no more than $\bar{T}$ balls. Consider the kth batting partnership facing the tth ball of the innings with R_{t-1} runs having been accumulated. We suppose that the control variable is the *intended* run rate r_t^*. The effect of run scoring rates on the probability of dismissal is captured in a dismissal hazard $h_t = \eta_{k,t}(r_t^*)$. We assume $\eta_{k,t}(\cdot)$ to be positive, increasing and convex, which assures a well-behaved solution to the problem. This is clearly a simplification. The hazard may depend, for instance, on the number of balls since the beginning of the partnership as well as the number of balls since the beginning of the innings and the relationship between the two depends on the uncertain timing of dismissals. To capture this in the notation would be extremely cumbersome and we hope that nothing is lost in ignoring it.

Actual runs achieved if not dismissed are r_t. In the simplest case, there is no uncertainty in run scoring other than that associated with the possibility of dismissal and hence $r_t = r_t^*$. More generally, if we wish to allow for the runs scored to depend uncertainly on the intended run rate even if not dismissed then r_t can be a random variable, the distribution of which depends on r_t^*. The probability of eventually achieving the target, which is in this context the value function, is denoted $V_{k,t} = \Omega_{k,t}(R_{t-1})$.

By Bellman's equation (see, for instance, Bertsekas 1987)

$$V_{k,t} = \max_{r_t^*}[\{1 - \eta_{k,t}(r_t^*)\}E\{\Omega_{k,t+1}(R_t)\} + \eta_{k,t}(r_t^*)\Omega_{k+1,t+1}(R_{t-1})]. \tag{19.1}$$

Assuming for heuristic purposes that we can regard the scoring rate as a continuous variable and that first-order conditions characterize the optimum, then

$$0 = \{1 - \eta_{k,t}(r_t^*)\}\frac{\partial}{\partial r_t^*}E\{\Omega_{k,t+1}(R_t)\} - \eta'_{k,t}(r_t^*)[E\{\Omega_{k,t+1}(R_t)\} - \Omega_{k+1,t+1}(R_{t-1})]. \tag{19.2}$$

The first term is the expected marginal gain from run accumulation whereas the second is the expected marginal loss from the increased likelihood of dismissal. We can reformulate this in the form

$$\psi_t \equiv \frac{\partial}{\partial r_t^*}\ln\{1-\eta_{k,t}(r_t^*)\} = \frac{\partial E\{\Omega_{k,t+1}(R_t)\}/\partial r_t^*}{E\{\Omega_{k,t+1}(R_t)\}-\Omega_{k+1,t+1}(R_{t-1})},$$

defining thereby ψ_t which can be seen to be increasing in the chosen run rate r_t^* and which might therefore be interpreted as an indicator of optimal lack of caution.

The case with no uncertainty is the easiest to analyse. For conciseness of exposition, let $V'_{k,t}$ denote the value of the derivative $\Omega'_{k,t}(R_{t-1})$ and let $V_{k+1,t}$ and $V'_{k+1,t}$ denote $\Omega_{k+1,t}(R_{t-2})$ and $\Omega'_{k+1,t}(R_{t-2})$, respectively, in each case as R_t evolves along the optimal path of run rates for the kth partnership. Then we can use equation (19.1) to show that

$$\begin{aligned}\psi_t &= \frac{V'_{k,t+1}}{V_{k,t+1}-V_{k+1,t+1}}\\ &= \frac{(1-h_{t+1})V'_{k,t+2}+h_{t+1}V'_{k+1,t+2}}{(1-h_{t+1})V_{k,t+2}+h_{t+1}V_{k+1,t+2}-V_{k+1,t+1}}\\ &= \psi_{t+1}\beta_{t+1}+\frac{h_{t+1}V'_{k+1,t+2}}{\Delta V_{k+1,t+2}}(1-\beta_{t+1})\end{aligned}$$

where

$$\beta_{t+1} \equiv \frac{(1-h_{t+1})\Delta V_{k,t+2}}{(1-h_{t+1})\Delta V_{k,t+2}+h_{t+1}\Delta V_{k+1,t+2}}.$$

Thus

$$\Delta\psi_{t+1} = \left\{\psi_{t+1}-\frac{h_{t+1}V'_{k+1,t+2}}{\Delta V_{k+1,t+2}}\right\}(1-\beta_{t+1}). \tag{19.3}$$

Now consider the last partnership ($k = 10$). The problem takes a particularly special form since there are no future partnerships to achieve the target. Therefore $V_{k+1,t} \equiv 0$ for all t and $\beta_{t+1} = 1$. It follows immediately from equation (19.3) that ψ_t is constant throughout the partnership. If the hazard is not duration dependent, i.e. $\eta_{10,t}(\cdot) \equiv \eta_{10}(\cdot)$ for all t, then the optimum strategy is to aim at the required run rate throughout the duration of the partnership. For the final partnership the proper balance between accumulating runs and safeguarding the wicket in such circumstances implies neither speeding up nor slowing down in intended run rates.

Of course, it may be more complicated than this in practice for various reasons. Most particularly, certain forms of duration dependence in the hazard may provide batsmen with a good case to take time to assess the bowling conditions. Note, however, the form that this would need to take—a mere decline in the hazard with duration t would not suffice. Rather, a good case for 'playing oneself in' would have to involve a decline in the dependence of the hazard on the scoring rate. Variations in the quality of the bowling over the course of the innings and fielding restrictions such as the 15-overs rule could provide other arguments for a non-constant intended rate.

The solution to the last partnership problem can be treated as an input to the penultimate partnership problem and by backward recursion we can work back ultimately to a solution of the opening batsmen's problem. If, as suggested earlier, partnerships with no succeeding batsmen should aim at or close to the required rate, then intuition suggests that earlier partnerships ought to be aiming above it because of the security offered by the availability of further partnerships if early batsmen are dismissed. This can be confirmed for the case of the penultimate partnership ($k = 9$) on the penultimate ball of the innings. It is easy to see from equation (19.2) that this partnership should choose a higher run rate if a succeeding partnership is available than if not. Establishing the generality of this conjecture would require a more involved argument and we rely on simulations for corroboration.

How are these arguments affected by uncertainty over run scoring if batsmen are not dismissed? It is not obvious whether the response will be precautionary run scoring to build up a buffer of runs to guard against ill luck later in the innings or wicket retention to enable an onslaught on the target late in the innings if there is ill luck earlier. Considering only the effect of uncertainty on the optimal run rate on the penultimate ball makes plain that the answer depends on the concavity properties of the marginal hazard $\eta'_{k,t}(r_t)$. Again we address these issues through simulation.

The first-innings problem

The aim for the team batting first is to minimize the probability of victory for the second innings. This will correspond to the optimized probability derived in the previous section if, but only if, the opponent can be assumed to be behaving optimally.

Formally the problem has considerable similarity with that for the second innings with a similar Bellman's equation, for instance, governing the evolution of the probability of victory. The difference is only in the less precipitous shape of that probability at the cessation of the innings. The probability of victory is no longer equal to 0 or 1 depending on whether a target has been achieved. Rather it is a steadily increasing function of runs accumulated and therefore set as the target for the opponent.

There are strong grounds for thinking that there is no completely general result on whether optimal run trajectories should increase or decrease. This can be illustrated with a simple example. Suppose that the last batsman has two balls left in the innings, there is no run scoring uncertainty and

$$h_{k,t}(r) = \min\left(\frac{r^2}{100}, 1\right),$$
$$V_{k+1,t}(R) = \Phi(R - \gamma)$$

where $\Phi(\cdot)$ denotes the standard Gaussian cumulative distribution function. The optimal run rate trajectory is increasing if $\gamma \leqslant 2.526$ and decreasing otherwise.

However, there are plausible arguments establishing a presumption in favour of an increasing trajectory. Suppose that the probability of victory at the end of the innings depends linearly on the number of runs scored, which may be a reasonable approximation over a fair range of accumulated runs. This is compatible with the approach in Clarke (1988), where maximization of runs scored is simply assumed to be the objective of the first innings. Consider the last partnership ($k = 10$) and adopt the same notation as that used for the second innings. Given the formal similarity to the second-innings problem, equation (19.3) continues to hold with $V_{k+1,t+1} \equiv R_t$. Hence

$$\Delta\psi_{t+1} = \left(\psi_{t+1} - \frac{h_{t+1}}{r_t}\right)(1 - \beta_{t+1}).$$

Assuming that $r_t \simeq r_{t+1}$, this will be positive given the assumptions on $\eta_{k,t}(\cdot)$, which imply that $\psi_{t+1}r_{t+1} - h_{t+1} > 0$. If there is no dependence on duration this implies a run rate increasing over the course of the partnership. It is easy to see why this should be. Under the assumed objective, the loss from dismissal diminishes over the course of the innings since it depends on the potentially forgone overs if the innings ends early. This may give an incentive for pursuing a strategy of increasing recklessness over the course of the innings. The numbers in Table 19.2 of Clarke (1988) appear to support this contention, even though the interpretation in the text claims that 'the generally accepted view of scoring slowly at the beginning of the innings is not optimal'. The essence of this argument may be expected to have some validity even where the objective is not linear in runs.

This conclusion may also shed light on batting behaviour in the second innings. Notice the implied conflict between the personal objective of a batsman interested in his own batting score and the objectives of the team if this is not simply to maximize expected runs, as in the second innings. Batsmen pursuing their own score would be expected to follow increasing run rates independently of whether that suited the interest of the team.

As regards earlier partnerships, the intuitive argument for presuming that earlier partnerships will aim ahead of the rates that would be chosen if there were no succeeding batsmen seems equally applicable and we would expect this to be evident in simulations.

Simulations

We explore some of the issues that were unresolved in the previous section through numerical solution. We assume that the match lasts for 40 overs and, for simplicity, take the over rather than the ball as the unit of decision.

We specify the hazard so that it is multiplicatively separable in components depending on the intended run rate and on the number of wickets lost. The former is assumed increasing and convex whereas the latter is stepwise increasing. The need to bound the dismissal probability from above at 1 leads to eventual non-convexity but, given chosen parameter values, this would come into play only for unrealistically high initial required run rates:

$$\eta_k(r^*) = \min\{f(k)g(r^*), 1\},$$

$$f(k) = \begin{cases} 0.5 & \text{if } k \leqslant 5, \\ 1 & \text{if } 6 \leqslant k \leqslant 7, \\ 1.5 & \text{if } 8 \leqslant k, \end{cases}$$

$$g(r^*) = \frac{1}{120} r^{*2}.$$

The assumption of declining skill down the batting order seems reasonable though we appreciate that, in practice, batsmen differ in more than one dimension of scoring ability. Our main conclusions above about the general shape of run rate trajectories were not dependent on this assumption, though the extent to which earlier batsmen should score differently from later batsmen will be affected by this sort of detail. We abstain completely from questions involving tactics on the bowling side.

To allow for uncertainty in run scoring we allow for a probability p that the batsmen fail to score in any particular over (independently of the intended scoring rate). By varying p from 0 upwards we can investigate numerically the importance of run scoring uncertainty. Runs scored if not dismissed are

$$
\begin{aligned}
r &= [r^*] && \text{with probability } (1-p)([r^*] - r^* + 1),\\
r &= [r^*] + 1 && \text{with probability } (1-p)(r^* - [r^*]),\\
r &= 0 && \text{with probability } p,
\end{aligned}
$$

where $[\cdot]$ denotes the greatest integer function, introduced as a convenient way of recognizing the discreteness that is inherent in run scoring while still allowing the intended run rate to vary continuously. One consequence of this is that there remains a small uncertainty for non-integer values of r^* even if $p = 0$. This could in principle upset the result that optimum scoring rates are flat for the final partnership and create a somewhat artificial preference for integer values of r^* in the optimum strategy. This does not seem important in the simulations reported here.

The optimal strategy in each innings was then found by dynamic programming methods for different values of the run scoring probability p. First-innings strategies are calculated as best responses to the assumed optimal behaviour by teams batting second. Run rate trajectories for each undismissed partnership were averaged over 100 simulations and run rate trajectories for whole innings, including dismissals, were averaged over 1000 simulations. Run targets for second-innings simulations were set at average achieved totals in the first innings. The results for individual partnerships given values $p = 0$ and $p = 0.5$ are illustrated in Fig. 19.1. These are

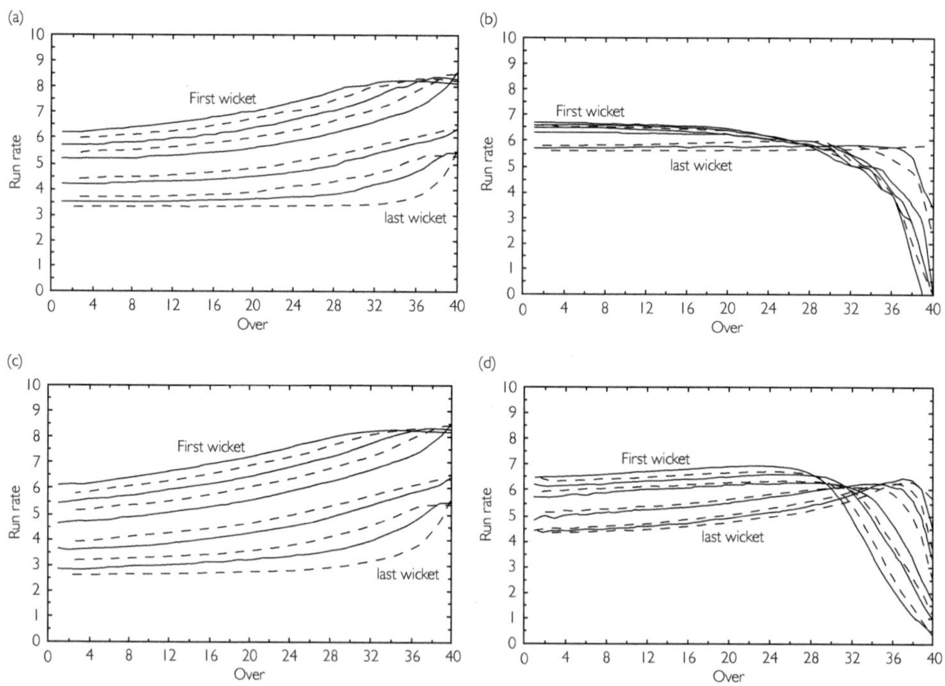

Figure 19.1 Simulated partnerships: (a) first innings, $p = 0$; (b) second innings, $p = 0$; (c) first innings, $p = 0.5$; (d) second innings, $p = 0.5$.

arranged in a 2×2 array with first-innings partnerships in the first column and second-innings partnerships in the second. Figs 19.1(a) and 19.1(b) show results with no uncertainty in run scoring, $p = 0$, and Figs 19.1(c) and 19.1(d) show those with $p = 0.5$.

The earlier discussion is supported by the simulations for the case without uncertainty in run scoring. The last partnership in the second innings pursues a flat scoring rate and earlier partnerships do, as conjectured, aim ahead of the rate required and pursue a declining scoring rate. Note also that optimal scoring rates increase over the course of first-innings partnerships, as conjectured. All first-innings simulations start with 0 runs and 40 overs left. For later partnerships this would be an extremely bleak beginning and one in which early dismissal would lead to extremely likely defeat. In these sorts of circumstances linearity of the victory probability in runs scored might be a poor approximation.

The main unresolved issue regards the effect of uncertainty. In these simulations (and others unreported here) uncertainty seems to encourage the retention of wickets rather than precautionary run accumulation although this does not upset the conclusion that early partnerships in the second innings aim ahead of the rate required.

Empirical specification: the second innings

Model

We concentrate empirical attention on second-innings strategies. This is mainly because the need to achieve a known run target provides a useful empirical anchor on the level of intended run rates, which is absent in the first innings. Although it would be possible to model an overall aspired run rate for the first innings there are no obvious covariates in the data and modelling the way in which this might change with the loss of wickets raises forbidding problems.

To simplify the notation consider the position at the beginning of a partnership. Let $\bar{r}$ be the required run rate, T be the remaining number of balls at the beginning of a partnership and Z be other factors which could include wickets lost, among other things.

The intended run rate after t balls (which need not but could be the optimum strategy) is

$$r^* = r^*(\bar{r}, T, Z, T - t, t).$$

It seems helpful to represent r^* as depending both on the number of balls faced t (because of initial nerves etc.) and on the potential number of balls remaining $T—t$ (for strategic reasons).

Without wishing to impose optimality, it still seems sensible to require that the strategy pursued involves eventually attaining the required target. Modelling, for convenience, in continuous time

$$\int_0^T r^*(\bar{r}, T, Z, T - t, t)\, \mathrm{d}t = \bar{R}.$$

If the hazard function is $\eta(r^*, Z, t)$ then the hazard along the chosen path is

$$\begin{aligned} h^* &= \eta\{r^*(\bar{r}, T, Z, T - t, t), Z, t\} \\ &= \eta^*(\bar{r}, T, Z, T - t, t). \end{aligned}$$

Note the potentially important restriction that $\bar{r}$ and T enter only through r^*.

Suppose that both the intended run rate and the hazard are multiplicatively separable (in the spirit of the proportional hazard models—see, for instance, Cox and Oakes 1985) with

$$r^*(\bar{r}, T, Z, T-t, t) = r_0(\bar{r}, T, Z)\phi(T-t, t),$$
$$\eta(r^*, Z, t) = r^{*a} f(Z)\psi(t).$$

Then the proportional hazard feature is preserved in the chosen hazard specification

$$\eta^*(\bar{r}, T, Z, T-t, t) = r_0(\bar{r}, T, Z)^\alpha f(Z)\phi(T-t, t)^\alpha \psi(t).$$

In the spirit of the Weibull model, we set

$$\phi(T-t, t) = (T-t+1)^\lambda t^\mu,$$
$$\psi(t) = \kappa\rho t^{\rho-1}$$

so that

$$r^*(\bar{r}, T, Z, T-t, t) = r_0(\bar{r}, T, Z)(T-t+1)^\lambda t^\mu,$$
$$\eta^*(\bar{r}, T, Z, T-t, t) = r_0(\bar{r}, T, Z)^\alpha f(Z)\kappa\rho(T-t+1)^{\lambda\alpha} t^{\mu\alpha+\rho-1}.$$

By compatibility with the target we have that

$$r_0(\bar{r}, T, Z)\int_0^T (T-t+1)^\lambda t^\mu \mathrm{d}t = \bar{r}T$$

and therefore

$$r_0(\bar{r}, T, Z) = \frac{\bar{r}}{(T+1)^{\lambda+\mu}B(\mu+1, \lambda+1)},$$

where $B(\cdot, \cdot)$ denotes the beta function and we have assumed that $\mu + 1, \lambda + 1 > 0$ and that T is large.

Then the survival function S is given by

$$\begin{aligned}
-\ln\{S(\bar{r}, T, Z, s)\} &= \int_0^s \eta^*(\bar{r}, T, Z, T-t, t)\,\mathrm{d}t \\
&= r_0(\bar{r}, T, Z)^\alpha f(Z)\kappa\rho \int_0^s (T-t+1)^{\lambda\alpha} t^{\mu\alpha+\rho-1}\,\mathrm{d}t \\
&= \bar{r}^\alpha f(Z)\kappa\rho(T+1)^\rho \frac{B_{s/(T+1)}(\mu\alpha+\rho, \lambda\alpha+1)}{B(\mu+1, \lambda+1)^\alpha}
\end{aligned}$$

where $B_x(\cdot, \cdot) \equiv I_x(\cdot, \cdot)B(\cdot, \cdot)$ and $I_x(\cdot, \cdot)$ is the incomplete beta function and we have assumed that $\mu\alpha + \rho, \lambda\alpha + 1 > 0$.

In recognition of the uncertainty in run scoring, assume that the actual run rate r follows $r^* + \epsilon$ where ϵ is a mean zero independent and identically distributed error term, assumed to be distributed independently of the event of dismissal. Then the average achieved run rate over the duration of the partnership r_s is

$$\begin{aligned} r_s &= \frac{1}{s}\int_0^s r^*(\bar{r}, T, Z, T-t, t)\,dt + \frac{1}{s}\int_0^s \epsilon\,dt \\ &= \frac{1}{s} r_0(\bar{r}, T, Z)\int_0^s (T-t+1)^{\lambda} t^{\mu} dt + \upsilon_s \\ &= \frac{T+1}{s}\bar{r}\, I_{s/(T+1)}(\mu+1, \lambda+1) + \upsilon_s. \end{aligned}$$

υ_s must be heteroscedastic with a variance which diminishes with the length of spell. Thus it would be feasible to estimate an achieved run rate equation by, say, non-linear least squares on the homoscedastic equation

$$s^{1/2} r_s = s^{1/2}(T+1)\bar{r}\, I_{s/(T+1)}(\mu+1, \lambda+1) + s^{1/2}\upsilon_s.$$

Estimation

We estimate the model by maximum likelihood assuming the error term ϵ to be Gaussian. Given the assumption that ϵ is distributed independently of dismissal we can factorize the likelihood for a partnership surviving s balls and achieving a run rate of r_s given $\bar{r}$, T and Z into a form

$$l_1(s, \bar{r}, T, Z; \mu, \lambda, \alpha, \rho, \theta_1)\; l_2(r_s, \bar{r}, T, Z; \mu, \lambda, \sigma, \theta_2),$$

where θ_1 and θ_2 denote parameters other than those already defined. μ, λ and σ are thus identified from the information on runs scored alone, whereas durations of partnerships provide identifying information on the remaining parameters as well as further information on μ and λ. Our estimates maximize the complete likelihood over the whole parameter space but we exploit this structure to derive starting values, firstly estimating μ, λ, σ and θ_1 by maximizing $l_2(\cdot)$ alone, then fixing those parameters at the estimated values and estimating α, ρ and θ_1 by maximizing $l_1(\cdot)$. The theory suggested that the slope of optimal trajectories should differ for differing partnerships so in certain estimations we allowed λ to differ between upper, middle and lower order batsmen.

There is a potentially serious source of bias due to omitted variables if we do not allow for match conditions, since a poor pitch is likely to lead to low targets and high second-innings hazards for reasons which are not connected with the choice of strategy. We choose to do this by including a full set of match effects (in the logarithm of the hazard) when estimating—i.e. we estimate a different intercept for each match. This, of course, precludes the inclusion of any other variables which do not vary within matches, such as team effects, though this should not cause worries since these effects will be completely absorbed in estimation (by the match effects).

Data and results

We have data from all 145 Axa Equity & Law Sunday League matches in the 1996 season. The 18 competing county teams would each have played each other once if the weather had allowed, but eight matches for various reasons lack adequate data for the purpose. The data cover 959 second-innings partnerships. Data on lengths of partnerships and runs scored were typed in from the original score-cards, because the conventional abbreviated published score-cards do not contain all the detail required. Sunday league innings last for 40 overs and conveniently there are no fielding restrictions which apply early in the innings.

Results with λ constant across partnerships are reported in Table 19.1. We produce columns for estimates without and with match effects, though we see strong grounds to prefer the latter. Match effects are jointly strongly significant and the increased coefficient on the required run rate when they are included accords with the expected direction of bias from their omission.

The estimated value of 1.591 for α, significantly different from 1 at the 5% level, offers evidence of a convex hazard as assumed in the theory. The estimated values of μ and λ, which are significantly different from 0 only in the latter case, suggest a strategy of increasing run rates. As the earlier simulations showed, this might be optimal with uncertainty in run scoring but not without. The positive estimated value for ρ suggests dismissal hazards which

Table 19.1 ***Model estimates with fixed λ.***[†]

Parameter	Estimates for model without match effects		Estimates for model with match effects	
	Coefficient	Estimated standard error	Coefficient	Estimated standard error
Constant	−3.526	−14.874	−4.705	−8.254
Required run rate	0.450	4.608	1.597	7.548
1st wicket	−0.818	−3.112	0.794	2.050
2nd wicket	−0.888	−3.375	0.835	2.163
3rd wicket	−0.694	−2.646	0.814	2.166
4th wicket	−0.534	−2.074	0.782	2.172
5th wicket	−0.529	−1.999	0.588	1.667
6th wicket	−0.245	−0.973	0.922	2.762
7th wicket	−0.113	−0.449	0.728	2.221
8th wicket	0.204	0.815	0.908	2.777
9th wicket	−0.002	−0.010	0.532	1.697
$\ln(\rho + \mu\alpha)$	0.201	6.439	0.292	6.898
$\ln(\sigma)$	1.747	50.594	1.766	48.973
$\ln(1 + \mu)$	−0.044	−1.082	0.070	1.678
$\ln(1 + \lambda)$	−0.484	−8.946	−0.304	−5.319
M	−0.044	−1.107	0.073	1.620
Λ	−0.383	−11.505	−0.262	−6.21
A	0.450	4.608	1.597	7.548
P	1.242	13.030	1.222	5.110
Σ	5.736	28.966	5.845	27.737
Log–likelihood	−5129.630		−5029.524	
Sample size	959		959	

[†] Likelihood ratio test for significance of match effects $\chi^2_{139} = 200.212, p = 0.001$.

Table 19.2 ***Model estimates with wicket-dependent*** λ.[†]

Parameter	Estimates for model without match effects		Estimates for model with match effects	
	Coefficient	Estimated standard error	Coefficient	Estimated standard error
Constant	−3.043	−12.415	−3.541	−5.839
Required run rate	0.512	6.030	1.099	7.227
1st wicket	−1.135	−4.473	−0.672	−2.125
2nd wicket	−1.209	−4.764	−0.606	−1.894
3rd wicket	−1.034	−4.070	−0.608	−1.921
4th wicket	−0.894	−3.536	−0.620	−2.006
5th wicket	−0.807	−3.095	−0.530	−1.666
6th wicket	−0.531	−2.099	−0.181	−0.590
7th wicket	−0.436	−1.668	−0.314	−0.962
8th wicket	0.303	1.200	0.872	2.689
9th wicket	0.034	0.150	0.507	1.636
$\ln(\rho + \mu\alpha)$	0.226	7.555	0.324	8.639
$\ln(\sigma)$	1.423	50.167	1.427	46.924
$\ln(1 + \mu)$	0.147	3.789	0.173	3.880
$\ln(1 + \lambda_{1-4})$	0.051	0.830	0.083	1.133
$\ln(1 + \lambda_{5-7})$	−0.211	−3.763	−0.185	−2.984
$\ln(1 + \lambda_{8-10})$	−1.265	−21.869	−1.213	−18.591
M	0.159	3.523	0.188	3.563
λ_{1-4}	0.052	0.809	0.087	1.087
λ_{5-7}	−0.190	−4.188	−0.169	−3.279
λ_{8-10}	−0.718	−43.952	−0.703	−36.221
A	0.512	6.030	1.099	7.227
P	1.172	11.208	1.175	5.923
Σ	4.150	35.253	4.165	32.890
Log–likelihood	−4965.867		−4868.359	
Sample size	959		959	

† Likelihood ratio test for significance of match effects $\chi^2_{139} = 195.015$, $p = 0.001$.

increase with the duration of partnerships (even given the intended run rate) and is suggestive of increasing carelessness or adventurousness as the duration of an innings is prolonged.

Table 19.2 reports results from allowing λ to vary with the number of wickets lost. The gain in the likelihood is sufficiently large to suggest a rejection of constancy for λ. Furthermore the differences accord with what theory suggests to be reasonable in an optimization context where the slopes to run trajectories are steeper for later partnerships. Indeed the picture of a negligible slope for high order partnerships and a more pronounced upward slope for those in the lower

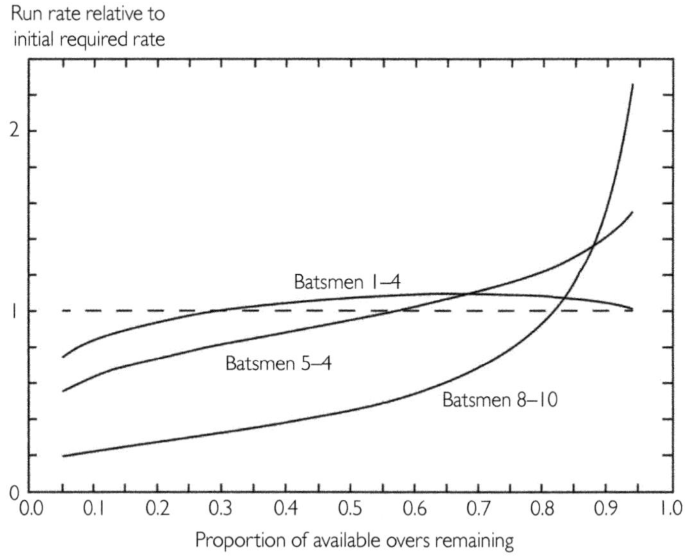

Figure 19.2 Estimated batting strategies.

order, illustrated in Fig. 19.2, has some similarity with what is seen in the Fig. 19.1(d), particularly when it is remembered that the data for higher order partnerships are concentrated in early overs and for lower order partnerships in later overs.

Discussion

We have established several results concerning the nature of the optimum batting strategy for limited overs cricket and pointed to fundamental and possibly poorly appreciated differences in the optimum strategies for the two innings. We have also estimated the behaviour of actual cricketers in the second innings of such matches. Interestingly, actual trajectories share some of the features of what we believe to be optimal behaviour though we would be reluctant to suggest that we have any strong evidence for saying that cricketers are actually optimizing. It might be a worthwhile exercise to apply simulation-based methods to estimate an explicitly optimizing model using data from the same score-cards (in the fashion of Rust 1995, 1996). Other strategic aspects in cricket may also deserve and repay attention, including bowling strategy.

ACKNOWLEDGEMENTS

We are grateful for the assistance of Tim Lamb and Richard Lockwood in gaining access to the data which were made available by the England and Wales Cricket Board and the Press Association. The responsibility for the interpretation of the data rests solely with the authors. We have benefited from discussions with or suggestions from the Joint Editor, Frank Duckworth,

Jeremy Edwards, Tony Lewis, Hamish Low, Malcolm Pemberton, James Symons, Richard Vaughan, Frank Windmeijer, participants in seminars in Birmingham, Glasgow and London and a referee.

REFERENCES

Bairam, E. I. and Howells, J. M. (1991a) The probability of winning and maximising behaviour for one-day international cricket. *Economics Discussion Paper 9120*. University of Otago, Dunedin.

—— (1991b) The probability of winning and maximising behaviour in cricket: the New Zealand experience. *Economics Discussion Paper 9121*. University of Otago, Dunedin.

Bairam, E. I., Howells, J. M. and Turner, G. M. (1990a) Production functions in cricket: the Australian and New Zealand experience. *Appl. Econ.*, **22**, 871–879.

—— (1990b) Production functions in cricket: one-day and three-day games in New Zealand. *J. Aust. Soc. Sports Hist.*, **6**, 203–217.

Bertsekas, D. (1987) *Dynamic Programming: Deterministic and Stochastic Models*. New York: Prentice Hall.

Clarke, S. R. (1988) Dynamic programming in one-day cricket—optimal scoring rates. *J. Oper. Res. Soc.*, **39**, 331–337.

Cox, D. and Oakes, D. (1985) *Analysis of Survival Data*. New York: Chapman and Hall.

Duckworth, F. C. and Lewis, A. J. (1998) A fair method of resetting the target in interrupted one-day cricket matches. *J. Oper. Res. Soc.*, **49**, 220–227.

Kimber, A. C. and Hansford, A. R. (1993) A statistical analysis of batting in cricket. *J. R. Statist. Soc.* A, **156**, 443–455.

Marylebone Cricket Club (1998) *Cricket: Know the Game*. London: Black.

Morrison, I. (1998) *Play the Game: Cricket*. London: Ward Lock.

Rust, J. (1995) Do people behave according to Bellman's principle of optimality? University of Wisconsin, Madison.

—— (1996) Numerical dynamic programming in economics. In *Handbook of Computational Economics* (eds H. Amman, D. Kendrick and J. Rust). Amsterdam: North-Holland.

PART V

Golf

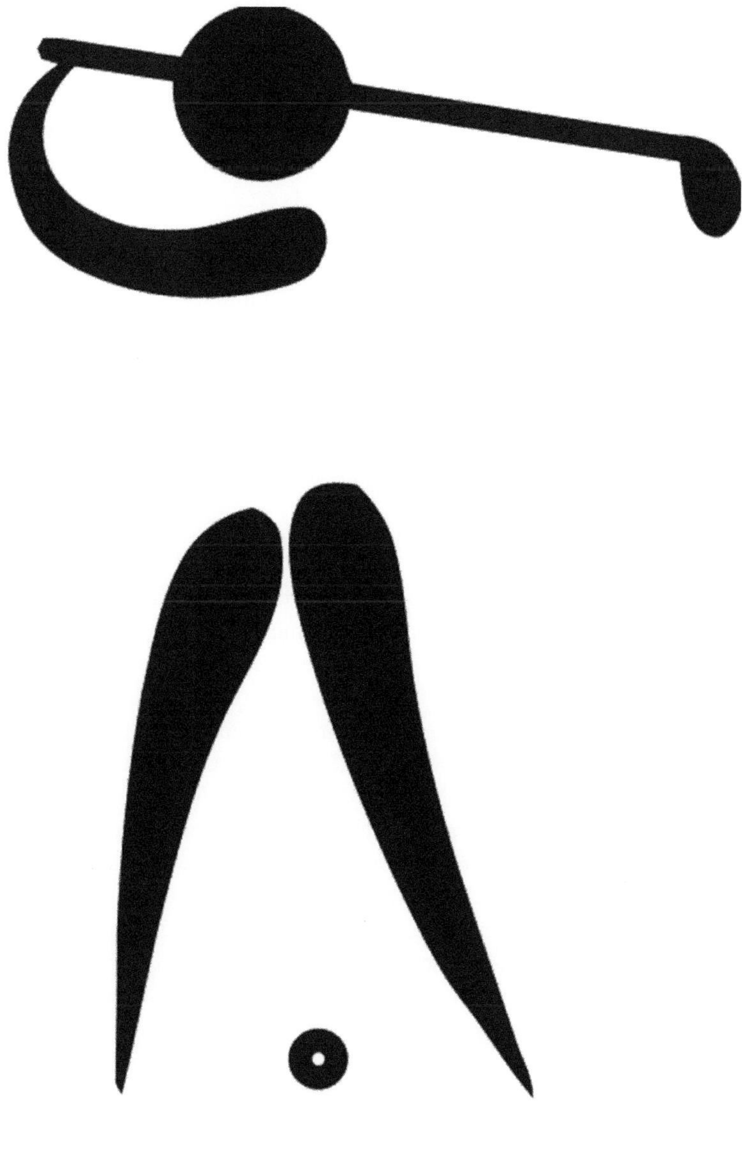

Introduction

SCOTT E. GRASMAN

Golf is a popular sport dating back many centuries. While similar games existed in central Europe earlier, a Scottish version from the 1400s is widely accepted as the origin of modern golf. Today, golf has spread throughout the world with amateurs and professionals playing for both fun and livelihood. The overarching premise of golf is to use clubs to move the ball from the tee to the hole in the minimum number of strokes possible while adhering to the principle of 'play the ball as it lies and, if you can't, do what is fair'. The Rules of Golf, dating back to the 1700s, are maintained by the United States Golf Association (USGA) and the Royal and Ancient (R&A) and form the basis for doing what is fair.

Each hole on a golf course has an associated number referred to as the hole's par, where par refers to the predetermined number of strokes that a golfer should require to complete the hole. A player who moves the ball from the tee to the hole in the number of strokes that is par for the hole is said to have shot par on the hole. A player who moves the ball from the tee to the hole in one shot less than par is said to have shot a birdie on the hole, and a player who moves the ball from the tee to the hole in two shots less than par is said to have shot an eagle on the hole. A player who moves the ball from the tee to the hole in one shot more than par is said to have shot a bogey on the hole, a score that is two shots over par is referred to as a double bogey, and so on. The number of shots for a hole is primarily a function of the distance from the tee to the hole. The United States Golf Association has established guidelines for the par ratings of holes by gender of player; the current guidelines are:

Par	Men	Women
3	Up to 250 yards	Up to 210 yards
4	251 to 470 yards	211 to 400 yards
5	471 yards to 690 yards	401 to 575 yards
6	691 yards or more	575 yards or more

The difficulty of a hole can be increased by narrowing the fairway (the path of well-maintained grass between the tee and the hole), adding water and/or sand hazards between the tee and the hole, and adding a dog leg or sharp bend in the fairway. Figure P V.1 provides a diagram of a typical golf hole.

There are many forms of golf competitions for individuals and for teams. Medal play, match play, and skins are the most popular individual formats, while scrambles, best-ball, and alternate-shot are often used for team play.

Medal play, or stroke play, is the traditional form of golf competition where one wins by accumulating the lowest number of strokes over a set of holes. In match play, two players compete with individual holes won based on the lower number of strokes; the overall winner is based on total holes won. If players have the same number of strokes on a particular hole, then the hole is simply halved. A variation on match play (normally with more than two players) is referred to as skins. In skins, the player recording the lowest score on an individual hole wins that hole plus all previous halved holes subsequent to the last non-halved hole (unless there is

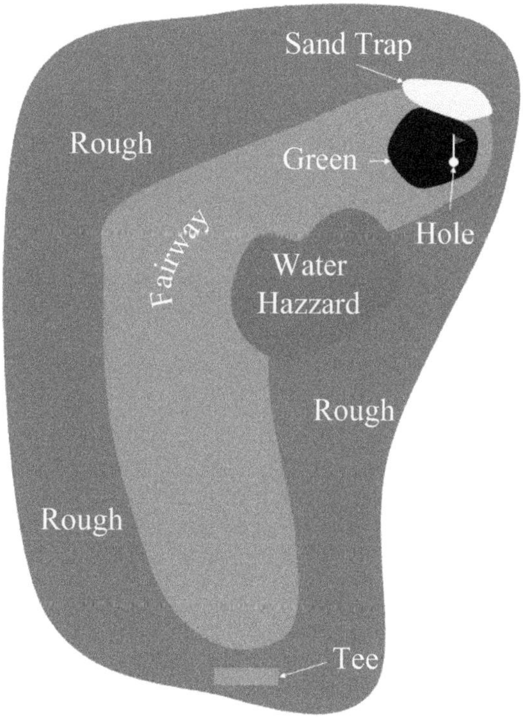

Figure P V.1 Layout of a typical golf hole with a dogleg right.

no carryover). In some cases, skins have to be 'validated' by winning, or perhaps simply tying, the next hole as well.

Scrambles, best-ball, and alternate-shot are popular for team play. In a scramble, two or more players each play a tee shot and then each plays from the better/best resulting lie (as deemed by the team) until the ball is holed; thus only one score is recorded for each team. In best-ball, each player on the team plays his/her own ball on each hole, and the lower/lowest score of the team is recorded as the team score. Alternate-shot, is a format where two players alternate hitting the same ball until the ball is holed. Players alternate hitting tee shots so that the same player does not start every hole. Scrambles, best-ball, and alternate-shot can all be played as stroke play (by counting the total strokes), match play (by counting the total holes won), or skins. Other less common formats also exist for team play.

There are two main streams of statistical research in golf: the first addresses handicapping systems, and the second addresses collecting and analysing strokes and associated outcomes. It is interesting to note that statistical research in golf does not always reach consistent conclusions and, in some cases, finds contradictory results. While this may be common due to modelling assumptions and data availability of empirical research, it is clear that opportunities exist for more statistical research using superior data that are now available.

In order to make competition more interesting and perhaps more equitable, it is common for amateur players to use a handicapping system when competing against one another. A handicap, a measure of a golfer's ability, is roughly expressed as the number of strokes above par at which the player usually plays. In medal play, the handicap is subtracted from the raw score before the lowest overall score is determined. In match play, the difference between the handicaps is calculated and strokes are subtracted from the score of the higher-handicapped player on an equivalent number of holes.

The USGA Handicapping System (http://www.usga.org/Handicapping/handicap-manual.html) calibrates a player's *Handicap Index* as the player's game changes, while disregarding outlier scores. The Handicap Index compares the potential scoring ability of an individual to that of a scratch golfer (zero handicap, or a player who is expected on average to shoot par) on a standard course. A Handicap Index is portable from course to course, as well as from sets of tees on the same course, by adjusting for the playing difficulty of a course under normal conditions (using *Course Rating* and *Slope Rating*). The R&A uses a similar approach to handicapping (http://www.randa.org/en/Playing-Golf/Handicapping.aspx).

There is debate as to whether handicapping systems provide a level playing field for individual and team competitors. As early as the 1970s, operations researchers have addressed the fairness of various schemes. The popular golfing periodical, *Golf Digest*, was perhaps the first to ask the question. Other papers presented early evaluations of the USGA handicapping system, and generally concluded that the handicapping system provides the lower handicap golfer with a competitive advantage. Some of these results and suggested modifications are partially responsible for changes to handicapping systems, but the debate continues. For example, some continue to argue that the current handicapping system provides higher probabilities of winning to more consistent and lower handicap golfers in both stroke and match play. While the handicapping system is well-established for individual play, it is less so for team play. A number of unofficial procedures have been used for various formats, but no official system for handicapping teams has been adopted. Some methodologies have been proposed, such as assigning a linear combination (e.g., the sum) of individual handicaps as the team handicap. However, some argue that these methods do not lead to teams with equal win probabilities.

Two articles in this chapter address handicapping systems. Kupper et al. (2001) provides an overview of the USGA Handicap System and presents an order statistics argument to show that a more consistent golfer has an advantage over a less consistent golfer with the same handicap. The paper includes insights based on empirical data providing intuition into this effect. Of note is the implication that handicapping systems should include variability in the calculation of handicaps. This result has been supported by recent papers. Bingham and Swartz (2000) present a rationale for an alternative handicapping method based on the assumption that scores are more variable as the handicap increases. Additionally, they concluded both from theoretical and empirical results that the handicapping system actually favours the higher handicap player in medal play when both players are playing well. While they did not directly consider match play, they conjectured that the same results would hold true.

With modern technology, and many volunteers, nearly every shot of an amateur or professional player can be captured and used for data analysis. While amateurs can use mobile apps, professional tours collect and disseminate scoring and statistical data on every shot by every player in real-time. Common statistics relate to driving distance and accuracy, greens in regulation, proximity to hole, sand save percentage, putting average and distribution, birdie/eagle percentage, and many others. Statistical research dating back to the 1980s has shown useful correlations between these statistics and scoring average.

Most earlier works attribute a statistic to a unique ability; however, as Ketzscher and Ringrose (2002) point out, greens in regulation is a function of both driving and iron play, while putting is a combination of proximity to the hole and other factors, thus more complex statistics may be influenced by playing style. Other metrics include variations of strokes gained, scrambling, and others. And while there is much anecdotal evidence related to how these metrics relate to competition, prediction based on such metrics is a significant challenge. Thus, the aim of golf analytics is to see the usefulness of these data in predicting the outcome of competitions. Two of the articles in this chapter analyse such data.

Berry (2003) discusses improvements in golf equipment and the impact on three standard driving statistics—driving distance, driving accuracy, and combined driving. The article presents boxplots for driving distance and driving accuracy that show a strong increasing trend for driving distance and weaker improving trend for driving accuracy. A linear model is created for the average driving distance. The article also presents contour plots that show values of the driving distance and driving accuracy that lead to the same mean score. Finally, box plots show the mean score based on driving, concluding that mean scores have decreased, first due to improved accuracy and second due to improved distance. While the article only looks at driving statistics, it conjectures that similar results would be seen for other statistics.

Ketzscher and Ringrose (2002) look at a pool of standard statistics (driving distance and accuracy, greens in regulation, total putts, total putts per green in regulation, and sand saves) to predict scoring average. Since the relationship between these statistics and scoring average is not surprising, the authors develop new statistics that may lead to better prediction. Regression and ANOVA models are created for a variety of statistics. The conclusion is that current statistics are insufficient for prediction and call for a greater level of detail in the statistics collected. Ironically, the data are currently available, for example, through the US PGA ShotLink® systems.

REFERENCES

Berry, S.M. (2003). Game of Which I Am Not Familiar. *Chance*, 16(3), 45–48.

Bingham, D.R. and Swartz, T.B. (2000). Equitable Handicapping in Golf. *The American Statistician*, 54(3), 170–177.

Ketzscher, R. and Ringrose, T.J. (2002). Exploratory Analysis of European Professional Golf Association Statistics. *The Statistician*, 51(2), 215–228.

Kupper, L.L., Hearne, L.B., Martin, S.L., and Griffin, J.M. (2001). Is the USGA Golf Handicap System Equitable? *Chance*, 14(1), 30–35.

CHAPTER 20

Is the USGA Golf Handicap System equitable?

LAWRENCE L. KUPPER, LEONARD B. HEARNE, SANDRA L. MARTIN, AND JEFFREY M. GRIFFIN

GOLF IS ARGUABLY the fastest growing sport in the United States, and it is played by many thousands of people with widely varying skill levels. The United States Golf Association (USGA), a nonprofit organization that has served as the national governing body of golf since its formation in 1894, has developed a handicap system with the expressed purpose of "making the game of golf more enjoyable by enabling golfers of differing abilities to compete on an equitable basis" (see the Web site www.usga.org/handicap/manual/index.html for details about the manual describing the current USGA Handicap System). The issue that we address in this article is whether or not the USGA Handicap System is, in fact, "equitable."

To borrow some terminology from an interesting 1988 commentary in *Golf Digest* by Robert Carney, suppose that "Steady Eddie" and "Wild Willie" are two golfers competing against each other in an 18-hole net medal play match. By net medal play, we mean that each golfer records a "gross score" for 18 holes (namely, the sum of the number of actual strokes plus any penalty strokes made on each of the 18 holes), and then each gross score is converted to a "net score" by subtracting the number of (USGA-based) handicap strokes that each golfer receives for that particular match. The winner of this net medal play competition between Eddie and Willie is that golfer with the lower net score. As an example, if Eddie shoots a gross score of 80 with a USGA handicap of 6, and if Willie shoots a gross score of 90 with a USGA handicap of 17, then Willie's net score of 73 bests Eddie's net score of 74.

No handicap needed

In what follows, we will consider the USGA Handicap System to be "equitable" if both Eddie and Willie have approximately the same chance of winning in a two-person net medal play match. Because the probability of a tie is reasonably small, this is essentially the same as saying that Eddie and Willie should each have roughly a 50–50 chance of winning the match when

This image was originally posted to Flickr by Jim Epler at http://flickr.com/photos/71707869@N00/2566168152. Reproduced under a Creative Commons licence

USGA handicaps are used. As Carney conjectured in his commentary and as we will show analytically, there are many realistic situations in which one golfer will have a decided advantage over another under the current USGA Handicap System. As an example, suppose that Steady Eddie and Wild Willie both shoot about the same average gross score for 18 holes, but (as their names are meant to imply) Eddie is much more consistent (i.e., is less variable or erratic) in the scores that he shoots than is Willie. In such a situation, Eddie will generally have a significantly greater than 50% chance of defeating Willie in a two-person net medal play competition. As we will show, one of the main reasons for this inequity in the USGA Handicap System is that the method used for computing handicaps does not appropriately account for the basic fact that golfers are not equally consistent.

Order statistics

The USGA's golf handicap system assigns a handicap based on a golfer's lowest 10 of 20 handicap differentials. To evaluate the consequences of the USGA handicap system, the authors of the accompanying article make use of statistical results concerning "order statistics." Given a sample of observations, the first order statistic is the smallest observation in the sample, the second order statistic is the next smallest observation, and so on. A typically assumed situation is that the observations in question are independently selected values from a common underlying distribution (a normal distribution in the article).

The statistical properties of ordered data are different from those of the original unordered observations. For instance, the probability that the smallest observation in a random sample

This image was originally posted to Flickr by Keith Allison at http://flickr.com/photos/27003603@N00/3608157023. Reproduced under a Creative Commons licence

is less than some fixed number is typically much higher than the corresponding probability for any unordered observation in that same sample. As another example, suppose that intelligence-test scores are normally distributed with mean 100 and standard deviation 15. For any randomly selected person taking this test, the expected value of that person's score is 100. If 20 randomly selected people take this test, however, then the expected value of the smallest score among these 20 people is about 72. In general, statistical methods involving order statistics are more complicated than those based on the use of unordered data.

For more information, refer to H. A. David's book *Order Statistics* (2nd ed., 1981, Wiley) and to other listed references.

The USGA Handicap System

A golfer's handicap is based on the most current 20 gross scores that the golfer has submitted for handicap determination purposes, so that handicaps continually change as new scores replace old scores. To compute a USGA handicap, each gross score (S) is first converted to a so-called "handicap differential" (HD), which is a standardized measure reflecting the difficulty of the course that was played. In particular, the formula relating HD to S is $HD = (S - CR)(113/SR)$, where CR is the USGA "Course Rating" and SR is the USGA "Slope Rating" for the particular course and set of tees played.

More specifically, CR is an index that reflects the average 18-hole gross score for an expert (or "scratch") golfer who shoots around par. Par is the score that an expert golfer would be "expected" to make on any given hole; to quote the USGA, "par means errorless play without flukes under ordinary weather conditions, allowing two strokes on the putting green." So, a typical value for CR would be about 72 for a course with 4 par three holes, 10 par four holes, and 4 par five holes; CR values are commonly in the low 70s, with higher CR values indicating harder courses. *SR* is an index that ranges between 55 and 155 and reflects a course's relative playing difficulty for players who are not scratch golfers. In the preceding formula, the value 113 is the USGA slope rating for a course of "standard playing difficulty," so the higher the *SR* value relative to 113, the more difficult is the course for the nonexpert golfer. The USGA golf handicap manual contains further details about how *CR* and *SR* are determined for any specific course and set of tees.

As an example, if a golfer shoots a gross score *S* of 93 on a course and tees with $CR = 73.1$ and $SR = 130$, then $HD = (93 - 73.1)(113/130) = 17.3$. So, although this gross score is about 20 strokes above the course rating, the resulting handicap differential is less than 20 after a slope rating adjustment for the increased difficulty of this course relative to one of standard playing difficulty. Note also that *HD* is negative, zero, or positive according to whether *S* is less than, equal to, or greater than *CR*. And, for $SR = 113$, $HD = (S - CR)$ is simply the difference between the score and the course rating. For a scratch golfer, the arithmetic mean of 20 representative *HD*s would be expected to have a value somewhat close to zero.

Once the 20 gross scores are converted to 20 handicap differentials, the 20 *HD*s are ordered from smallest to largest. Then, the golfer's USGA Handicap Index (*HI*) is computed as 0.96 times the mean of the smallest 10 *HD*s (with deletion of all digits after the tenths position). A golfer uses his or her current *HI* to determine the number of strokes (i.e., the Handicap *H*) that he or she "gets" for a particular net play competition via the formula $H = (HI)(SR/113)$, where *SR* is the slope rating for the particular course and tees to be played. For example, if a golfer has $HI = 10.5$ and if the course and tees to be played have $SR = 135$ (indicating a fairly difficult course), then $H = (10.5)(135/113) = 12.54$. This golfer would then get 13 strokes (since values ending with 0.5 or more are rounded upward to the next highest integer) so that this golfer's net score for this particular golf match would be 13 strokes lower than his or her gross score. As a special case, when $SR = 113$, then *H* is essentially equal to *HI*. For a course with *SR* greater than 113, *H* is clearly larger than *HI*, with the disparity increasing as *HI* itself increases. Because golfers don't typically carry calculators in their golf bags, they convert their *HI*s to *H*s using appropriate USGA course handicap tables available at the course to be played.

Rationale for the USGA Handicap System

A basic premise underlying the USGA Handicap System is that every player will try to make the best possible score on each hole in every round played, regardless of where the round is played, and will post every acceptable round for handicap determination purposes. Given this important assumption, the USGA's primary rationale for dropping the highest 10 of a golfer's 20 most recent *HD*s when assigning handicaps is that such "high" *HD*s bear little relation to that golfer's "potential" ability. As we will show, however, the USGA's decision to use only the lowest 10 *HD*s for handicap determination purposes is exactly the reason there is an inequity problem with the current USGA Handicap System.

Statistical developments

Suppose that Steady Eddie and Wild Willie are playing each other in a net medal play competition on a course with course rating *CR* and slope rating *SR*. Clearly, golf scores are subject to random variation, and hence so are handicap indexes and handicaps. To model this randomness, we let S_{se} and S_{ww} be independent random variables denoting, respectively, the gross scores that Steady Eddie and Wild Willie shoot during this golf match. And, we let H_{se} and H_{ww} be independent random variables representing the corresponding handicaps for Eddie and Willie based on their most recently submitted sets of 20 gross scores. Now, recalling that Willie wins if his net score ($S_{ww} - H_{ww}$) is lower than Eddie's net score ($S_{se} - H_{se}$), we can express the probability that Willie defeats Eddie in a net medal play match as

$$\Pr[\text{Willie defeats Eddie}] = \Pr[(S_{ww} - H_{ww}) < (S_{se} - H_{se})].$$

Using the equations that relate *HD* to *S* and *HI* to *H* and performing some algebra, we can rewrite this probability directly in terms of *HD*s and *HI*s as

$$\Pr[\text{Willie defeats Eddie}] = \Pr[(\text{HD}_{ww} - HD_{se}) < (HI_{ww} - HI_{se})].$$

In words, this last result says that the probability that Willie defeats Eddie in a net medal play match is equivalent to the probability that the difference in handicap differentials posted by Willie and Eddie for that match is smaller than the corresponding difference in their current handicap indexes.

To proceed further, we need to say something about the behavior of the four random quantities HD_{se}, HD_{ww}, HI_{se}, and HI_{ww}. Based on some empirical evidence, we claim that, at least to a reasonable degree of approximation, HD_{se} and HD_{ww} are each roughly normally distributed. To support this claim, we examined available sets of handicap differentials posted by male members of the University of North Carolina's Finley Golf Course from April 1, 1997, through October 31, 1997. For a set of 130 golfers each with at least 10 recorded *HD*s during this seven-month period, we found no strong statistical evidence against this approximate normality assumption. For illustrative purposes, Figure 20.1 provides relative frequency histograms of *HD* values for three of these 1997 golfers. In addition, there was no indication of significant autocorrelation among a golfer's set of handicap differentials (i.e., the *HD*s for a golfer seemed to vary fairly randomly over this seven-month period). The same general conclusions were obtained from a similar analysis of data on 60 golfers over the same seven-month period in 1994. Thus, it seems reasonable to treat these four random variables as being mutually independent.

Now, what can we say about the distributions of the random variables HI_{se} and HI_{ww}? Given that HD_{se} and HD_{ww} can each be taken to be approximately normally distributed (with, say, respective means μ_{se} and μ_{ww} and variances σ^2_{se} and σ^2_{ww}), it follows that HI_{se} and HI_{ww} are functions of the lowest 10 of 20 ordered normal variates. The properties of "order statistics" based on random samples from normal populations have been well studied (see sidebar). Using these known properties of normal order statistics, it can be shown (we omit the details) that the probability that Willie defeats Eddie in a net medal play match can be reasonably approximated by the formula

$$\Phi\{-0.715(\sigma_{ww} - \sigma_{se}) + 0.039(\mu_{ww} - \mu_{se})]/(\sigma^2_{ww} + \sigma^2_{se})^{1/2}\}$$

where $\Phi\{\cdot\}$ is the standard normal cumulative distribution function (CDF) for which extensive tables are widely available. Hence, for given values of the means (μ_{se} and μ_{ww}) and variances

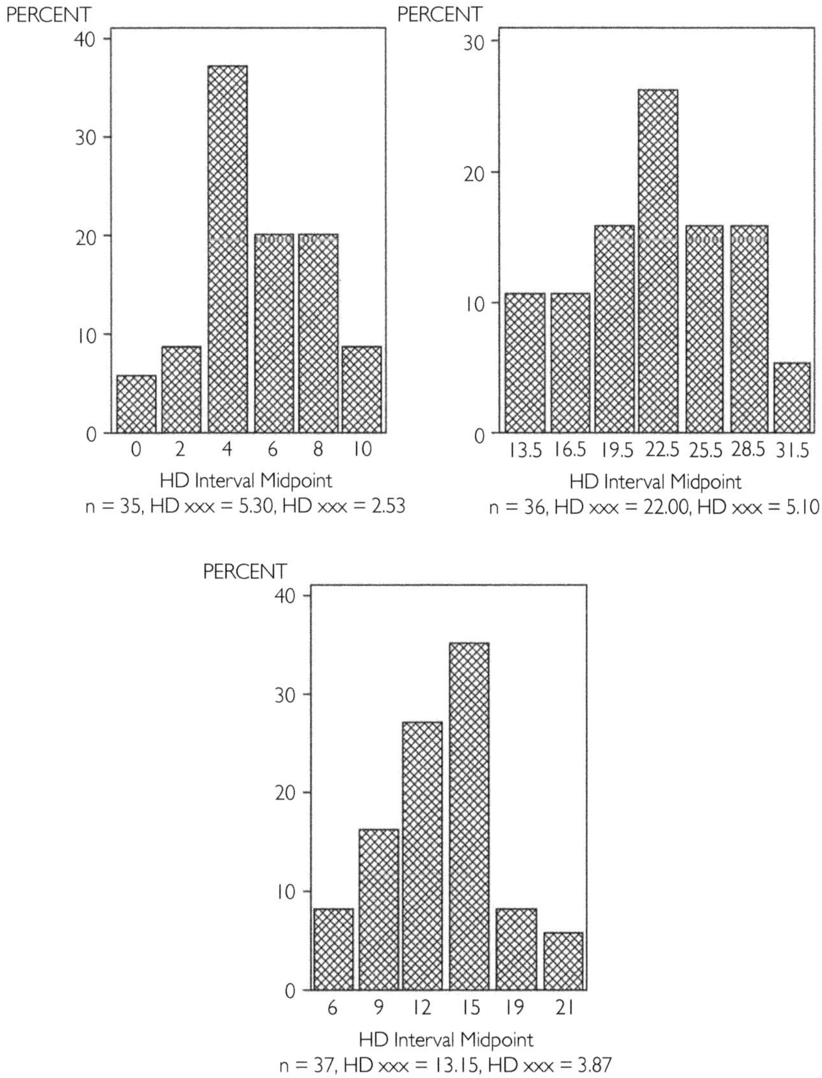

Figure 20.1 Histograms of handicap differential (HD) values for three 1997 golfers.

(σ_{se}^2 and σ_{ww}^2) of the approximate normal distributions for HD_{se} and HD_{ww}, we can compute the approximate probability that Willie defeats Eddie in a net medal play competition.

As a numerical illustration, suppose that $\mu_{se} = 5$, $\mu_{ww} = 20$, $\sigma_{se}^2 = 3$, and $\sigma_{se}^2 = 10$ so that Steady Eddie, on average, shoots lower scores and is also more consistent (i.e., is less variable) than Wild Willie. Then, substituting these four specified values into the preceding formula gives Pr[Willie defeats Eddie] $\approx \Phi(-0.446) \approx 0.33$ using standard normal CDF tables.

Now, what is the appropriate interpretation for the numerical value 0.33 just computed? As discussed earlier, over a series of net medal play matches between Eddie and Willie, both the gross scores that they shoot, and hence their handicaps, will vary from match to match. Hence, the probability that Willie defeats Eddie will also vary from match to match as a function of these changing handicaps. Given these particular choices for μ_{se}, μ_{ww}, σ_{se}^2, and σ_{ww}^2 (i.e., 5, 20, 3, and 10, respectively) that characterize the (normal) distributions of possible handicap differentials for Eddie and Willie, the value 0.33 should be interpreted as the average of all these match-specific probabilities, or, equivalently, as the approximate probability that Willie will defeat Eddie in any randomly occurring net medal play match. To put it another way, if Willie and Eddie play a large number of net medal play matches under the stated conditions, then Willie would be expected to win only about 1/3 of these matches under the current USGA Handicap System. It is very important to emphasize here that our results should *not* be used to compute the probability that Willie defeats Eddie in a specific match involving *fixed* handicaps; this probability could be either somewhat greater than or somewhat less than 1/3, the actual value depending on the current handicap indexes for Willie and Eddie at the time of the match.

In the preceding numerical example, Steady Eddie is clearly a better golfer than Wild Willie in both average score and consistency, and we find that Willie is at a distinct disadvantage under the current USGA Handicap System. For the USGA Handicap System to be equitable, both Eddie and Willie should each be expected to win roughly half of a series of net medal play matches that they play against each other; equivalently, both Eddie and Willie should each have about a 50–50 chance of winning any randomly occurring net medal play match. An inspection of our approximate formula for Pr[Willie defeats Eddie] quickly reveals that a sufficient condition for the USGA Handicap System to be equitable is that both the equalities $\mu_{se} = \mu_{ww}$ and $\sigma_{se} = \sigma_{ww}$ must simultaneously hold—namely, that the normal distributions of possible handicap differentials for Eddie and Willie must be identical. To put it another way, the current USGA Handicap System is equitable if Eddie and Willie tend to shoot about the same average score and are about equally consistent when playing under the same conditions. In essentially all two-person golf matches, however, the two golfers involved are not equally consistent and they do not have the same average score.

To proceed further, let us take advantage of the generally recognized phenomenon that golfers with low average scores tend to be more consistent (i.e., are less erratic or variable) than golfers with high average scores. As empirical evidence in support of this phenomenon, Figure 20.2 (see previous section) plots, for each of the 130 golfers mentioned earlier, the sample mean of that golfer's *HD* values versus the sample standard deviation of these same *HD* values. Clearly, there is solid evidence that the sample standard deviation tends to become large as the sample mean becomes large. Thus, when comparing Steady Eddie to Wild Willie (and, more generally, when comparing a good golfer to a not-so-good golfer), we would expect that both the inequalities $\mu_{se} < \mu_{ww}$ and $\sigma_{se} < \sigma_{ww}$ would typically hold. For any choices of μ_{se}, μ_{ww}, σ_{se}, and σ_{ww} that satisfy these two inequalities, the probability that Willie defeats Eddie is less than 1/2 based on our formula.

Thus, a good golfer like Steady Eddie who, on average, shoots lower scores and is also more consistent than a not-so-good golfer like Wild Willie *always* has an advantage under the current USGA Handicap System. Of course, the "size" of this advantage will depend on the specific values of μ_{se}, μ_{ww}, σ_{se}, and σ_{ww}. For a reasonable range of values for $(\mu_{ww} - \mu_{se})$, σ_{se} and σ_{ww} suggested by Figure 20.2, Table 20.1 provides values for Pr[Willie defeats Eddie]. As can be seen from our formula, even if Eddie and Willie have about the same average score so that $(\mu_{ww} - \mu_{se}) \approx 0$, the more consistent golfer (Eddie) will always have an advantage under the current USGA Handicap System.

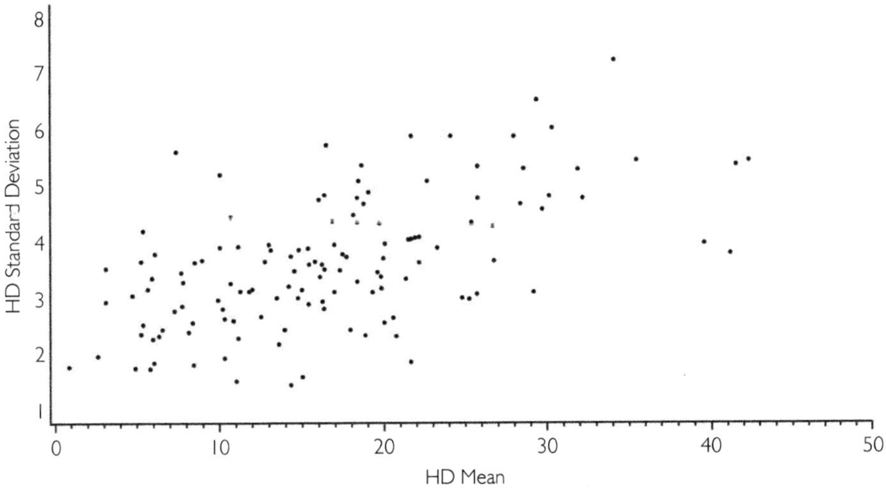

Figure 20.2 Graph of HD standard deviation versus HD mean for 130 players in 1997.

Table 20.1 ***Values of Pr[Willie Defeats Eddie] as a Function of (μ_{ww} – μ_{se}), σ_{se}) and σ_{ww}.***

$(\mu_{ww} - \mu_{se})$*	σ_{se}	σ_{ww}	Pr[Willie defeats Eddie]
0	2	3	0.42
0	2	4	0.37
0	2	6	0.33
5	2	3	0.40
5	2	4	0.36
5	2	6	0.32
10	2	3	0.38
10	2	4	0.34
10	2	6	0.31
20	2	3	0.34
20	2	4	0.31
20	2	6	0.28
30	2	3	0.30
30	2	4	0.28
30	2	6	0.26

*NOTE: Based on order statistics theory, the expected (or average) difference in handicap indexes between Wild Willie and Steady Eddie can be approximated as

$$E(HI_{ww} - HI_{se}) \approx 0.96(\mu_{ww} - \mu_{se}) - 0.7368(\sigma_{ww} - \sigma_{se})$$

For example, if $(\mu_{ww} - \sigma_{se}) = 5$ and $(\sigma_{ww} - \sigma_{se}) = 2$, then $E(HI_{ww} - HI_{se}) \approx 3.33$.

Our quantitative results regarding the importance of considering a golfer's consistency (or variability) when developing and evaluating a golf handicap system agree closely with the intuitive arguments given in Carney's *Golf Digest* article. Carney introduced the concept of an "anti-handicap" (i.e., a handicap index based on the worst 10, rather than the best 10, of a golfer's most recent set of 20 handicap differentials) to conjecture that a golfer like Steady Eddie would defeat a golfer like Wild Willie well more than half the time when these two golfers have about the same USGA handicaps.

Discussion and recommendations

Based on an empirically supported normality approximation, we have provided some quantitative arguments that the current USGA Handicap System is biased in favor of the better golfer in a two-person net medal play competition. Furthermore, it is possible to provide an intuitive explanation as to why such a bias is present. Given that the underlying distribution of a golfer's handicap differentials is approximately normal, then the expected (positive) difference between this normal distribution's mean (a measure of the golfer's average performance level) and the golfer's USGA handicap index (a measure designed to reflect the golfer's "potential" performance level) is directly proportional to the value of the standard deviation σ for this distribution. In other words, the larger σ is, the larger is the expected difference between a golfer's average performance level and his or her (USGA-defined) potential performance level. Thus, since $\sigma_{ww} > \sigma_{se}$, it follows that the expected difference between Wild Willie's average performance level and his potential performance level is larger than the corresponding expected difference for Steady Eddie. In other words, Willie's handicap index is less representative of the scores that he typically shoots than is Eddie's handicap index. Since Willie's handicap is thus generally too low when he competes against Eddie in a two-person net medal play competition, the result is that Eddie tends to win much more often than would be the case with an unbiased golf handicap system.

Given the preceding findings, it seems that the USGA should not continue to ignore the impact of variability in golfers' scores on its currently recommended golf handicap system. In particular, we feel that it is desirable to have a golf handicap system that has two important properties: (1) It involves a single numerical index for each golfer and (2) it is equitable despite the fact that golfers are quite different in their levels of consistency or variability. Fortunately, under the USGA assumption that all scores posted by a golfer for handicap purposes are valid, it is easy to suggest some simple modifications to the current USGA Handicap System that will result in a single-index handicap system with the property that each golfer will have roughly an equal chance of winning in a two-person net medal play competition.

More specifically, we suggest that the USGA's handicap index be modified so that it reflects a golfer's current average scoring ability (or average performance level) rather than a golfer's "potential" for shooting a score well below average. Given a golfer's current 20 handicap differentials, one modified handicap index that immediately comes to mind is the simple arithmetic mean of all 20 handicap differentials. Another modified handicap index that would be less sensitive to uncharacteristically low or high handicap differentials is a so-called "trimmed mean." For example, given the assumption of approximate normality (or, more generally, symmetry), the two lowest and the two highest handicap differentials could be discarded, and the mean of the remaining 16 handicap differentials would then serve as an appropriate modified handicap index. Of course, larger trimming proportions could be used if deemed appropriate.

Given that the USGA Handicap System will be in effect for the foreseeable future, can two golfers like Steady Eddie and Wild Willie make handicap adjustments for a particular two-person net medal play match so that the competition will be essentially equitable? The approach that we endorse requires both Eddie and Willie to use a modified handicap index like one of those suggested previously. Of course, this means that both golfers would have to keep up-to-date and accurate records on scores, course ratings, and slope ratings and then convert these scores to handicap differentials. As an additional benefit, keeping such records would be quite helpful to any golfer interested in monitoring his or her performance over time.

Finally, although our work focused specifically on two-person net medal play, our findings can be reasonably extrapolated to other types of golfing competitions in which USGA handicaps are used. For example, in a two-person net match play competition, scoring is based on the number of holes won, with handicap strokes assigned to holes in their order of difficulty. Because the current USGA Handicap System does not give Wild Willie enough handicap strokes when playing Steady Eddie, Willie would also be at a disadvantage in a two-person net match play competition.

Another type of popular golfing event is a net medal play tournament involving many golfers, with the winner being that golfer with the lowest net score. In such an event, it is very often the case that a high-handicap golfer is the winner, the main reason being the large range in possible net scores for high-handicap golfers as compared to low-handicap golfers. In particular, the larger the number of high-handicap golfers participating, the more likely it is that at least one of these high handicappers will post an extremely low net score. Having golfers compete within subgroups (or "flights") based on their current USGA handicaps is one way to make such tournaments more equitable.

ACKNOWLEDGMENTS

We wish to thank Dr. Forrest Council, Mr. Robert Druhan, and Dr. Carl Henley, all long-time members of a golfing foursome that includes the first author, for helpful discussions and insights about the game of golf.

REFERENCES AND FURTHER READINGS

Bingham, D. R., and Swartz, T. B. (2000), "Equitable handicapping in golf," *The American Statistician*, 54, 170–177.

Carney, R. (1988), "What's your anti-handicap?" *Golf Digest*, November, 70–71.

David, H. A. (1981), *Order Statistics* (2nd ed.), New York: Wiley.

Harter, H. L. (1961), "Expected values of normal order statistics," *Biometrika*, 48, 151–165.

Jorgensen, T. P. (1994), *The Physics of Golf*, New York: American Institute of Physics (AIP) Press.

Sarhan, A. E., and Greenberg, B. G. (1956), "Estimation of location and scale parameters by order statistics from singly and doubly censored samples (Part I: The normal distribution up to samples of size 10)," *The Annals of Mathematical Statistics*, 27, 427–451.

Stigler, S. M. (1969), "Linear functions of order statistics," *The Annals of Mathematical Statistics*, 40, 770–788.

USGA Handicap System, With USGA Course Rating System and USGA Handicap Decisions (1999), published by the United States Golf Association.

CHAPTER 21

"A game of which I am not familiar"

SCOTT M. BERRY

The new game of golf

One of the most famous quotes in golf history—and maybe sports history—is what Bobby Jones said after watching Jack Nicklaus win the 1965 Masters, "Nicklaus played a game of which I am not familiar." Bobby Jones, who himself was one of the greatest players in golf history, was commenting on Nicklaus' overpowering a golf course. Nicklaus' combination of distance and accuracy was something never before seen in the game. Nicklaus made this same comment when he played a practice round with Tiger Woods. When Woods was an "unknown" amateur player, Nicklaus said that Tiger would win more Green Jackets than he and Arnold Palmer combined (10). While sometimes a player comes along who hits the ball better and farther than anyone else, I think if Bobby Jones were alive today he would say that all Professional Golfers of America (PGA) Tour players play a game of which he was not familiar. In fact, almost all Tour players play a game of which anyone who played before 1990 is not familiar.

The last 20 years have seen amazing improvement in golf equipment. The advent of metal-headed drivers, improved irons, and juiced golf balls has apparently improved the distance and accuracy of modern golfers. In 1980 the leader in driving distance on the PGA Tour, Dan Pohl, averaged 274.1 yards. In 2002 that value would have placed Dan Pohl 158th on the PGA Tour. In this column I analyse data from the PGA Tour on driving statistics. I look at the changes in driving distance and accuracy over the last 24 years, and the impact any changes have made on scoring.

The PGA Tour web site, *pgatour.com*, provides data on PGA Tour players since 1980. The average driving distance and the driving accuracy percentage are reported. At each tournament, two holes are selected to measure driving distance. They are selected to be in opposite directions, to minimize the effects of wind, to be fairly flat, and to be holes where most players will hit drivers. On these two holes the distance of the drives for each of the players is recorded.

The average of these drives is considered the player's average driving distance. For all par-4 and par-5 holes, whether the drive is in the fairway is recorded. The percentage of times a player "hits the fairway" is recorded as that player's driving accuracy. I don't know the minimum-round requirements for inclusion in the yearly data sets, but there are between 150 and 200 players for each of the seasons. I downloaded the driving distance and accuracy for each of these qualifying players. There are 735 different players who have qualified over the 24 years' worth of data (I include the 2003 data through the 2003 Masters).

Robert Tyre "Bobby" Jones Jr. greets lack Nicklaus, 20, at the Metropolitan Golf Writers Association dinner in New York City on Jan. 26, 1960. Jones is the only golfer to win four major championships in one year, at age 28 in 1930, and Nicklaus is the youngest player in 50 years to win the U.S. Amateur Golf tournament. (AP Photo/Marty Zimmerman)

Average driving distance

Figure 21.1 shows a boxplot of the driving distances of each player for each golf season from 1980 through 2003. For each boxplot the box represents the middle 50% of the data, with the top of the box representing the 75th percentile and the bottom of the box the 25th percentile. The middle line represents the median. The "whiskers" extend to the largest and smallest observations that are not deemed outliers. Any outliers are shown with open circles. The average driving distances stay fairly flat during the 1980s, with a slight increase at the end of the decade. A slight decrease in driving distances in the early 1990s is followed by a steady, strong increase. From 1993 through 2003 the median driving distance increases every year (strangely, there was a decrease for three consecutive years from 1990 to 1993). In the current golf season,

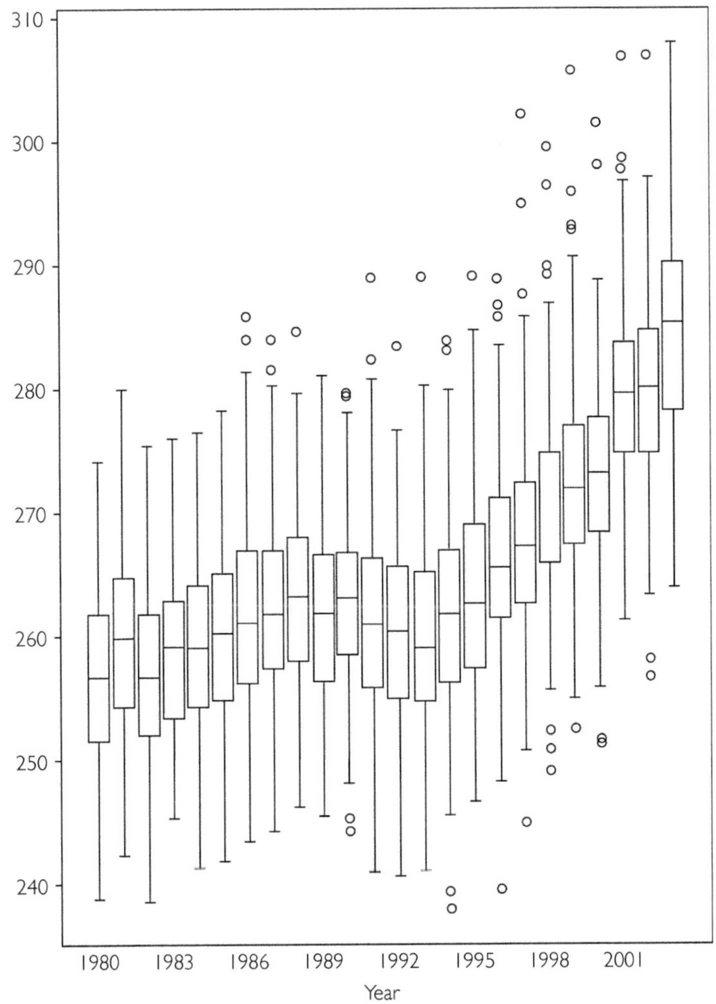

Figure 21.1 Boxplots for driving distances from 1980 through 2003.

every single PGA Tour player has an average driving distance larger than the median player's average driving distance in 1993. The average driving distance of the median Tour player has increased 27.3 yards from 1993 to 2003.

One difficulty in interpreting the effects of the equipment is that the player pool changes across the years. It may be that players are just getting stronger through time. There are many players included in multiple years. I want to know if the trend of increased distance observed in the late 1990s and the early 2000s is due to the equipment or to a stronger pool of players. Figure 21.2 shows the yearly average driving distance for the seven players included in at least 23 of the 24 seasons of data. The same trend appears to be holding for these players–a constant level until the early 1990s, when the average driving distance steadily increases. This is an interesting trend, especially because these golfers, all having been Tour players for at least 20 years, are getting older. They are all at least 40, and hitting the ball farther than they ever have!

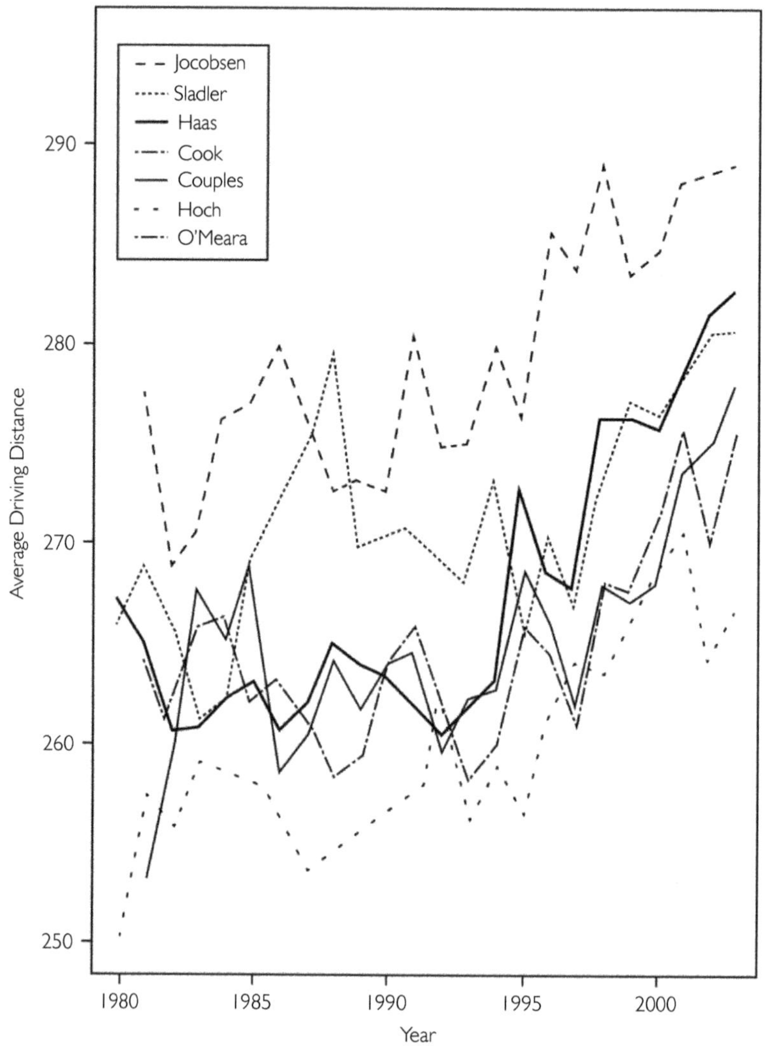

Figure 21.2 The yearly average driving distance for the seven players with 23 or more years of driving data.

In order to model the effects for each player, I present an additive linear model to address the effects each year has on individual players. Let Y_{ij} be the average driving distance for player i in year j. We model the average driving distance as a function of the player and the year:

$$Y_{ij} = \theta_i + \gamma_j + \epsilon_{ij}$$

where θ_i, represents the mean driving distance for player i in 1980, γ_j represents the difference in average driving distance for a PGA Tour player in year j as compared to 1980, and the ϵ_{ij} are independent normally distributed random variables with a mean of 0 and a standard deviation of σ. This model is rather simple in that it assumes players have the same driving distance across the years, except for the technology of the year. It assumes there is no aging effect. To

Table 21.1 ***Posterior means and standard deviations of the yearly effects from the normal linear model for the distance and accuracy cases.***

Year	Distance		Accuracy	
	Mean of γ	S.D. of γ	Mean of γ	S.D. of γ
1980	0	0	0	0
1981	2.36	0.49	–0.76	0.45
1982	–0.50	0.50	0.08	0.46
1983	0.74	0.51	–0.17	0.46
1984	1.44	0.51	–0.17	0.47
1985	2.02	0.52	1.52	0.48
1986	2.72	0.52	1.37	0.48
1987	3.30	0.52	2.20	0.48
1988	4.40	0.52	0.81	0.48
1989	2.01	0.53	1.74	0.49
1990	3.57	0.52	1.92	0.48
1991	1.27	0.53	3.97	0.49
1992	0.75	0.54	5.09	0.49
1993	–0.26	0.54	5.66	0.49
1994	1.41	0.54	5.93	0.50
1995	2.36	0.54	5.96	0.50
1996	5.17	0.54	4.95	0.50
1997	6.46	0.55	5.34	0.50
1998	9.09	0.55	6.44	0.51
1999	9.99	0.55	5.74	0.51
2000	10.38	0.56	5.76	0.51
2001	15.95	0.57	5.98	0.52
2002	16.02	0.57	5.10	0.52
2003	20.39	0.59	2.54	0.54

keep the comparisons to 1980, we restrict γ_{1980} to be 0. We use a hierarchical model for the distribution of players. Let the θ_i be normally distributed with a mean of μ and a standard deviation of τ. The prior distributions for the γs, μ, σ, and τ are non-informative, with little effect on the results. Table 21.1 presents the posterior means and standard deviations for each of the γs.

The posterior means for the γs demonstrate a very similar pattern to what is seen in the boxplots: a fairly constant level through the 1980s, with no difference larger than 5 yards until 1996. From the mid-1990s until 2003 there is a steady constant increase in average driving distance. The posterior mean in 2003 is a more than 20-yard increase in average driving distance. These estimates are probably conservative in estimating the true additive effects of technology. There is probably a decreasing ability of players to hit the ball a long way as they get older. This ability would be partially masked by the γs.

Driving accuracy

The previous section demonstrates that technology has a clear effect on driving distance in the last 10 years. Has technology also aided the ability to hit drives more accurately? It seems counterintuitive that it could do both. The error in the angle to which you have to hit the ball to keep it in the fairway gets smaller the farther the ball goes. Figure 21.3 shows the boxplots for the driving accuracy over the same 1980–2003 time span. There is not nearly as strong a pattern in the accuracy boxplots as there was in the distance boxplots. Amazingly, it appears that accuracy is also improving. In 1980 the median accuracy is 62.4%, and in 1981 it is 61.8%. This increases to a maximum of 69.8% in 1998. The accuracy has decreased in the last 5 years, with the 2003 median of 65.7% the lowest since 1990.

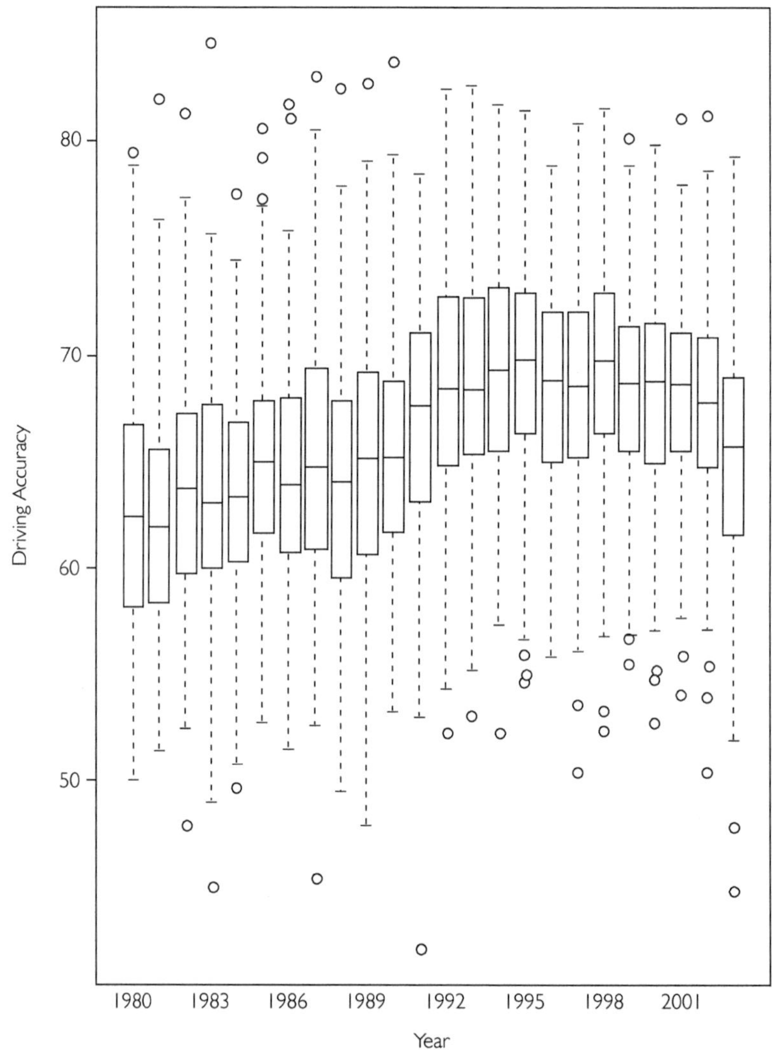

Figure 21.3 Boxplots for driving accuracy from 1980 through 2003.

We use the same linear model as in the previous section (which technically is not well-suited for "percentage" data, but due to the sample sizes and the percentages not near 0% or 100% this will work very well). The posterior means and standard deviations are presented in Table 21.1. Before the 2003 season, it appears that there has been about a 5% increase in accuracy percentage in the last 20 years. The technology of the last 20 years has helped pro golfers hit the ball not only farther but also straighter.

Combining distance and accuracy

The correlation between distance and accuracy is clearly negative. The correlation coefficient between them is –0.16. Thus, in general, the farther a player hits the ball, the less often it is in the fairway. In my column, "Drive for Show and Putt for Dough," in the Fall 1999 issue of *Chance*, I modeled the effect on scoring of driving distance and driving accuracy (as well

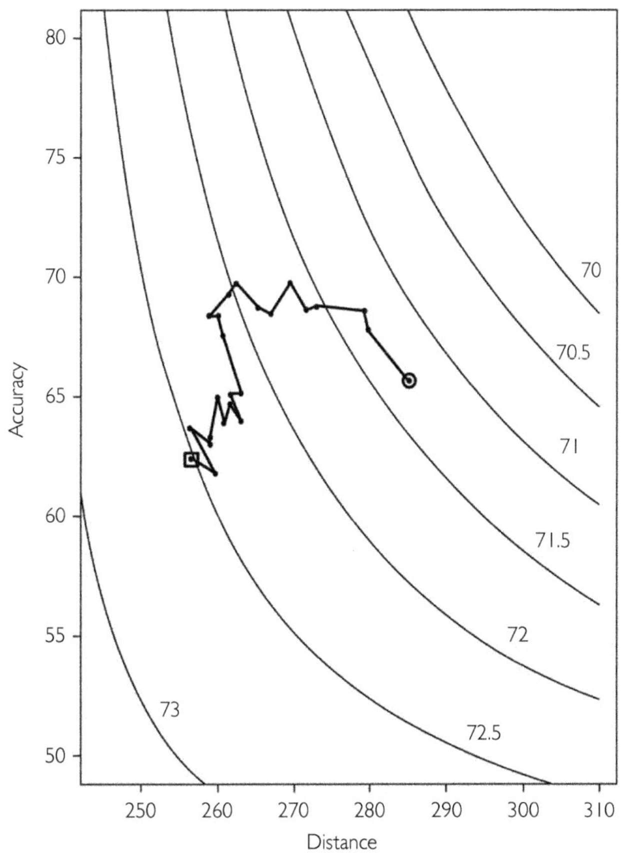

Figure 21.4 The contour plots for different values of distance and accuracy. Each contour is the mean score based on driving ability for a player of average putting, sand save, scrambling, and iron play ability. The points represent the median distance and accuracy for each year from 1980 to 2003. The 1980 median pair is shown with an open square and the 2003 median pair is shown with an open circle. The line follows the median pair through time.

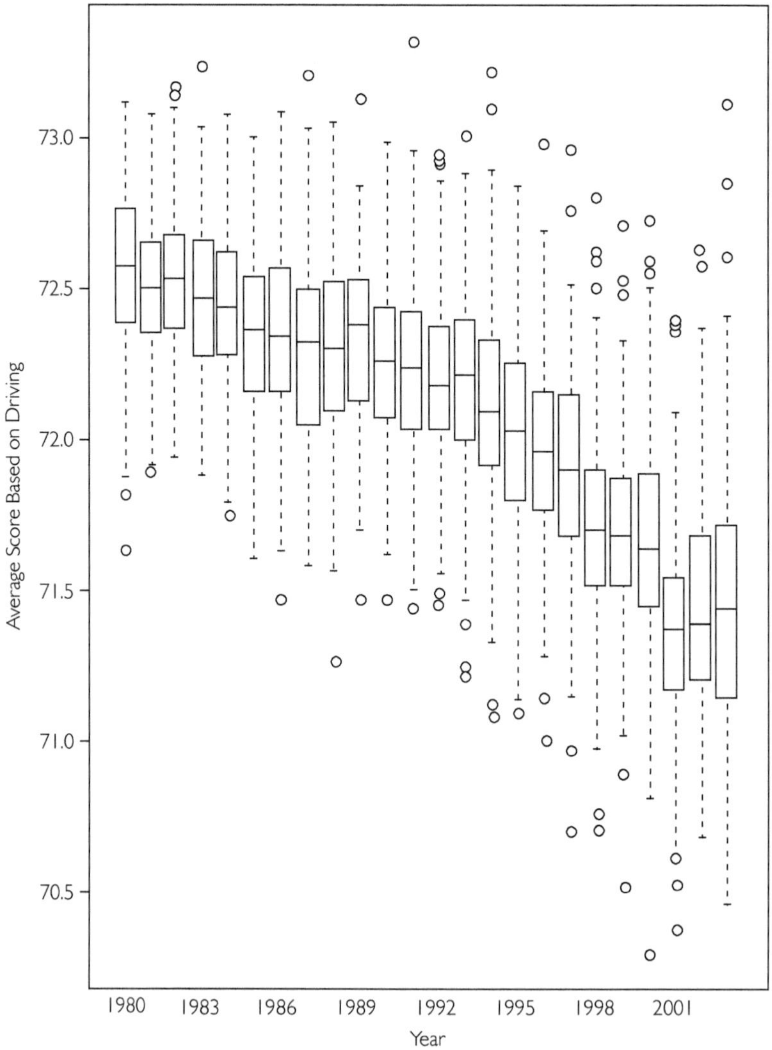

Figure 21.5 Boxplots for the mean score based solely on driving ability.

as other variables) using data from the 1999 season. That column was one of my favorites, as it showed the intuitive notion that there is a trade-off between distance and accuracy. For a player of average ability at putting, sand saves, scrambling, and iron play, the estimated scoring during the 1999 season, based on distance and accuracy, was estimated to be

$$57.70 + 0.067(\text{Distance}) + 0.373(\text{Accuracy})$$
$$- 0.0016(\text{Distance} * \text{Accuracy}).$$

Figure 21.4 shows the contour plots for this function. The contour lines show the distance and accuracy values that result in a constant mean score. For example, someone who averages 300 yards per drive but has only a 50% accuracy will have the same mean score as someone who averages 250 yards with a 72% accuracy, as long as the rest of their golling abilities are the same.

On Figure 21.4 I plotted the median distance and accuracy values through time. The open square is the median pair for 1980. The open circle is the median pair for 2003. The line follows the median pairs from 1980 through 2003. The median driver on the PGA Tour has improved, consistently through time, with about a 1.25 shot advantage, just from driving ability, in the last 24 years.

Figure 21.5 presents boxplots for the average score for a player that is average at each aspect of golf, except driving. Based on their accuracy and distance, I calculate the mean score, based on only driving ability, for each player in each season. This "driving scoring average" has a fairly constant decrease through time. The decrease through the 1980s appears to be from an improved accuracy, and the decrease through the 1990s appears to be mostly from improved distance, but also from not losing accuracy.

SUMMARY

Many of the best golf courses in the world have had to renovate their design to defend from golfers hitting the ball too straight and too far. It may be that actual scores on the PGA Tour have gone up (I don't know), but playing the same course as players did in 1980, scores would be about 1.25 shots per round better, if the rest of the golf attributes remained the same. In fact, chipping, iron play, and even putting have also probably improved. With new wedges, new putters, and more forgiving irons, there has surely been improvement in these areas.

With today's improved technology, the median player drives the ball better than the best drivers of 1980. The best drivers of today are almost 1.5 shots per round better than the best drivers of 1980, resulting in a 6-shot difference in a 4-round tournament. In the highly competitive sport of professional golf a 6-shot difference in a tournament is huge, or put more eloquently, it is a game of which the 1980 golfer is not familiar.

CHAPTER 22

Equitable handicapping in golf

DEREK R. BINGHAM AND TIM B. SWARTZ

Abstract

Previous studies on handicapping have suggested that in matches between two golfers, the better golfer has an advantage. This article considers medal play between two golfers when they are both playing well. The study uses the new slope system for handicapping. In this context we argue that it is actually the weaker golfer who has an advantage. The conclusions are based on both the analysis of actual golf scores and the analysis of theoretical models. We suggest an alternative scoring formula which leads to "fairer" competitions.

Keywords: *data analysis, order statistics, statistical modeling*

Introduction

The purpose of handicapping in the game of golf is to allow golfers of varying skill levels to compete fairly. The fairness of handicap systems has been addressed by a number of authors. For instance, Tallis (1994) investigated handicapping in various team competitions and found that certain handicap systems can be extremely unfair. This article focuses on matches between two golfers. In such matches, the general consensus is that handicap systems tend to favor the better golfer. For example, Scheid (1977) looked at 1,000 scores obtained from members of the Plymouth Country Club in Plymouth, MA. Using simulated matches based on these scores, he calculated the winning proportion for the better golfer in medal play using handicaps. He found that with handicap differentials greater than three, the winning proportion for the better golfer exceeded 60% and was as high as 85%. Pollock (1977) used a theoretical golf model based on the normal distribution and found that under reasonable variance assumptions, the better golfer has a competitive advantage in both medal play and match play. In a nice overview of the statistical literature pertaining to golf, Larkey (1998) explained that the pioneering work of Francis Scheid, beginning with Scheid (1971), was instrumental in initiating changes to the handicap system.

We address a slightly different problem in the context of medal play using the current handicap system. We are not concerned with the probability that golfer A defeats golfer B where golfer A is the better golfer. Rather, we are interested in the conditional probability that golfer A defeats golfer B when they are both playing well. Under a fair handicap system, the conditional probability that golfer A defeats golfer B should be 0.5. From a tournament perspective, this is a most practical issue. For a golfer does not expect to win a prize when he plays poorly, but when he plays well, at the very least, he expects a fair chance of winning a prize.

The central idea of this article is that golfers with higher handicaps (i.e., weaker golfers) tend to have more variability in their golf scores. Consequently, the weaker golfer is more likely to post results well below (or above) their typical scores. Under the current handicap system (i.e., the slope system), this greater variability is not taken into consideration. We demonstrate that this gives an advantage to the weaker golfer when both golfers play well.

In the next section, we briefly describe handicapping and indicate changes that have taken place in the handicap system since the articles of Scheid (1977) and Pollock (1977). Some real data are analysed in the section after that. We see that in medal play using handicaps, the opposite effect to that observed in Scheid (1977) and Pollock (1977) takes place. That is, when both golfers are playing well, it is actually the weaker golfer who has an advantage in winning a match. The advantage is most dramatic when the better golfer happens to be a low-handicapper. In the section "Analysis of a normal model," we observe the same phenomenon by approaching the problem from a theoretical perspective. In the section "An alternative to net scores," we describe a simple alternative scoring formula for use in tournament play and suggest that it is a more equitable method for prize distribution. We then conclude with some caveats and closing remarks.

Handicapping in golf

Since the publication of the articles by Scheid (1977) and Pollock (1977), there have been changes to the United States Golf Association (USGA) handicap system. Consider a golfer who has completed at least 20 rounds of golf and wishes to update his handicap. We refer to a gross score as a golfer's actual score obtained by summing his strokes over 18 holes. Using the old handicap system, a differential D_{Old} is defined as

$$D_{\text{Old}} = X - R,$$

where X is the adjusted gross score modified according to equitable stroke control (see Table 22.1) and R is the course rating for the course being played. Using the old handicap system, the golfer's handicap is calculated by averaging the 10 lowest differentials in the last 20 rounds, multiplying by 0.96, and then rounding to the nearest integer value. When playing on any course, a golfer's net score in medal play is then obtained by subtracting his handicap from his gross score. We note that only a golfer's lowest differentials (i.e., best rounds) are used in the handicap calculation. This is because the best rounds provide a better indication of ability; when a golfer is playing poorly, it is easy to lose concentration and have scores deteriorate. Using only the lowest differentials also provides a deterrent to the "sandbagger" who attempts to artificially inflate his handicap by intentionally playing poorly.

Table 22.1 ***Equitable Stroke Control. The maximum score that can be reported on a hole for a golfer of a given handicap.***

Handicap	Maximum score
≤9	double bogey
10–19	7
20–29	8
30–39	9
≥40	10

Since the early 1990s, the slope system has been gradually introduced to golf courses across North America as an enhancement to fair handicapping. The slope system generalizes the old handicap system by further considering course difficulty. Consider then a golfer who has completed at least 20 rounds where none of the resultant scores are deemed to be "eligible tournament" scores. The definition of the differential D is modified from $D_{\text{Old}} = X - R$ to

$$D = \frac{113(X - R)}{S},$$

where S is the slope rating for the course. A handicap index I is then obtained by averaging the 10 lowest differentials in the most recent 20 rounds, multiplying by 0.96 and truncating to 1 decimal place. When playing on a given course with slope rating S^*, the number of handicap strokes is now determined to be $I(S^*/113)$ rounded to the nearest integer value. If the competition is net medal play between two golfers, then the golfer's net score is obtained by subtracting the handicap strokes from his gross score. At this point we emphasize the distinction between the handicap index I and the number of handicap strokes for a given match; this distinction is important throughout the remainder of the article.

In Table 22.2, we provide an example of the calculation of differentials under both handicap systems. Using the old system, the golfer receives $0.96(158.3/10) \rightarrow 15$ handicap strokes independent of the course and the set of tees. Using the slope system, the golfer has a handicap index of $0.96(154.8/10) \rightarrow 14.8$, and, for example, is allotted $14.8(120)/113 \rightarrow 16$ handicap strokes when playing from a set of tees on a course whose slope rating is 120.

Consider then two golfers who play most of their rounds on courses of average difficulty (i.e., $S \approx 113$) and then play an easy course (i.e., $S^* < 113$). From the definitions above, it is clear that they receive proportionally fewer handicap strokes under the slope system than under the old system. Similarly, they receive proportionally more handicap strokes under the slope system than under the old system when playing a difficult course (i.e., $S^* > 113$). If we accept the results of the existing literature that suggest that the better golfer always has an advantage, then it follows that the advantage on an easy course will be even more pronounced under the slope system.

For detailed information on the calculation of handicaps, eligible tournament scores, equitable stroke control, and so on, see the USGA handicap Web page at www.usga.org/handicap/manual. For information on the theory underlying the slope system, see Scheid (1995) and the references therein.

Table 22.2 ***An example showing the calculation of differentials under the old handicap system (D_{Old}) and under the slope system (D).***

	X	R	S	D_{Old}	D
	90	70.1	116	19.9	19.4
	91	70.1	116	20.9	20.4
	94	72.3	123	21.7	19.9
	88	70.1	116	17.9*	17.4*
	89	70.1	116	18.9	18.4
	90	72.3	123	17.7*	16.3*
	91	72.3	123	18.7*	17.2*
	91	70.1	116	20.9	20.4
	91	70.1	116	20.9	20.4
	86	68.7	105	17.3*	18.6
	90	70.1	116	19.9	19.4
	92	72.3	123	19.7	18.1*
	85	68.0	107	17.0*	18.0*
	78	68.7	105	9.3*	10.0*
	82	70.1	116	11.9*	11.6*
	84	70.1	116	13.9*	13.5*
	94	72.3	123	21.7	19.9
	93	72.3	123	20.7	19.0
	89	72.3	123	16.7*	15.3*
	88	70.1	116	17.9*	17.4*
Sum				158.3	154.8

NOTE: Here *X* is the adjusted gross score, *R* is the course rating and *S* is the slope rating. The asterisks indicate the differentials that are included in the handicap calculations.

Data analysis

Data were collected from the computer handicap system at the Pemberton Valley Golf and Country Club in Pemberton Valley, British Columbia, during the 1997 golf season. Only full data on rounds played at Pemberton Valley were available. From the white tees, Pemberton Valley is a 5,972-yard golf course with a course rating of 68.5 and a slope rating of 120. From the blue tees, Pemberton Valley measures 6,407 yards with a course rating of 70.5 and a slope rating of 124. We demonstrate that in a competition between two golfers, the better golfer is disadvantaged in medal play when both golfers are playing well.

In this section, all reference to net scores should actually be interpreted as adjusted net scores where two adjustments are made. The first adjustment is due to equitable stroke control. Unfortunately, members enter adjusted gross scores into the computer database and there is no method of retrieving the (unadjusted) gross scores. However, we do not anticipate this to be a problem as we consider only the very best rounds of golf. Presumably, when a golfer is playing very well, the adjusted gross score will be equal or nearly equal to the gross score. The second adjustment is a standardization that transforms an adjusted gross score X_1 on a course with course rating R_1 and slope rating S_1 to an adjusted gross score $R_2 + S_2(X_1 - R_1)/S_1$ corresponding to a course with course rating R_2 and slope rating S_2. The standardization is based on the recognition that the differential $113(X_1 - R_1)/S_1$ is equivalent to the differential $113(X_2 - R_2)/S_2$. In our analysis, we transform scores from the blues tees to scores from the white tees at Pemberton Valley.

We limit our analysis to the 49 male members who completed 40 or more rounds during the year. We use the first 20 rounds as a tune-up period to allow the golfer to reach "mid-season" form and to allow the handicap index to settle. We also restrict our study to the immediate 20 rounds following the tune-up period. We hope that by using a shortened period, golfers will not experience dramatic changes in their skill levels. Each golfer will also have completed the same number of rounds of golf. Therefore our data analysis is based on 49(20) = 980 scores.

For each of the 49 golfers, we calculate net scores for each of the 20 rounds subsequent to the initial 20 rounds of golf. We then take each golfer's best net score and use this as a measure of when the golfer is playing well. In Figure 22.1, we plot the minimum net score versus the handicap index I where the handicap index is calculated immediately after the tune-up period. We see that the plot has a negative regression line (slope = −0.10) indicating that the weaker golfer has a better chance of defeating the stronger golfer when they both play well. A statistical test for the slope gives a marginally significant p value of 0.049, where of course it is more difficult to establish an effect when the effect size is small. To put the plot into perspective, we would expect a 25-handicapper to beat a 5-handicapper by (25 − 5)(0.10) = 2 net strokes in medal play when they are both playing their best games. This highlights the considerable inequity facing the low-handicapper in tournaments where only a few players win prizes.

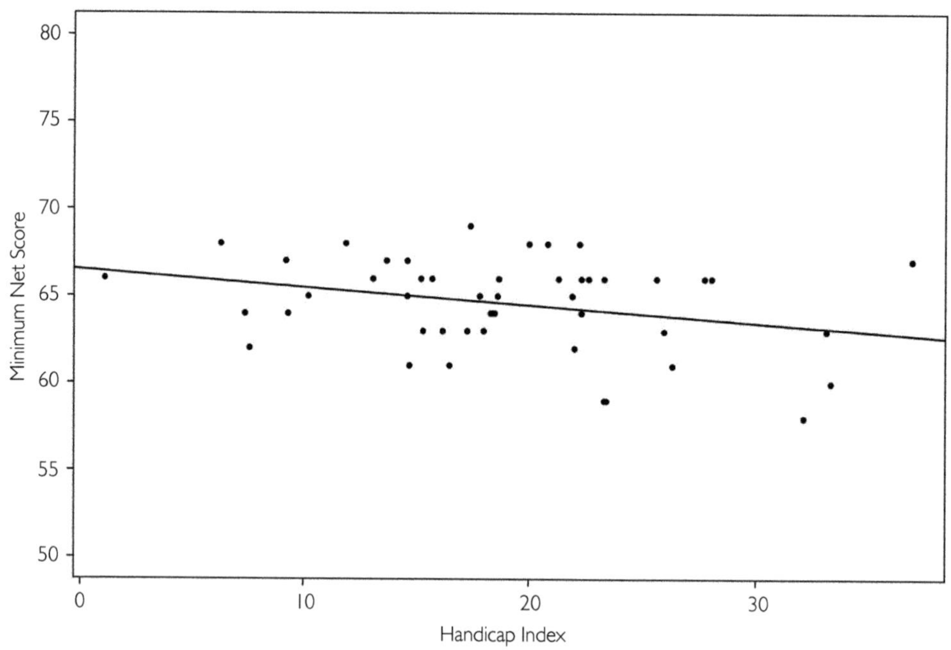

Figure 22.1 Minimum net score versus handicap index.

Using the same dataset, we again consider the 49 golfers and choose their best m net scores among the 20 rounds immediately following the initial tune-up period. Again, the net scores are based on the handicap index I determined at the end of the tune-up period. With m scores for each golfer, there are $\binom{49}{2}m^2$ possible matches between two golfers that can be simulated. The matches are simulated in the sense that the two golfers have not directly competed against one another. We consider $m = 2, 3, 4$ as this represents the best 10%, 15%, and 20% of net scores (i.e., occasions when the golfers play well). We exclude from the analysis the five pairs of golfers that have the same handicap index. In Table 22.3, we give the results of the simulated matches, and again, we observe that the weaker golfer enjoys an advantage when both golfers are playing well. For example, with $m = 2$, the weaker golfer wins or ties 65.6% of the matches.

In Figure 22.2, we plot the sample standard deviation of the 20 net scores following the tune-up period versus the handicap index established at the end of the tune-up period. The least squares line has a slope of 0.07 and a corresponding p value of 0.001. In this plot we see

Table 22.3 ***Simulated matches between two golfers based on their best m out of 20 net scores. The percentages refer to matches won, lost, and tied by the lower handicap (i.e., better) golfer.***

M	Matches	Wins	Losses	Ties
2	4,684	34.3%	52.8%	12.8%
3	10,539	35.4%	51.1%	13.5%
4	18,736	36.5%	50.4%	13.1%

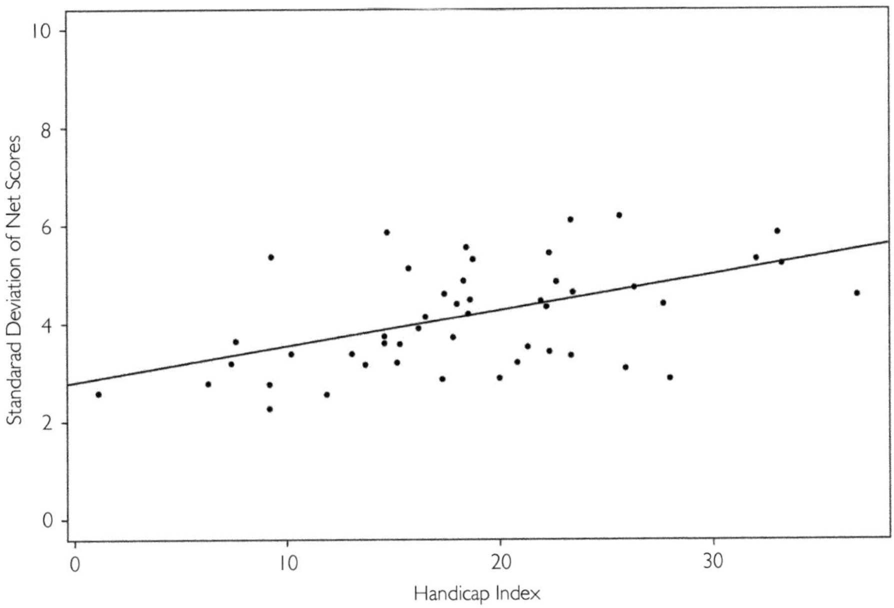

Figure 22.2 Standard deviation of net scores versus handicap index.

that the low handicap golfer is generally more consistent than the high handicap golfer. This result is not surprising. However, it does have implications for the fairness of matches when both golfers are playing well. This result, together with Figure 22.1, suggests that the very low handicap golfer has little chance of winning a tournament based on net scores.

We note that these results are striking and contradict the USGA literature. From section 10–2 of the USGA Handicap Formula manual on the USGA web page, we quote, "As your Handicap Index improves (gets lower), you have a slightly better chance of placing high or winning a handicap event."

Analysis of a normal model

The use of the normal distribution in modeling golf scores has been previously considered in the literature. For example, Pollock (1977) used a normal model to investigate theoretical properties of handicapping. More notably, Scheid (1990) carried out an extensive analysis on the distribution of golf scores. Except for a slightly longer right tail, Scheid (1990) found the normal approximation to be satisfactory.

In this analysis, we define X_{ij} as the gross score corresponding to the ith golfer on the jth course with course rating R_j and slope rating S_j. We assume that scores are independent and that

$$D_{ij} = \frac{113(X_{ij} - R_j)}{S_j} \sim \text{Normal}[\theta_i, \tau_i^2]. \tag{22.1}$$

The normality assumption can be motivated by the central limit theorem by noting that gross scores are obtained by summing the number of strokes taken over 18 holes. The assumption of the constancy of the distribution of D_{ij} over courses and tees is essentially the rationale for the slope system. Note that D_{ij} is not quite a differential since X_{ij} has not been modified according to equitable stroke control.

To test the adequacy of model (22.1), we calculate the Cramér-von Mises statistic W^2 and the corresponding p value (D'Agostino and Stephens 1986, p. 122) for each of the 49 golfers using the differentials obtained over the 20 post tune-up rounds. The results are consistent with the hypothesis of normality as only 2 of the 49 p values (0.046 and 0.047) are significant at level 0.05. When using the Anderson–Darling statistic A^2, only 1 p value (0.039) is found to be significant. With respect to the Scheid (1990) study, we observe that the slight departures from normality appear to be the result of a shorter than expected left tail. Although we recognize deficiencies in the model—such as the approximation of a discrete distribution by a continuous distribution—it provides insight on a number of handicap issues as well as motivation for a new scoring formula as developed in the section "An alternative to net scores."

Now according to model (22.1), consider two golfers playing the same course so that we can drop the subscript j. Let H_i be the number of handicap strokes for the ith golfer, as described in the section "Handicapping in golf," and without loss of generality let $H_1 < H_2$ such that the first golfer is the better player. Then

$$X_i \sim \text{Normal}[\mu_i, \sigma_i^2],$$

where $u_i = R + S\theta_i/113$ and $\sigma_i = S\tau_i/113$. Our interest lies in the investigation of

$$P_k = \Pr(\text{the better golfer wins} \mid \text{both golfers play well})$$
$$= \Pr(X_1 - H_1 < X_2 - H_2 \mid X_i < \mu_i - k\sigma_i, \quad i = 1, 2),$$

where $k > 0$. Here, $X_i - H_i$ represents the net score of golfer i and we condition on both golfers playing better than k standard deviations below their average gross score. Since $\mu_i - H_i$ represents the average net score of golfer i, it follows from the results of Scheid (1977) and Pollock (1977) that

$$\Delta = (\mu_2 - H_2) - (\mu_1 - H_1) > 0.$$

When $\sigma_1 < \sigma_2$ and k $> \Delta/(\sigma_2 - \sigma_1)$, it is then easily shown that

$$P_k = \frac{\int_{z=-\infty}^{(\Delta-k\sigma_2)/\sigma_1} \phi(z)\left[\Phi(-k) - \Phi\left(\frac{\sigma_1 z-\Delta}{\sigma_2}\right)\right] dz}{\Phi^2(-k)},$$

where φ is the density and Φ is the cumulative distribution function of the standard normal. To get a sense of P_k for realistic values of k, consider typical values $\Delta = 1, \sigma_1 = 2$, and $\sigma_2 = 3$. Integrating via Simpson's rule, we obtain the decreasing probabilities $P_{1.0} = 0.39$, $P_{1.5} = 0.22$, $P_{2.0} = 0.10$, and $P_{2.5} = 0.03$.

Now let $G(z)$ be such that $G'(z) = \phi(z)\left[\Phi(-k) - \Phi\left(\frac{\sigma_1 z-\Delta}{\sigma_2}\right)\right]$ and $G(-\infty) = 0$. Then

$$\lim_{k\to\infty} P_k = \lim_{k\to\infty} \frac{G\left(\frac{\Delta-k\sigma_2}{\sigma_1}\right)}{\Phi^2(-k)} = 0,$$

where the second equality follows from an application of l'Hospital's rule. Therefore, as both golfers play better (i.e., $k \to \infty$), it becomes impossible for the better golfer to win the match. This conclusion is in the same direction as the empirical results from the section on "Data analysis."

Mosteller and Youtz (1992) considered the scores of professional golfers during the final two rounds of PGA tournaments under ideal weather conditions. They found that adjusted scores could be well approximated by a base score plus a Poisson variate. Whereas the Mosteller and Youtz (1992) analysis involved the most homogeneous of conditions, we are faced with data involving golfers of varying skill levels playing under various conditions. Furthermore, little is at stake for our golfers and we therefore do not expect their effort to be constant over all rounds. In an earlier version of this article (which is available upon request), we extended the Mosteller and Youtz (1992) model such that the net score of the ith golfer $X_i - H_i$ is given by

$$X_i - H_i = B_i + W_i,$$

where B_i is the idealized or perfect base score for the ith golfer and $W_i \simeq$ Poisson (λ_i). This model leads to the same theoretical result that the better golfer is disadvantaged when two golfers play well.

An alternative to net scores

In this section we propose a simple method for promoting fairness in matches. It is based on the normal model from the previous section "Analysis of a normal model" and consideration of the slope system. From model (22.1), it follows that

$$T = \frac{D_{ij} - \theta_i}{\tau_i}$$

has the same distribution (i.e., standard normal) for all golfers and ought to be a "fairer" yardstick than using the traditional net score. Moreover, T ought to work well generally (i.e., matches based on T should be fair unconditionally and matches based on T should be fair conditionally when golfers play well/average/poorly). Of course, T cannot be calculated since θ_i and τ_i are unknown. For simplicity of notation, we now drop the subscripts i and j. Our strategy then is to estimate θ and τ with $\hat{\theta}$ and $\hat{\tau}$, respectively, and then rate golfing performances using the quantity

$$T^* = \frac{D - \hat{\theta}}{\hat{\tau}}. \tag{22.2}$$

In obtaining $\hat{\tau}$, we refer to Figure 22.3 which plots the standard deviation of the differentials versus the handicap index I for each of the 49 golfers over the 20 post tune-up rounds. The handicap index I is established immediately after the tune-up period. Using a least squares fit in Figure 22.3, we have that $\hat{\tau} = \hat{\tau}(I)$, where

$$\hat{\tau}(I) = 2.74 + .053\, I. \tag{22.3}$$

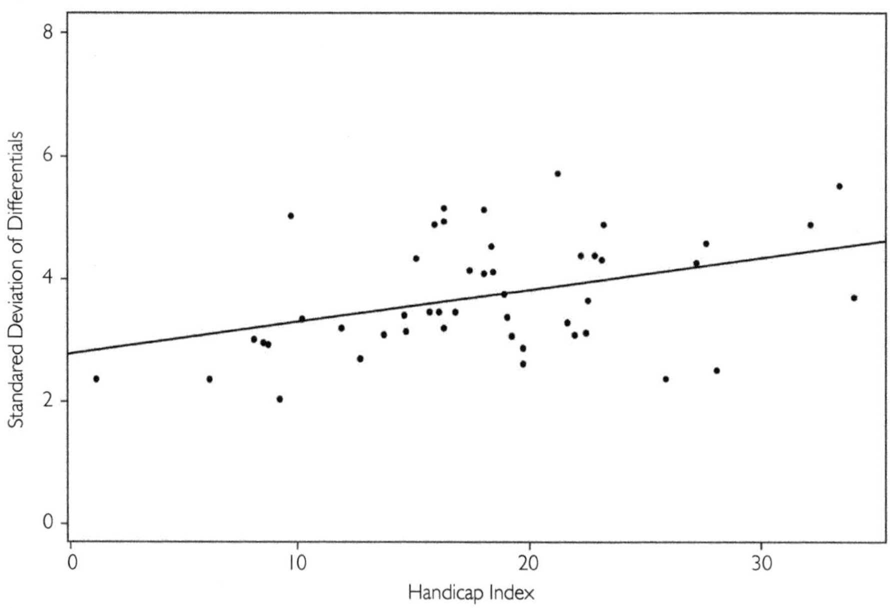

Figure 22.3 Standard deviation of differentials versus handicap index.

Note that Figure 22.3 involves adjusted gross scores and therefore $\hat{\tau}(I)$ may slightly underestimate τ.

In obtaining $\hat{\theta}$, we recall from the section "Handicapping in golf" that the handicap index I is calculated by

$$I = .96\left(\frac{\sum_{i=1}^{10} D_{i:20}}{10}\right), \qquad (22.4)$$

where $D_{i:20}$ is the ith lowest differential in the last 20 rounds. We refer to $D_{i:20}$ as the ith order statistic. Now a differential is only included in the calculation of the handicap index I when a golfer plays well. Presumably, under these conditions, a golfer's adjusted gross score equals his gross score where the adjustment is due to equitable stroke control. Therefore, using (22.4) and the normal model (22.1) without subscripts,

$$I = .096 \sum_{i=1}^{10} (\theta + \tau Z_{i:20}),$$

where $Z_{i:20}$ is the ith order statistic of the standard normal distribution. We therefore estimate θ by

$$\begin{aligned}\hat{\theta} = \hat{\theta}(I) &= \left[I/.096 - \hat{\tau}(I)\sum_{i=1}^{10} \mathrm{E}(Z_{i:20})\right]/10 \\ &= I/.96 + .7674\ \hat{\tau}(I), \qquad (22.5)\end{aligned}$$

where the moments of the standard normal order statistics were obtained from the tables in Harter (1961).

Putting (22.2), (22.3), and (22.5) together, we propose the quantity

$$T^* = \frac{113(X - R)/S - 2.10 - 1.082\,I}{2.74 + .053\,I}$$

as a new scoring formula where small values indicate good rounds of golf.

We investigate the performance of T^* in the same way that we exposed the inadequacy of traditional net scores in Figure 22.1 and Table 22.3. For each of the 49 golfers, we calculate T^* for each of the 20 rounds subsequent to the initial 20 rounds of golf. Again, the handicap index I is calculated prior to each of the rounds of golf. We then take each golfer's best T^* and use this as a measure of when the golfer is playing well. In Figure 22.4, we plot the minimum T^* versus the handicap index. We see that the points are scattered in a horizontal band. A straight line regression gives an insignificant p value of 0.27 for the slope. The lack of a pattern suggests that there are no handicap pairings for which a systematic advantage exists when using the new quantity T^*.

Using the 49 golfers, we also choose their m best T^* scores among the 20 rounds immediately following the initial tune-up period. Here the T^* scores are based on the handicap index I determined at the end of the tune-up period. We consider the $\binom{49}{2}m^2$ simulated matches between two golfers where $m = 2, 3, 4$ and exclude from the analysis the five pairs of golfers that have the same handicap index. In Table 22.4, we see that the outcomes of the simulated matches are far more balanced than when using traditional net scores. For example, with $m = 4$, the better

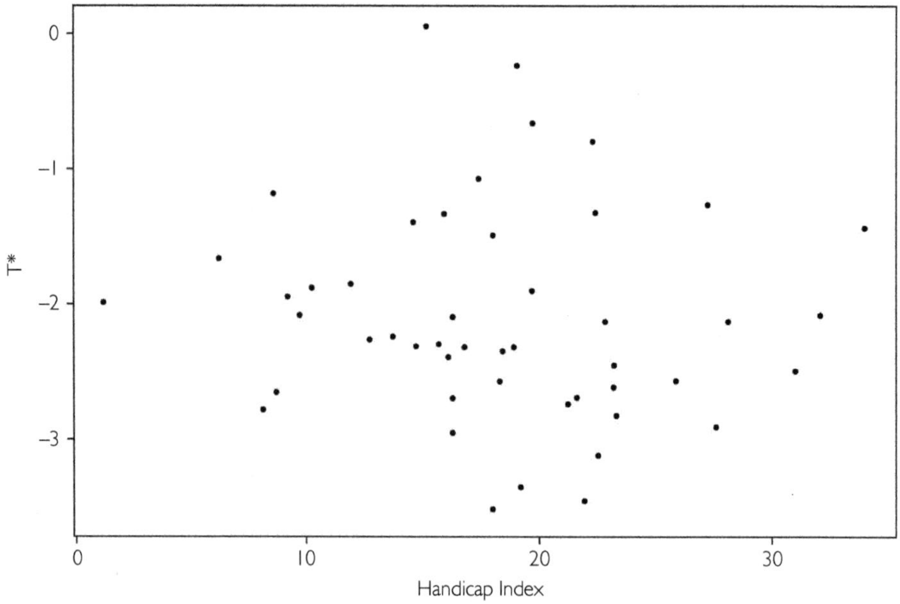

Figure 22.4 Minimum T* versus handicap index.

Table 22.4 ***Simulated matches between two golfers based on their best m out of 20 scores using the statistic T*. The percentages refer to matches won, lost, and tied by the lower handicap (i.e., better) golfer.***

m	Matches	Wins	Losses	Ties
2	4,684	47.9%	52.1%	0%
3	10,539	49.0%	50.9%	0%
4	18,736	50.3%	49.7%	0%

golfer wins 50.3% of the matches. This is much closer to the idealized value 50% than the value 36.5% + (1/2)13.1% = 43.1% obtained using traditional net scores.

Unfortunately, the average golfer may not find the quantity T^* as intuitive and appealing as the traditional net score. In tournament play, however, we do not envision the individual golfer actively calculating T^*. This would be the responsibility of the prize committee. The situation is no different from tournaments where the prize committee privately selects random holes to establish individual handicaps and then determines net scores (e.g., the Peoria system).

We wish to stress that T^* is a first attempt at improving fairness in tournament play. The coefficients in T^* may be modified through a more comprehensive study of golf data. On the other hand, we have seen in Table 22.4 that T^* offers dramatic improvements over the traditional net scores and allows comparisons of scores from different courses. It is notable that the use of T^* does not require an overhaul of the handicap system; the ingredients in T^* are the course rating R, the slope rating S and the handicap index I, all of which are the same as before.

Concluding remarks

We have demonstrated that a good golfer is disadvantaged in a tournament when prizes are distributed strictly according to net scores. It appears that this phenomenon has not gone unnoticed. As a result, there are typically various ways to win prizes in a large golf tournament. For example, the field of golfers is sometimes divided into flights so that only golfers of comparable abilities are competing against one another. Prizes are also often distributed according to gross scores within flights. Although these modifications are helpful, inequities are still bound to exist. For example, the first flight may group golfers in the 0 (scratch) to 8 handicap range, where the 8 handicapper still has a distinct advantage over the scratch golfer when they both play well. Also, some tournaments are so small that it is not sensible to divide the field into flights.

We have focused on the case when golfers play well and, consequently, may win a tournament prize. In the less interesting case when both golfers play poorly, we expect the opposite results. That is, due to the greater variability in the scores of the weaker golfer, we expect that the better golfer is advantaged. Also, we conjecture that the same general conclusions would be obtained in an analysis of match play using handicaps. To study match play, additional data are required. In particular, one needs the scores on individual holes.

In the section "An alternative to net scores," we presented a simple scoring formula T^* that is more equitable than the calculation of traditional net scores. The basic idea underlying the approach is that golf scores become more variable as the handicap increases. However, there may be golfers who are consistent but are not very good. For example, certain senior golfers who hit the ball short distances but with high accuracy may fall into this class. For them, the proposed scoring formula may be penalizing. We suggest a more extensive collection of golf scores to further assess variability. Perhaps what is needed is a variability index, V, for every golfer. Like the handicap index I, V could be calculated from a golfer's best 10 differentials from their most recent 20 rounds. Following the section "An alternative to net scores," a gross score X could then be transformed to a new net score $[113(X-R)/S-\hat{\theta}(I)]/V$. Of course, unlike T^*, such a proposal would require a complete overhaul of the USGA handicap system.

FURTHER READING

Balakrishnan, N., and Cohen, A. C. (1991), *Order Statistics and Inference*, New York: Academic Press.

REFERENCES

D'Agostino, R. B., and Stephens, M. A. (1986), *Goodness-of-Fit Techniques*, New York: Marcel Dekker.

Harter, H. L. (1961), "Expected values of normal order statistics," *Biometrika*, 48, 151–165; Correction, 48, 476.

Larkey, P. D. (1998), "Statistics in golf," in *Statistics in Sport*, ed. J. M. Bennett, London: Arnold, pp. 121–140.

Mosteller, F., and Youtz, C. (1992), "Professional golf scores are Poisson on the final tournament days," in *Proceedings of the Section on Statistics in Sports*, Alexandria, VA: American Statistical Association, pp. 39–51.

Pollock, S. M. (1977), "A model of the USGA Handicap System and "fairness" of medal and match play," in *Optimal Strategies in Sports*, eds. S. P. Ladany and R. E. Machol, Amsterdam: North Holland, pp. 141–150.

Scheid, F. (1971), "You're not getting enough strokes," *Golf Digest*, June; reprinted in *The Best of Golf Digest*, 1975, 32–33.

—— (1977), "An evaluation of the handicap system of the United States Golf Association," in *Optimal Strategies in Sports*, eds. S. P. Ladany and R. E. Machol, Amsterdam: North Holland, pp. 151–155.

—— (1990), "On the normality and independence of golf scores, with applications," in *Science and Golf: Proceedings of the First World Scientific Congress of Golf*, ed. A. J. Cochran, London: E. and F.N. Spon, pp. 147–152.

—— (1995), "Adjusting golf handicaps for the difficulty of the course," in *Proceedings of the Section on Statistics in Sports*, Alexandria, VA: American Statistical Association, pp. 1–5.

Tallis, G. M. (1994), "A stochastic model for team golf competitions with applications to handicapping," *Australian Journal of Statistics*, 36, 257–269.

CHAPTER 23

Exploratory analysis of European Professional Golf Association statistics

ROBERT KETZSCHER AND TREVOR J. RINGROSE

Abstract
Various summary statistics collected by the Professional Golf Association European tour for 1998–2001 are investigated, in particular to see how well these predict golfing success. The paper tries to avoid predicting something by itself, as previous papers have in effect done, concentrating on key parts of the game only. We conclude that the game of golf is too complex to characterize by the relatively small set of variables that are currently used, and we suggest further quantities which, if measured, would allow a more informative analysis.

Keywords: *correlation, European Professional Golf Association, golf, linear regression*

Introduction

The Professional Golf Association (PGA) in both Europe and the USA collect summary statistics for each of the players on their respective tours. A few papers (see 'Literature review and comments') have looked at the US PGA data for a small number of years, but surprisingly little has been done with these data, despite their popularity with golf journalists, especially in the USA. These previous papers have concentrated on trying to assess the importance of various elements of playing style and ability in predicting golfing success, and we largely follow this approach.

We look at the European tour performance data for the years 1998–2001. The 'Variables' section describes the data collected, the section 'Literature review and comments' reviews previous work, the subsequent section introduces two new variables derived from the existing data, the sections that follow present some simple analyses and discussions of the data and the final section discusses possible future work.

We now discuss why using data for every round of each player should be representative of a player's true abilities and hence lead to comparatively more useful results than, for example, analysing the match statistics of all football games over a season. Essentially it is because the nature of the game and its reward system means that a player has an incentive to try his best with every shot, whereas in many sports the result of a game or championship may be clear long before the end, so players or teams may not try as hard as normal.

With the 'cut-off' system players who are far adrift are not allowed to participate in the final two rounds. In football a team which is 4–0 down at half-time will often have given up and not play to their true potential in the second half. Hence data for the second half may not be representative of the team's true abilities. In golf the cut-off prevents this. This does not imply biased sampling; it ensures that only competitive data are being used.

Prize-money is available for everybody who exceeds the cut-off and is staggered depending on the final total score. If a player is 10 shots off the lead with nine holes to play it would be extremely unlikely for him to win the tournament. However, by maintaining this margin or even decreasing it over the last nine holes he will still ensure a respectable payment, which depends on every single shot and putt that he takes. If in a football match a team is winning or losing by three clear goals with 10 minutes to go, the remaining 10 minutes will not be as representative as earlier parts of this game.

Finally, in league games of most sports, frequently the league champions and/or relegated teams are already known before the last game(s). Hence these teams may not try as hard as teams who have something to play for. In golf each tournament presents a new chance for the golfer to do well, all players starting from zero each time.

Playing in a league system, teams must play their matches at a given frequency. This is sometimes worsened by progress in various cup competitions and the resulting extra matches. Consequently teams must play even when they are not at their best. Golf in contrast is different. Generally players participate in only about 60% of tournaments on the PGA European tour, which enables them to select tournaments and courses in which they hope to do well. This also enables them to withdraw from tournaments because of injuries. In team league games such as football, injuries to key players imply that an understrength team is fielded.

These comments only apply to medal play, where players compete as individuals and their finishing position is determined by the total number of shots taken. As pointed out by Larkey (1998), there are problems in using the data from other types of competition.

Variables

The standard figures collected from the PGA European tour and the PGA tour are driving distance and accuracy, greens in regulation GIR, total putts, putts per GIR and sand saves. These are all usually presented as averages or percentages for each player over a whole season and are assumed to be so throughout the next few sections. In the section 'Per-round results for 2001' we discuss less aggregated data, where the variables are recorded for each player for each round. We now define and discuss these variables in greater detail.

A drive is a tee shot taken on a par 4 or 5 hole. Generally the purpose of this shot is to achieve a good position from which a player will reach the green with one or two more shots. Each drive has two main components. Firstly the driving distance DrivDist is how far the ball has travelled from the tee to its final resting position. Unlike the other figures, this is based on a sample of two holes per round, specifically those where it is thought that players will try to hit the drive as far as possible. The Tournament Director chooses two holes where drivers will be used, which face different directions (to negate the effect of wind), and with few contour changes (so that measurement is easier). The distance is taken from course measurements and does not (at present) use the global positioning system.

Secondly, the driving accuracy DrivAccu is the percentage of drives which finish on the fairway or on the green, which is the ideal finishing position. Less favourables outcomes are bunkers (i.e. in sand), the rough on either side of the fairway or even worse among trees, water hazards or 'out of bounds'. This includes the results of all tee shots on all holes of par 4 or 5, not just those used to calculate DrivDist.

The variable greens in regulation GIR represents the percentage of holes in which the player reached the green in not more than the 'expected' number of shots, i.e. in $n - 2$ shots on a par n. However, players can reach par 5 holes in two shots rather than the regulation three shots and this difference is not incorporated in GIR. However, because in such cases the second shot is from a greater distance, even if it finds the green it usually results in a lengthy first putt. The potential situation after this putt has been taken (i.e. the regulation third shot) is similar to the case when the green is reached with the third shot by means of a short iron shot, so that not being able to distinguish whether par 5 holes are reached in two or three shots does not seem crucial. Similarly, if the green is not hit in regulation there is no distinction between $n - 1$, n, $n + 1$ etc. shots.

The variable average putts AvPutts is simply the average number of putts per round. As noted above, the green in a par n should be reached in $n-2$ shots. This leaves two putts to complete the hole 'to par'. Each hole always has such an allocation of two putts, so consequently players are expected to take 36 putts each round.

The variable putts per GIR PPGIR is the average number of putts per hole, for all holes where the green was hit in regulation. This implies that the previous stroke was usually an approach shot from distance, so this variable should characterize putting skill slightly better than AvPutts, which includes more cases resulting from short chips onto the green, where resulting putts are generally shorter.

The variable sand saves SandSave represents the percentage of holes where the player achieved par out of all the holes where he found a bunker. Bunkers are placed on courses to complicate the play of the hole significantly without automatically implying penalty strokes, as opposed to water hazards.

There are various ways to compare the golfing success of players. The 'money list' compares players on the total amount of prize-money won in each season. Prize-money is staggered depending on the finishing position in a tournament. The actual amounts differ from tournament to tournament. All the other variables are unweighted averages over the whole season and hence the money list does not seem to be an appropriate comparison.

Another possible way is to compare the average finishing positions. Although this seems a much better unweighted indicator, it still does not relate to the actual golf shots. A player finishing runner-up six shots adrift is treated the same as a player losing in a play-off.

Therefore it seems that the best indicator is the stroke average StrAver, the average number of strokes taken per round over the entire season. This is also not perfect since some courses have a higher par, or higher difficulty with the same par, than others and players enter different combinations of tournaments, but this effect should be small over a whole season.

We define a further variable, average non-green shots NonGreen, to be the difference between the stroke average and average putts, since this is an exact measure of the number of shots that a player takes to reach the green, whereas GIR is less exact for the reasons noted above. However, the one advantage of GIR over NonGreen is that it accounts for the differences in course lengths (pars).

The same course may differ in difficulty between different times of the day in the same tournament, potentially affecting all the above variables, so again we hope that, over a full season, this will average out.

Literature review and comments

It seems that so far there have been only a few simple investigations of the standard variables described in the previous section.

Belkin et al. (1994) reviewed some earlier work and calculated means, standard deviations and correlations for the above variables for the years 1986–1988 for the US PGA tour. For each year significant (Pearson) correlations were found (in the obvious directions) between the stroke average and each of the other variables, apart from drive distance in 1986. Regression with stroke average as the response and the other variables as explanatory gave similar results, with drive accuracy in 1986 and sand saves in 1987 and 1988 the only non-significant terms. The variables together had R^2-values of at least 0.8 in each year. For those three years there was a 'trend' for the coefficients for drive distance and drive accuracy to increase over time, which Belkin et al. (1994) ascribed to changes in location of the pin on greens making driving relatively more important. Engelhardt (1995) looked at this further, using 1993 and 1994 US PGA data. It is not entirely clear what Engelhardt (1995) was attempting, but it claims to calculate (Spearman) correlations between money rank with driving and GIR ranks. However, since only the top 10 money earners were used, this does not seem to make sense. The claim was made, though, that driving had become more important than GIR over time. Engelhardt (1997, 1999) used Mann–Whitney's U-test to assess the 'significance' of the difference between the above variables for the top 10 and bottom 10 earners on the PGA tour in 1995. Despite such *post hoc* testing only the two driving variables gave significant results. A similar look at the PGA European tour for 1998 by Jiminez and Fierro-Hernandez (1999) found drive distance, GIR and sand saves to be significant.

Those references seemed to assume that GIR represents a player's ability in iron play whereas AvPutts represents their putting ability. However, this is clearly not true. A player's GIR is a function of their driving and iron play, since the better their driving the easier their iron shots will be and the more likely they are to reach the green quickly. Similarly, the better their driving and iron play the closer their first putt is likely to be to the hole. Anecdotally, we suspect that the players with the lowest AvPutts may be those with the best short iron games. Alternatively, players who aggressively aim for the pin rather than for the centre of the green will often leave short putts but will often miss the green and leave short chips (sometimes bunker shots) onto it, both cases tending to lead to low AvPutts. In contrast, a very good putter may play conservatively, aiming for the centre of the green and leaving long putts, thus concentrating on his strength but resulting in a high AvPutts. Hence, an aggressive player with good short iron skills may have a lower AvPutts than a conservative player who putts well.

This is all speculation, but it illustrates that the relationship between the numbers collected and players' abilities is not as straightforward as has been assumed, since on any hole every shot has an effect on every subsequent shot. Drive distance and drive accuracy describe a player's ability at tee shots on par 4 and 5 holes, but the other variables are mixtures of players' abilities with woods, long and short irons and (for AvPutts) the putter.

Previous papers have seemed to regard the relationship between these variables and stroke average as a genuine 'finding', but surely it is hardly surprising. By definition the stroke average is the sum of AvPutts and NonGreen, so since GIR is just a recoding of the latter it is hardly surprising that the stroke average is predicted very well by a regression model including AvPutts and GIR. If NonGreen were used instead of GIR then the prediction would of course be exact. The same comments apply to money earned as a response, since this can be regarded as a recoding of stroke average.

Hence, rather than just trying to predict the stroke average by using components of itself, the present paper attempts to predict it by using other quantities. This in turn leads to a consideration of other variables which would need to be collected to achieve genuine insight into how different abilities and styles relate to the stroke average.

New variables

It will be interesting to compare the standard figure PPGIR, the average number of putts taken on holes where the green was reached in regulation ($n - 2$ shots or fewer for a par n hole) with a new figure PPNIR, the average number of putts taken on holes where the green was not reached in regulation. This can be calculated from the existing data as follows:

$$\text{PPNIR} = \frac{\text{total putts for greens not hit in regulation}}{\text{numberof greens not hit in regulation}}$$

$$= \frac{\text{AvPutts} - 18(\text{GIR}/100)\text{PPGIR}}{18(1 - \text{GIR}/100)}.$$

Figure 23.1 shows plots of PPGIR against PPNIR for the 2000 season (plots for other years are very similar), with the line of equality shown. For all golfers and seasons PPGIR is much larger than PPNIR, so if a player finds the green in regulation he will on average take more putts than if he does not. This can be explained by the fact that, at the professional level, even if the green is missed by the approach shot the ball will still often be fairly close to the green, so the next shot is usually reasonably short (e.g. a chip or pitch). Hence in this case the first putt will be from closer to the pin than if the green had been reached with the, often quite long, approach shot.

If the values of PPGIR and PPNIR are reasonably close, we could assume that, if a player finds a green in regulation, he is quite close to the pin, i.e. not much further than if he had chipped on. Hence the greens hit in regulation are hit very well. This could be interpreted as a measure of aggression, i.e. the player risks not hitting the green at all, but when he does the shot is close to the hole. Hence we tentatively define Aggression Aggr as

$$\text{Aggr} = \frac{\text{PPNIR}}{\text{PPGIR}}.$$

However, in Figure 23.2 we see that Aggr depends primarily on PPNIR and is positively correlated with both its numerator and its denominator (again, plots for other years are very

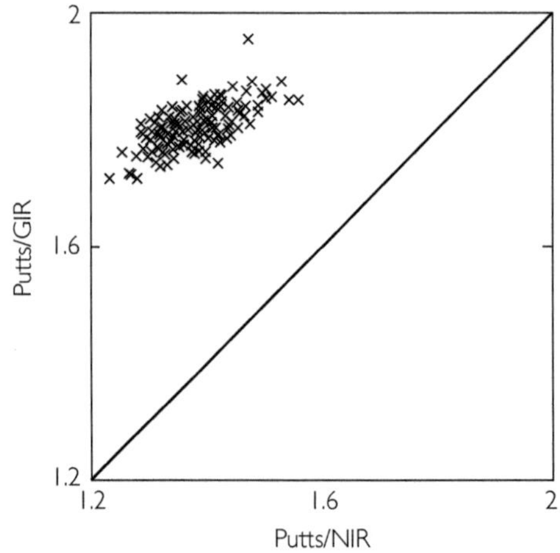

Figure 23.1 PPGIR versus PPNIR.

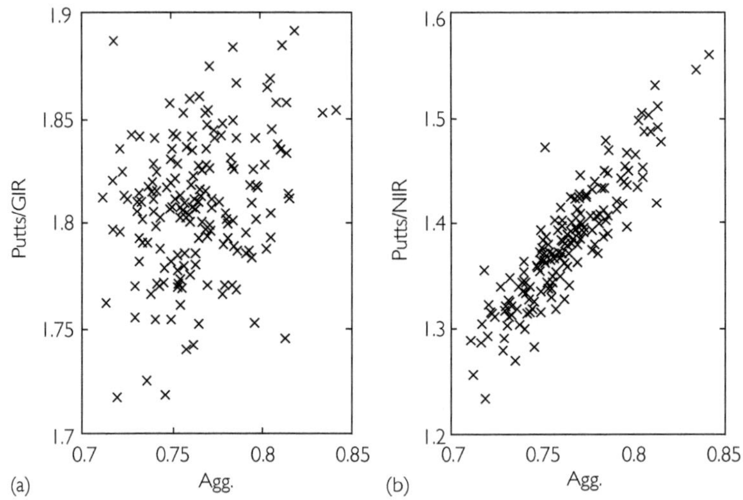

Figure 23.2 Aggr versus (a) PPGIR (Pearson correlation $r = 0.224$) and (b) PPNIR (Pearson correlation $r = 0.825$).

similar). Hence another interpretation is that high values represent players who take almost as many putts when not reaching a green in regulation as when they do. This might be because their earlier drive or fairway shots have gone badly wrong so their approach shot onto the green started from almost as far away as a normal GIR approach shot would have done.

Another variable which may lead to new insight is course management Man, defined as

$$\text{Man} = 1 - \frac{\text{sand shots per round}}{\text{storke average} - \text{total putts} - 18}.$$

This is the percentage of shots which were not taken from bunkers, but which potentially could have been. We may find that players overcome disadvantages in skill, such as poor accuracy or bunker play, by choosing routes to greens which avoid bunkers.

Per-season results for 1998–2001

We looked at data available for the European tour spanning 4 years from 1998 until 2001 (copyright is held by the European PGA). These data are averages over a full season of the variables described in the 'Variables' section, with the exception of 2001 where it covers only the period until the Omega European Masters, consequently leaving out the last nine scheduled events. The summary statistics are shown in Table 23.1.

From this we see that most variables vary little between seasons. The driving distance is increasing with time, which can be attributed to the development of both balls and clubs. The average number of putts is also decreasing slightly, which may be because the quality of the greens is improving (Nick Metcalf, personal communication). Apart from this, all seems to be random variation. Note that, contrary to the notional 36 putts per round, implying 36 putts and 36 non-green shots on a par 72 course, in reality these top players average 30 putts and 42 non-green shots.

Looking at histograms of variables, such as the stroke averages as shown in Figure 23.3, we find that the distributions are reasonably symmetric, albeit with slightly longer tails to the right. However, for the 2000 season there are a few players with outlying values. For example Derrick Cooper had a stroke average of 77.32 shots, nearly two shots worse than the second worst player, whereas Seve Ballesteros's driving accuracy of 32.0% was 16.8% lower than the second worst player in that category.

From the discussion in 'Literature review and comments', DrivDist and DrivAccu are the only true 'input' variables recorded whereas we regard StrAver as the best overall output measure. However, we can also view AvPutts, NonGreen and/or GIR as (component) outputs, which seems more sensible than viewing them as inputs as in earlier references. It is interesting to examine the breakdown of StrAver into AvPutts and NonGreen, and Table 23.2 shows the variances and covariances. The negative covariance between AvPutts and NonGreen is not surprising, since certain styles of play may increase one and decrease the other, whereas wood or iron and putt shots are so different that it is unlikely that the best (of the best) players in each category will contain many names in common. The variance of StrAver can be decomposed into the sum of its covariances with each of the other two, showing that in this sense NonGreen accounts for 3–4 times more of the total variability than AvPutts does. It is noticeable that from 1999 onwards the variance of NonGreen was larger than that of StrAver. The variances are rather lower for 2001, though this may be an artefact of the absence of the last nine tournaments.

Alternatively, results of weighted (by number of rounds played) linear regression with StrAver as the response, given in Tables 23.3 and 23.4, show that NonGreen produces much larger R^2-values than AvPutts when predicting SrtAver, though in both cases the R^2-values decrease over time.

This all suggests that, on the PGA European tour at least, off-green shots are the more important component of overall success.

We note here that all plots of variables and residuals considered were well behaved, so Pearson correlations and linear regression are appropriate throughout this section. Table 23.5 shows that GIR and NonGreen are very strongly related, as expected. Players counting their putts who also count their GIR-value are gaining nothing, since they must keep track of the score anyway, so they could just count the number of putts and deduce the number of non-green shots. GIR itself does not store any extra information, since different course par values

Table 23.1 ***Summary data for all variables.***

Variable	Unit	Results for the following years:							
		1998, $n = 169$		1999, $n = 176$		2000, $n = 181$		2001, $n = 184$	
		Mean	Standard deviation	Mean	Standard deviation	Mean	Standard deviation	Mean	Standard deviation
StrAver	shots	72.24	1.01	72.28	1.02	72.29	1.04	71.71	0.90
NonGreen	shots	42.14	0.98	42.24	1.05	42.35	1.10	42.06	1.04
AvPutts	shots	30.10	0.67	30.04	0.73	29.94	0.76	29.65	0.74
DrivAccu	%	63.00	6.03	64.15	6.28	62.32	6.29	63.00	5.81
DrivDist	yards	267.3	7.37	269.8	7.35	273.5	7.21	280.3	7.65
GIR	%	66.57	4.05	67.00	4.44	65.82	4.24	66.49	4.14
PPGIR	putts	1.81	0.03	1.81	0.03	1.81	0.03	1.65	0.05
PPNIR	putts	1.40	0.06	1.39	0.06	1.38	0.06	1.36	0.06
SandSave	%	52.61	9.27	53.22	8.42	52.09	8.56	53.35	9.21
Aggr	%	83.62	2.31	82.98	2.26	83.10	2.06	82.29	2.26
Man	%	94.90	1.21	94.42	1.15	94.07	0.96	94.94	1.09

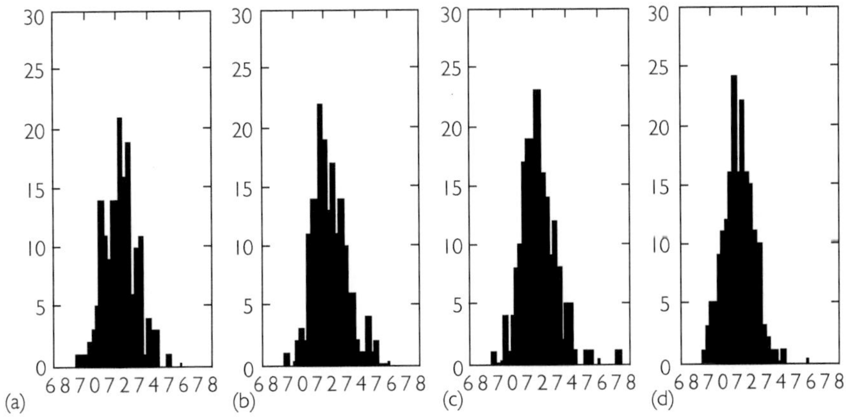

Figure 23.3 Histograms of stroke averages for (a) 1998, (b) 1999, (c) 2000 and (d) 2001.

Table 23.2 ***Variances and covariances for StrAver, AvPutts and NonGreen.***

Variable	Year	StrAver	AvPutts	NonGreen
StrAver	1998	1.0119		
	1999	1.0315		
	2000	1.0844		
	2001	0.81561		
AvPutts	1998	0.2577	0.4579	
	1999	0.2301	0.5370	
	2000	0.2259	0.5722	
	2001	0.1476	0.5538	
NonGreen	1998	0.7541	–0.2002	0.9544
	1999	0.8014	–0.3069	1.1083
	2000	0.8585	–0.3463	1.2048
	2001	0.6680	–0.4062	1.0742

Table 23.3 ***Coefficients for weighted regression of StrAver on NonGreen.***

Year	NonGreen	Constant	R^2 (%)
1998	0.755	40.4	57.6
1999	0.702	42.6	55.2
2000	0.679	43.5	53.4
2001	0.615	45.8	50.3

Table 23.4 ***Coefficients for weighted regression of StrAver on AvPutts.***

Year	AvPutts	Constant	R^2 (%)
1998	0.489	57.5	11.6
1999	0.390	60.4	8.3
2000	0.366	61.2	7.9
2001	0.261	63.9	4.7

Table 23.5 ***Coefficients for weighted regression of GIR on NonGreen.***

Year	NonGreen	Constant	R^2 (%)
1998	−4.001	235.1	91.9
1999	−4.082	239.4	92.0
2000	−3.842	228.5	90.3
2001	−3.769	225.0	88.9

can be easily incorporated into a player's non-green shots total by using $n-36$ as the par for off-green shots on a par n course. Though GIR is a nice summary statistic for journalists, it has no real analytic value.

From Table 23.6 we can see that both DrivAccu and DrivDist are useful predictors for NonGreen, with the driving accuracy coefficients remaining approximately the same over time. The decrease in the coefficient for driving distance is not as severe since the actual distance values increase as noted in Table 23.1. Despite the fact that, even together, the two variables are a fairly crude measure of driving ability, R^2 is moderately high in all cases. This suggests that either drives have a large effect on subsequent shots or that good drivers tend to be good iron players. From Table 23.6 we can estimate a relationship between accuracy and distance. For example for 1999 we have

$$\text{NonGreen} = 67.45 - 0.1\ \text{DrivAccu} - 0.07\ \text{DrivDist}.$$

Very approximately, a difference of 10 yards in driving distance is equivalent to a difference of 7% in driving accuracy. However, an extreme example where DrivAccu seems to be far

Table 23.6 ***Coefficients for weighted regression of non-green shots on driving accuracy and distance.***

Year	DrivAccu	DrivDist	Constant	R^2 (%)
1998	−0.1094	−0.0706	67.89	49.1
1999	−0.0994	−0.0699	67.45	42.4
2000	−0.0986	−0.0614	65.20	36.8
2001	−0.1080	−0.0451	61.45	42.1

Table 23.7 ***Coefficients for weighted regression of stroke average.***

Year	DrivAccu	DrivDist	Aggr	Constant	R^2 (%)
1998	−0.0473	−0.0471	0.23753	67.909	52.0
1999	−0.0342	−0.0342	0.26360	61.786	49.0
2000	−0.0491	−0.0410	0.24201	67.409	50.0
2001	−0.0377	−0.0320	0.19345	67.115	41.3

more important than DrivDist is the performance of the Swedish golfer Pierre Fulke. In 1999 his driving accuracy was an excellent 76.6% which together with a rather mediocre driving distance of 256.8 yards resulted in taking an average of 41.42 off-green shots. In the following season of 2000, he improved his driving distance to 270.4 yards, while losing 1% of accuracy (down to 75.6%), which meant that his average NonGreen still increased to 42.01.

It is of interest to see whether SandSave and Man are related, in particular whether there is any suggestion that players who go into bunkers more often are better at saving par from them. The correlation between these two variables decreased from 0.31 (1998) to 0.21 (1999) to 0.09 (2000), from clear significance to clear non-significance. However, it then increases to 0.21 (2001), illustrating the pitfalls of trying to infer temporal changes with few time points. Hence it is not clear that there is a relationship, but if there is then it appears to be positive, so players who avoid bunkers better also escape from them better. The picture is similar for the correlations of Man with other variables in different years, with a few moderately large (and hence 'significant') values, but this is probably just a consequence of the fact that we are data trawling.

Table 23.7 summarizes stepwise weighted (by number of rounds) regressions of StrAver on the other variables defined above, DrivAccu, DrivDist, Aggr, Man and SandSave, though not AvPutts, NonGreen or GIR for the reasons outlined earlier. In each case, the best model, under the usual F-test criteria, is that containing just DrivDist, DrivAccu and Aggr, whereas Man and SandSave were clearly non-significant. In each case this accounted for over 40% of the total variation in StrAver, with Aggr clearly the best single predictor.

The main feature of these results is that Aggr seems to be the most important factor (of those considered). Hence, if the interpretation of this ratio as aggression is valid, this suggests that players who try ambitious approach shots are penalized and score higher on average than safety first players.

Per-round results for 2001

The data investigated in the previous section are clearly highly aggregated, being averages over a full season, and this may mask interesting features. Therefore, in this section we consider data for the 2001 season on a per-round basis, so that each variable is recorded for each player for each round. We have these data only for 2001, and for some tournaments in this season the full per-round data are not available, so in this section we use only the 25 tournaments for which data on all rounds are available.

For these data the meanings are slightly different. For example, the per-season data can show that a player of middling ability in all variables measured will have middling results on average, whereas the per-round data may show whether his best performances have been those where he has driven most accurately. Similarly, if we had a genuine measure of putting ability, we could see whether his best performances were those where he putted best.

We now perform analyses similar to those in the previous section, comparing the results and usefulness of per-season and per-round data. Previously, treating StrAver over a season as normally distributed seemed reasonable, and all the model checking in 'Per-season results for 1998–2001' reinforced this. However, the score on a single round is clearly not normally distributed. The distribution tends to have a longer tail to the right since it is easier to drop large numbers of shots than to gain them. However, there is no standard discrete distribution which appears to be appropriate, in particular not the Poisson, which is also nearly symmetric and constrains the mean and variance to be equal. Hence we appeal to the central limit theorem, using the fact that the score is a sum of individual shots, and use normal-based regression as before, albeit with more caution. In some of the analyses performed, the residual plots showed that the normal distribution assumption was clearly invalid, but in those reported here there was some indication of non-normality but nothing serious. Hence, since in most cases the F-tests were so unequivocal, this should not materially affect the conclusions reached below. As an aside, finding an appropriate distributional form for the number of shots per round is an interesting question for the future.

Earlier we noted that, although different courses and conditions will have an effect on the scores for each round, these differences should be less significant when considering seasonal averages (i.e. StrAver). However, for analysing round-by-round data, this is clearly not the case. Hence Tournament is included as a factor (with 25 levels), in all regression models, to correct for the fact that different courses have different par values and competitions will have been played in different conditions. For a single year this is equivalent to the course used, though with data for multiple years tournament would be nested within a variable Course.

Table 23.8 shows the decomposition of the variance of round score, similarly to Table 23.2. This shows that, for these per-round data, putts contribute more to the overall variation than with per-season data.

Tables 23.9–23.11 give results for the regression of round score on NonGreeen and AvPutts separately, similarly to the analysis given in Tables 23.3 and 23.4. For NonGreen the results are similar, but for AvPutts the prediction is much better, and the coefficient larger, for the per-round case. Combined with the conclusions in 'Per-season results for 1998–2001', this suggests that, whereas over a season a player's NonGreen value is a much better predictor of his overall success than his AvPutts value, for an individual round this difference is less pronounced (at least in the 2001 season).

Table 23.8 ***Variances and covariances for round score, AvPutts and NonGreen for individual rounds.***

Variable	Round score	AvPutts	NonGreen
Round score	10.0217		
AvPutts	3.4515	6.5538	
NonGreen	6.5702	–3.1022	9.6724

Table 23.9 ***Analysis-of-variance table for regression of round score on NonGreen with Tournament as a factor.***

Source	Degrees of freedom	Sum of squares	F
NonGreen	1	31,494.2	6003.12
Tournament	24	2725.0	21.64
Residual	8276	43,418.1	
Total	8301	83,189.9	

Table 23.10 ***Analysis-of-variance table for regression of round score on AvPutts with Tournament as a factor.***

Source	Degrees of freedom	Sum of squares	F
AvPutts	1	15,931.5	2235.45
Tournament	24	9119.8	53.32
Residual	8276	58,981.1	
Total	8301	83,189.9	

Table 23.11 ***Coefficients for regression of round score on NonGreen and AvPutts separately with Tournament as a factor.***

Variable	Coefficient	Constant	R^2 (%)
NonGreen	0.677	43.3	47.8
AvPutts	0.561	55.0	29.1

Table 23.12 can be compared with Table 23.6 and we see that both coefficients and the R^2-value are rather lower in the per-round case. This is perhaps surprising, especially in comparison with the results above and below.

Finally, we regress round score on DrivAccu, DrivDist, Aggr and Man. The variable SandSave is not very meaningful on a per-round basis (the denominator is often zero) and so was omitted. In addition to Tournament, we now include Round number (four levels) and Player (184 levels) as factors. The latter attempts to assess how much of the overall variation in scores can be explained simply by the identity of the players. The analysis-of-variance table and coefficients of the covariates are given in Table 23.13, whereas the coefficients of the factors are omitted for obvious reasons. Of the terms reported, the interaction between Tournament and DrivAccu is clearly non-significant, the effect of Round is equivocal (given the imperfect assumption of normality) and all the other terms are clearly significant. Other interaction terms not reported were either non-significant or were considered inappropriate (for example an interaction between Player and Tournament would not be informative, given the objectives of the analysis).

Table 23.12 ***Analysis-of-variance table for regression of non-green shots on driving accuracy and distance.***†

Source	Degrees of freedom	Sum of squares	*F*
DrivAccu	1	7067.62	969.45
DrivDist	1	1168.04	160.22
Tournament	24	10,480.62	59.90
Residual	8275	60,327.41	
Total	8301	80,290.54	

† DrivAccu, −0.0626; DrivDist, −0.0253; constant, 53.04; R^2, 24.9%.

Table 23.13 ***Analysis-of-variance table and coefficients of covariates for regression of round score.***

Source	Degrees of freedom	Sum of squares	*F*	Coefficient	Standard error
DrivAccu	1	2900.20	389.37	−0.0463	0.00235
DrivDist	1	506.57	68.01	−0.0220	0.00267
Aggr	1	1202.69	161.47	2.43	0.1914
Man	1	339.23	45.54	−5.16	0.7645
Tournament	24	354.50	1.98		
Round	3	75.55	3.38		
Player	183	5252.91	3.85		
Tournament.DrivAccu	24	214.29	1.20		
Tournament.DrivDist	24	562.05	3.14		
Tournament.Man	24	424.14	2.37		
Tournament.Aggr	24	801.32	4.48		
Residual	7945	59,177.86			
Total	8255	82,613.54			

The main differences between these results and those in Table 23.7 are that all (covariate) coefficients except that for DrivDist have increased, and that Man is now clearly significant.

The difference between the effects of DrivAccu and DrivDist in the per-year and the per-round cases are interesting. For DrivAccu the relationship appears stronger in the per-round case, since in a particular round consistently finding, or failing to find, the fairway from the tee is a reasonably good predictor of the round score. However, when DrivAccu and round score are averaged over the season, some of the strength of this relationship is lost, since all players have good and bad rounds. Similar comments apply to Aggr and Man. For DrivDist, in contrast, the value for each round is based on only two observations, so this has much less predictive value. Averaging DrivDist over a season may indeed be preferable, since this then makes it a more reliable measure of a player's driving power.

The factor Round, though borderline significant, shows that if the effect is genuine then it is due to scores being higher in the fourth round (possibly because of more difficult pin positions) and much higher in the first round (possibly because of lesser familiarity with the course). Not surprisingly, the factor Player is significant, but like all the other terms it still only explains a small proportion of the variability, probably because it is rare in golf for any player to do consistently well in tournaments. In some sports, such as tennis, the top-ranked few players in any tournament usually do very well, whereas in golf it is much more common for top players to perform no better than most others in the tournament, while less fancied players are at the top of the leaderboard. This is because of the greater variation in scores and the potential to lose several shots very quickly.

It is reassuring that the variables defined above have much larger F-ratios than the factors used here, suggesting that these stylistic variables are not swamped by simple between-tournament and between-player variability. This leads to the hope that more meaningful variables will have even greater predictive value.

Conclusions and proposals for further work

We have attempted to investigate the components of golfing success in a slightly more subtle way than previous references, trying to avoid simply predicting that players who reach the green quickly and take few putts will score well, since this is obvious. However, as we have seen, it is very difficult to tease out measures of playing style or component skills from the data that are currently collected. Only the figures for driving accuracy and distance can be regarded as measures of this type, since all other figures represent mixtures of skills.

Therefore, if people wish to break golf down into its constituent skills more effectively then more data need to be collected, and in particular more different variables. The driving data are relatively easy to collect since, at a given hole, every player starts in virtually the same place. Collecting similar data for iron shots would need to record start as well as finish locations, not to mention positions relative to the green, since all analyses would need to be conditional on where the drive ended to distinguish between driving and iron skills. The same comments apply to putting, if we wish to separate out putting skill from iron skill.

A very interesting figure, if collected, would be to record the distance from the hole both before and after every putt a player makes, since this would be close to being a direct measure of a player's putting ability. Lorensen and Yamron (1992) illustrated the detection of golf-balls from television cameras, stressing the large times required for surveying the greens, although with current hardware this should be much easier. However, the results obtained could still be biased by the very different natures of different greens, since the difficulty of a putt is a function of the topography of the green as well as distance from the hole. Similarly, a record of final distances from the hole, and whether on the green or not, for iron shots of given distances and locations, would give some chance of disentangling the various effects. However, as we move further from the hole the numbers become more difficult to interpret as the different natures of the particular holes may make averaging figures over them meaningless.

In addition, these figures need to be available at a finer level than simply averages over the season, as for the variables that are currently recorded. Although such averaging is often sensible, as with the case of DrivDist noted above, having data on individual tournaments, rounds or even holes clearly gives more information on variability and links between variables. More detailed measurements, such as those proposed above, available per hole, per round or at least per tournament, should help to uncover the components of golfing success. For example, does success in a given competition tend to go to those who putt best, those who pitch onto the green most accurately or those who drive furthest?

In summary, current measurements are clearly insufficient if they are to be used for predictions. Although, compared with other sports, statistics of golf seasons are a true reflection of what they are supposed to represent, a greater variety of data, at a greater level of detail, needs to be collected to achieve practically useful results.

ACKNOWLEDGEMENTS

The authors would like to thank the European PGA for the provision of, and permission to use, the data. In particular we thank Nick Metcalf for many clarifications and much useful information.

REFERENCES

Belkin, D. S., Gansneder, B., Pickens, M., Rotella, R. J. and Striegel, D. (1994) Predictability and stability of Professional Golf Association tour statistics. *Percept. Motor Skills*, **78**, 1275–1280.

Engelhardt, G. M. (1995) 'It's not how you drive, it's how you arrive': the myth. *Percept. Motor Skills*, **80**, 1135–1138.

—— (1997) Differences in short-making skills among high and low money winners on the PGA Tour. *Percept. Motor Skills*, **84**, 1314.

—— (1999) Is American PGA Tour more competitive than European Tour?: Reply to Jiminez and Fierro-Hernandez. *Percept. Motor Skills*, **89**, 1028.

Jiminez, J. A. and Fierro-Hernandez, C. (1999) Are European and American golf players different?: Reply to Engelhardt (1997). *Percept. Motor Skills*, **89**, 417–418.

Larkey, P. D. (1998) Statistics in golf. In *Statistics in Sport* (ed. J. Bennett), pp. 121–140. London: Arnold.

Lorensen, W. E. and Yamron, B. (1992) Golf green visualization. *IEEE Comput. Graph. Applic.*, **12**, 35–44.

PART VI

Soccer

Introduction

PHIL SCARF

The rules of soccer (which is referred to as Football outside of North America) are simple. Two teams compete to kick a ball into the opposition goal, and the team with the most goals after 90 minutes of play wins. Handling of the ball is not permitted and it is therefore a game that requires great skill to play well. This is a large part of its fascination and popularity; this and the uncertainty of outcome due to the paucity of goals. Only hockey comes close in terms of low scoring rate among the team ball-sports. Soccer is not a sport that is popular in North America or in the Indian sub-continent, and in Asia and Australasia it has a moderate following. In contrast, South Americans, Africans, and Europeans are fanatical about the game, and the World Cup tournament is the most popular sporting event on the planet. Soccer is the number one sport globally for spectator interest.

In terms of statistics research interest, soccer takes third place after baseball and cricket. Cricket researchers are fascinated with matters that fall into five categories. First and foremost is match outcome prediction, which can be further categorized into modelling match outcomes (win, draw, loss) using, for example, ordinal logistic or ordered probit regression and modelling scores using, for example, the double Poisson distribution and derivatives thereof. The seminal paper on modelling scores is Maher (1982). In the double Poisson model, the goal-scoring intensity of a team depends on its attacking strength and the defensive weakness of the opposition. Each team i possesses an attacking strength parameter α_i and a defensive weakness parameter β_i, and the expected score of team i in a match against team j is $\alpha_i\beta_j$, and vice versa. Many extensions of this basic model of Maher have been published that allow dependence in the bivariate Poisson distribution, extra-Poisson variation, weighted estimation schemes that emphasize recent results, and evolution of team strengths both deterministically and stochastically. In the paper by Crowder et al. (2002) in this anthology, teams' attacking and defensive strengths evolve through time according to a latent bivariate stochastic process. The contribution of this paper was not only the model but also the approximate methods used for computation of the parameter estimates—a non-trivial task for the 92 teams in the English football league with eight parameters per team. The likelihood for the model of Crowder et al. cannot be evaluated numerically because the model is non-Gaussian and non-linear. Instead they develop an approximating model for which the likelihood can be computed. More recently, Koopman and Lit (2015), in what is possibly the last word on the double Poisson model, use MCMC methods to evaluate this likelihood.

Modelling match outcome is the subject of Forrest and Simmons (2000) who investigate the match forecast accuracy of a set of experts, the 'Pools Panel'. The Pools is an anachronistic gambling instrument in the U.K. whereby punters bet on outcomes that are score-draws—tied match outcomes that exclude 0–0. A punter who correctly predicts eight score-draws among the matches played in a weekend round of the English and Scottish leagues shares the jackpot with others who do likewise. When a round of fixtures is disrupted and results are not available, due to adverse weather for example, the Pools Panel sits to determine the missing outcomes. Since the introduction of the national lottery in the U.K., the popularity of the Pools has waned. Given the focus of the Pools Panel on outcomes, Forrest and Simmons necessarily use an outcome model and one that is commonly used by econometricians—the ordered probit with a latent stochastic strength variable that determines the outcome category (win, draw, loss) and whose expectation depends on covariates that quantify the performance of the teams in recent rounds. Their model predicts Pools Panel outcomes better than it does real outcomes, and their insight therefore is that soccer is less predictable than the experts themselves perceive it to be. This unpredictability of the game is arguably fundamental to its popularity—this and its simplicity to the untrained observer.

The notion of unpredictability brings us naturally to the second category of research papers in soccer—those concerned with uncertainty of outcome or competitive balance and whether policy interventions influence competitive balance for the better or worse. This is the topic of our third featured paper by Koning (2000). While it focuses on competitive balance in Dutch soccer, its approach is the maturation of the methodology in general, whereby competitive balance in a tournament is simply (represented by) the collection of strengths of the competitors—unknown but estimated using some explanatory or predictive model of match outcomes. Then questions about competitive balance are equivalent for practical purposes to questions about the estimated competitors' strengths, both within-season (short term competitive balance) and between seasons or tournaments (persistence in competitive balance). Koning takes the econometricians approach, using ordered probit to estimate team strengths. In the ordered probit or logistic models, teams have a single strength parameter, whereas in the double Poisson model strengths are vectors. In the Koning approach, ordering the strength estimates gives a ranking and their relative sizes the ratings. Given estimated strength vectors, rankings, and ratings of teams, particular competitive balance measures can be obtained by simulating multiple tournaments or leagues (Scarf and Yusof, 2011). Competitive balance measures and their estimates allow the effect on interventions by administrators to be studied, although the value of retrospective analyses is debatable.

Strategic interventions at match level—coach and player decisions and their modelling and optimization—is the third strand of research in soccer. Determining optimal in-play strategy given the state of the match is an open problem. This is challenging partly because the game is continuous and partly because the type of play at any point is difficult to categorize. Authors have attempted to discretize the game using possession transitions, and relating possession to scoring is one possible way to consider the strategy question. Hirotsu and Wright (2003) make an important contribution here, although the impact of their work has been less significant. Nonetheless they also pioneered the use of big data in soccer research using counts of player actions such as passes and shots.

The fourth strand of research concerns strategic interventions by administrators and our fifth featured paper (Fernández-Cantelli and Meeden, 2003) exemplifies this. There is much scope for the statistician to ask 'what if' questions through modelling and therefore to aid the decision-making of administrators, although there is scant evidence that administrators take notice. Tournament design changes are easily investigated with simulation analysis for example. Lenten et al. (2013) present a fine example of the use of an econometric model to ask what should be the format of the penalty shoot-out in matches where a draw or tie is not permitted. The fifth strand is player performance measurement and the quantification of individual player effects (e.g. McHale et al., 2012). Work here is less well developed. Tournament scheduling is a further important research topic but outside the scope of this overview, this being the domain of the OR community. Returning to the work of Fernández-Cantelli and Meeden, this work is interesting for two reasons: firstly, they conclude that the change in the rules for the awarding of league points (from 2–1–0 for win-draw-loss, respectively, to 3–1–0) had no noticeable effect on the stated aim of this change—to incentivize more attacking play, more goals, and fewer tied matches—and secondly they demonstrate that statistical analysis of the effect of such interventions can be considered 'in-silico'. On the other hand, their suggested points modification is offbeat, because even though it scores on the incentivizing attacking play criterion it concedes on the simplicity criterion and is at best 0–0 on the fascination criterion. That said, other sports have adopted new, radical scoring schemes; for example, Rugby Union in the form of a try bonus. This sport might do better to simplify its scoring rules, making a try the only means of scoring (a point), so that uncertainty of outcome increases and the game is made more exciting. Administrators will, however, baulk at this. Thus, while other team sports fiddle with their rules, Soccer, the simple game, remains the beautiful game.

REFERENCES

Crowder, M., Dixon, M., Ledford, A., and Robinson, M. (2002), Dynamic Modelling and Prediction of English Football League Matches for Betting, *Journal of the Royal Statistical Society: Series D (The Statistician)*, 51(2), 157–168.

Fernández-Cantelli, E., and Meeden, G. (2003). An Improved Award System for Soccer. *Chance*, 16(1), 23–29.

Forrest, D. and Simmons, R. (2000). Making up the Results: The Work of the Football Pools Panel, 1963–1997. *Journal of the Royal Statistical Society: Series D (The Statistician)*, 49(2), 253–260.

Hirotsu, N. and Wright, M. (2003). An Evaluation of Characteristics of Teams in Association Football by Using a Markov Process Model. *Journal of the Royal Statistical Society. Series D (The Statistician)*, 52(4), 591-602.

Koning, R.H. (2000). Balance in Competition in Dutch Soccer. *Journal of the Royal Statistical Society. Series D (The Statistician)*, 49(3), 419–431.

Koopman, S.J. and Lit, R. (2015). A Dynamic Bivariate Poisson Model for Analysing and Forecasting Match Results in the English Premier League. *Journal of the Royal Statistical Society: Series A (Statistics in Society)*, 178(1), 167–186.

Lenten, L., Libich, J., and Stehlık, P. (2013). Policy Timing and Footballers' Incentives: Penalties Before or After Extra Time? *Journal of Sports Economics*, 14(6), 629–655.

Maher, M. (1982). Modelling Association Football Scores. *Statistica Neerlandica*, 36(3), 109–118.

McHale, I.G., Scarf, P.A., and Folker, D.E. (2012). On the Development of a Soccer Player Performance Rating System for the English Premier League. *Interfaces*, 42(4), 339–351.

Scarf, P.A. and Yusof, M.M. (2011). A Numerical Study of Tournament Structure and Seeding Policy for the Soccer World Cup Finals. *Statistica Neerlandica*, 65(1), 43–57.

CHAPTER 24

An improved award system for soccer

EDUARDO FERNANDEZ-CANTELLI AND
GLEN MEEDEN

Historically, say, prior to 1980, soccer leagues awarded two points to the winning team in a match and zero points to the losing team. In the case of a tie, each team was awarded one point. The final standings for a league were based on the total number of points earned by each team during the season. Under that system, there was concern with the lack of goal scoring and with the proportion of games that end in a tie. In part, these outcomes result from the weaker team playing a very conservative or defensive strategy. To try to overcome these problems, several important European leagues replaced the 2–1–0 award system with the 3–1–0 award system, in which the winning team was awarded three points instead of two. The 3–1–0 system is now common throughout the world, including the quadrennial World Cup competition.

England was the pioneer in adopting the 3–1–0 award system. It was first used there in the 1981–82 season. In the following years it was taken up in other countries. For 10 different countries, we collected data from leading leagues that changed from the 2–1–0 award system to the 3–1–0 award system. For example, for England we collected data for the 17 seasons preceding 1981–82 for the 2–1–0 award system and the subsequent 19 seasons under the 3–1–0 award system. In the 10 countries for the data collected the smallest number of games played under one of the two award systems was 1,230 while the largest number was 9,616. For each of the 10 countries, we calculated the average number of goals scored per game and the proportion of ties under the two award systems. The results are given in the Table 24.1.

From the table we see that there is little evidence to support either the hypothesis that the 3–1–0 award system led to increased scoring, or the hypothesis that it led to fewer ties. Half the countries' teams showed an increase in scoring after the change, but the others had a decrease in scoring. Six of the 10 had fewer ties, but only teams in Turkey and Italy showed noteworthy decreases. Here we will explain why the switch to the 3–1–0 award system did not have the desired results and propose a new award system that should increase scoring.

Table 24.1 ***The average number of goals scored per game and the proportion of games that end in tie for ten countries with n_2 seasons under the 2–1–0 award system and n_3 subsequent seasons under the 3–1–0 award system.***

Country	2–1–0 Award system			3–1–0 Award system		
	n_2	Average # of goals	Proportion of ties	n_3	Average # of goals	Proportion of ties
England	17	2.70	0.28	19	2.64	0.27
Iceland	20	2.93	0.26	16	2.97	0.22
N. Ireland	22	3.40	0.20	14	2.89	0.24
Norway	23	2.83	0.25	13	3.24	0.22
Turkey	23	2.01	0.34	13	2.85	0.24
Sweden	26	2.84	0.27	10	2.81	0.26
Finland	27	3.00	0.24	9	2.70	0.25
Italy	30	2.12	0.37	6	2.64	0.28
Spain	31	2.48	0.26	5	2.68	0.28
Germany	31	3.18	0.26	5	2.86	0.29

Mexico's Cuauhtemoc Blanco, center, tries to dribble between USA's Pablo Mastroeni, left, and Cobi Jones during their 2002 World Cup second round soccer game at the Jeonju World Cup stadium in Jeonju, South Korea, Monday June 17, 2002. The USA won 2–0 to advance to the quarterfinals.

Computing expected winnings

The main tool that we use to discuss team strategy is to compute a team's expected winnings, that is, the number of points they expect to earn from a game. Suppose that Barcelona and Madrid are to play a game in which we know that on average Barcelona scores λ (2 in our example) goals per game and Madrid scores μ (1.5 in our example) goals per game. Several researchers have found that the distribution of scores obtained by each team can be well approximated by the Poisson distribution (see sidebar). If we let B and M represent the number of goals each will score, then a reasonable model for the final score is to assume that B and M have independent Poisson distributions with mean parameters λ and μ respectively. With these assumptions, we can compute the probability of any possible outcome of the game, e.g., the probability that Barcelona wins 2–1 or the probability that Madrid wins 3–0. This is done using the Poisson distribution. By considering all possible scores we can evaluate the probability that each team wins and that the teams play to a draw. Hence before the game begins we can compute the expected winnings for each team. Under the 2–1–0 award system, Barcelona's expected winnings are just two times the probability that Barcelona scores more goals than Madrid plus one times the probability they score the same number of goals. The answer just depends on the values of λ and μ. With $\lambda = 2$ and $\mu = 1.5$, Barcelona wins with probability 0.49, loses with probability 0.29 and draws with probability 0.22; the expected number of league points earned by Barcelona is $0.49 \times 2 + 0.22 = 1.20$.

Suppose that at the end of the first half neither team has scored. If Barcelona scores on the average λ goals per game, we would expect that on the average it would score $\lambda/2$ goals in the second half. Similarly, we would expect Madrid to score $\mu/2$ goals in the second half. Assuming that neither team scored in the first half, we can compute the expected winnings of each team at the beginning of the second half. As we shall soon see, their expected winnings at halftime will be different from their expected winnings at the start of the game. More generally, let $0 \leq t \leq 1$ represent the proportion of the game that has been played. If neither team has scored by time t, then under our Poisson model, the number of goals that Barcelona and Madrid will score in the remainder of the game follow independent Poisson distributions with mean parameters $\lambda(1 - t)$ and $\mu(1 - t)$ respectively. We can compute the expected winnings for each team at this point in the game. In much the same way, we can calculate the expected winnings of Barcelona if they are ahead by one goal or two goals at time t or behind by one goal or two goals and so on. The same is true for Madrid. However, under the 2–1–0 award system at any stage of the game, one only needs to compute the expected winnings of one of the teams, since the sum of their expected winnings must always be two (because two points are allocated to the two teams in each game).

As an example, we calculated the expected winnings of Barcelona over the course of the game under two different scenarios, one in which the score is tied and the other when they are ahead by one. In each case, we assumed that on the average Barcelona scores 2.0 goals per game and Madrid scores 1.5 goals. The results are given in Figure 24.1. Note that as the game progresses, Barcelona's expected winnings decrease from 1.2 at the beginning to 1.0 at the end when the score is tied. If Barcelona is ahead by one goal, the expected winnings increase from 1.6 (if they score early in the match) to 2.0 (if they lead when the match ends). This is typical under the 2–1–0 award system. The stronger team's expected winnings always decrease when the score is tied or they are behind and increases when they are ahead.

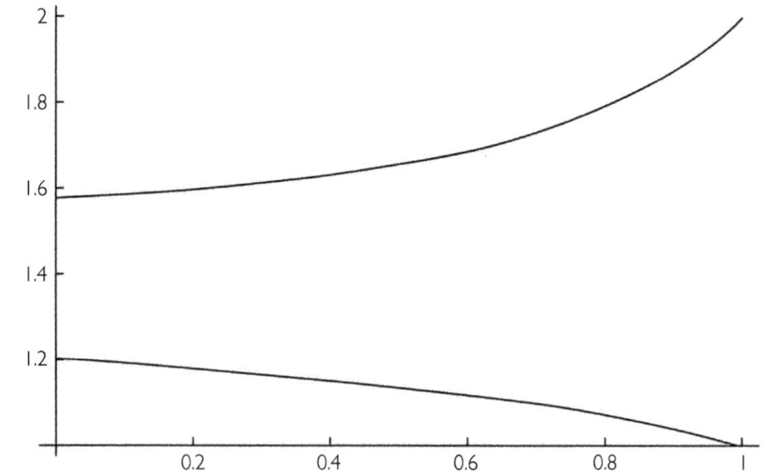

Figure 24.1 Expected winnings of Barcelona under the 2–1–0 award system over the duration of the game when they are ahead by one goal (upper curve) and when the score is tied (lower curve), where on the average they score 2.0 goals per game and Madrid scores 1.5 goals.

So far, we have assumed that the 2–1–0 award system was being used, but similar calculations can be done for the 3–1–0 award system. In fact for any award system we can compute the expected winnings of either team at any stage of the game as long as we know the two teams' original Poisson parameters, the time remaining and the current score. Next, we will study teams' expected winnings under both the 2–1–0 and 3–1–0 award systems.

Strategy under the standard systems

We begin with the 2–1–0 award system and consider the behavior of the expected winnings of Barcelona as the game progresses. We first consider the situation in which the score is tied. In Figure 24.1 we saw that when $\lambda = 2.0$ and $\mu = 1.5$, Barcelona's expected winnings always decrease as t moves from zero to one. This remains true whenever Barcelona is the better team, that is, when $\lambda > \mu$. Hence, the longer the game goes on without a goal being scored, the greater the disadvantage for the better team. Conversely, Madrid's expected winnings increase as the game progresses. The longer the game continues without a score, the more advantageous it becomes for the weaker team. This helps to explain the popular strategy of the weaker team playing very conservatively from the beginning of the game. By playing very defensively the weaker team decreases the scoring rates for both teams to values closer to zero. When both of the Poisson mean parameters are quite small, both teams have expected winnings close to one. This is better for the weaker team than playing normally; playing defensively or conservatively in effect shortens the game. This makes sense intuitively since the shorter the game, the less chance the better team has to demonstrate its superiority.

More generally, anytime a team's expected winnings increase as the game progresses, its players should use a defensive or conservative strategy which limits goal scoring and, in effect, shortens the game. On the other hand, if a team's expected winnings are decreasing with time, they should continue to play normally. The team could even consider using a more aggressive or more offensive strategy which increases goal scoring and, in effect, lengthens the game. This observation is basic to understanding how a team should determine their style of play during the game. In order to maximize their payoff or expected winnings, they should play defensively when their expected winnings increase and play normally when their expected winnings decrease. This underlies much of the argument that follows.

Next, we consider the situation in which Barcelona, the stronger team, has scored first. When $\lambda = 2.0$ and $\mu = 1.5$, we saw in Figure 24.1 that their expected payoff increases as t moves up to one. That is, once the stronger team gets ahead, it is in their interest to play defensively, decrease both teams' chances of scoring and shorten the game. On the other hand, the weaker team should play more aggressively once they get behind.

Suppose the weaker team happens to score first. Then calculations demonstrate, as one would expect, that it is in their interest to play defensively and decrease the likelihood of any goal being scored. All these results show that, no matter what the state of the game, it is always in the interest of one of the teams to play conservatively, except when they are of equal strength and the score is tied. This fact helps to explain the lack of scoring under the 2–1–0 award system.

In the 3–1–0 award system, the sum of the expected winnings for the two teams is not constant, as it is for the 2–1–0 award system. However, calculations reveal that the optimal strategies for the two teams are the same as in the 2–1–0 award system. The weaker team should play conservatively until they get behind. The stronger team should play normally until they get ahead. Once either team gets ahead, they should play conservatively, while the opponent should play aggressively.

A new award system

The problem with the standard systems is that, no matter what the state of game and the relative strengths of the teams, it is always in one team's best interest to try to limit the likelihood of a goal. The only exception is when the teams are of equal ability and the score is tied. In light of this, we wish to find an award system that encourages both teams to play normally throughout a portion of the game. As a practical matter any new system should not be too different from the standard systems.

Table 24.2 ***The number of points awarded under the two standard systems and the new system for the various possible outcomes of a game.***

Game outcome	Points awarded		
	New	2–1–0	3–1–0
Win by ≥ 3	3	2	3
Win by 1 or 2	1	2	3
Tie	0	1	1
Lose by 1 or 2	−0.5	0	0
Lose by ≥ 3	−1.5	0	0

We now propose such an award system. In this system each team receives zero points for a tie. If a team wins by three or more goals, they receive three points. If a team wins by one or two goals, they receive one point. The losing team receives −1.5 points if they lose by three or more goals and receives −0.5 points if they lose by one or two goals. The payoffs for the three systems are summarized in Table 24.2. Under this system, teams near the bottom of the standings at the end of the year would have a negative point total for the season. Just as with the 3–1–0 award system, in the new system the sum of the expected winnings is not constant, and the sum of the total payoffs for any game depends on its outcome. But, as before, one can compute the expected winnings for either team at any point in the game.

Strategy under the new system

To see how the new system could affect teams' strategies we need to find the optimal strategy of each team for different game situations and different choices of the Poisson parameters. We return to the example discussed earlier in which Barcelona scores on the average $\lambda = 2.0$ goals per game and Madrid $\mu = 1.5$ goals.

First, we assume that no goals have been scored and we plot the expected winnings of each team as the game progresses. This is Graph I in Figure 24.2. The upper curve is Barcelona's expected winnings, which decrease as the game continues with no score. This comes as no surprise since Barcelona is the stronger team. The decreasing expected winnings means that Barcelona will continue to play their normal style. Note, however, that Madrid's expected winnings (the lower curve) increase until about 25% of the game remains and then start to decrease. Recall that under the old award system Madrid's expected winnings increased throughout the game under this scenario. Hence, Madrid's optimal strategy under the proposed award system is to play conservatively until about one-fourth of the game remains, and then play normally. What is the reason for this welcome change? When there is only a small part of the game left, it is most likely that no goals or just one goal will be scored in the remainder of the game. When the skill levels of the two teams are similar, the probability that the weaker team scores first, although less than one-half, will be close enough to one-half that it is in their interest to play for a win because the reward for winning by one goal (1 point) is twice the penalty for losing by one goal (−0.5 points). If the weaker team's probability of scoring first is greater than one-third, they should stop playing conservatively and go for the goal. The point at which the weaker team should switch from conservative to normal strategy depends on the relative strengths of the two teams. But when the two teams are similar in ability, the new scoring system encourages both to play for the win when the score is tied near the end of a game.

Next, we assume that Barcelona is ahead by one goal. Under the standard award systems their optimal strategy would be to play conservatively for the rest of the game. We see in Graph *II* of Figure 24.2, however, that their expected winnings (the upper curve) continues to decrease until less than 20% of the game remains, so they should only switch to a conservative style near the end of the contest. Why is that? Since they are the better team, it is in their interest to stay aggressive and try to win by three goals since they expect to score more goals than Madrid if both teams play their normal games. Only near the end should Barcelona play

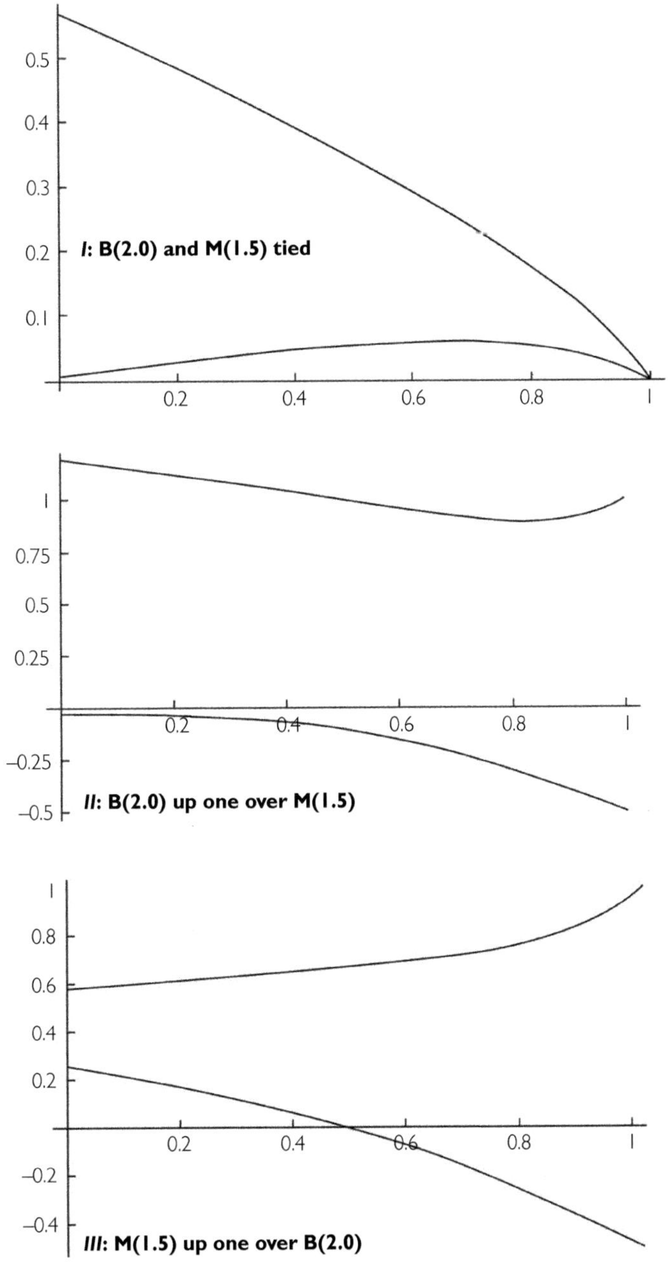

Figure 24.2 Over the duration of the game, the expected winnings of Barcelona and Madrid under the new award system for three different situations. In Graphs *I* and *II* the upper curve represents Barcelona's expected winnings, and the lower curve Madrid's. In Graph *III*, Barcelona's is the lower curve, and Madrid's the upper curve.

conservatively when it becomes unlikely that they can score two more goals before Madrid scores one. How near the end they should start to be conservative depends on the values of λ and μ. On the other hand, Madrid should also continue to play normally because their expected winnings (the lower curve) in Graph II are also decreasing. This is because losing by two is no worse than losing by one and they have a reasonable chance to score the next goal.

In Graph *III* of Figure 24.2 we assume that Madrid is ahead by one goal. Their expected winnings (the upper curve here) increase over the entire range. Since they are the weaker team, their optimal strategy is to forget about trying to play for a win by at least three goals and play defensively to conserve their one goal lead. Barcelona, of course, should continue to play normally.

In Figure 24.3, we assume that the two teams are equal with $\lambda = \mu = 2$ and that Barcelona is ahead by one goal. Madrid trails and should thus play normally. Barcelona's expected winnings (the upper curve) is decreasing until about 27% of the game remains; here again Barcelona plays normal at first to try for a big win and only begins to play conservatively when the chances of a big win are small. Again, the point at which the expected winning curve turns and the optimal strategy changes depends on the scoring rates. If each team only scores on the average one goal a game, then the first team to score will turn to conservative play earlier. Other game situations can be analysed in a similar manner.

These examples suggest the following differences between the proposed new award system and the standard systems:

- Under the proposed new system, when the score is tied the weaker team should switch from a conservative to a normal style of play near the end of the game. Under the standard systems the weaker team should always play conservatively.
- Under the proposed new system, once the stronger team gets ahead by one goal they should continue to play normally until the later stages of the game. Under the standard systems the stronger team should switch to a conservative strategy once it gets the lead.

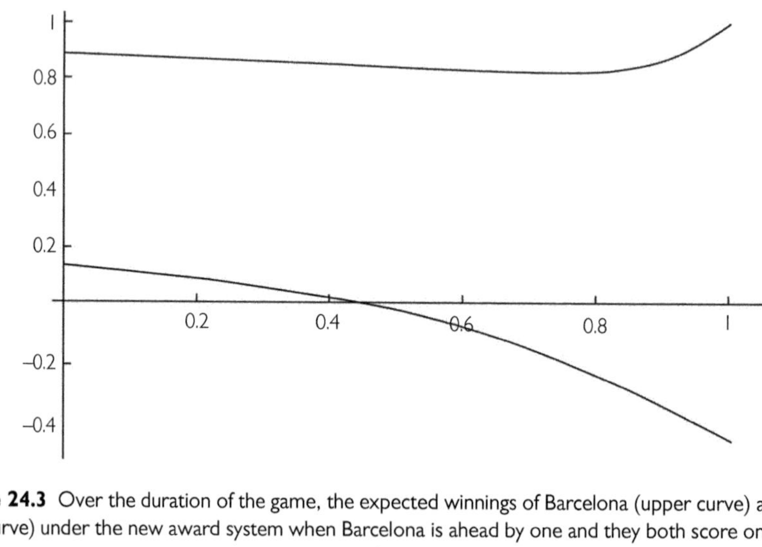

Figure 24.3 Over the duration of the game, the expected winnings of Barcelona (upper curve) and Madrid (lower curve) under the new award system when Barcelona is ahead by one and they both score on the average two goals per game.

- Under the proposed new system, once a significantly weaker team gets behind by one goal they should play conservatively until near the end of the game. Under the standard systems they should always play normally once it falls behind.
- Under the proposed new system, once the superior team gets ahead by two goals they should continue to play normally and the inferior team, conservatively. (The teams reverse roles relative to the standard system because of the presence of a reward for big wins.)

In summary, the two main advantages of the new system are that there can be significant portions of the game in which both teams should play normally and, except in the later portions of the game, the stronger team should continue to play normally once they get ahead. The length of these intervals of normal play depends on the relative strengths of the two teams. But games in which the better team continues to try to score until it is up by three goals should be more exciting than those in which the better team begins to play conservatively after it gets ahead by one goal.

Final remarks

One possible objection to our approach and to the new award system is that teams likely do not have the ability or even an interest in mathematically computing their expect winnings. Even if they did, selecting the appropriate Poisson parameters for any given game would be very difficult. These depend on many factors, including where the game is being played, the weather, the playing style of the opponent, the team's place in the league standings, the strategies of the coaches, etc. In practice teams cannot gauge precisely their relative strengths, and we are not claiming that they always follow their optimal strategies, especially when they are about evenly matched. But, as we argued earlier, the practice of the weaker team often playing for a tie and the negligible effect the change to the 3–1–0 award system seemed to have on the number of ties, as well as the overall level of scoring suggest that teams can and do compute their expected winnings when selecting their game strategies. We believe that this would remain true if the new system were adopted. In particular, when the teams are roughly equal and not low scoring and the first goal is scored early in the game, it is in neither of their interests to adopt a conservative style of play. Note that this state of affairs can be expected to hold in many actual games and no complicated computations need to be done for each team to decide that it is their best interests to continue to play normally. A real advantage of the new award system is that it encourages the better team to play normally throughout a great proportion of the game, a fact which should lead to more interesting games.

The Poisson Distribution in Soccer

The Poisson distribution models random events happening over time, for example, goals in a soccer game. If $\lambda > 0$ is the average number of goals scored by team B during a typical game, then in a particular game the probability that team B scores k goals under the Poisson distribution is

$$\Pr(B = k) = \frac{\exp^{-\lambda} \lambda^k}{k!}.$$

If team M scores on the average $\mu > 0$ goals per game and the goal scoring of the two teams is independent, then the probability that B scores k goals and M scores l goals is

$$\Pr(B = k, M = l) = \frac{\exp^{-\lambda} \lambda^k}{k!} \frac{\exp^{-\mu} \mu^l}{l!}.$$

Though the assumption of two teams scoring independently can't be right since only one team can score at a time, the results of soccer games appear to be consistent with the independent Poisson model. Using these expressions it is possible to find the probability of any outcome. For example, with $\lambda = 2$ and $\mu = 1.5$ (the case considered in the accompanying paper), the probability of a scoreless draw is 0.030, the probability of a 2–0 Barcelona win is 0.060, and the probability of a 0–2 Madrid win is 0.034, etc. By adding over all possible outcomes we find the distribution of outcomes for the game as shown in Table 24.3.

Table 24.3 ***Probability of different outcomes under the Poisson model for Barcelona (2 goals per game) versus Madrid (1.5 goals per game).***

Game outcome	Probability
Barcelona wins by 5 or more	0.019
Barcelona wins by 4	0.036
Barcelona wins by 3	0.081
Barcelona wins by 2	0.149
Barcelona wins by 1	0.209
Tie game	0.216
Madrid wins by 1	0.157
Madrid wins by 2	0.084
Madrid wins by 3	0.034
Madrid wins by 4	0.011
Madrid wins by 5 or more	0.004

Another possible objection is that the proposal is too complicated. We considered some other award systems as well. They all awarded zero points for a tie, positive points for the winner and negative points for the loser where the points awarded could depend on the size of the win and the loss. It is easy to find others that work similarly to the one we have proposed. After considerable calculations, we selected this one because it is not so different from the standard systems and we could not find another that seemed to work any better. Furthermore, the new system is at most marginally more sophisticated than the 3–1–0 award system, and it would take just a short time for both the teams and the fans to adapt. A related objection might be that the 3–1–0 system was also advertised as offering to provide more offense and failed to do so. However, our approach here shows that this is not surprising. Award systems that differentiate big wins and small wins are needed to provide an incentive for more attacking play during games.

The recent 2002 World Cup competition yields some suggestive evidence on this point. In the preliminary round the field of 32 teams was divided into eight groups of four. Within each group a round robin tournament was held to determine which two teams advanced to the final round. For teams with the same record, the ties were broken by rules that depended on the number of goals scored. The final 16 teams then competed in a single elimination tournament. In the 15 elimination games in the second stage, there were 26 goals scored for an average of 1.73 per game. Of these, two came in overtime. Many commentators noted the paucity of scoring in these games. On the other hand, in the 30 first-round games involving at least one team that made it into the round of 16 there were 118 goals scored for an average of 3.93 per game. In the six first-round games that involved two teams that made it into the final round of 16, there were 21 goals scored for an average of 3.5 goals per game. Clearly, there was more scoring in the preliminary round than in the final round. One possible explanation is the presence of weaker teams in the original field of 32. But more important, we believe, were the rules for breaking ties. In the preliminary round it was not necessarily in the interest of the better team to become conservative if they got the lead, yet while in the final round, in which winning was the only thing that mattered, once a team got ahead it was clearly in their interest to become conservative.

Of the other major team sports, hockey is the one that is most similar to soccer. This is true for how the teams interact during the game and in the determination of the standings. There are important differences, however. In particular, in hockey it is much more difficult for a weaker team to play defensively for a tie. This is because there are many more scoring opportunities in a typical hockey game than in a typical soccer game. One way to increase scoring in soccer would be to change the rules so that there are more scoring opportunities. For example, one could change the offside rule or increase the size of the goal. In contrast to American professional sports such as baseball, basketball, and football, which are constantly tinkering with the rules in an attempt to produce a more entertaining product, soccer officialdom has been much more conservative about making changes. In lieu of making rule changes that affect how the game is played (eliminating off sides, changing the size of the goal), the award system we propose is the simplest way to achieve more entertaining games with more scoring.

ACKNOWLEDGMENTS

We wish to thank Hal Stern for his many helpful comments during our writing of this piece. Research for Glen Meeden was supported in part by NSF Grant DMS 9971331.

ADDITIONAL READING

Keller, J. (1994), "A characterization of the Poisson distribution and the probability of winning a game," *The American Statistician*, 48, 294–298.

Lee, A. (1997), "Modeling scores in the Premier League: Is Manchester United really the best?" *Chance*, 10, 15–19.

CHAPTER 25

Balance in competition in Dutch soccer

RUUD H. KONING

Abstract

We estimate an ordered probit model for soccer results in the Netherlands. The result of a game is assumed to be determined by home ground advantage and differences in quality between the opposing teams. The parameters of the model are used to assess whether the balance in competition in Dutch professional soccer has changed over time. Contrary to popular belief, we find that the balance has not changed much since the mid-1970s.

Keywords: *balance in competition, home advantage, ordered probit model, professional soccer*

Introduction

Professional soccer is a big business today. Broadcasting rights for 4 years have been sold in England for approximately US $1 billion, a typical sponsor contract for a top European team is valued at US $6 million (yearly) and the annual salary of a top striker is rumoured to be US $6 million. Demand, as measured by attendances in stadiums or numbers of spectators watching live broadcasts, has increased as well in recent years. In other words, the soccer business is becoming a major amusement industry (*Economist*, 1997).

The two main reasons for interest in a particular soccer game, or in any sports contest, are the quality of the play in absolute terms and the uncertainty about the outcome. The home games of weak teams are usually sold out when the top teams visit. Strong teams tend to win their games but sometimes they are taken by surprise by a weaker team. Most soccer *aficionados* can recall stories about a leader of a league who lost unexpectedly against a team that was at the bottom of the league. In fact, soccer results are more random than the results of games in other sports as only a few goals are scored each game and chance may be quite influential in determining the outcome. (The relationship between predictability and scores in different sports has been examined by Stefani, 1983.)

Schemes such as sponsorship contracts, proceeds from the lucrative Champions League competition (a European competition in which only a selected number of teams participate), merchandizing and television rights allow wealthy teams to lure players away from poorer teams, even to sit on the substitutes' bench. As a result, weak teams are concerned that an increasing inequality in the income distribution of clubs leads to a decrease in the odds of beating strong teams. Poorer teams used to receive revenues from transferring players with great talent to top teams, but this source of income has vanished after the Bosman ruling. (According to the Bosman ruling by the European Court of Justice, a soccer player from the European Community is a free agent after his contract has expired.) This income could be used to improve training facilities for weaker teams or to increase the quality of the team by hiring players. The increased demand for top players across Europe has sent salaries sky high with obvious repercussions for the salary demands of mediocre players in average teams. These developments may cause a breakdown in balance in competition between teams and hence may decrease interest in soccer in the long run. Some weaker teams use these arguments to call for a redistribution of the proceeds of the sale of television rights (both of the national competitions and of the Champions League).

This paper examines the development of the balance in competition in Dutch professional soccer. Our aims are modest: we shall measure the balance in various ways, and we shall discuss its development over time. The structure of the paper is as follows. The next section discusses some theory on balance in competition. In the section entitled 'A model to analyse soccer results' we develop a simple statistical model that can be used for this purpose. The balance in competition and its evolution over time are discussed in the subsequent section. We end with some conclusions and directions for further research in the last section.

Theory

Sports contests are interesting when there is not much difference in the quality of the contenders. As Quirk and Fort (1992), page 243, put it:

> 'One of the key ingredients of the demand by fans for team sports is the excitement generated because of uncertainty of outcome of league games.... In order to maintain fan interest, a sports league has to ensure that teams do not get too strong or too weak relative to one another so that uncertainty of outcome is preserved.'

In fact, this is cited as the reason why some sports organizations in the USA are exempted from anti-trust regulation. Two teams engage in a joint production when they play a game. The outcome and the quality of the game are the good that is sold to the public. The public is worse off when the outcome of a game is easily predicted than if the game is tight. Therefore, collusion between teams to increase the quality of the game may be in the public's interest. This view neglects the absolute quality of the play. In fact, one of the important instruments to maintain balance in competitions in the USA is the inverse draft system, where lower ranked teams can pick talented new players before higher ranked teams can. In soccer leagues, there is no such balancing regulation.

According to the view cited above, an important task for sport bodies like the Union of European Football Associations or the Dutch Soccer Association is to maintain balance in competition because it is needed to ensure long-term interest in the league. The instruments that are available to achieve balance are limited, however. In the Netherlands, a court has decided in a preliminary ruling that individual teams are the owners of the broadcasting rights and not the organizing body. Each team can therefore sell its broadcasting rights individually and take the proceeds of this transaction. An implicit subsidy from wealthy teams to poor teams by the organizing body, to maintain balance, is no longer possible. Moreover, in contrast with baseball and football in the USA, gate receipts are not split between both teams. This may favour teams with big stadiums, even though a completely balanced competition is played (each pair of teams meets twice, once at each venue). In addition, there are no salary caps, either for the teams in total or for individual contracts in European soccer.

Balance in competition and regulations that intended to change it have been studied in the American context but not in the context of European soccer. Neumann and Tamura (1996) studied the balance in the National Football League in the USA. It is measured as the spread of parameters in a non-linear regression model. These parameters capture the quality of the teams. Bennett and Fizel (1995) examined the effect of telecast deregulation on balance in competition in US college football. They measured it by comparing actual performances in a league with the performances that would be found if all teams were of equal strength (an approach developed by Noll 1991 and Scully 1989). Empirical results have also been published in Quirk and Fort (1992). They measured the long-term development in five American professional sports leagues. They measured the balance by comparing the percentages of wins or losses for each league for each year with the percentages we would expect to find if all teams were equally strong. Each of the five leagues that they analysed showed a significant imbalance, though the imbalance in both baseball leagues has been decreasing in the last 20 or 30 years.

In the American literature, balance is usually defined as a win percentage of 50%. Such a definition is not useful for soccer, because of the prevalence of draws (unless we consider a draw to be a half-win). Draws are quite common in soccer. Over the period from 1956–1957 to 1996–1997, 26% of all league games ended in a draw, 48% ended in a win by the home team and the remaining 26% ended in a win for the away team. We define a soccer league to be in perfect balance for a certain year if the probability that any team wins a home game does not vary with the opponent or with the team. We assume that the home ground advantage is equal for each team, so in a balanced competition two teams would have equal probability of winning if the game were played on a neutral ground. In a balanced competition the probability that a team wins its home game may exceed the probability of a loss of a home game because of the home advantage. This definition allows for home advantage that changes over time, while the league is still in complete balance.

A model to analyse soccer results

The statistical model

In this section we propose a simple statistical model to analyse the outcome of soccer games. The model is an extension of the model of Neumann and Tamura (1996) in that we allow for an advantage for the home team. The strength of team i in the league is measured by a single parameter α_i. This parameter is independent of the opponent and venue of the game, and it is assumed to be constant during the season. If we assume that team i plays at home and team j is the away team, the difference in strength is $\alpha_i - \alpha_j$. To allow for unmeasured characteristics (i.e. those not captured by α), chance events during a game that influence the score etc., we assume that the outcome of the game is determined by the random variable D^*_{ij}:

$$D^*_{ij} = \alpha_i - \alpha_j + h_{ij} + \eta_{ij}, \quad i,j = 1,\ldots,18, \quad j \neq i. \tag{25.1}$$

In equation (25.1), h_{ij} is the home ground advantage of team i over team j which is assumed to be normally distributed with mean h. η_{ij} is a mean 0 random variable that captures other determinants of the result of the game. If D^*_{ij} is positive, team i is stronger than j, and D^*_{ij} is negative if team j is stronger than i. We do not observe the actual difference in strength; we only observe the outcome of the game. In fact, we observe whether team i has won, has played a draw or lost against team j. The latent difference in strength is transformed into an observed outcome of the game by

$$D_{ij} = \begin{cases} 1 & D^*_{ij} > c'_2, \\ 0 & c'_1 < D^*_{ij} \leq c'_2, \\ -1 & D^*_{ij} \leq c'_1 \end{cases} \tag{25.2}$$

with $D_{ij} = 1$ if team i wins, $D_{ij} = 0$ if team i plays a draw and $D_{ij} = -1$ if team j (the away team) wins the game. If we assume that h_{ij} and η_{ij} in equation (25.1) are independent normally distributed ($\epsilon_{ij} = h_{ij} + \eta_{ij} \sim \mathcal{N}(h, \sigma^2)$), then the probabilities of the possible outcomes of a game are

$$\begin{aligned} \Pr(D_{ij} = 1) &= 1 - \Phi\{(c_2 - \alpha_i + \alpha_j)/\sigma\}, \\ \Pr(D_{ij} = 0) &= \Phi\{(c_2 - \alpha_i + \alpha_j)/\sigma\} - \Phi\{(c_1 - \alpha_i + \alpha_j)/\sigma\}, \\ \Pr(D_{ij} = -1) &= \Phi\{(c_1 - \alpha_i + \alpha_j)/\sigma\} \end{aligned} \tag{25.3}$$

with $\Phi(\cdot)$ the standard normal distribution function and $c_1 = c'_1 - h$ and $c_2 = c'_2 - h$.

The statistical model in equation (25.2) allows for a constant home ground advantage. Consider two (hypothetical) teams of equal strength so that $\alpha_i - \alpha_j = 0$. The probability that the home team wins is $1 - \Phi(c_2/\sigma)$ and the probability that the home team loses is $\Phi(c_1/\sigma)$. These two probabilities are not constrained to be equal. In fact, we would expect that $\Phi(c_1/\sigma) < 1 - \Phi(c_2/\sigma)$ and this is confirmed by the results of our estimation. The existence of a home advantage can be examined formally by testing the hypothesis $c_1 = -c_2$.

It is not possible to identify all the parameters of this model. First, we need to fix the location of the quality parameters α. We impose the identifying restriction $\Sigma_i \alpha_i = 0$ so that the parameters α can be interpreted as deviations from a hypothetical average team with quality 0. A positive α_i implies that the quality of team i is better than average; a negative α_i implies the opposite. In addition, we fix the scale of the model by imposing the standard normalization $\sigma^2 = 1$.

Model (25.3) resembles models used by Clarke and Norman (1995), Stefani (1980) and Kuk (1995) to model soccer results. In Stefani (1980) and Clarke and Norman (1995), the dependent variable is the goal difference, and their models are estimated by least squares techniques. By using the least squares method, the dependent variable is assumed to follow a normal distribution. However, the observed goal difference takes only integer values with a limited range. They allow home ground advantage to vary between teams. The model in

Kuk (1995) resembles the model above in that the dependent variable is the result of the game (and not the goal difference), and that an ordered probit model is used to derive the probability that a game is won, drawn or lost. In his model, the quality of a team differs between home and away games, and the home advantage varies between teams and over games. He estimated his model by using methods of moments using only the final ranking at the end of the season. An alternative approach could be based on Poisson-like models for the exact score in a game; see for instance Maher (1982), Dixon and Coles (1997) and Dixon and Robinson (1997). The reason that we prefer the ordered probit model is the simplicity of model (25.2): the quality of each individual team is captured by a single parameter. Moreover, this model allows for a simple separation between the measurement of quality of the teams and home advantage. In addition, the values that the dependent variable take are consistent with the stochastic specification of the model. Poisson-like models are usually more complex and have more parameters. For instance, in Maher (1982) at least two parameters per team had to be estimated and these parameters are difficult to interpret. Maher also assumed that the numbers of goals scored by the home team and the away team in a particular game are statistically independent. This may be too strong an assumption. He also assumed that games are mutually independent, an assumption that we shall make as well.

Description of the data and estimation results

The parameters were estimated by using the complete history of the Premier League of professional soccer in the Netherlands. (The data were obtained from Michael Koolhaas, private correspondence, and http://www.noord.bart.nl/~kammenga/soccer.) Organization of the competition as we know it today was introduced in the 1955–1956 season. In the 1962–1963, 1963–1964, 1964–1965 and 1965–1966 seasons only 16 teams participated in the Premier League. In all the other seasons 18 teams participated. Rules for relegation to the first division have changed over time. In the last few seasons the team finishing last was relegated automatically to the first division. The teams ranked 16 and 17 had to play additional games against teams in the first division. However, in earlier seasons, the teams finishing in the 17th and 18th places were relegated without having to play additional games. In total, 54 different clubs have played in the Premier League since the start of the competition in 1955–1956. Each year a couple of new teams entered the competition, either because of mergers or because of promotion. (A list of all the relevant mergers is given in Appendix A.) In each season, any combination of two teams meet twice: once at each venue. Therefore, a competition with 18 teams consists of 306 games; in total the data set comprises 12,155 games. However, data from only 179 games are recorded from the 1996–1997 season.

We begin by estimating the parameters α_i, c_1 and c_2. It is assumed that the quality of the teams (measured by α_i) and the home advantage (measured by c_1 and c_2) were constant during the history of professional soccer. We use these results to calculate an all-time ranking of Dutch soccer teams by pooling the data over all seasons. The parameters were estimated by maximization of the log-likelihood function

$$l(\theta) = \sum_{\tau} \sum_{(i,j)\in\mathcal{J}_\tau} [I_{(D_{ij}=1)} \ln\{1 - \Phi(c_2 - \alpha_i + \alpha_j)\} + I_{(D_{ij}=0)} \ln\{\Phi(c_2 - \alpha_i + \alpha_j) - \Phi(c_1 - \alpha_i + \alpha_j)\} + I_{(D_{ij}=-1)} \ln\{\Phi(c_1 - \alpha_i + \alpha_j)\}]. \quad (25.4)$$

In this equation, τ is the index indicating the season and $\mathcal{J}_\tau$ is the index set of teams playing in the Premier League in season τ. The point estimates and their standard errors are given in Table 25.3 in Appendix A. The ordering of the parameters α_i indicates an all-time ranking. The best three teams have been Ajax (0.963), Feyenoord (0.750) and PSV Eindhoven (0.735), and the teams performing least well have been Dordrecht (−0.581), Fortuna SC (−0.498) and SVV (−0.433). This ranking is not necessarily equal to the standard ranking in which two or three points are awarded for each win and one for a draw. Teams that are relegated during some seasons do not earn any points in this ranking and would be at the bottom of the standard ranking. In our approach, there are no observations on a team if it does not participate in the Premier League during a season. Hence, the estimated α_i of a team that has participated in the Premier League for two seasons only can exceed the estimated α_i of a team that has played in the Premier League for, say, three seasons, if that first team played well during these two seasons. Indeed, we find that the teams with least points are not those with the smallest α_i: these are Fortuna SC, SHS and Alkmaar.

Second, we estimate the parameters for each year. This approach allows for variation in team-specific quality over time. Fig. 25.1 presents estimates of the home advantage. Detailed information on the individual estimates for each year is available on request. Home advantage is measured as the difference between the probability that the home team wins if both teams are of equal quality (i.e. $\alpha_i = \alpha_j$) and the probability that the away team wins. The circles depict the probability that the home team wins and the triangles the probability that the away team wins. If there had been no home advantage both sets would coincide (apart from sample variation). However, we see that there is a clear home advantage which has increased markedly during the second half of the 1960s to 33% in 1970–1971. Since the early 1970s, the probability that the home team wins against an opponent of equal strength is approximately 45–50%. The corresponding probability for the away team appears to have increased since then from

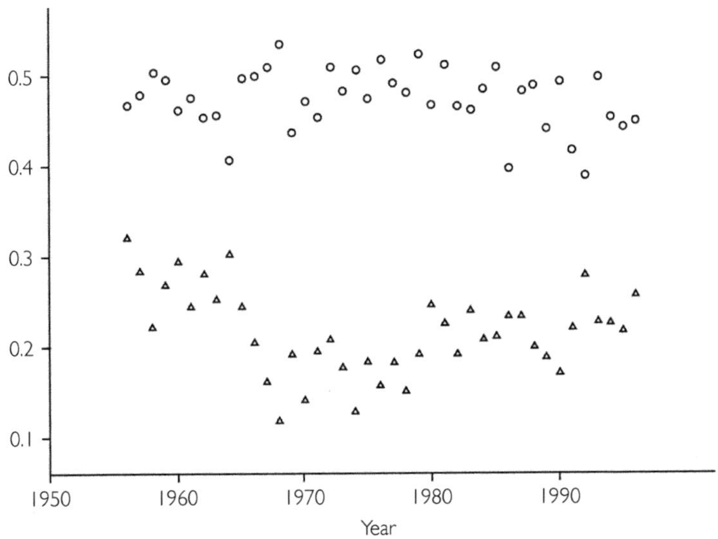

Figure 25.1 Probabilities of winning of home (◯) and away (Δ) games when the teams are of equal strength.

approximately 15% to 20%. Therefore, home advantage decreased over that period. Indeed, over the 1990–1996 period, home advantage averaged 21% compared with 31% in the period 1970–1975. Home advantage in the league varies from year to year. This was also found by Clarke and Norman (1995). A test of whether $c_1 = -c_2$ is rejected at any reasonable level of significance for all years.

In Table 25.1, we present some summary statistics of the estimation results for each year. For each season we list the strongest and the weakest teams, and the standard deviation of the estimated α_is. Instead of giving the point estimates for the α_i-parameters, which are difficult to interpret, we give a transformation of these estimates. If team i has quality α_i in a given year, then the probability that this team wins a home game against the hypothetical team with quality 0 is $1 - \Phi(\alpha_i - c_2)$. This probability is given in the third and fifth columns of Table 25.1, where $\pi_{(n)}$ denotes the probability that the worst team in a given year beats the average team in a home game. $\pi_{(1)}$ is calculated similarly for the best team in that year. We have set $c_2 = 0.060$, the value obtained when estimating the model for the whole sample period. Hence, the variation in the probabilities reflects variations in quality, not changes in home advantage.

Note that the maximum likelihood estimate for an α_i would diverge to $-\infty$ if a team loses all its games during a season or to ∞ if a team wins all its games. No such teams are to be found even though Ajax came close in the 1971–1972 season by losing only one game, drawing in a mere three games and winning all the other games.

Specification tests

The model used in the previous sections is estimated by using maximum likelihood. The parameters are estimated inconsistently if the distributional assumption of normality is not correct. We tested whether we should reject the assumption of normality against the more general alternative that the distribution of the error terms belongs to a member of the Pearson family of distributions (for details of the test statistic we refer to Glewwe 1997 and Weiss 1997). Members of the Pearson family include the normal, t- and Γ-distributions. Essentially, we test whether the third moment of ϵ is 0 and the fourth moment of ϵ is 3. The test statistic was calculated for each year that the model was estimated. The null hypothesis of normality was not rejected in any season at a 5% level of significance. This conclusion is not at odds with the Poisson assumption that is often made when analysing soccer scores. The score difference is estimated by summing over a large number of Poisson-distributed scores, and hence in the limit it can be approximated by a normal distribution. Clarke and Norman (1995) also found that the residuals were approximately normally distributed when they estimated a model similar to equation (25.1) by least squares.

We tested whether or not restrictions on the parameters could be imposed. First, for each year all estimated α_i were found to be jointly significantly different from 0. Second, the hypothesis that the home advantage is constant over time had to be rejected. We also rejected the hypothesis that the quality of a given team does not vary over time.

If we assume that the quality parameters α_i remain constant over time, it is possible to test for variation in the variance of the error term in model (25.2). We imposed this restriction, and re-estimated the model for the period from 1991–1992 to 1996–1997 with unrestricted variances. We could not reject the null hypothesis that the variance of ε_{ij} is constant over time.

Table 25.1 ***Best and worst team for each year.***

Year	Worst team	$\pi_{(n)}$	Best team	$\pi_{(1)}$	Standard deviation
1956	Eindhoven	0.259	Ajax	0.714	0.320
1957	Den Bosch	0.260	SC Enschede	0.679	0.273
1958	SHS	0.212	Sparta	0.755	0.412
1959	Sittardia	0.196	Ajax	0.735	0.363
1960	Noad	0.212	Feyenoord	0.776	0.386
1961	Rapid JC	0.263	Feyenoord	0.730	0.308
1962	Volewijckers	0.112	PSV	0.700	0.399
1963	Blauw-Wit	0.265	DWS	0.706	0.362
1964	NAC	0.335	Feyenoord	0.747	0.270
1965	Heracles	0.297	Ajax	0.867	0.466
1966	Willem II	0.143	Ajax	0.863	0.537
1967	Sittardia	0.212	Ajax	0.898	0.529
1968	Fortuna SC	0.175	Feyenoord	0.894	0.628
1969	SVV	0.144	Ajax	0.918	0.600
1970	AZ	0.128	Feyenoord	0.904	0.739
1971	Volendam	0.168	Ajax	0.959	0.725
1972	Den Bosch	0.179	Ajax	0.931	0.678
1973	Groningen	0.233	Feyenoord	0.865	0.595
1974	Wageningen	0.261	PSV	0.852	0.520
1975	Excelsior	0.216	PSV	0.810	0.501
1976	DE Graafschap	0.273	Ajax	0.794	0.429
1977	Telstar	0.165	PSV	0.795	0.443
1978	VVV	0.134	Ajax	0.841	0.539
1979	Haarlem	0.323	Ajax	0.749	0.352
1980	Wageningen	0.257	AZ	0.888	0.486
1981	De Graafschap	0.122	Ajax	0.852	0.552
1982	NEC	0.259	Ajax	0.887	0.514
1983	DS79	0.163	Feyenoord	0.860	0.560
1984	Pec Zwolle	0.200	Ajax	0.813	0.446
1985	Heracles	0.123	PSV	0.896	0.550
1986	Excelsior	0.223	PSV	0.894	0.484
1987	DS79	0.138	PSV	0.877	0.476
1988	Veendam	0.306	PSV	0.808	0.370
1989	Haarlem	0.167	Ajax	0.736	0.403
1990	NEC	0.292	PSV	0.812	0.440
1991	VVV	0.120	PSV	0.875	0.576
1992	Fortuna SC	0.252	Feyenoord	0.792	0.467
1993	Cambuur	0.233	Ajax	0.820	0.409
1994	Dordrecht	0.238	Ajax	0.914	0.556
1995	Go Ahead Eagles	0.197	Ajax	0.872	0.550
1996	AZ	0.237	PSV	0.795	0.422

Empirical evidence of balance in competition

In this paper we measure the balance in competition in three different ways.

(a) If the standard deviation σ_P of the number of points in the final ranking of a competition is small, there is not much spread in the points gained at the end of the season and the competition has been tight.
(b) Since α_i is the extent to which team i is better than a hypothetical team with quality 0, it is natural to measure the balance in competition by the total deviation from average quality, $\Sigma_i \alpha_i^2$. This is proportional to the standard deviation σ_α of the quality parameters of the statistical model of the previous section. Again, if this number is small, the quality of the teams does not vary much.
(c) The concentration ratio CR_K is defined as the number of points obtained by the top K teams divided by the number of points that they could have gained. If there are J teams in a competition the team winning the competition could have obtained $2W(J-1)$ points where W is the number of points awarded for a game won. We denote the number of points obtained by the kth-best team by $P_{(k)}$. The concentration ratio is formally defined as

$$CR_K = \sum_{k=1}^{K} P_{(k)} \Big/ KW(2J - K - 1), \tag{25.5}$$

the number of points actually obtained by the K best teams divided by the maximum number of points that they could have obtained. If the concentration ratio is high, the top K teams did not lose many points to weaker teams.

These three measures are to some extent 'static' as they refer to balance *within* a particular season. An advantage of the second measure compared with the first measure is that it does not require that the season be complete. The concentration ratio is not a measure of balance in the whole competition; it applies to the quality of the top teams. This measure is interesting though as it is commonly believed that the gap between top teams and the rest has increased over time. It is not possible to address this issue with the first two measures.

The first two measures are not completely equivalent: a crucial drawback of σ_P as a measure of balance in competition is that it is not invariant under changes in home advantage. As our statistical model separates home advantage and team quality, σ_α does not suffer from this drawback. Moreover, in the 1995–1996 season the number of points obtained for a win was raised from 2 to 3. In contrast with the standard deviation of the number of points, the standard deviation of the α_i is invariant to changes in the number of points awarded for a win or a draw. Finally, we can estimate the α_i and their standard deviations even if a season is not finished completely. Note that σ_P and σ_α may rise if the average scoring frequency of all teams increases. In this case, skilled teams accumulate more wins whereas less skilled teams accumulate more losses, leading to more dispersion in the point and quality distributions. This is of little practical importance as goal scoring does not vary much over the years.

Estimates of σ_P and σ_α are graphed in Fig. 25.2. The standard deviation of the α_i varies between 0.25 and 0.75, and the standard deviation of the number of points varies between 5.9 and 13.5. To interpret the level of σ_P and σ_α we have simulated two hypothetical competitions and calculated σ_P and σ_α for each simulated competition. In the first competition, each team is

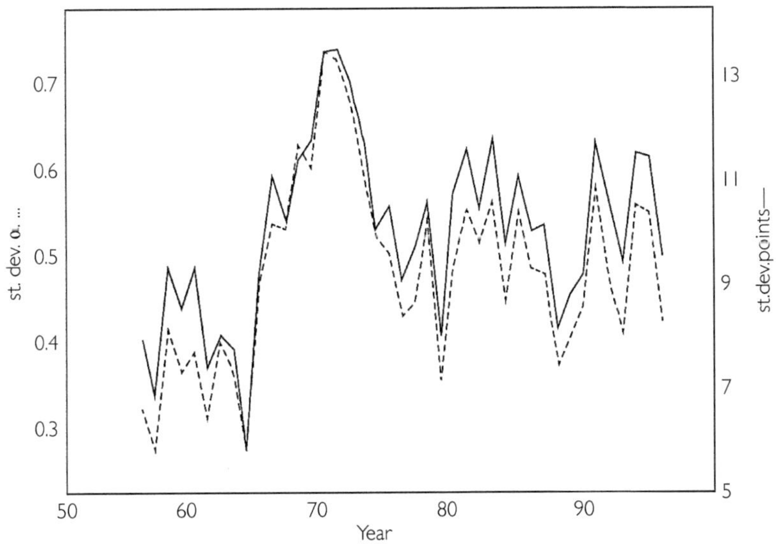

Figure 25.2 Standard deviations of the estimated α_i (· · · · · · · · ·) and of the numbers of points (——).

equally strong and has a 50% probability of winning a home game, a 25% probability of playing a draw and 25% of losing a home game. For this case, the average value of σ_P is 4.92, and the average value of σ_α is 0.20.

In the second simulation the probabilities are the same, except for three strong teams. These three strong teams win games against each of the remaining 15 weaker opponents with 90% probability (and a 5% probability of a draw and a 5% probability of a loss). In this second case, the average value of σ_P is 9.87, and the average value of σ_α is 0.46. Therefore, a competition with three strong teams and 15 identical weaker teams gives the same values for σ_P and σ_α as we find in Fig. 25.2.

The balance in competition has not changed systematically from the early start of professional soccer in 1955 until the mid-1960s. We then see a marked decrease between 1965 and 1970 followed by an increase in between 1970 and 1976. Coincidentally (or not) it was in the period 1966–1970 that Dutch professional soccer caught up with the best teams in Europe. Ajax was the first Dutch finalist in a European tournament in the spring of 1969 and Feyenoord was the first Dutch winner in a European tournament in 1970. Dutch (and European) soccer was dominated by Ajax from 1970 until 1973, a period when the balance in competition in Dutch soccer increased sharply.

From the mid-1970s to the end of the century there is no clear trend in the balance in competition. One year, competition is tighter than in another year, but no clear trends are discernible from the data. The spread from year to year is considerable; years with a tight competition are interspersed with years with a one-sided competition. This is especially noteworthy because it was feared in the early 1980s that competition would become less tight because of shirt sponsorship which in Dutch soccer has been allowed from the 1981–1982 season onwards. At first, it seemed that criticism of shirt sponsorship was justified, since in the 1981–1982 season two teams could not find a sponsor at all. This was only temporary though: even amateur teams now have shirt sponsors. Some teams have had better sponsoring deals than others, resulting in a more unequal distribution of income. This increase in income inequality is claimed to lead to a decrease in balance in competition. However, we do not find any evidence for this hypothesis. Less successful teams use the same arguments as those used against shirt sponsorship to oppose television contracts that give a larger share of the revenue to more successful teams. According to them, an unequal distribution of television revenues will lead to an unequal distribution of quality and this leads to a decrease in general interest in soccer. However, the balance in competition has not decreased significantly since the introduction of shirt sponsorship. In fact, shirt sponsorship may have enabled most semiprofessional players to become full professionals and this may have increased the overall quality of soccer.

To examine the robustness of our results, we have estimated two other models. First, we estimated the quality parameters of the teams by using the model of Clarke and Norman (1995). In this model the dependent variable was the difference in goals scored, and the parameters were estimated by using least squares. In the second model, the ordered probit structure of equation (25.3) was retained, but two additional categories were added. The dependent variable now makes a distinction between a win (or loss) by a three-goal difference (or more). In both cases, the variation in the estimated quality parameters was examined, and it turned out that the results were very similar to those in Fig. 25.2. Hence, our conclusions about the development of the balance in competition are robust with respect to the statistical model used.

As a third indicator of the balance in competition, we look at the concentration ratios for the first and fourth place. The results are given in Fig. 25.3. Qualitatively, we see the same picture as in the previous graphs: until the mid-1960s the top team in each year captured only 75% of the number of points that it could have obtained at the end of the season. This percentage increases during the second half of the 1960s, to a maximum of 94% for the top team in 1971–1972. Then an increase in the balance of competition sets in, followed by an irregular period

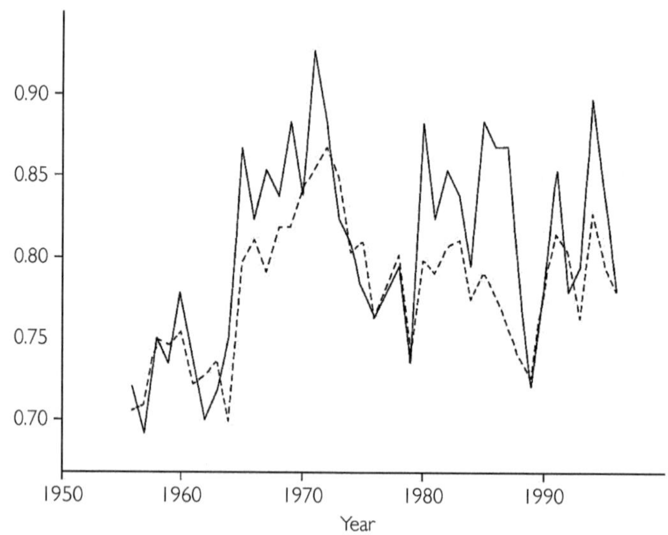

Figure 25.3 Concentration ratios for the first and fourth places:——, first place; · · · · · · · · · · ·, fourth place.

with no clear trends. The picture is slightly different for the top four teams: a slight upward trend of the concentration ratio during the 1980s and 1990s is visible. (CR_4 exceeds CR_1 in a particular year if the number of points obtained by the teams that end second, third and fourth does not differ much from the number of points obtained by the team that ended first (i.e. if $P_{(2)} + P_{(3)} + P_{(4)} > (45/17)P_{(1)}$). This happens in 13 seasons.) Contrary to common opinion and despite the slight upward trend, the value of CR_4 is not high when compared with typical values encountered during the 1960s.

Conclusion

In this paper we have discussed the balance in competition in Dutch professional soccer. We used a simple model to estimate the quality of the teams participating in the Premier League. We find that the balance decreased markedly during the second half of the 1960s and increased during the first half of the 1970s but that there has been no clear trend since. We also find that the introduction of shirt sponsorship did not lead to a noticeable significant decrease in the balance. These conclusions are borne out by three different measures of balance and hence are not sensitive to model specification.

This paper provides only a starting-point for a more structural economic analysis of the balance in competition. The 'superstar' model of Rosen (1981) may provide insight into why an increase in income inequality that may have taken place did not lead to a decrease in the balance of competition. Another issue to be resolved is to examine whether the lack of

Table 25.2 ***Mergers in Dutch professional soccer.***†

Year	New team	Merged teams
1958	DWS	Amsterdam, *DWS*
1962	Roda JC	*Roda Sport*, Rapid JC
1963	Telstar	*Stormvogels, VSV*
1965	Twente	SC Enschede, *Enschedese Boys*
1967	AZ	Alkmaar, *Zaanstreek*
1967	Den Bosch	*Den Bosch, Wilhelmina*
1967	Xerxes/DHC	*DHC*, Xerxes
1968	Fortuna SC	Fortuna '54, Sittardia
1970	Utrecht	Dos, Elinkwijk, *Velox*
1971	Den Haag	ADO, Holland Sport
1972	FC Amsterdam	DWS, Blauw-Wit
1974	FC Amsterdam	FC Amsterdam, De Volenwijckers
1991	Dordrecht '90	*Dordrecht '90*, SVV

† Teams in italics have never played in the Premier League.

Table 25.3 ***Estimation results of the full model.***

Team	$\hat{\alpha}$	Standard deviation	Team	$\hat{\alpha}$	Standard deviation	Team	$\hat{\alpha}$	Standard deviation
Ado	0.322	0.032	Fortuna 54	0.159	0.029	SHS	–0.546	0.174
Ajax	0.951	0.031	Fortuna SC	0.009	0.022	Sittardia	–0.212	0.040
Alkmaar	–0.219	0.043	Go Ahead Eagles	0.057	0.028	Sparta	0.262	0.029
Amsterdam	0.034	0.033	Graafschap	–0.046	0.030	SVV	–0.443	0.047
AZ	0.215	0.032	Groningen	0.123	0.021	Telstar	–0.082	0.033
Blauw-Wit	0.023	0.031	Haarlem	–0.032	0.032	Twente	0.389	0.028
Cambuur	–0.297	0.040	Heerenveen	0.105	0.041	Utrecht	0.127	0.023
De Graafschap	–0.446	0.054	Helmond	–0.446	0.044	Veendam	–0.255	0.038
Den Bosch	–0.067	0.022	Heracles	–0.192	0.051	Vitesse	0.252	0.031
Den Haag	–0.002	0.025	Holland	0.031	0.031	Volendam	–0.050	0.021
Dordrecht	–0.335	0.050	MVV	0.048	0.032	Volewijckers	–0.353	0.035
DOS	0.156	0.019	NAC	0.045	0.033	VVV	0.004	0.021
DS79	–0.829	0.113	NEC	–0.085	0.029	Wageningen	–0.356	0.045
DWS	0.141	0.023	Noad	–0.235	0.040	Willem II	–0.047	0.022
Eindhoven	–0.283	0.057	PEC Zwolle	–0.103	0.041	Xerxes	0.212	0.085
Elinkwijk	–0.136	0.051	PSV	0.724	0.029	Xerxes/DHC	0.272	0.116
Excelsior	–0.236	0.039	Rapid JC	0.111	0.020	c_1	–0.721	0.012
FC Amsterdam 1	0.280	0.049	RKC	0.107	0.019	c_2	0.060	0.010
FC Amsterdam 2	–0.152	0.026	Roda JC	0.267	0.034			

recent trends in the balance of competition is specific to Dutch soccer or whether it is a more international phenomenon.

ACKNOWLEDGEMENTS

The author thanks Marco Haan, Peter Hopstaken, Bas van der Klaauw, Geert Ridder, seminar participants at the University of Mannheim, Concordia University, Queen's University and participants at the Statistische Dag and the European Economic Association meeting in Berlin for helpful discussions and comments. The exposition and the content of the paper have benefited from detailed comments by three referees.

Appendix A: Estimation results of the complete model

In this appendix we discuss the construction of our data set and give estimation results of the model with the α_i constant during the sample period. First, in Table 25.2 we give a list of mergers in Dutch professional soccer. (The list is based on information provided in Verkammen and Vermeer 1994.) Some teams have played under several names. For instance, FC Dordrecht changed its name in 1979 to DS '79 and in 1990 again to Dordrecht '90. In other cases, professional soccer teams merged with other professional soccer teams (for example, in 1991 Dordrecht '90 merged with SVV to form SVV/Dordrecht '90 which changed its name again in 1992, reverting to Dordrecht '90). We have treated each team that resulted from a merger as a new team, so we distinguish between DS '79 (a predecessor of Dordrecht '90 that played in the Premier League in 1987–1988) and Dordrecht '90 that resulted from a merger with SVV. In the same vein, FC Amsterdam before 1974 is considered to be a different team from FC Amsterdam after the year when it merged with De Volenwijckers.

In Table 25.3 we give the estimation results of the model estimated over the period 1956–1996 in which all parameters are constant over time. The number of cases is 12,155, and the mean log-likelihood is −0.980507.

REFERENCES

Bennett, R. W. and Fizel, J. L. (1995) Telecast deregulation and competitive balance: NCAA Division I football. *Am. J. Econ. Sociol.*, **54**, 183–199.

Clarke, S. R. and Norman, J. M. (1995) Home ground advantage of individual clubs in English soccer. *Statistician*, **44**, 509–521.

Dixon, M. J. and Coles, S. G. (1997) Modelling association football scores and inefficiencies in the football betting market. *Appl. Statist.*, **46**, 265–280.

Dixon, M. J. and Robinson, M. E. (1997) A birth process for association football matches. Lancaster University, Lancaster.

Economist (1997) Golden goals. *Economist*, May 31st, 63–65.

Glewwe, P. (1997) A test of the normality assumption in the ordered probit model. *Econometr. Rev.*, **16**, 1–19.

Kuk, A. Y. C. (1995) Modelling paired comparison data with large numbers of draws and large variability of draw percentages among players. *Statistician*, **44**, 523–528.

Maher, M. J. (1982) Modelling association football scores. *Statist. Neerland.*, **36**, 109–118.

Neumann, G. R. and Tamura, R. F. (1996) Managing competition: the case of the national football league. University of Iowa, Iowa City.

Noll, R. G. (1991) Professional basketball: economic and business perspectives. In *The Business of Professional Sports* (eds P. D. Staudohar and J. A. Mongan), pp. 18–47. Chicago: University of Illinois Press.

Quirk, J. and Fort, R. D. (1992) *Pay Dirt*. Princeton: Princeton University Press.

Rosen, S. (1981) The economics of superstars. *Am. Econ. Rev.*, **71**, 845–858.

Scully, G. W. (1989) *The Business of Major League Baseball*. Chicago: University of Chicago Press.

Stefani, R. T. (1980) Improved least squares football, basketball, and soccer predictions. *IEEE Trans. Syst. Man Cyber.*, **10**, 116–123.

—— (1983) Observed betting tendencies and suggested betting strategies for European football pools. *Statistician*, **32**, 319–329.

Verkammen, M. and Vermeer, E. (1994) *Om 't Spel en de Knikkers*. Bilthoven: Mundt.

Weiss, A. A. (1997) Specification tests in ordered logit and probit models. *Econometr. Rev.*, **16**, 361–391.

CHAPTER 26

Dynamic modelling and prediction of English Football League matches for betting

MARTIN CROWDER, MARK DIXON,
ANTHONY LEDFORD AND MIKE ROBINSON

Abstract

We focus on modelling the 92 soccer teams in the English Football Association League over the years 1992–1997 using refinements of the independent Poisson model of Dixon and Coles. Our framework assumes that each team has attack and defence strengths that evolve through time (rather than remaining constant) according to some unobserved bivariate stochastic process. Estimation of the teams' attack and defence capabilities is undertaken via a novel approach involving an approximation that is computationally convenient and fast. The results of this approximation compare very favourably with results obtained through the Dixon and Coles approach. We note that the full model (i.e. the model before the above approximation is made) may be implemented using Markov chain Monte Carlo procedures, and that this approach is vastly more computationally expensive. We focus on the probabilities of home win, draw or away win because these outcomes constitute the primary betting market. These probabilities are estimated for games played between any two of the 92 teams and the predictions are compared with the actual results.

Keywords: *attack and defence strengths, football results; Poisson models, State space modelling*

Team talk

Consider the match between team i (home) and team j (away) played at time t, in which the goals scored are x_{it} (by team i) and y_{jt} (by team j). According to the independent Poisson model, the probability of this result is

$$\{\exp(-\eta\alpha_{it}\beta_{jt})(\eta\alpha_{it}\beta_{jt})^{x_{it}}/x_{it}!\}\{\exp(-\alpha_{jt}\beta_{it})(\alpha_{jt}\beta_{it})^{y_{jt}}/y_{jt}!\}, \qquad (26.1)$$

where α_{it} and α_{jt} represent the attack rates (strengths) of teams i and j at time t, β_{it} and β_{jt} their defence rates (weaknesses) and η is the home advantage factor. With increasing η, α_{it} and β_{jt}, more goals will tend to be scored by team i against team j; likewise, there will be more goals at the other end when $\alpha_{jt}\beta_{it}$ increases. Though the model describes goals scored by the teams, this is simply a route to estimating the probabilities of win, draw or lose for the primary football betting market.

Maher (1982) took the αs and βs to be constant over time, so each team is assigned a pair of fixed parameters, α_i and β_i. Dixon and Coles (1997) adjusted the model to reflect a dependence between x_{it} and y_{jt} at the lower end (when x_{it} and y_{jt} are both less than 2). Also, with prediction in mind, they tapered the likelihood function to give greater weight to more recent results. This is in recognition of the probable need to allow α_i and β_i to vary through time, not to be constrained to stay constant.

Recently, three papers have appeared with a theme that is similar to ours. Where we take the goals scored as the data, the observations in Knorr-Held (2000) and Koning (2000) comprise only the win, lose or draw results. In addition, their models are based on a single team strength, rather than separating attack and defence, and are based on thresholds in cumulative distribution functions, in the usual way for ordered categorical data. Both Knorr-Held (2000) and Koning (2000) derived point estimates for the parameters, Koning applying maximum likelihood and Knorr-Held using an extended Kalman filter together with an *ad hoc* method for a variance parameter.

Rue and Salvesen (2000) used a modified independent Poisson model, as proposed by Dixon and Coles (1997). They allowed for separate attack and defence strengths but incorporated these in a further-modified way that involves a psychological factor to reflect the overall difference in the opposing teams' strengths. Also, they truncated the numbers of goals at 5, so a score of 9–6 is interpreted as 5–5, for example. In a final modification, a mixture model was proposed in which the previously described form is mixed with a similar form based on average scores. There was no allowance for a home ground advantage, however.

In both Knorr-Held (2000) and Rue and Salvesen (2000) the team strength parameters were allowed to vary over time according to random walks with independent normal increments. This aspect is similar to our formulation, given in the section 'A stochastic process tactic', but in other respects our treatment and aims are quite different from those of Knorr-Held and Rue and Salvesen.

Constraints on team behaviour

The αs and βs only appear in the likelihood function (26.1) in combinations $\alpha_{it}\beta_{jt}$. In consequence, there is a lack of identifiability; for example, the as could be doubled and the βs halved without altering their products. A standard approach in such cases is to apply one or more constraints to the parameters. However, any such constraint has the undesirable effect of tying their values together in some way. Thus, if one team's (α, β) changes, as a result of some local event, say, other teams' (α, β)s must change to preserve the constraints, even though the event in question does not affect them. This consideration suggests that any particular set of (α, β)s does not itself necessarily constitute the 'true' team strengths: the values are merely the projections of the strengths onto the constraint surface along trajectories of constant likelihood.

It is not immediately obvious how many constraints will be needed on the (α, β)s to obtain identifiability, though it is clearly at least one because of the scaling ambiguity mentioned above. With m teams, there are $2m$ αs and βs, and $m(m-1)$ combinations $\alpha_i\beta_j$. Thus, with enough data, i.e. sufficient hypothetical replication of the matches, to fix the combinations, there are $m(m-1)$ equations in $2m$ unknowns. For example, with $m = 4$ this is 12 equations in eight unknowns, which would seem to imply inconsistency even without any constraints. However, the 12 equations are not necessarily independent, and we must identify the extent of this dependence to calculate the number of constraints required. It is shown in Appendix A.1 that a single constraint will suffice for identifiability.

We shall take the single constraint to be

$$\sum_{i=1}^{m} \log(\alpha_{it}\beta_{it}) = 0, \tag{26.2}$$

i.e. that the geometric mean of the $(\alpha_{it}\beta_{it})$ is 1. Note that $\alpha_{it}\beta_{it}$ is the average scoring rate of team i's attack against its own defence per game. Constraint (26.2) will be imposed by expressing the α_{it} and β_{it} in terms of a more basic set of unconstrained parameters. Let these underlying parameters be $\gamma_{it} = (\gamma_{\alpha_{it}}, \gamma_{\beta_{it}})^{\mathrm{T}}$, and take

$$\begin{aligned} \log(\alpha_{it}) &= \gamma_{\alpha_{it}} - \bar{\gamma}_t, \\ \log(\beta_{it}) &= \gamma_{\beta_{it}} - \bar{\gamma}_t, \end{aligned} \tag{26.3}$$

where

$$\bar{\gamma}_t = (2m)^{-1} \sum_{j=1}^{m} (\gamma_{\alpha_{jt}} + \gamma_{\beta_{jt}}).$$

Thus, constraint (26.2) is automatically satisfied.

A stochastic process tactic

Dixon and Coles (1997) recognized the probable need for the αs and βs to vary over time. One way of tackling such a drift, alternative to theirs, is to model the processes explicitly. The changes over time may be small, for consistently performing teams, or larger, reflecting more substantial events, planned or not. We shall implement this by adopting an autoregressive AR(1) process for $\gamma_{it} = (\gamma_{\alpha_{it}}, \gamma_{\beta_{it}})^{\mathrm{T}}$:

$$\gamma_{it} - \gamma_{i0} = R(\gamma_{i,t-1} - \gamma_{i0}) + u_{it}. \tag{26.4}$$

Thus, the stochastic process moves in the γ-space, of $2m$ dimensions, and then, by equation (26.3), this is projected into the (α, β) space, of $2m - 1$ dimensions. In equation (26.4) R is a 2×2 matrix of autoregression parameters,

$$R = \begin{pmatrix} \rho_{\alpha\alpha} & \rho_{\alpha\beta} \\ \rho_{\beta\alpha} & \rho_{\beta\beta} \end{pmatrix},$$

the u_{it} are independent $N_2(0, \Sigma)$ innovations (independent of all previous outcomes and all other u_{jt}s) and γ_{i0} is a base-line value towards which γ_{it} is drawn if R is small in some sense (e.g. $\text{tr}(R^T R) < 1$, which ensures that $|Rx| < |x|$ for all x).

Because of the independence of the u_{it}, the γ_{it}-processes for different teams evolve over time independently of one another. However, dependence between the αs and βs for different teams is introduced via equation (26.3). These processes are not directly observable, but their effects are seen in the match results. Special cases of equation (26.4) include the following. If the γ_{i0} are all 0, the base-line attraction aspect is absent and the processes are purely autoregressive. If R and Σ are both diagonal the processes $\gamma_{\alpha_{it}}$ (attack) and $\gamma_{\beta_{it}}$ (defence) evolve independently. If $R = I$, the unit matrix, the γ_{it}-process is just a two-dimensional random walk with steps u_{it}. If $R = I$ and $\Sigma = 0$, the zero matrix, γ_{it} is constant over time as in the Maher (1982) model.

Even if we take the γ_{i0} that appear explicitly in equation (26.4) as 0 we still need to specify initial values because for $t = 1$ the formula involves γ_{i0}. These values can be defined with reference to the previous season's results, or perhaps as random effects from some suitable distribution, or retained as fixed effects parameters to be estimated.

An overall likelihood function can be constructed for matches up to and including time t as follows. The vector of unknown parameters is $\theta = (\eta, R, \Sigma)$, of dimension $1 + 4 + 3 = 8$. Let G_t comprise the goal results of all the matches played at time t, i.e. all the x_{it} and y_{jt} for i and j ranging over all the teams involved, and let Γ_t represent the set of unobserved γ_{it} at time t for these teams. Then

$$p(\Gamma_1, \ldots, \Gamma_t | \Gamma_0, \theta) = \prod_{s=1}^{t} p(\Gamma_s | \Gamma_{s-1}, \theta) = \prod_{s=1}^{t} \prod_{i} p(\gamma_{is} | \gamma_{i,s-1}, \theta) \tag{26.5}$$

and

$$p(G_1, \ldots, G_t | \Gamma_0, \ldots, \Gamma_t, \theta) = \prod_{s=1}^{t} p(G_s | \Gamma_s, \eta) = \prod_{s=1}^{t} \prod_{ij} p(x_{is}, y_{js} | a_{is}, a_{js}, \eta) \tag{26.6}$$

where $a_{is} = (\log(\alpha_{is}), \log(\beta_{is}))^{\mathrm{T}}$ and the product $\prod_{ij}$ in equation (26.6) is over all the games played at time s. It has been assumed in equation (26.6) that, given $(\alpha_{is}, \beta_{is}, \eta)$, the result (x_{is}, y_{js}) is independent of all others. Hence, the likelihood function for θ based on the matches up to time t, conditional on Γ_0, is

$$\begin{aligned} L_t(\theta) &= p(G_1, \ldots, G_t | \Gamma_0, \theta) \\ &= \int p(G_1, \ldots, G_t, \Gamma_1, \ldots, \Gamma_1 | \Gamma_0, \theta) d\Gamma_1 \ldots d\Gamma_t \\ &= \int p(G_1, \ldots, G_t | \Gamma_0, \ldots, \Gamma_t, \theta) p(\Gamma_1, \ldots, \Gamma_t | \Gamma_0, \theta) d\Gamma \ldots d\Gamma_t \\ &= \int \prod_{s=1}^{t} \left\{ \prod_{ij} p(x_{is}, y_{js} | a_{is}, a_{js}, \eta) \prod_{i} p(\gamma_{is} | \gamma_{i,s-1}, \theta) \right\} d\Gamma_1 \ldots d\Gamma_t, \end{aligned} \tag{26.7}$$

using equations (26.5) and (26.6). The two core ingredients in the integrand of equation (26.7) are straightforward to compute: $p(x_{is}, y_{js} | a_{is}, a_{js}, \eta)$ is given as the Poisson product in expression (26.1) and $p(\gamma_{is}|\gamma_{i,s-1}, \theta)$ as the density of $N\{\gamma_{i0} + R(\gamma_{i,t-1} - \gamma_{i0}), \Sigma\}$, from equation (26.4). However, the dimension of the integration is $2mt$, m being the number of teams involved. For instance, with $m = 92$, the number of teams in the English Football League, and $t = 40$, roughly the number of games played in one season, $2mt = 7360$. This makes equation (26.7) difficult to deal with directly.

Markov chain Monte Carlo (MCMC) methods were initially applied to the present problem but, for brevity, they are not reported here. In short, a form of MCMC sampling was employed using a Gibbs-type method with the components, γ_{is} and γ_{js} of $(\Gamma_1, \ldots \Gamma_t, \theta)$, being updated one at a time via the usual Metropolis acceptance formula. By virtue of the Markov property of the γ_{it}-processes, seen in equation (26.5), and of the dependence of (x_{is}, y_{js}) on $(\Gamma_1, \ldots, \Gamma_t, \theta)$ only through (a_{is}, a_{js}, η), γ_{is} and γ_{js} each appear in only three individual factors of the full posterior expression. Thus, a significant short-cut can be made in computing the acceptance probability for proposal values of γ_{is}. On this basis the computations for predictions for a single week become feasible. However, for comparisons over whole seasons, as required for the present study, the MCMC method is still too slow to be used routinely. For this reason a faster method was sought, and such an approach is described next.

An approximation

The model here is essentially a non-normal, non-linear state space model in which the states are the Γ_{it} and the observed quantities are the x_{it} and y_{it}. Many references have presented approximate methods for filtering and smoothing for such models, e.g. Kitagawa (1987), Carlin et al. (1992), Carter and Kohn (1994) and Shephard and Pitt (1997). Some of the methods proposed are based on numerical integration and others on simulation, but what they all have in common is the retention of the underlying model formulation. The approach to be outlined in this section differs in that the original model is replaced by a derived model that is easier to handle. The attitude here is that no model is likely to be 'true', however carefully constructed, and the only true worth of any model lies in its predictive performance: in the present context this is its ability to predict the outcomes of matches so that useful bets can be identified and placed.

It is shown in Appendix A.1 that, in consequence of equations (26.4) and (26.3), the $(\alpha_{it}, \beta_{it})$ process is given by

$$a_{it} - a_{i0} = R(a_{i,t-1} - a_{i0}) + (RJ - JR)(\gamma_{t-1} - \gamma_0) + u_{it} - (2m)^{-1}\sum_{j=1}^{m} Ju_{jt}, \tag{26.8}$$

where $a_{it} = (\log(\alpha_{it}), \log(\beta_{it}))^{\mathrm{T}}$ and J is a 2×2 matrix of 1s. Note that

$$RJ - JR = \begin{pmatrix} \rho_{\alpha\alpha} + \rho_{\alpha\beta} & \rho_{\alpha\alpha} + \rho_{\alpha\beta} \\ \rho_{\beta\alpha} + \rho_{\beta\beta} & \rho_{\beta\alpha} + \rho_{\beta\beta} \end{pmatrix} - \begin{pmatrix} \rho_{\alpha\alpha} + \rho_{\beta\alpha} & \rho_{\alpha\beta} + \rho_{\beta\beta} \\ \rho_{\alpha\alpha} + \rho_{\beta\alpha} & \rho_{\alpha\beta} + \rho_{\beta\beta} \end{pmatrix}$$

$$= \begin{pmatrix} \rho_{\alpha\beta} - \rho_{\beta\alpha} & \rho_{\alpha\alpha} - \rho_{\beta\beta} \\ \rho_{\beta\beta} - \rho_{\alpha\alpha} & \rho_{\beta\alpha} - \rho_{\alpha\beta} \end{pmatrix}.$$

Thus, $RJ - JR$ is 0 if $R = I$, as for the random walk model. However, in that case equation (26.4) would represent a non-stationary autoregressive model. More generally, $RJ - JR$ is 0 if R is symmetric about both diagonals, i.e. $\rho_{\alpha\alpha} = \rho_{\beta\beta}$ and $\rho_{\alpha\beta} = \rho_{\beta\alpha}$, and a stationary autoregressive model obtains if $|R| < 1$, i.e. $\rho_{\alpha\alpha}^2 + \rho_{\beta\beta}^2 + \rho_{\alpha\beta}^2 + \rho_{\beta\alpha}^2 < 1$.

Suppose that $RJ - JR = 0$, and omit the $O_p(m^{-1})$ contribution in equation (26.8). Then we obtain an autoregressive model for the a_{it}:

$$a_{it} - a_{i0} = R(a_{i,t-1} - a_{i0}) + u_{it}. \tag{26.9}$$

This model might have been adopted at the outset, but it does not incorporate the constraint explicitly. However, it can lead to a useful approximate approach as follows.

Note that equation (26.9) gives

$$p(a_{it}|a_{i0}, a_{i1}, \dots a_{i,t-1}) = \det(2\pi\,\Sigma)^{-1/2} \exp\left\{-\frac{1}{2}(a_{it} - m_{it})^{\mathrm{T}}\Sigma^{-1}(a_{it} - m_{it})\right\}, \tag{26.10}$$

with

$$m_{it} = a_{i0} + R(a_{i,t-1} - a_{i0}). \tag{26.11}$$

Also, the appropriate distribution for updating a_{it} and a_{jt}, as a result of the unobserved changes from time $t - 1$ together with the evidence from the match result (x_{it}, y_{jt}), is

$$\begin{aligned} &p(a_{it}, a_{jt}|a_{i,t-1}, a_{j,t-1}, x_{it}, y_{jt}) \\ &\quad = p(x_{it}, y_{jt}|a_{it}, a_{jt}, a_{i,t-1}, a_{j,t-1})p(a_{it}, a_{jt}|a_{i,t-1}, a_{j,t-1})/p(x_{it}, y_{jt}|a_{i,t-1}, a_{j,1-1}) \\ &\quad = p(x_{it}, y_{jt}|a_{it}, a_{jt})p(a_{it}|a_{i,t-1})p(a_{jt}|a_{j,t-1})/p(x_{it}, y_{jt}|a_{i,t-1}, a_{j,t-1}), \end{aligned} \tag{26.12}$$

where $p(x_{it}, y_{jt}|a_{it}, a_{jt})$ is given by equation (26.1) and $p(a_{it}|a_{i,t-1})$ and $p(a_{jt}|a_{j,t-1})$ are given by equation (26.10); the denominator does not involve (a_{it}, a_{jt}).

We seek some summary of the information in equation (26.12) that will enable us to avoid the integration in equation (26.7). A natural suggestion is to adopt, for the updated versions of a_{it} and a_{jt}, their maximum probability estimators based on equation (26.12); see, for example, Fahrmeir and Kaufman (1991) and Fahrmeir (1992). These are derived in Appendix A.2 as

$$\begin{aligned} a_{it} &= m_{it} + \Sigma r_{i\,jt}, \\ a_{jt} &= m_{jt} + \Sigma r_{jit}, \end{aligned} \tag{26.13}$$

where

$$r_{i\,jt} = (x_{it} - \eta\alpha_{it}\beta_{jt}, y_{jt} - \alpha_{jt}\beta_{it})^{\mathrm{T}}$$

is the vector of 'Poisson residuals', i.e. the discrepancies between the home and away goals and their expected values; r_{jit} is just r_{ijt} with the components interchanged. The updated version of a_{it} is thus expressed as its autoregressive mean m_{it} plus an adjustment Σr_{ijt}, to take account of the match result, and likewise for a_{jt}. Unfortunately, however, r_{ijt} itself involves the components of a_{it} and a_{jt}, so equations (26.13) give a_{it} and a_{jt} implicitly. An iterative solution, described in Appendix A.2, has been implemented and works reasonably well. At least one solution to equations (26.13) exists because the probability density (26.12) must have at least one maximum. If teams i and j do not play at time t, the appropriate updating distribution is just $p(a_{it}, a_{jt}|a_{i,t-1}, a_{j,t-1})$. On the same basis as above, this leads to the reduced updating formulae

$$\begin{aligned} a_{it} &= m_{it}, \\ a_{jt} &= m_{jt}. \end{aligned} \tag{26.14}$$

We now propose that the updating formulae (26.13) and (26.14) be adopted in place of equation (26.9) as the formal model for $(\alpha_{it}, \beta_{it}, \alpha_{jt}, \beta_{jt})$. In effect, the $(\alpha_{it}, \beta_{it})$ become interlinked stochastic processes driven by the innovations (x_{it}, y_{jt}) with initial values $(\alpha_{i0}, \beta_{i0})$ and parameter set $\theta = (\eta, R, \Sigma)$. In equation (26.9) the $(\alpha_{it}, \beta_{it})$ evolve independently over time and information would just be gathered about their progress from the match results. Now, in equations (26.13) and (26.14), the processes $(\alpha_{it}, \beta_{it})$ are actually driven by the match results and are thereby interdependent.

The nature of the approximation made in this section can be described as follows: the original process (26.8) has been replaced by the approximation (26.9), and then equation (26.9) has been replaced by a different, but closely related, process.

A likelihood function based on this scheme can be constructed for the matches up to and including time t as

$$L_t(\theta) = p(G_1, \ldots, G_t|A_0, \theta) = \prod_{s=1}^{t} p(G_s|G_1, \ldots, G_{s-1}, A_0, \theta),$$

where $A_t = (a_{1t}, \ldots, a_{mt})$. Assume that $(G_1, \ldots, G_{s-1}, A_0)$ and $(A_0, \ldots, A_{s-1})$ are equivalent, i.e. that expression (26.13) has a unique solution for a_{it} and a_{jt}. Then,

$$\begin{aligned} p(G_s|G_1, \ldots, G_{s-1}, A_0, \theta) &= p(G_s|A_0, \ldots, A_{s-1}, \theta) \\ &= \int p(G_s|A_0, \ldots, A_s, \theta)p(A_s|A_0, \ldots, A_{s-1}, \theta)\mathrm{d}A_s \\ &= p(G_s|A_0, \ldots, A_{s-1}, A_s = M_s, \theta) = p(G_s|A_s = M_s, \theta) \end{aligned} \tag{26.15}$$

because, according to equations (26.14), $p(A_s|A_0, \ldots, A_{s-1}, \theta)$ is concentrated at the single point $A_s = M_s$, where M_s is the complete set of m_{is}-values for all the teams. Hence,

$$L_t(\theta) = \prod_{s=1}^{t} p(G_s|A_s = M_s, \theta) = \prod_{s=1}^{t}\prod_{ij} p(x_{is}, y_{js}|a_{is} = m_{is}, a_{js} = m_{js}, \theta), \tag{26.16}$$

where $\prod_{ij}$ is the product over all the matches played at time s and $p(x_{is}, y_{js}|a_{is} = m_{is}, a_{js} = m_{js}, \theta)$ is given by equations (26.1) and (26.11). Note that $L_t(\theta)$ in equation (26.16) is implicitly conditioned on Γ_0 because the m_{i1} depend on the a_{i0}. This likelihood function is very simply computed, unlike that in equation (26.7).

Prediction

Because of the steadily accruing information on the a_{it} we might hope that predictions would become more useful as time goes on. However, the inherent variability of low count Poisson variables limits the attainable accuracy of forecasting match results. Even if the model were correct, and the parameters precisely known, there would still be considerable uncertainty in the predictions.

For prediction one step ahead, the relevant distribution is $p(G_t|G_1, \ldots, G_{t-1}, \Gamma_0)$. This gives the probabilities of various scores (x_{it}, y_{jt}) for particular games, and so the probabilities of home or away wins can be computed. It may be evaluated as

$$p(G_t|G_1, \ldots, G_{t-1}, \Gamma_0) = \int p(G_t|G_1, \ldots, G_{t-1}, \Gamma_0, \theta)p(\theta|G_1, \ldots, G_{t-1}, \Gamma_0)d\theta, \qquad (26.17)$$

where $p(\theta|G_1, \ldots, G_{t-1}, \Gamma_0)$ is the posterior distribution of θ given the match results up to time $t - 1$.

For the approximation method we have, from equation (26.15),

$$p(G_t|G_1, \ldots, G_{t-1}, \Gamma_0, \theta) = p(G_t|A_t = M_t, \theta),$$

which is given by equations (26.1) and (26.11). Provided that $p(\theta|G_1, \ldots, G_{t-1}, \Gamma_0)$ is sufficiently peaked, the integral (26.17) can be well approximated by a plug-in estimate, $p(G_t|A_t = M_t, \theta = \hat{\theta}_{t-1})$, where $\hat{\theta}_{t-1}$ is the maximum likelihood estimate based on $(G_1, \ldots, G_{t-1}, \Gamma_0)$. The integral could be evaluated by simulation, e.g. by MCMC sampling from the posterior, but we have not pursued this alternative because our focus is on a computational approach that avoids MCMC methods: see the discussion in the last section.

Full results round-up

The methods described above were applied to the Football Association League fixtures over the years 1992–1997, and home win, draw and away win probabilities were calculated accordingly. It is these probabilities that are of chief concern in the primary betting market. Some summary statistics examining the performance of each approach were evaluated and are now presented.

Table 26.1 shows summaries of the actual outcomes given the most likely predicted outcome for the approximation method and the Dixon and Coles (1997) model. If a model is performing well then the elements in the leading diagonal of these tables should dominate. In practice though, since a draw is so seldom predicted as the most likely outcome, only the results for a home win and an away win are of significant interest. The two approaches appear to have roughly the same predictive ability for home wins (in the first row of the table), with just under 50% of games predicted as home wins actually resulting in home wins. For away wins, the methods perform less well, with the approximation method slightly better.

The indicators examined above, although informative, are prone to considerable variability since, for example, a home win could be the most likely outcome for any probability greater than $\frac{1}{3}$. As an alternative, consider the predictive ability of each method when only strong

Table 26.1 *Comparison of predicted and actual outcomes.*

Most likely predicted outcome	Approximation results for the following actual outcomes:			Dixon–Coles results for the following actual outcomes:		
	Home win	Draw	Away win	Home win	Draw	Away win
Home win	0.48	0.28	0.24	0.49	0.28	0.23
Draw	–	–	–	0.33	0.33	0.33
Away win	0.33	0.28	0.39	0.33	0.31	0.36

Table 26.2 *Comparison of predicted and actual outcomes.*

Approximation	0.51	0.43
Dixon–Coles	0.52	0.49

favourites are considered. Table 26.2 gives a summary of the empirical outcomes conditional on strong favourites, 'strong' in the sense that the predicted probability is 0.5 or more. Again, for both home and away wins the two approaches perform very similarly.

Tables 26.1 and 26.2 provide summary information averaged over the entire time horizon of the study. Of potentially greater interest is the performance of the approaches through time:

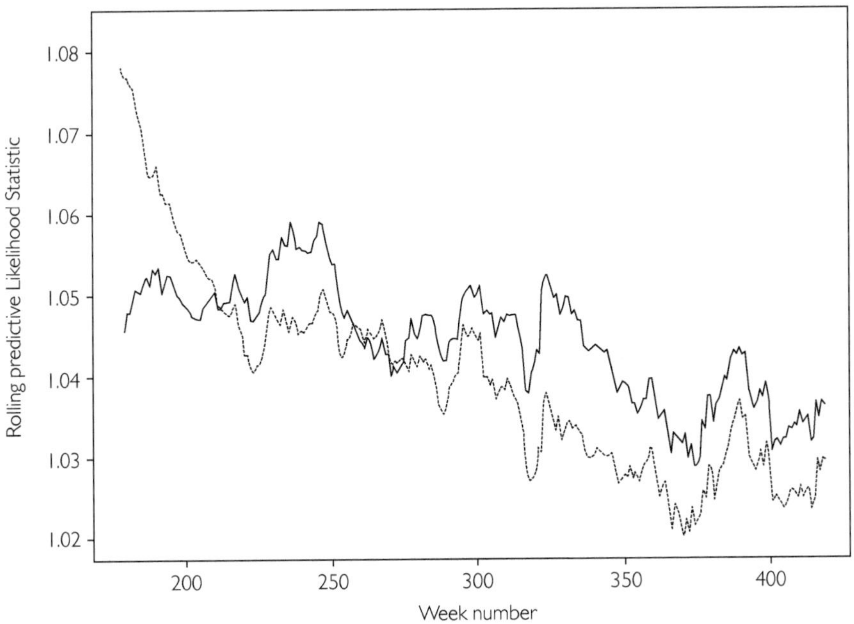

Figure 26.1 Rolling predictive likelihood statistic for the Dixon–Coles method (·······) and the approximation (——).

do the methods become better at predicting the outcomes of games as time goes on? Fig. 26.1 depicts a quantity that is related to this issue, namely, a rolling predictive likelihood defined as

$$\mathrm{RPL}(s,t) = -N(s,t)^{-1} \sum_{r=s}^{t} \sum_{ij} \log\{\mathrm{pr}(o_{ijr})\},$$

where o_{ijr} denotes the outcome of the match between teams i and j played at time r, Σ_{ij} denotes summation over those games and $N(s,t)$ denotes the total number of games played at times $s, s+1, \ldots, t$; $\mathrm{pr}(o_{ijr})$ is the predicted probability of outcome o_{ijr}. A good performance is indicated by a small value of this statistic. Fig. 26.1 shows a plot, for each method, of $\mathrm{RPL}(t-69, t)$ *versus* t; the window length, 70 time periods, is roughly one season. The Dixon–Coles method starts poorly but improves rapidly and betters the approximation around week 210. However, the approximation, which is a more general, likelihood-based approach, is at least competitive throughout.

Conclusion

The purpose of the work described here is to develop a fast computational approach for predicting the results of football matches with a view to betting in the primary market on win, draw or lose outcomes. For this, a novel method has been developed in which the original stochastic process model is replaced by an approximation that yields more tractable computation than MCMC methodology in this case, but which does not lose too much predictive power. The approximation method clearly has wider potential for tractable computations in the context of non-linear state space models.

ACKNOWLEDGEMENTS

We thank the Joint Editor and referees for their helpful comments.

Appendix A

A.1 The identifiability constraint

Suppose that $\alpha_i\beta_j$ is fixed as $\exp(\nu_{ij})$, i.e. $\log(\alpha_i)+\log(\beta_j) = \nu_{ij}$. Then the whole set of equations can be written as $Ax = \nu$, where x is the $2m \times 1$ vector

$$(\log(\alpha_1), \ldots, \log(\alpha_m), \log(\beta_1), \ldots, \log(\beta_m))^{\mathrm{T}},$$

ν is the $m(m-1) \times 1$ vector of ν_{ij}s and A is the appropriate $m(m-1) \times 2m$ incidence matrix. With $m = 4$, for instance,

$$
A = \begin{pmatrix}
1000 & 0100 \\
1000 & 0010 \\
1000 & 0001 \\
0100 & 1000 \\
0100 & 0010 \\
0100 & 0001 \\
0010 & 1000 \\
0010 & 0100 \\
0010 & 0001 \\
0001 & 1000 \\
0001 & 0100 \\
0001 & 0010
\end{pmatrix}
$$

The identifiable parametric functions are the elements of Ax, and so the number of independent identifiable functions is rank(A). Thus, the appropriate number of constraints will be $2m$ – rank(A). We can determine the dimension of the kernel of A quite easily by seeing that the only solutions of $Ay = 0$ are y-vectors proportional to $(1, \ldots, 1, -1, \ldots, -1)$; this can be checked for the example with $m = 4$ given above, and thence generalized. Note that this y-vector corresponds to adding a constant to each $\log(\alpha_i)$ and subtracting the same constant from each $\log(\beta_i)$, which amounts to precisely the rescaling of the αs and βs referred to in in the section 'Constraints on team behaviour'. Thus, $\dim\{\ker(A)\} = 1$, so $\text{rank}(A) = 2m - 1$ and so only a single constraint is needed.

The implications of the autoregressive model (26.4) for the αs and βs defined by equation (26.3) are as follows. We have

$$
\begin{pmatrix} \log(\alpha_{it}) \\ \log(\beta_{it}) \end{pmatrix} = \begin{pmatrix} \gamma_{\alpha it} - \gamma_t \\ \gamma_{\beta it} - \gamma_t \end{pmatrix} = \gamma_{it} - (2m)^{-1} \sum_{j=1}^{m} J\gamma_{jt},
$$

where J is a 2×2 matrix of 1s. Hence,

$$
\begin{aligned}
\begin{pmatrix} \log(\alpha_{it}) - \log(\alpha_{i0}) \\ \log(\beta_{it}) - \log(\beta_{i0}) \end{pmatrix} &= \gamma_{it} - \gamma_{i0} - (2m)^{-1} \sum_{j=1}^{m} J(\gamma_{jt} - \gamma_{j0}) \\
&= R(\gamma_{i,t-1} - \gamma_{i0}) + u_{it} - (2m)^{-1} \sum_{j=1}^{m} J\{R(\gamma_{j,t-1} - \gamma_{j0}) + u_{jt}\} \\
&= R(\gamma_{i,t-1} - \gamma_{i0}) - (2m)^{-1} \sum_{j=1}^{m} RJ(\gamma_{j,t-1} - \gamma_{j0}) \\
&\quad + (2m)^{-1} \sum_{j=1}^{m} (RJ - JR)(\gamma_{j,t-1} - \gamma_{j0}) + u_{it} - (2m)^{-1} \sum_{j=1}^{m} Ju_{jt} \\
&= R \begin{pmatrix} \log(\alpha_{i,t-1}) - \log(\alpha_{i0}) \\ \log(\beta_{i,t-1}) - \log(\beta_{i0}) \end{pmatrix} + (RJ - JR)(\gamma_{t-1} - \gamma_0) + u_{it} - (2m)^{-1} \sum_{j=1}^{m} Ju_{jt},
\end{aligned}
$$

which is equation (26.8).

A.2 Equations and numerical solution for the approximation method

Equation (26.13) is derived together with a suggested numerical method of solution. The relevant part of the logarithm of equation (26.12) is

$$M = -\gamma\alpha_{it}\beta_{jt} + x_{it}\log(\gamma\alpha_{it}\beta_{jt}) - \alpha_{jt}\beta_{it} + y_{jt}\log(\alpha_{jt}\beta_{it})$$
$$-\frac{1}{2}(a_{it} - m_{it})^{\mathrm{T}}\Sigma^{-1}(a_{it} - m_{it}) - \frac{1}{2}(a_{jt} - m_{jt})^{\mathrm{T}}\Sigma^{-1}(a_{jt} - m_{jt}),$$

and this has derivatives as follows:

$$\frac{\partial M}{\partial \alpha_{it}} = -\gamma\beta_{jt} + \alpha_{it}^{-1}x_{it} - \alpha_{it}^{-1}[\{\log(\alpha_{it}) - m_{it1}\}\Sigma^{11} + \{\log(\beta_{it}) - m_{it2}\}\Sigma^{12}];$$

$$\frac{\partial M}{\partial \beta_{it}} = -\alpha_{jt} + \beta_{it}^{-1}y_{jt} - \beta_{it}^{-1}[\{\log(\beta_{it}) - m_{it2}\}\Sigma^{22} + \{\log(\alpha_{it}) - m_{it1}\}\Sigma^{12}];$$

$$\frac{\partial M}{\partial \alpha_{jt}} = -\beta_{it} + \alpha_{jt}^{-1}y_{jt} - \alpha_{jt}^{-1}[\{\log(\alpha_{jt}) - m_{jt1}\}\Sigma^{11} + \{\log(\beta_{jt}) - m_{jt2}\}\Sigma^{12}];$$

$$\frac{\partial M}{\partial \beta_{jt}} = -\gamma\alpha_{it} + \beta_{jt}^{-1}x_{it} - \beta_{jt}^{-1}[\{\log(\beta_{jt}) - m_{jt2}\}\Sigma^{22} + \{\log(\alpha_{jt}) - m_{jt1}\}\Sigma^{12}];$$

here, m_{it1} and m_{it2} are the components of m_{it}, and the Σ^{jk} are the elements of Σ^{-1}. On equating these derivatives to 0, and rearranging, we obtain

$$\{\log(\alpha_{it}) - m_{it1}\}\Sigma^{11} + \{\log(\beta_{it}) - m_{it2}\}\Sigma^{12} = x_{it} - \gamma\alpha_{it}\beta_{jt},$$
$$\{\log(\alpha_{it}) - m_{it1}\}\Sigma^{12} + \{\log(\beta_{it}) - m_{it2}\}\Sigma^{22} = y_{it} - \alpha_{jt}\beta_{it},$$
$$\{\log(\alpha_{jt}) - m_{jt1}\}\Sigma^{11} + \{\log(\beta_{jt}) - m_{jt2}\}\Sigma^{12} = y_{jt} - \alpha_{jt}\beta_{it},$$
$$\{\log(\alpha_{jt}) - m_{jt1}\}\Sigma^{12} + \{\log(\beta_{jt}) - m_{jt2}\}\Sigma^{22} = x_{it} - \gamma\alpha_{it}\beta_{jt}.$$

These equations may be 'solved' to give

$$\{\log(\alpha_{it}) - m_{it1}\}\{\Sigma^{11}\Sigma^{22} - (\Sigma^{12})^2\} = \Sigma^{22}(x_{it} - \gamma\alpha_{it}\beta_{jt}) - \Sigma^{12}(y_{jt} - \alpha_{jt}\beta_{it}),$$
$$\{\log(\beta_{it}) - m_{it2}\}\{\Sigma^{11}\Sigma^{22} - (\Sigma^{12})^2\} = -\Sigma^{12}(x_{it} - \gamma\alpha_{it}\beta_{jt}) + \Sigma^{11}(y_{jt} - \alpha_{jt}\beta_{it}),$$
$$\{\log(\alpha_{jt}) - m_{it1}\}\{\Sigma^{11}\Sigma^{22} - (\Sigma^{12})^2\} = -\Sigma^{12}(x_{it} - \gamma\alpha_{it}\beta_{jt}) + \Sigma^{22}(y_{jt} - \alpha_{jt}\beta_{it}),$$
$$\{\log(\beta_{jt}) - m_{jt2}\}\{\Sigma^{11}\Sigma^{22} - (\Sigma^{12})^2\} = \Sigma^{11}(x_{it} - \gamma\alpha_{it}\beta_{jt}) - \Sigma^{12}(y_{jt} - \alpha_{jt}\beta_{it}).$$

Using the identity $(\Sigma^{-1})^{-1} = \Sigma$, in component form, these equations yield equations (26.13):

$$\log(\alpha_{it}) = m_{it1} + \Sigma_{11}(x_{it} - \gamma\alpha_{it}\beta_{jt}) + \Sigma_{12}(y_{jt} - \alpha_{jt}\beta_{it}),$$
$$\log(\beta_{it}) = m_{it2} + \Sigma_{12}(x_{it} - \gamma\alpha_{it}\beta_{jt}) + \Sigma_{22}(y_{jt} - \alpha_{jt}\beta_{it}),$$
$$\log(\alpha_{jt}) = m_{jt1} + \Sigma_{12}(x_{it} - \gamma\alpha_{it}\beta_{jt}) + \Sigma_{11}(y_{jt} - \alpha_{jt}\beta_{it}),$$
$$\log(\beta_{jt}) = m_{jt2} + \Sigma_{22}(x_{it} - \gamma\alpha_{it}\beta_{jt}) + \Sigma_{12}(y_{jt} - \alpha_{jt}\beta_{it}).$$

For numerical solution we first reduce these four equations to two, by adding the first and fourth, and the second and third, to obtain

$$\log(\alpha_{it}\beta_{jt}) = m_{it1} + m_{jt2} + (\Sigma_{11} + \Sigma_{22})(x_{it} - \gamma\alpha_{it}\beta_{jt}) + 2\Sigma_{12}(y_{jt} - \alpha_{jt}\beta_{it}),$$

$$\log(\alpha_{jt}\beta_{it}) = m_{it2} + m_{jt1} + 2\Sigma_{12}(x_{it} - \gamma\alpha_{it}\beta_{jt}) + (\Sigma_{11} + \Sigma_{22})(y_{jt} - \alpha_{jt}\beta_{it}).$$

We now have two non-linear equations in the two quantities, $\alpha_{it}\beta_{jt}$ and $\alpha_{jt}\beta_{it}$, so an iterative scheme for solution is called for. The first attempt was one in which the right-hand sides contained the previous estimates, but the successive values tended to oscillate. This suggested that convergence could be accelerated by updating the estimates as the average of the current and previous values, and this scheme was found to work quite well. The technique bears a strong resemblance to Aitken's δ^2 acceleration method (Dixon 1974, section 7.6). Once the process has converged, the previous set of equations is used to compute α_{it}, β_{it}, α_{jt} and β_{jt} separately.

REFERENCES

Carlin, B. P., Polson, N. G. and Stoffer, D. S. (1992) A Monte Carlo approach to nonnormal and nonlinear state-space modeling. *J. Am. Statist. Ass.*, **87**, 493–500.

Carter, C. K. and Kohn, R. (1994) On Gibbs sampling for state space models. *Biometrika*, **81**, 541–553.

Dixon, C. (1974) *Numerical Analysis*. London: Blackie.

Dixon, M. J. and Coles, S. G. (1997) Modelling association football scores and inefficiencies in the football betting market. *Appl. Statist.*, **46**, 265–280.

Fahrmeir, L. (1992) Posterior mode estimation by extended Kalman filtering for multivariate dynamic generalised linear models. *J. Am. Statist. Ass.*, **87**, 501–509.

Fahrmeir, L. and Kaufman, H. (1991) On Kalman filtering, posterior mode estimation and Fisher scoring in dynamic exponential family models. *Metrika*, **38**, 37–60.

Kitagawa, G. (1987) Non-Gaussian state-space modelling of nonstationary time series (with discussion). *J Am. Statist. Ass.*, **82**, 1032–1063.

Knorr-Held, L. (2000) Dynamic rating of sports teams. *Statistician*, **49**, 261–276.

Koning, R. H. (2000) Balance in competition in Dutch soccer. *Statistician*, **49**, 419–431.

Maher, M. J. (1982) Modelling association football scores. *Statist. Neerland.*, **36**, 109–118.

Rue, H. and Salvesen, Ø. (2000) Prediction and retrospective analysis of soccer matches in a league. *Statistician*, **49**, 399–418.

Shephard, N. and Pitt, M. K. (1997) Likelihood analysis of non-Gaussian measurement time series. *Biometrika*, **84**, 653–667.

CHAPTER 27

An evaluation of characteristics of teams in association football by using a Markov process model

NOBUYOSHI HIROTSU AND MIKE WRIGHT

Abstract

The paper proposes a statistical model of a football match that is useful in providing insights into the characteristics of teams. We evaluate the characteristics by means of maximum likelihood estimators for factors such as home advantage, offensive and defensive strength, and their interactions, by taking account of not only goals but also possession of the ball. We build the model on the basis of the real data from the 1999–2000 season of the English Premier League and illustrate the characteristics of the teams in ways that allow us to visualize offensive and defensive strength, preference for playing at home and relative success against particular opposition teams.

Keywords: *Football, Markov process, Maximum likelihood method, Poisson model, soccer, sports*

Introduction

The evaluation of teams' strength in association football (soccer) is a topic of interest not only to football fans but also to researchers (e.g. Norman 1998). Maher (1982) analysed three consecutive seasons of four English football league divisions starting from the 1971–1972 season and estimated the offensive strength for scoring goals and defensive strength against conceding goals for each team in his Poisson regression model by means of the maximum likelihood method. Lee (1997) also analysed the 1995–1996 season of the English Premier League in a similar manner. Dixon and Coles (1997) evaluated the strengths of teams for the purpose of making profits in the football betting market, using English league and cup football data from

1992 to 1995. They introduced a time-dependent effect into a Poisson regression model based on Maher's model. Dixon and Robinson (1998) developed a more complicated statistical model by incorporating the rate of scoring goals, given the current time and score during the match, using English league and cup football data from 1993 to 1996.

Though there are various statistical models, from simple to complicated, for evaluating teams' strength, until now their analyses have been based on the data of scoring and conceding goals. This information of course provides us with an aspect of the characteristics of the strengths of teams, but we can estimate this aspect simply by looking at the league table, which shows the final position of teams and the total number of goals scored and conceded by each team. However, the information in relation to goals may not be enough to understand fully the characteristics of teams. Information about possession of the ball, which is a very important consideration in any football match, may show other illuminating characteristics of teams.

In Hirotsu and Wright (2002) we described a Markov process model of a football match, together with a log-linear model representing not only the transition rates of scoring and conceding goals but also the rates of gaining and losing possession, using explanatory variables of home advantage, offensive strength and defensive strength. In this paper, we build a statistical model which incorporates these factors, based on real data from the English Premier League in the 1999–2000 season, and then we illustrate various characteristics of the teams involved.

Statistical modelling

A model of a football match

A football match can be seen as progressing through a set of stochastic transitions due to a change of possession of the ball or the scoring of a goal. In Hirotsu and Wright (2002) we assumed a Markov property in these transitions and proposed a Markov process model, which seems appropriate to a football match as a first approximation. Under this assumption the transition time until a goal is scored or possession changes follows an exponential distribution and the number of these transitions in any given time interval follows a Poisson distribution.

Fig. 27.1 shows a graphical representation of the model. This model consists of the following four states:

(a) state AG is the state in which team A scores a goal;
(b) state A is the state in which team A is in possession of the ball;
(c) state B is the state in which team B is in possession of the ball;
(d) state BG is the state in which team B scores a goal.

In Fig. 27.1, N_{AG} represents the total number of goals scored by team A in a match against team B. N_{AB} represents the total number of changes of possession from team A to team B. Here, under the assumption of the Markov property, N_{AG} and N_{AB} follow Poisson distributions whose means are proportional to T_A, the total possession time for team A. N_{BG}, N_{BA} and T_B are also defined in the same manner for team B.

To check on the validity of the Markov assumption, we used a video player and stop-watches to record possession times for each team in two half-matches of the 1999–2000

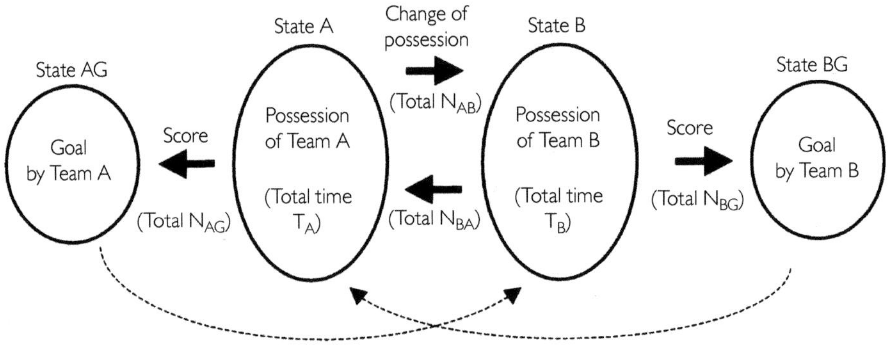

Figure 27.1 Four-state model of a football match.

English Premier League. There was no evidence against the hypothesis that these times follow independent exponential distributions.

The data

To evaluate the characteristics of teams for goals and possession based on this model, we need to obtain N_{AG}, N_{AB}, N_{BA} and N_{BG} with the total possession time T_A and T_B for each match. Table 27.1 lists part of the data from the 380 matches that were played. The goals data can easily be taken from Rollin and Rollin (2000) for example, but N_{AB} and N_{BA} are inferred from data collected by Opta Index Ltd, part of which are reported in Opta Index (2000). Plainly, the difference between N_{AB} and N_{BA} should not exceed one more than the difference in goals, so these data in Table 27.1 are not error free.

The possession times T_A and T_B are also inferred from the Opta Index data. To do this, let N_{AA} be the number of 'activities' which result in possession staying with team A. Similarly to the count of N_{AB}, we count N_{AA} by summing the number of occurrences of the following: 'pass to own player', 'cross-pass to own player', 'goal assist', 'dribble, the ball retained' and 'distribution to own player by the goalkeeper'. In the same manner, N_{BB} is the number of 'activities' which result in possession staying with team B.

Here, it seems natural to think that, if the total number of activities relating to possession of team A is larger than that of team B, the total possession time of team A would be longer than that of team B in the match, i.e. there will be a relationship between the ratio of the possession time T_A/T_B in the match and the ratio of the total number of activities $(N_{AB} + N_{AA})/(N_{BA} + N_{BB})$. Thus we make the assumption that the ratio of total possession time for team A and team B is equal to the ratio of the total number of activities starting with possession by team A and team B, i.e.

$$\frac{T_A}{T_B} = \frac{N_{AB} + N_{AA}}{N_{BA} + N_{BB}}. \tag{27.1}$$

Clearly this is an approximation. To test the adequacy of this approximation, we measured T_A and T_B, as well as N_{AB} etc., for 10 matches (Table 27.2). The t-statistic to test equality is not significant. The values of T_A and T_B that are shown in Table 1 have been estimated by using equation (27.1).

Table 27.1 ***Part of the statistics of the 1999–2000 season of the English Premier League.***

Home team	Away team	N_{AG}	N_{BG}	N_{AA}	N_{AB}	N_{BA}	N_{BB}	T_A	T_B
Arsenal	Aston Villa	3	1	439	145	156	176	57.4	32.6
Arsenal	Bradford City	2	0	368	127	130	251	50.9	39.1
Arsenal	Chelsea	2	1	256	146	143	483	35.2	54.8
Arsenal	Coventry City	3	0	374	156	178	344	45.3	44.7
Arsenal	Derby County	2	1	364	193	201	250	49.7	40.3
Arsenal	Everton	4	1	456	165	168	283	52.1	37.9
Arsenal	Leeds United	2	0	323	150	159	256	47.9	42.1
Arsenal	Leicester City	2	1	471	167	144	224	57.1	32.9
Arsenal	Liverpool	0	1	546	198	157	201	60.8	29.2
⋮	⋮	⋮	⋮	⋮	⋮	⋮	⋮	⋮	⋮
Wimbledon	Manchester United	2	2	280	202	199	393	40.4	49.6
Wimbledon	Middlesbrough	2	3	238	138	129	276	43.3	46.7
Wimbledon	Newcastle United	2	0	279	205	198	406	40.0	50.0
Wimbledon	Sheffield Wednesday	0	2	279	198	199	275	45.1	44.9
Wimbledon	Southampton	1	1	187	147	158	188	44.2	45.8
Wimbledon	Sunderland	1	0	209	193	187	288	41.3	48.7
Wimbledon	Tottenham Hotspur	1	1	153	141	158	303	35.0	55.0
Wimbledon	Watford	5	0	247	178	176	241	45.4	44.6
Wimbledon	West Ham United	2	2	238	194	164	236	46.7	43.3

Table 27.2 ***Ratios for the 10 matches.***

Match	T_A/T_B	$(N_{AB} + N_{AA})/(N_{BA} + N_{BB})$
1	1.15	1.14
2	0.93	0.95
3	1.07	1.11
4	1.53	1.79
5	1.07	0.93
6	0.94	0.93
7	1.12	1.28
8	0.94	1.03
9	1.16	1.18
10	0.90	0.80
t		–0.84

Hierarchy of models

We now apply a generalized linear model to the data in Table 27.1 in a manner similar to Lee (1997) and estimate the following factors for each team by means of the maximum likelihood method: λ, the intercept for scoring goals; λ_{home}, the home team advantage for scoring goals; $\lambda_{\text{score}}(X)$, the strength of team X for scoring goals; $\lambda_{\text{concede}}(X)$, the strength of team X against conceding goals; $\lambda_{\text{scorehome}}(X)$, the strength of team X for scoring goals at home; $\lambda_{\text{concedehome}}(X)$, the strength of team X against conceding goals in a home match of the opposing team; μ, the intercept for gaining possession; μ_{home}, the home team advantage for gaining possession; $\mu_{\text{gain}}(X)$, the strength of team X for gaining possession; $\mu_{\text{lose}}(X)$, the strength of team X against losing possession; $\mu_{\text{gainhome}}(X)$, the strength of team X for gaining possession at home; $\mu_{\text{losehome}}(X)$, the strength of team X against losing possession in a home match of the opposing team; $\mu_{\text{gainversus}}(X, Y)$, the strength of team X for gaining possession against team Y. Here, we set the possession times T_{A} and T_{B} as offsets in the models.

We started with a simple model and then added factors to test their significance in improving the model. Following this procedure, we obtained the hierarchy of models for goals and possession as shown in Fig. 27.2.

Fig. 27.2(a) shows the hierarchy of the model for goals. Here, we note that, as there are 20 teams in the league, the change in degrees of freedom caused by the inclusion of the offensive

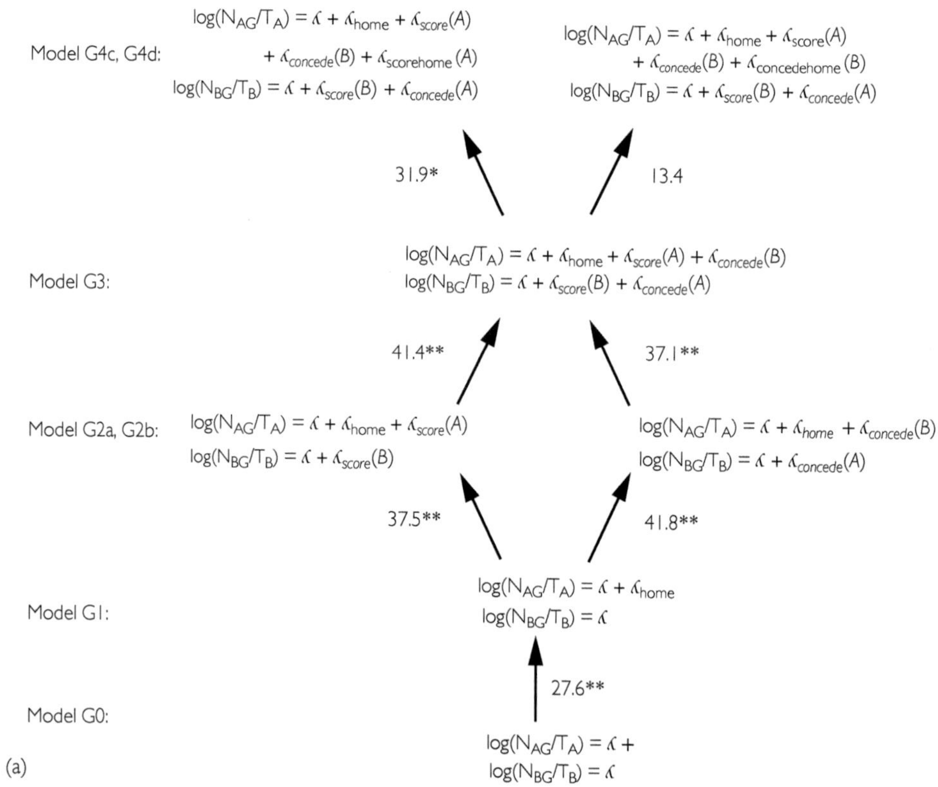

Figure 27.2 Hierarchy of models (a) for goals and (b) for possession, with changes in the value of twice the maximized log-likelihood (*, value significant at the 5% level; **, value significant at the 1% level).

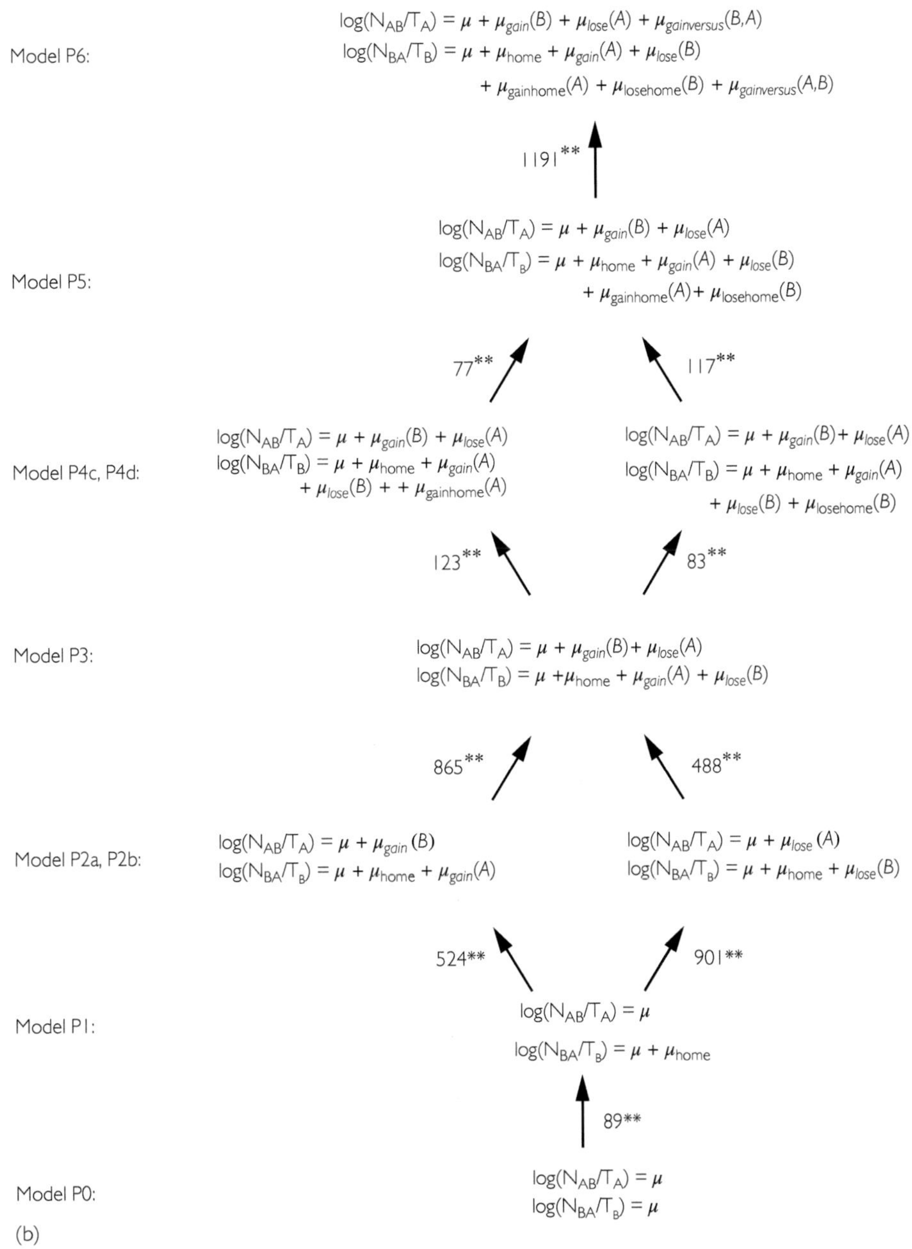

Figure 27.2 (Continued)

factor λ_{score} or the defensive factor $\lambda_{concede}$ is 20 – 1 = 19. In Fig. 27.2(a), we finally reach model G4c, which includes the interaction between home and offensive strength for scoring.

Similarly, we build the hierarchy of the model for possession as shown in Fig. 27.2(b). In this case, as a change in possession occurs around 150 times in an average match, which is much larger than the number of goals that are scored (usually no more than 5), each move

Table 27.3 ***Observed and expected frequencies of scores in goals and gains in possession.***

Goals			Possession		
Number of scores	Observed	Expected	Number of gains	Observed	Expected
Home					
0	81	84.1	≤ 110	13	11.1
1	112	115.7	111–130	64	55.1
2	101	88.8	131–150	112	118.0
3	44	50.7	151–170	100	117.2
4	28	24.2	171–190	62	56.3
≥ 5	14	16.6	191–210	24	17.9
			≥ 211	5	4.4
χ^2		3.80			7.31
Away					
χ^2		2.94			1.57

in Fig. 27.2(b) is highly significant compared with the case for goal scoring that is shown in Fig. 27.2(a). In Fig. 27.2(b) we reach model P6, which includes the interaction between home and offensive strength for possession, the interaction between home and defensive strength, and the interaction between offensive and defensive strengths of teams. As the result of this statistical modelling, model G4c for goals and model P6 for possession are considered to be the most suitable models for explaining these data.

Goodness-of-fit-tests

We tested the Poisson assumption in the statistical models by calculating the expected frequency of goals and gains in possession by using the maximum likelihood estimators of the factors in models G4c and P6 in Fig. 27.2, and comparing with the observed frequency. As shown in Table 27.3, the observed and expected frequencies in the 1999–2000 season are roughly equal. We tested the goodness of fit of the distribution in Table 27.3 and found that the resulting χ^2-statistics are not significant at the 5% level. Thus there is no evidence in favour of rejecting the Poisson assumption in this analysis.

Evaluation of characteristics of teams

Offensive and defensive strengths for goals and possession

Now we evaluate the characteristics of teams in the Premier League by using the models that are shown in Fig. 27.2. Firstly, the parameters for offensive and defensive strengths for goals and possession are evaluated by the maximum likelihood estimators using model G3 and P3

Table 27.4 *Offensive and defensive strengths of the English Premier League teams (with standard errors in parentheses).*†

Rank	Team	Results for goals		Results for possession	
		Offence (λ_{score})	Defence ($\lambda_{concede}$)	Offence (μ_{gain})	Defence (μ_{lose})
1	Manchester United	0.494 (0.155)	0.063 (0.214)	0.075 (0.0187)	−0.143 (0.0184)
2	Arsenal	0.263 (0.198)	−0.070 (0.198)	0.123 (0.0187)	−0.055 (0.0187)
3	Leeds United	0.071 (0.176)	−0.143 (0.216)	0.029 (0.0184)	−0.028 (0.0184)
4	Liverpool	−0.091 (0.183)	−0.505 (0.238)	0.052 (0.0184)	−0.075 (0.0185)
5	Chelsea	−0.075 (0.181)	−0.316 (0.230)	0.104 (0.0183)	−0.095 (0.0184)
⋮	⋮	⋮	⋮	⋮	⋮
16	Derby County	−0.096 (0.191)	0.032 (0.202)	−0.125 (0.0187)	−0.011 (0.0187)
17	Bradford City	−0.194 (0.200)	0.165 (0.195)	−0.106 (0.0184)	0.088 (0.0185)
18	Wimbledon	0.018 (0.186)	0.257 (0.192)	−0.054 (0.0182)	0.179 (0.0181)
19	Sheffield Wednesday	−0.263 (0.200)	0.263 (0.194)	−0.001 (0.0182)	0.050 (0.0183)
20	Watford	−0.273 (0.206)	0.297 (0.191)	−0.015 (0.0181)	0.172 (0.0181)

† Intercept for goals, $\lambda = -3.71(0.20)$; intercept for possession, $\mu = 1.21(0.0187)$; home advantage for goals, $\lambda_{home} = 0.330(0.06)$; home advantage for possession, $\mu_{home} = 0.0557(0.0059)$.

in Fig. 27.2 respectively. The results of the offensive and defensive strengths of the top five and the bottom five teams, following the order of ranking in the 1999–2000 season, are represented in Table 27.4.

We have plotted these results in Figs 27.3(a) and 27.3(b), in which the top five and bottom five teams are represented by large black and white dots respectively. The teams ranked 6–10 and 11–15 are represented by small black and white dots respectively. Note that the direction of the y-axis representing defensive strength is reversed because, if the value is negatively larger, it means stronger in the sense of conceding fewer goals and losing less possession. Therefore the teams that are plotted in the upper right-hand area are relatively stronger than the teams that are located in the lower left-hand area.

Fig. 27.3(a) shows that Manchester United, league champions for the 1999–2000 season, has easily the highest goal scoring parameter—not surprisingly, since Manchester United scored 97 goals, which is far more than any other team. However, Manchester United's defensive parameter is not particularly good; the best is Liverpool's, again not surprisingly since Liverpool conceded only 30 goals, which is fewer than any other team.

Fig. 27.3(b) shows that Chelsea is the best team at gaining possession, with Derby County worst. It also shows that Manchester United is the best team at not losing possession, with Wimbledon worst. These results are of particular interest since they give insights which could not be ascertained simply by examining the league tables.

Table 27.4 shows that the standard errors of the estimates of teams' goal and possession parameters are very similar across the whole league, whereas the parameters themselves are very different, and consistent with what we would expect, given the league positions.

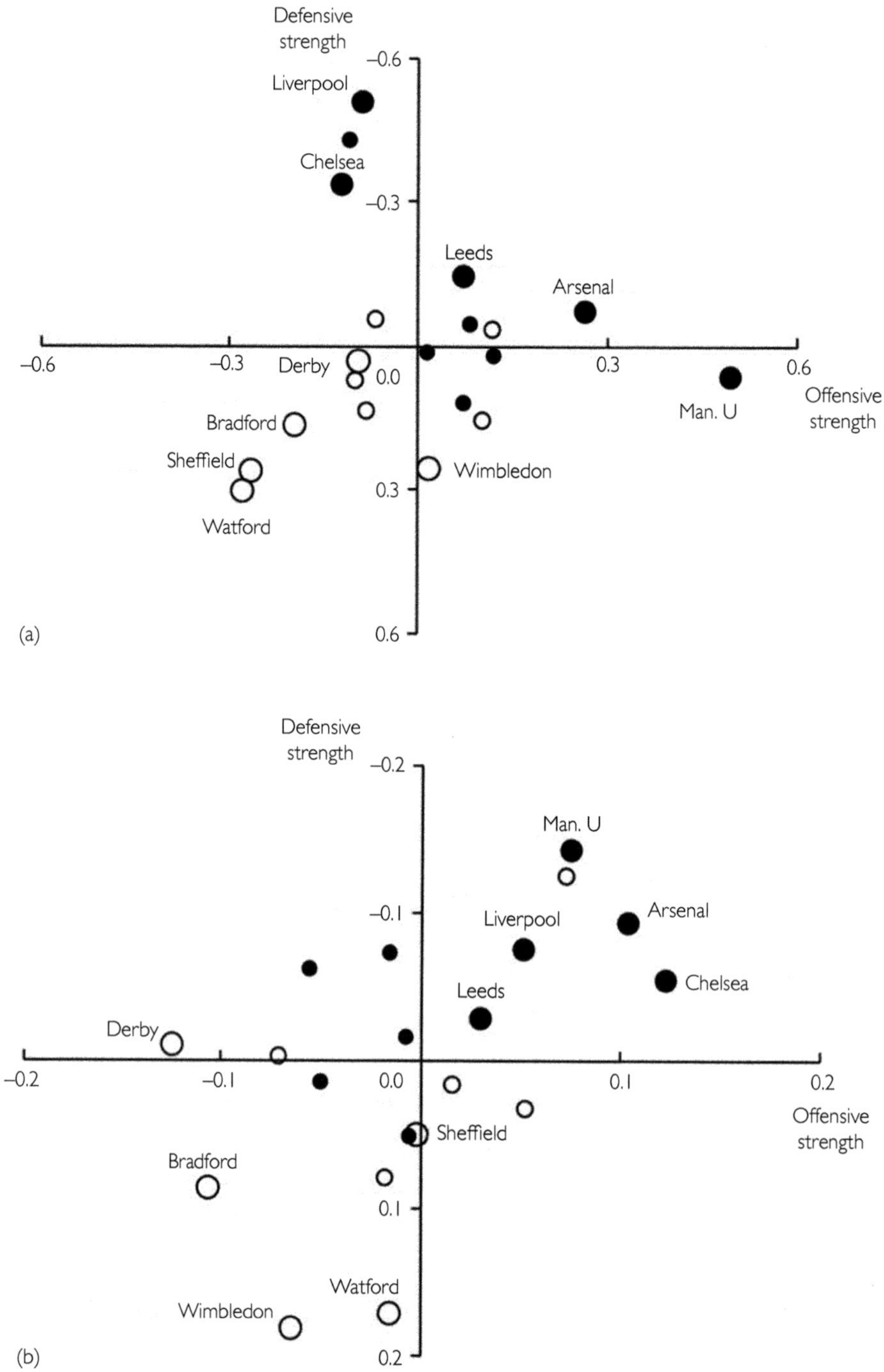

Figure 27.3 Offensive and defensive strengths (a) for goals and (b) for possession.

Table 27.5 ***Relative success for scoring goals at home (with standard errors in parentheses).***

Team	$\lambda_{scorehome}$	Home goals–away goals
Coventry City	1.13 (0.44)	29
Tottenham Hotspur	0.44 (0.37)	23
Watford	0.34 (0.43)	13
Bradford City	0.31 (0.42)	14
Chelsea	0.28 (0.37)	17
⋮	⋮	⋮
Derby County	− 0.44 (0.38)	0
Leeds United	− 0.44 (0.35)	0
Sunderland	− 0.44 (0.36)	−1
Aston Villa	− 0.46 (0.38)	0
Middlesbrough	− 0.46 (0.38)	0

Relative success for scoring goals at home

We can also evaluate the interaction parameter $\lambda_{scorehome}(A)$, which represents the propensity for team A to score a goal when playing at home, by using model G4c in Fig. 27.2(a). The average value of this parameter is constrained to be 0. This parameter can be used to give an idea of a team's 'home preference' with regard to scoring goals. Table 27.5 lists the top five and the bottom five teams ranked by the value of this interactive parameter in descending order.

This shows that Coventry City has the largest value of the interaction parameter. This means that Coventry City has the strongest tendency to score goals at home rather than away, compared with other teams. This is not surprising since Coventry City scored 38 goals at home but only 9 goals away during the 1999–2000 season, i.e. the value of home goals minus away goals is 29 as also shown in Table 27.5. In contrast, Middlesbrough has the lowest value of the parameter, indicating a tendency to score fewer goals at home compared with other teams. This again is not surprising since Middlesbrough scored 23 goals both at home and away during the season.

Relative success in gaining or losing possession at home

Similarly, we can evaluate the parameters $\mu_{gainhome}(A)$ and $\mu_{losehome}(A)$, which represent team A's relative success in gaining possession at home and its increased tendency to lose possession in a home match of the opposing team (i.e. away) respectively, by using model P5 in Fig. 27.2(b). Here the sign of the latter parameter should be changed to represent its increased tendency to lose possession at home, so Table 27.6 shows the value of $-\mu_{losehome}(A)$, which is a negative value for those teams that have relative success against losing possession at home.

These results are plotted in Fig. 27.4, in which the direction of the y-axis representing the relative success against losing possession at home is also reversed—thus a team in the upper right-hand area has a stronger tendency to gain more possession and to lose less possession

Table 27.6 ***Relative success in gaining or losing possession at home (with standard errors in parentheses).***

Rank	Team	$\mu_{gainhome}$	$-\mu_{losehome}$
1	Manchester United	−0.05 (0.037)	− 0.01 (0.037)
2	Arsenal	−0.03 (0.026)	0.00 (0.0026)
3	Leeds United	−0.05 (0.037)	− 0.01 (0.037)
4	Liverpool	−0.03 (0.038)	− 0.09 (0.037)
5	Chelsea	−0.08 (0.037)	− 0.02 (0.037)
⋮	⋮	⋮	⋮
16	Derby County	−0.06 (0.037)	− 0.03 (0.037)
17	Bradford City	0.12 (0.037)	0.08 (0.037)
18	Wimbledon	0.03 (0.037)	0.03 (0.036)
19	Sheffield Wednesday	0.08 (0.036)	0.08 (0.037)
20	Watford	0.01 (0.036)	− 0.01 (0.036)

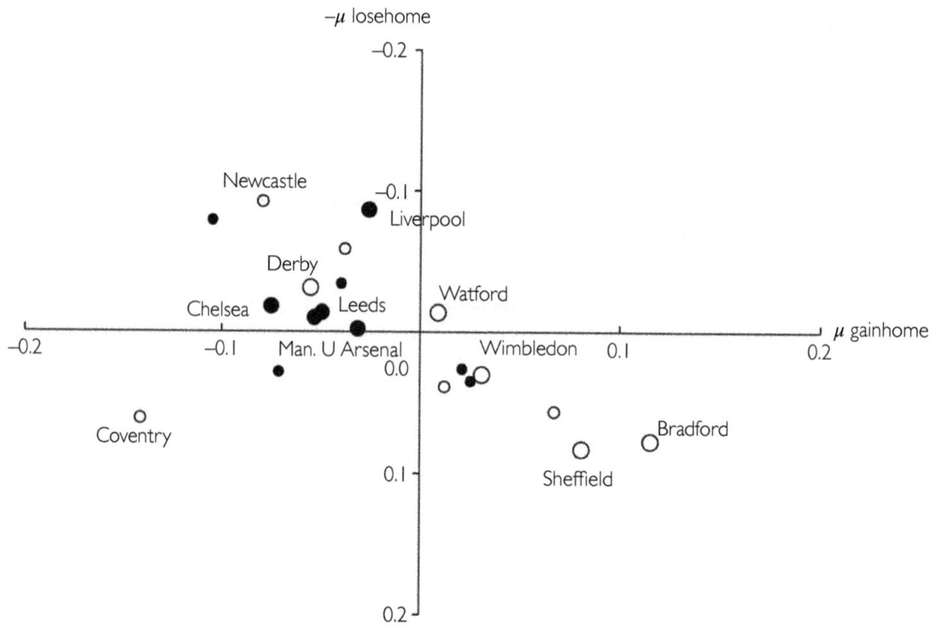

Figure 27.4 Relative success in gaining or losing possession at home.

at home rather than away, compared with other teams. Here, Watford is the only team in the upper right-hand area, and it is close to the origin corresponding to the average of all teams. This shows that no team has a particular overall propensity of relative success both for gaining and for the avoidance of losing possession, i.e. there can be no team which has a strong overall 'home preference' in possession.

We can also see that Coventry City has the weakest home preference with regard to gaining possession and almost the weakest home preference with regard to the avoidance of losing possession. Recall that in the previous section it was shown that Coventry City had the strongest home preference of all with regard to scoring goals and this result appears very surprising indeed.

This result derives partly from the fact that Coventry City had a higher estimated total possession time than its opponents away from home—883 minutes against 827 minutes—but a lower total possession time than its opponents in its home matches—852 minutes against 858 minutes. Moreover, the team gained more possessions (2971) in its away games during its opponents' possession time of 827 minutes than in its home games (2896) during its opponents' possession time of 858 minutes. This means that it was gaining possession at a rate of 3.59 times per minute of opponents' possession away, but only 3.38 times per minute at home. This puts Coventry City on the extreme left in the diagram. These findings appear to support comments made at the time by the team's manager, who frequently said that he could not understand why their home record was so much better than their away record when they were playing just as well, if not better, away from home! In all respects other than goal scoring, he appears to have had a valid point.

Relative success in gaining possession against different teams

Table 27.7 shows the estimates of the parameters $\mu_{\text{gainversus}}(A, B)$, when Manchester United is team A, for the other 19 teams as team B and vice versa. This parameter represents the strength of A to gain possession against B but, as it is based on only two matches over one season, little should be read into these values. But a similar analysis over several seasons might objectively identify 'bogey' teams. Fig. 27.5 plots these data; teams in the positive quadrant are those that were less likely to gain possession from Manchester United, and more likely to lose

Table 27.7 *Relative success of Manchester United in gaining and losing possession against different teams (with standard errors in parentheses).*

Rank	Team	$\mu_{\text{gainversus}}$ (Manchester United, *)	$\mu_{\text{gainversus}}$(*, Manchester United)
1	Manchester United	–	–
2	Arsenal	0.17 (0.10)	0.13 (0.10)
3	Leeds United	0.11 (0.14)	–0.05 (0.14)
4	Liverpool	–0.22 (0.14)	0.17 (0.14)
5	Chelsea	–0.02 (0.14)	–0.06 (0.14)
⋮	⋮	⋮	⋮
16	Derby County	0.12 (0.14)	–0.23 (0.14)
17	Bradford City	0.16 (0.14)	0.10 (0.14)
18	Wimbledon	–0.03 (0.14)	0.02 (0.11)
19	Sheffield Wednesday	–0.09 (0.14)	0.01 (0.14)
20	Watford	0.04 (0.14)	–0.07 (0.14)

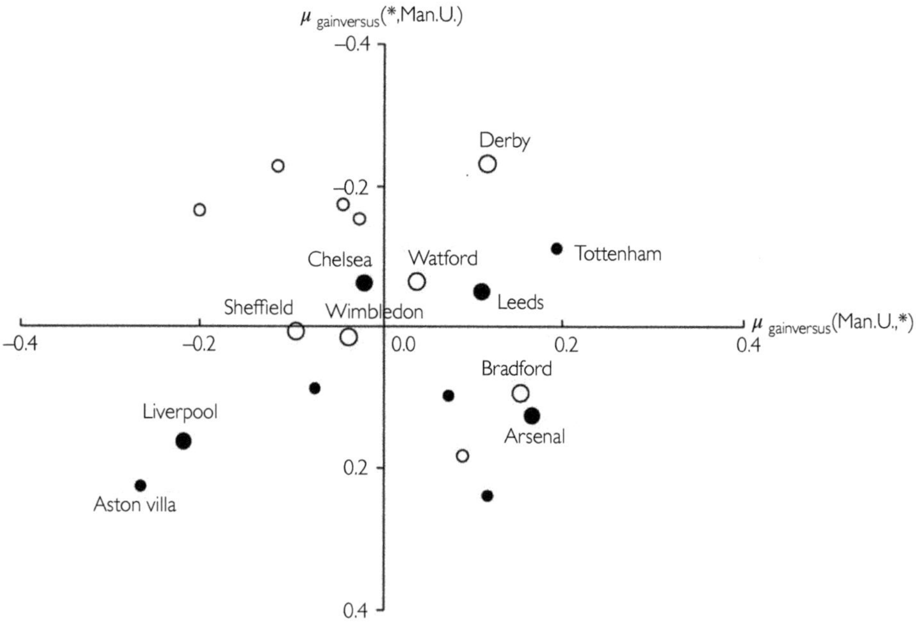

Figure 27.5 Relative success of Manchester United in gaining and losing possession against various teams.

such possession, compared with other teams. Conversely, Aston Villa and Liverpool did better than average under both criteria.

Conclusions

We have built a statistical model of a football match and evaluated the characteristics of the Premier League teams by means of parameters representing home advantage, offensive and defensive strengths and their interactions.

As we evaluate the strength of teams by using the data not only of goals but also of possession, our approach has produced more insight into the strength of teams than previous work has, in which only the data for goals scored were used in the models. Moreover, we have given an interpretation of these interactions and illustrated the characteristics of the teams in ways that allow us to visualize a preference for playing at home and relative success against particular opposition teams.

However, it should of course be borne in mind that many other factors, such as the availability of players and pitch conditions, could be involved in this type of evaluation. Further, an introduction of the time dependence of parameters into this analysis would also be of great interest (i.e. how they vary during the course of a game). If appropriate data can be obtained, they would enable us to test the significance of these factors and thus to improve our models for evaluating the characteristics of teams. Assumptions that were made in this paper, such as the Markov property and the ratio of total possession times, are supported by simple tests. However, it would be interesting to explore the validity of the assumptions more thoroughly by using more data sets.

ACKNOWLEDGEMENTS

This research uses real data provided by Opta Index Ltd, the official statisticians of the English Premier League. We thank Opta Index Ltd for allowing us to refer to these data and especially Mr James Dennis for supplying all the statistics and giving us practical advice. We also thank Dr Damon Berridge, in the Centre for Applied Statistics at Lancaster University, for giving us practical advice from the statistical point of view. We appreciate helpful comments and suggestions from the referees for improvement.

REFERENCES

Dixon, M. J. and Coles, S. G. (1997) Modelling association football scores and inefficiencies in the football betting market. *Appl. Statist.*, **46**, 265–280.

Dixon, M. J. and Robinson, M. E. (1998) A birth process model for association football matches. *Statistician*, **47**, 523–538.

Hirotsu, N. and Wright, M. (2002) Using a Markov process model of an association football match to determine the optimal timing of substitution and tactical decisions. *J. Ops Res. Soc.*, **53**, 88–96.

Lee, A. J. (1997) Modeling scores in the Premier League: is Manchester United really the best? *Chance*, **10**, 15–19.

Maher, M. J. (1982) Modelling association football scores. *Statist. Neerland.*, **36**, 109–118.

Norman, J. M. (1998) Soccer. In *Statistics in Sport* (ed. J. Bennett), pp. 105–120. London: Arnold.

Opta Index (2000) *The Opta Football Yearbook: 2000–2001*. London: Carlton.

Rollin, G. and Rollin, J. (eds) (2000) *Rothmans Football Yearbook 2000–2001*. London: Headline.

CHAPTER 28

Focus on sport

Making up the results: The work of the Football Pools Panel, 1963–1997

DAVID FORREST AND ROBERT SIMMONS

Abstract
Since 1963, a panel of experts has determined for gambling purposes the results of soccer matches in Britain which were included in the football pools weekly game but which could not be played because of adverse weather. The paper models its behaviour and highlights differences between the factors which appear to influence the panel's predictions and those that matter to the outcomes of real games. Panel results are more predictable than real results and the paper offers a specific recommendation to the Pools Panel to render its judgments less predictable. The panel also appears to overemphasize the importance of the recent away form of the away team. An explanation for this systematic difference in the statistical estimates of Pools Panel and actual results is found in the home advantage literature in sports statistics.

Keywords: *football pools, ordered logit, pools panel*

The football pools

The Pools Promoters' Association Pools Panel (henceforth, the Pools Panel) is a very peculiar British sporting institution. Currently, the treble chance is the most popular of several games offered by the pools industry. This *pari-mutuel* game invites bettors to select eight matches from a weekly list of 49 games. If all eight selected matches are drawn (tied), the bettor wins a share in the jackpot prize; near misses are rewarded with lesser prizes. If any of the 49 games are postponed, the Pools Panel adjudicates on the outcomes solely for the purposes of pools betting.

The Pools Panel was initiated in January 1963 by the pools companies in response to the loss of revenue incurred through the cancellation of fixtures, and indeed abandonment of the pools game, on several Saturdays owing to the exceptionally severe winter of 1962–1963. In early

January 1963, a Swedish pools company set a precedent by paying out prize money on the pools on the basis of allocating postponed matches outcomes determined by a random draw. The British Pools Panel followed a different approach. A panel was set up to determine the outcomes of postponed and abandoned matches which appeared on pools coupons. This panel comprised five experts, a mix of dignitaries and former players and referees, and a non-voting chairman. Its first meeting was on January 26th, 1963. In its earlier stages the panel met during Saturday afternoons when 30 or more matches were postponed, this being the number that had previously been sufficient to cancel the pools competition. Now, the Pools Panel adjudicates whenever even a single match on the pools coupon cannot take place for whatever reason.

The integrity of any gambling medium is important for its survival. A loss of public confidence in the fairness of the betting process could lead to long-term loss of turnover. The industry may therefore hope that its appointed experts would seek to avoid controversy. If they do, their results will be more predictable than those of real life. However, this approach has its own problems. As early as February 1963, the *Daily Mirror*'s pools tipster, Longsight, pointed out that it was

> 'not only possible but comparatively easy for anyone who knew a lot about soccer to predict most of the results given to unplayed matches by the Panel on a Saturday'.

This greater predictability of panel outcomes could reduce individual prizes because more people each week would select the 'correct' results. This would lead to lower revenue for the pools companies as the expected reduced jackpot prize per winner would deter bettors from entering the pools competition.

Our analysis in this paper, based on 35 years of Pools Panel activity, shows Longsight's remarks to be prescient. We compare Pools Panel football match decisions with actual results and demonstrate the relatively greater predictability of the former. We also find systematic behavioural differences in the statistical processes which determine Pools Panel and actual results. The paper is structured as follows. The next section introduces our data set and presents some summary statistics. The section 'An ordered logit model' sets out our proposed model and statistical methodology. The following section reports our results, drawing attention to key differences between the estimates of actual and invented results. The last section concludes with a precise recommendation for the future conduct of the Pools Panel.

The data

We collected from newspaper microfilm records samples of the real or manufactured results of (English) Football League and Scottish League fixtures scheduled between 1963 and 1997. There were three criteria for sampling matches:

(a) we only included Pools Panel verdicts on Saturdays when it turned out to predict the results of at least 10 matches;
(b) we only included actual match results on Saturdays where they coexisted with at least 10 Pools Panel verdicts;
(c) cup games were excluded.

Table 28.1 *Outcomes (proportions) of real and panel matches, 1963–1997.*

Match	Away win	Draw	Home win
Real	0.24	0.30	0.46
Panel	0.21	0.27	0.52

This gave us matched samples of 1384 actual results and 1483 Pools Panel results. That the samples are from the same Saturdays is useful for our purpose of comparing what determined outcomes in the two categories because different factors may influence outcomes at different times of the year (for example, shock results may occur when the minds of weak teams are concentrated by end-of-season relegation struggles) and we would not therefore want a different pattern of dates for the sample of real results from that for the panel's results. Again, where 10 or more matches are postponed, it is likely that there are bad weather conditions over much of the country and the results of the matches that survived the weather may be affected by the marginal conditions in which they will have had to be played (and in which the postponed matches would also have had to be played if the referees had not called them off).

Table 28.1 displays the proportions of away wins, draws and home wins in our two samples. Draws are less represented in the panel than in the real matches data but so also are away wins. What is evident from Table 28.1 is that the panel stressed home advantage more than was justified by the record of results on the field. 52% of panel decisions favoured the home team but only 46% of home teams that played succeeded in securing a win. These percentages differ statistically at the 0.1% significance level.

Longsight (1963) laid particular emphasis on the predictability of the panel when a team had both home advantage and a lead over its opponent in the league standings. In his early assessment of the Pools Panel, after just 2 weeks, he pointed out that, where the home teams were at least 10 places higher in the league tables than the visitors, the panel always decided on a home win. It seems that this behaviour pattern became ingrained over the next 34 years of Pools Panel operation. We grouped matches according to the difference in the current league standings between the position of the home team and the position of the away team. For all bands where the home team occupied the higher league place, the panel called a home win very much more often than data on real matches suggested that they should.

An ordered logit model

The Pools Panel must form a judgment about where each game lies on a scale between 'away win certain' and 'home win certain'. Some games which it thinks likely to yield a home victory may be viewed as 'safer' calls than others for which a home win is nevertheless the most probable outcome. However, we do not observe which home win calls are regarded as close and which lie further towards the home win certain extreme because the announced predictions merely place each game in one of three categories: away win, draw or home win. Essentially, there is an unobserved (or latent) continuous variable (how far along the away–home scale the panel puts the individual match) but we observe only three categories of outcome.

An appropriate empirical methodology for a situation such as this is provided by the ordered logit model proposed by Zavoina and McElvey (1975). The model is specified as

$$y^* = \mathbf{b}'\mathbf{x} + \epsilon \tag{28.1}$$

where y^* is the unobserved dependent variable (here the relative strength of the home team as judged by the Pools Panel) and $\mathbf{x}$ is a vector of explanatory variables (here those which affect Pools Panel behaviour); $\mathbf{b}$ is a vector of parameters to be estimated and ϵ is an error term.

There are three ordered measures of relative home team strength which map into three categories selected by the Pools Panel: away win, draw and home win.

Then:

$$y = \begin{cases} 0 & \text{if } y^* \leq 0, & (28.2a) \\ 1 & \text{if } 0 < y^* < \mu, & (28.2b) \\ 2 & \text{if } \mu \leq y^* & (28.2c) \end{cases}$$

where μ is a threshold level to be estimated by the model. A set of ordered variables is constructed for categories of Pools Panel selection, where $j = 1, 2, 3$ indexes the categories of away win, draw and home win respectively and where $i = 1, \ldots, n$ represents individual fixtures. Hence, for a particular fixture i, we create the dummy variable

$$z_{ij} = \begin{cases} 1 & \text{if } y \text{ falls in the } j\text{th category}, & (28.3a) \\ 0 & \text{otherwise.} & (28.3b) \end{cases}$$

The basic principle of the ordered logit regression model is that the logit of the probability of each underlying outcome is estimated as a linear function of the independent variables and a set of threshold levels. For any fixture i, the probability of observing outcome j corresponds to the probability that the estimated linear function is within the range of the threshold levels estimated for the outcome:

$$\Pr(z_{ij} = 1) = \Pr(\mu_{j-1} < b_1x_{1j} + b_2x_{2j} + \ldots + b_kx_{kj} \leq \mu_j). \tag{28.2}$$

Maximum likelihood estimation, across a sample of fixtures, generates the vector of coefficients $\mathbf{b}$ along with the threshold levels μ_1 and μ_2. With three possible outcomes, μ_0 is taken as $-\infty$, μ_1 as 0 and μ_3 as ∞, leaving just $\mu_2 = \mu$ as the reported threshold level in the following tables. Estimates of $\mathbf{b}$ from any ordered response model are difficult to interpret. Consequently, following Greene (1997), we shall also compute the marginal effects of $\mathbf{x}$, the vector of explanatory variables, on the three categories of Pools Panel outcome. Here, we propose ordered logit estimation but (unreported) ordered probit estimation gave similar results.

Our choice of explanatory variables was influenced by an announcement by the first chairman of the Pools Panel, Lord Brabazon, on the criteria that it would employ in reaching its decision on a match: the panel would take account of the two clubs' current league records (i.e. league standings), the recent home form and away form of the home and away team respectively and the clubs' goal averages (i.e. the ratio of goals scored to goals conceded). Corresponding data are published by most daily newspapers as part of their weekly guide for pools bettors and so presumably the panel's criteria reflect a consensus that these factors matter in the determination of outcomes of matches. Accordingly, we collected data for each fixture in our samples on

(a) the difference between the league position of the home team and the away team (DIFFPOS) and its square (DIFFPOS2) to allow for non-linearity,
(b) the proportion of its preceding six home matches that the home team had won (draws counting as half a win) (HOMEFORM),
(c) the proportion of its preceding six matches that the away team had won (draws counting as half a win) (AWAYFORM),
(d) for the home team, the ratio of goals scored at home so far that season to goals conceded at home in the same period (HOMEGOALRATIO) and,
(e) for the away team, the ratio of goals scored away so far that season to goals conceded away in the same period (AWAYGOALRATIO).

In addition to these variables, we experimented with some dummy variables but none had an estimated regression coefficient that was significant at the 5% level and they were all excluded from our reported analysis. These included dummy variables to represent the division of the English or Scottish league in which the fixture was scheduled, a dummy variable to represent fixtures taking place under revised league rules under which winning teams in actual matches were rewarded with 3 league points instead of 2 (1 point being awarded for a draw in either case) and dummy variables to represent different time periods.

Our experiments indicated that the panel was uninfluenced in its allocation of results by which division was being discussed, or by whether teams had an enhanced incentive to play for a win rather than a draw; nor did the panel reveal structural breaks in its behaviour.

Results

Table 28.2 displays the results from estimating our ordered logit model on the sample of Pools Panel verdicts. It also displays the corresponding marginal effects (calculated as in Greene (1997), pages 927–930). The marginal effects show how the probability of being in each Pools Panel category changes with a 1-unit increase in the value of the relevant explanatory variable. Without the computation of marginal effects, the sign of the coefficient on the variable would not allow us to conclude whether the probability of the middle category (a draw) is raised or lowered.

In Table 28.3 we show how the probability of being in each panel category is affected when the value of the relevant variable is raised by one standard deviation from its mean. Table 28.3 excludes the zero marginal effects obtained for DIFFPOS2. For example, consider the variable HOMEFORM with mean 0.609 and standard deviation 0.205. If the home form measure rises by one standard deviation from 0.609 to 0.814, all else equal, then the marginal effect derived from Table 28.2 shows that the probability that the panel selects a home win is raised by 8.0 percentage points. This is the 'standard deviation marginal effect'. In this case, the probability of a draw as well as that of an away win is also lowered.

Table 28.2 *Pools Panel results: ordered logit estimation.*

Variable	Coefficient	Absolute *t*-statistic	Marginal effects			Variable mean	Variable standard deviation
			Away (0)	Draw (2)	Home (2)		
Constant	1.522	7.37†					
HOMEFORM	1.553	4.47†	–0.213	–0.175	0.388	0.609	0.205
AWAYFORM	–1.561	4.20†	0.214	0.176	–0.390	0.385	0.191
HOMEGOALRATIO	0.282	3.31†	–0.039	–0.032	0.070	1.607	0.962
AWAYGOALRATIO	–0.806	5.34†	0.111	0.091	–0.201	0.814	0.513
DIFFPOS	–0.584	6.59†	0.122	0.100	–0.222	0.582	8.471
DIFFPOS2	–0.001	0.92	0.000	0.000	–0.000	77.96	93.47
MU	1.573	20.41†					
% correct	59.5						
Sample size	1483						

† Significant at the 0.1% level.

Table 28.3 *Pools Panel results: marginal effects from a one-standard-deviation change in explanatory variables.*

Variable	Standard deviation marginal effects		
	Away (0)	Draw (1)	Home (2)
HOMEFORM	–0.044	–0.036	0.080
AWAYFORM	0.041	0.034	–0.074
HOMEGOALRATIO	–0.038	–0.031	0.067
AWAYGOALRATIO	0.057	0.047	–0.103
DIFFPOS	1.033	0.847	–1.881

The coefficients in our ordered logit equation are all of the expected sign and most are highly significant, with the exception of the squared difference in position. By contrast, when we experimented with a multinomial logit model, where the categories of result are not ordered, few of our variables were significant in explaining the Pools Panel's behaviour. The implication is that the Pools Panel indeed makes its judgments as if it is placing matches on a scale ordered from almost certain away win to almost certain home win with the draw as an intermediate case.

Our results demonstrate that our publicly available indicators used in press coverage of forthcoming matches are indeed taken into serious consideration by the Pools Panel. This is underlined by the count R^2 of the estimated equation which indicates that our model correctly predicted 59.5% of Pools Panel decisions. A contrasting naïve model might take the overall (panel) totals of away, draw and home decisions and allocate these at random among all the

Table 28.4 ***Real results: ordered logit estimation.***

Variable	Coefficient	Absolute *t*-statistic	Marginal effects			Variable mean	Variable standard deviation
			Away (0)	Draw (1)	Home (2)		
Constant	1.129	5.49†					
HOMEFORM	0.852	2.70†	–0.148	–0.064	0.212	0.619	0.217
AWAYFORM	–0.139	0.44	0.024	0.010	–0.034	0.385	0.239
HOMEGOALRATIO	0.110	2.08§	–0.018	–0.008	0.026	1.738	1.143
AWAYGOALRATIO	–0.699	4.58†	0.121	0.052	–0.174	0.783	0.444
DIFFPOS	–0.020	2.56§	0.003	0.002	–0.005	–0.454	8.773
DIFFPOS2	–0.000	0.03	0.037	0.016	–0.053	77.96	100.53
MU	1.415	22.45†					
%correct	48.2						
Sample size	1384						

† Significant at the 0.1% level.
‡ Significant at the 1% level.
§ Significant at the 5% level.

Table 28.5 ***Real results: marginal effects from a one-standard-deviation change in explanatory variables.***

Variable	Standard deviation marginal effects		
	Away (0)	Draw (1)	Home (2)
HOMEFORM	–0.032	–0.014	0.046
AWAYFORM	0.006	0.002	–0.008
HOMEGOALRATIO	–0.021	–0.009	0.030
AWAYGOALRATIO	0.054	0.023	–0.077
DIFFPOS	0.026	0.018	–0.044

fixtures in the sample. This would result in a 38.5% success rate. Using a few widely available indicators of team strength therefore added substantially to our ability to predict panel decisions, compared with employing only a naïve model.

It is relevant to ask whether the model implicitly used by the Pools Panel resembles that which determines the outcomes of real matches. In Tables 28.4 and 28.5, we report the regression results, marginal effects and standard deviation marginal effects obtained from applying the same model used to represent the Pools Panel's behaviour to our sample of 1384 real matches. Although all variables apart from AWAYFORM and DIFFPOS2 have significant coefficients (at the 5% level), the model predicts less well in this new context. Only 48.2% of results are correctly predicted. Our estimates clearly demonstrate the greater predictability of Pools

Panel outcomes over real results; the latter are simply noisier. There is still, however, enough relevance in the variables included for this to be sharply higher than the success rate (36.5%) that would follow from allocating results at random within the constraint that the proportions of outcomes in each category should match those in the sample.

It is not surprising that the model for predicting actual results has less predictive power. Panel members know that shock results happen but it is doubtful whether they have inside information which gives insight into where these may occur. Therefore, they are prone to base their judgments on criteria that are known in advance, such as the relative league positions of the teams before the games postponed.

Our estimates also suggest that, whereas there is symmetry in the treatment by the Pools Panel of the importance of the HOMEFORM and AWAYFORM variables, with almost exactly equal and opposite signed coefficients and marginal effects, the real match outcomes appear not to be influenced at all by AWAYFORM. Most newspapers, like the Pools Panel, think that the recent away form of the away club helps to predict outcomes (they print the data for each match) but our finding is that it does not.

A possible reason for the lack of significance of AWAYFORM in the actual results estimates lies in the various sources of home advantage in sports listed in a survey by Courneya and Carron (1992): these include the surface and dimensions of the playing area which differ between venues (home teams can adapt their playing style to exploit the peculiarities of their particular stadium), the adverse effect on the away team of the fatigue and disruption to routine that are associated with travel and the partisan nature of crowds which may influence players' performances or key refereeing decisions in matches to the advantage of the home side. Some recent statistical work on UK soccer by Clarke and Norman (1995) draws attention to home advantage being more important as the distance of travel between the grounds increases whereas Nevill et al. (1996) found that referees' decisions appeared to favour the home team to a greater extent in those divisions which attracted larger crowds. Given the various sources of home advantage, it is understandable why the recent home form of the home team is more relevant than the recent away form of the away team in predicting actual match outcomes. The away team's form in recent away matches will reflect a wide variety of venues and opposing home crowds. In contrast, the home team will have played its last six home games in the same stadium and probably before a crowd of similar size and partisanship. How well the players respond to these conditions will be embodied in the last six home results. The last six results at home may therefore have some informational content to aid in predicting the result of their next appearance in the same stadium. Recent away form is observed over more heterogeneous playing conditions and is less relevant although cumulative away form, in which longer-term average conditions are captured, may help to predict a team's performance as revealed by the significant coefficient on AWAYGOALRATIO.

Our ordered logit estimates show that Pools Panel decisions can be more easily predicted than can real outcomes. We have also shown that details of their implicit model may be flawed and their misguided faith in the relevance of recent away form is the most striking case in point. At the inception of the Pools Panel, Longsight (1963) claimed that its results would be more predictable than on-the-field results and that therefore the number of pools winners would increase (and the size of individual prizes consequently decrease). On the basis of our findings, he was correct to assert that it would be 'comparatively easy to anyone who knew a lot about soccer (or who used our model) to predict most of the results given to unplayed matches by the Panel'.

Conclusions

The first chairman of the Pools Panel effectively established its constitution by stating the factors that it would take into account in its decision-making on the outcomes of football games that could not be played because of weather conditions or other circumstances. The criteria matched those for which there appeared to be a consensus between journalists about their relevance for results of games. Our study shows that the panel has continued to employ these criteria throughout its life and has therefore followed its unwritten constitution. In doing so, it has avoided controversy. If it instead wished to simulate the process by which real results are generated, it would need to give less attention than it seems to do to certain factors such as the recent away form of the away team.

Our regression results show that actual game outcomes contain considerable noise whereas Pools Panel verdicts are much more predictable. Adding some random noise to the findings of the expert panellists, when fixtures are postponed, would simulate real life more closely and may now be acceptable to the pools companies, particularly in the face of severe competition from the UK National Lottery and growing forms of football betting such as fixed odds results betting and index betting.

We suggest that the Pools Panel should add some random noise to its verdicts to simulate real results more closely. Our ordered logit estimates gave an 11% difference between the prediction rates in the Pools Panel and actual results regressions (59.5% from Table 28.2 minus 48.2% from Table 28.4). To make the pools competition both less predictable and more interesting, and to simulate real results more closely, we would therefore suggest that the Pools Panel should introduce approximately 10% of 'wild' predictions into their results, to reflect better the unpredictable nature of real games.

ACKNOWLEDGEMENTS

We wish to thank Pam Carroll for her excellent research assistance. Stephen Drinkwater, David Percy and a referee provided helpful comments although they are not to be held responsible for the contents of the paper.

REFERENCES

Clarke, S. R. and Norman, J. M. (1995) Home ground advantage of individual clubs in English soccer. *Statistician*, **44**, 509–521.

Courneya, K. S. and Carron, A. V (1992) The home advantage in sports competitions: a literature review. *J. Sport Exercise Psychol.*, **14**, 13–27.

Greene, W. H. (1997) *Econometric Analysis*, 3rd edn. New York: Macmillan.

Longsight (1963) The Longsight column. *Daily Mirror*, Feb. 7th, 1963.

Nevill, A. M., Newell, S. M. and Gale, S. (1996) Factors associated with home advantage in English and Scottish soccer matches. *J. Sports Sci.*, **14**, 181–186.

Zavoina, R. and McElvey, W. (1975) A statistical model for the analysis of ordinal level variables. *J. Math. Sociol.*, **4**, 103–120.

PART VII

Other Sports

Introduction

FRITS SPIEKSMA

What makes a good scientific paper in sports? Of course, some ingredients are indispensable: sound methodological approach, new insights, lucid writing... But to top it off, one needs a good story; a story that captures the imagination, a story we can relate to, a story with which we are (perhaps vaguely) familiar. Every sport has good stories—a sport cannot survive without them.

The six papers presented in this section represent research in specific sports that are not covered by the other sections of the anthology; these papers address five different sports. In addition to representing research on a wide variety of sports, the types of questions considered in these papers vary quite a bit. However, the papers in this section also have something in common: they each tell a good story. From the famous Holyfield versus Lewis boxing match to the continuing debate on the hot-hand effect, each of the papers in this section offers an interesting story that expands our understanding. Here, we introduce each of these papers and identify their innovative contributions.

Zaman (2001) addresses a well-known hockey issue known as the Pull-the-Goalie (PtG) problem: at what moment should the coach decide to pull the goalie (i.e., to change the goalkeeper for a regular skater) in order to maximize the chance to score a tying goal when trailing by one (or two) goals late in the game? Research on this question was initiated by Morrison (1976), and a recent paper on this topic has been published by Beaudoin and Swartz (2010). Most approaches to the PtG problem are based on the arrival of goals being modelled as a Poisson process, thereby assuming a constant probability over time of a goal occurring. Then, pulling the goalie increases the scoring rate of the team that pulls its goalie (as well the scoring rate of the opposing team!). In contrast to these traditional approaches, Zaman proposes a Markov chain model that includes the location of the puck in the field as a state. There are seven states: Goal A, Shot A, Zone B, Neutral, Zone A, Shot B, and Goal B. One advantage of this point of view is that the state of the game has an impact on the decision when to pull the goalie. Indeed, one should be more careful when the puck is in one's defensive zone compared to when the puck is in the offensive zone. It turns out that, pulling the goalie with 5–8 minutes left to play is the right decision when trailing by one goal depending upon the location of the puck. This outcome is consistent with outcomes from traditional approaches, and also shows that in practice coaches are too conservative.

The second paper (Dawson and Magee, 2001) also deals with the National Hockey League (NHL). Can we reliably predict performance of NHL players as a function of the player's position (goaltenders, defensemen, forwards), year, and overall rank in the draft (or pick number)? Performance is measured by the number of NHL games played. A recent contribution attributing a value to each pick number is found in Hurley *et al.* (2012). Dawson and Magee focus on three questions: (i) What is the expected difference in number of games played between a player drafted as the ith pick, versus a player drafted as the jth pick (for any i,j)? (ii) How well does a franchise draft its players? (iii) To what extent are European players undervalued in the draft? The authors propose a multiplicative model, and describe a so-called weighted pool-adjacent-violators algorithm to arrive at the estimates. They find for each of the three positions a relation between pick number and number of games played. They also find substantial variation among franchises with respect to the difference between the actual and predicted number of games played by the franchises' draftees. Finally, it is found that in a particular period (1988–1992), European draftees played more matches than expected.

One of the most intensely debated issues in sport statistics is the "hot-hand" question: are success probabilities of consecutive attempts independent? Clearly, this question turns up in many different sports: basketball, baseball, golf, darts, tennis, volleyball, to name a few; see Bar-Eli *et al.* (2006) for an overview of research on this topic. Some papers find the effect, others do not. In particular, an influential study in basketball by Gilovich *et al.* (1985) finds no evidence of the effect. The third paper (Dorsey-Palamateer and Smith, 2004) addresses the presence of the hot hand in bowling. It is argued that bowling has a number of properties that make it attractive for testing the existence of a hot hand effect. Indeed, the absence of interplay of opponents, the almost identical circumstances prior to each attempt, and the regular time-intervals between throws, make bowling a sport well-suited for studying the presence of the hot hand effect. By categorizing each throw as a strike or a non-strike, a binomial model can be used to model that the success probability is stationary, and that throws are independent: player i has a probability p_i to throw a strike. Analyzing the data reveals that a bowler's success probabilities are not independent of previous trials. More specifically, success breeds success, as the authors say; most bowlers have a higher strike proportion after j consecutive strikes than after j consecutive non-strikes, and this effect increases with the value for j.

Who won the boxing fight? This innocent-looking question is not always so easy to answer, as those who watched the match between Evander Holyfield and Lennox Lewis on March 13, 1999 may remember. This particular match was called a draw, even though many observers felt Lewis had the upper hand. Lee *et al.* (2002) looks at this fight and the September 1999 fight between Oscar de la Hoya and Felix Trinidad with respect to interrater agreement. A boxing match is divided into 12 rounds with each round scored individually by three judges who cannot communicate with each other. An individual scorecard then consists of a sequence of 12 symbols, one for each round, indicating whether a round was won by one opponent, by the other opponent, or was called even. The paper shows that it is possible to compute the probability that a particular scorecard is observed, and thereby offer a way to identify and quantify the variance in scoring behavior of different judges, both official ones and unofficial ones. Of course, this gives the opportunity to assess judges' behavior (see also Balmer *et al.* (2005) who focus on home advantage). The authors further argue that unfamiliarity with the precise scoring rules and the perhaps not well-understood inherent stochastic nature of scoring a boxing fight may cause the perceptual difference between general opinion and the actual outcome of a fight. It also may result in bad advice to a boxer while the match is in progress, as exemplified by the match between Oscar de la Hoya and Felix Trinidad.

More and more, the National Basketball Association (NBA) drafts players who are not seniors in college, but instead sophomores, juniors, or freshmen. Thus, a player in college, aspiring to play in the NBA is confronted with the difficult of decision whether to stay in college for another year or enter the draft. There are obvious pros and cons for each alternative. The fifth paper in this section (Bishop and Gajewski, 2004) provides several examples indicating there is no single good choice. The authors define a statistic (the "bling" number), which assesses a player's potential and can be seen as representing the draft likelihood. This number is computed using first principal components analysis and next performing a logistic regression. The principal components analysis identifies from all criteria two major factors, namely game (minutes played, points made, rebounds, and so on) and body (height, weight, blocks). The logistic regression was then used to predict whether a player was drafted. To balance the error rate among players predicted to be drafted and players predicted not to be drafted, a cutoff probability of 0.20 was chosen, leading to 64 players that were predicted to be drafted in the 2001 season. Individual analysis of some of these athletes reveals that other factors (heart for the game, leadership) also influence drafting decisions.

In the sixth paper in this section, Marcus (2001) proposes a new table tennis rating system. Notice that a rate represents the strength of a player whereas a ranking system only produces an ordinal list of players. Probably, the best-known rating system in sports is the chess rating system (see e.g., Glickman (1999)). In table tennis, due to the large variety in player characteristics (in number of matches played per year), a Bayesian model is deemed appropriate; such a model treats a player's strength as uncertain and uses conditioning to incorporate new results. After a precise and clear description of the current system in use, the new system is explained and the two systems are compared. The paper argues convincingly that a rating system should not use scores of matches to rate players. In the new approach, the strength of each player is modelled as a random variable. The paper describes how to initialize this strength, how strength evolves over time (by adding a random walk), and how new match results are processed to arrive at adjusted strengths. Resolving this latter issue can be computationally intensive, and the paper discusses ways to speed up the algorithm.

REFERENCES

Balmer, N., Nevill, A., and Lane, A. (2005). Do Judges Enhance Home Advantage in European Championship Boxing? *Journal of Sports Sciences*, 23, 409–416.

Bar-Eli, M., Avugos, S., and Raab, M. (2006). Twenty Years of "Hot Hand" Research: Review and Critique. *Psychology of Sport and Exercise*, 7, 525–553.

Beaudoin, D. and Swartz, T.B. (2010). Strategies for Pulling the Goalie in Hockey. *The American Statistician*, 64, 197–204.

Bishop, T. and Gajewski, B. (2004). Drafting a Career in Sports: Determining Underclassmen College Player's Stock in the NBA Draft. *Chance*, 17, 9–12.

Dawson, D. and Magee, L. (2001). The National Hockey League Draft, 1969–1995. *The American Statistician*, 55, 194–199.

Dorsey-Palmateer, R. and Smith, G. (2004). Bowlers' Hot Hands. *The American Statistician*, 58, 38–45.

Gilovich, T., Vallone, R., and Tversky A. (1985). The Hot Hand in Basketball: on the Misperception of Random Sequences. *Cognitive Psychology*, 17, 295–314.

Glickman, M. (1999). Chess Rating Systems. *American Chess Journal*, 3, 59–102.

Hurley, W., Pavlov, A., and Andrews, W. (2012). A Valuation Model for NFL and NHL Draft Positions. *Journal of the Operational Research Society* 63, 890–898.

Lee, H., Cork, D., and Algranati, D. (2002). Did Lennox beat Evander Holyfield? Methods for Analyzing Small Sample Interrater Agreement Problems. *The Statistician*, 51, 129–146.

Marcus, D. (2001). New Table-Tennis Rating System. *Journal of the Royal Statistical Society*, Series D, 50, 191–208.

Morrison, D.G. (1976). On the Optimal Time to Pull the Goalie: A Poisson Model Applied to a Common Strategy Used in Ice Hockey. *TIMS Studies in Management Sciences, Vol. 4*, North Holland, New York, 137–144.

Zaman, Z. (2001). Coach Markov Pulls Goalie Poisson. *Chance*, 14, 31–35.

CHAPTER 29

Coach Markov pulls goalie Poisson

ZIA ZAMAN

In the hockey world, very little is predictable. Of course there are statistics and analyses that can be used to help coaches decide who should take a face-off in which circumstance, but because of the impromptu and fluid flow of the game, there have been far fewer applications of math to solving hockey problems than there have been in other sports, say baseball, for example. One notable exception is the celebrated *Pull the Goalie* problem. For hockey virgins, pulling the goalie is a strategy that coaches have at their disposal to remove the goalkeeper from the goal and replace him with another skater. For years, coaches have been guessing that if a team is losing by one goal, somewhere near the end of the game, the team should remove or "pull" the goalie in favor of an extra attacker. The general thinking is that by adding a skater in the offensive end the team can perhaps augment its chances of scoring the tying goal. True, the chances that they will be scored upon also increases since they have no goalkeeper, but losing by two goals is also no worse than losing by one goal. Typically, NHL coaches tend to pull the goalie with one minute remaining. This practice started about 25 years ago and has not significantly changed much at all, despite some analyses that posited that the optimal time to pull the goalie was as much as five or ten minutes before the end of the game. "Hogwash!" one can imagine coaches saying in response to the malevolent mathematical results, "I know hockey, and I know you just have to hunker down and score instead of playing around with your calculators."

It seems clear that there does come a point in a close game when the trailing team's coach is better off risking a goal against by pulling the goalie than by continuing even-strength play. But, having played and watched hockey all my life, I too felt that the five minute result was unintuitive. So I went out to try and find a more representative and accurate model for the game of hockey.

The traditional approach

For many years now, the *Pull the Goalie* problem has been discussed using the assumption that goals arrive as a Poisson process. This means that a goal can happen at any time with constant probability and that goal arrivals are independent of how much time has occurred since the last goal. In this case, the parameters required to determine optimal strategy are the arrival rate of goals in the steady state and in the empty-net state for both the leading and trailing teams. When solved, the math yields a time at which the probability of the trailing team's tying the score is equivalent in the steady-state and empty-net cases. The results uniformly pointed to a time approximately five minutes before the end of the hockey game. This result was verified again recently in Scott Berry's Summer 2000 *Chance* column.

One problem with the preceding is that any fan knows that the arrival pattern of goals *is* dependent on the state of the hockey game, especially the location of the puck. This motivation led me to the conclusion that a better representation of the game of hockey would be a Markov chain.

A Markov approach

Hockey is a fluid sport. It has ingenuity and spontaneity and lacks the clear analysable definitions of, say, baseball. Still, there is one elemental aspect of the game that persists. Players chase a puck around an oval ice surface in an attempt to shoot it into a $6' \times 4'$ net with their sticks. The rink is loosely separated into three zones. These zones are the two endzones and the neutral zone. When the puck is in the neutral zone, the game takes on a very different shape than when the puck is in an attacking zone. Players are careful to advance an attack by either amassing forces to enter the attacking zone or play off the transition game by hoping to go through the neutral zone quickly to get to the attacking zone before defenders have a chance to return. "Gaining the zone" is an important concept because of the offside rule and because the feasibility of scoring is nil unless you are close to the goal. Once in an attacking zone, the offensive side often "sets up." The team takes a formation which allows them to pass the puck back and forth in an attempt to free a player in a shooting position on the ice. If the player is given the puck in a shooting position, within the blink of the eye, he will be able to release a shot. This shot may well be stopped by the goalie but usually results in the puck being returned to a dangerous position where perhaps another shot can be taken. "You can't score if you don't shoot," is another popular hockey expression.

Because the sport involves the flow of the puck back and forth between zones, I submit that the state of the game can be captured at any one given instant by the location of the puck. The progress from state to state corresponds to the flow of action in the game. Immediately, this view lends itself to using a Markov chain as a model. The Markov-chain model (see sidebar) specifies the probability of making a transition from one state to another state; this transition probability does not depend on how we reached the current state. Clearly under such a model, the probability of scoring a goal is dependent on the state of the system. The probability of scoring will be much higher for a team in its attacking zone than for a team in its defensive zone.

How the Markov model works

Figure 29.1 is a graphical representation of the Markov model. It superimposes the seven possible states of the Markov model—*Goal A, Shot A, Zone B, Neutral, Zone A, Shot B, Goal B*—on the image of a hockey rink. Consider an example in which the puck is dropped at center ice to start the game. The puck is in the *Neutral Zone* at the start and stays there for a random amount of time until it is taken into another zone, let's say *Zone A* (short for the defensive zone of Team A), where Team A is on defense. The transition probability for the passage from *Neutral* to *Zone A* is the probability that in the next second of play the puck moves across the blue line separating the zones; in the diagram it is labeled *b*3. Now that the puck is in *Zone A*, it is entirely possible that the attacking team, Team B, manages to create a situation in which one of its players receives the puck in a shooting position. This represents a transition from *Zone* <$$$>A to the state labeled *Shot B* (the probability of this transition is labeled *b*4). Once

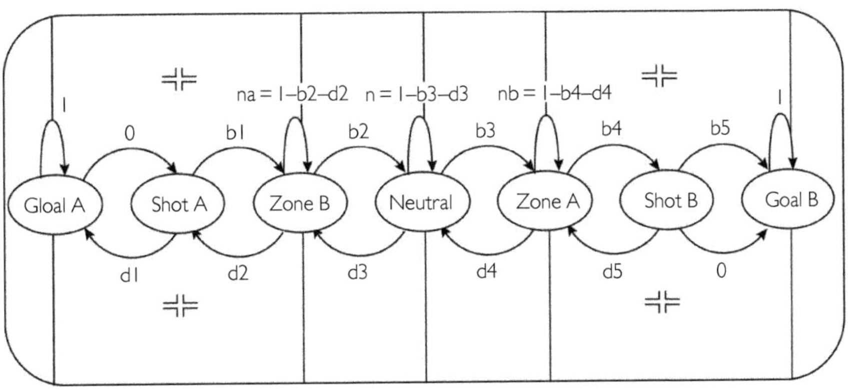

Figure 29.1 Graphical representation of the Markov model of hockey superimposed on a hockey rink. The seven ovals represent the seven states of the model with arcs representing transitions from one zone to another. Arcs are labeled with the probability of a transition.

the shot is taken, there are two possibilities: Either Team B scores (transition to *Goal B*) or it doesn't (return to *Zone A*). Let's assume that the puck does not go in. This corresponds to transition probability *d*5. The puck is in *Zone A* and play proceeds. If the puck ends up in state *Goal B*, then play is restarted at center ice.

For the empty-net case, there are a few changes. For one thing, all shots taken on an empty net end up scoring. Moreover, both teams benefit from an increased likelihood of scoring. To apply the Markov model to the question of determining the optimal time to pull the goalie, one computes the probability that the trailing team scores the next goal (to tie the game) under the *even strength* and *empty net* scenarios for a given amount of time remaining. When the two chains yield the same probability of success, we will have found the time at which we are indifferent between the two approaches. With this amount of time remaining before the completion of the game, it would be wise to withdraw the goalie from the goal.

An advantage of the new model is that it yields three results, dependent on the state of the game. Clearly, a team should be more reluctant to withdraw its goaltender when the puck is in its own defensive zone. It should also be more liberal when it has the puck in its opponents' end. These factors are taken into account by the nature of the Markov model. Therefore, a result is found for each of the three possible states of the system—namely, having the puck in the offensive, neutral, or defensive zone.

Estimating the probabilities

To make use of the Markov model one must estimate the transition probabilities. Most of the probabilities are estimated from published statistics in the NHL yearbook and Web site and from data collected by timing several NHL games to find out how long the puck stays in each zone. Some other parameters were found by using simulation to discover parameter values consistent with observed hockey results.

Markov-Chain Application

Markov chains are sequences of random variables in which the distribution of one variable depends only on the value of the preceding observation and not on any earlier information. In the accompanying article, the progress of the puck is modeled as a Markov chain. The transition probabilities pictured in Fig. 29.1 of the article can be displayed in a matrix as follows. This matrix gives the probability of going from one state to another in the next second of play. The columns are not labeled but they correspond to the row labels. For example, the second element in the third row gives the probability of a transition from Zone B to Shot A as being *d*2.

$$P = \begin{pmatrix} 1 & 0 & 0 & 0 & 0 & 0 & 0 \\ d1 & 0 & b1 & 0 & 0 & 0 & 0 \\ 0 & d2 & na & b2 & 0 & 0 & 0 \\ 0 & 0 & d3 & n & b3 & 0 & 0 \\ 0 & 0 & 0 & d4 & nb & b4 & 0 \\ 0 & 0 & 0 & 0 & d5 & 0 & b5 \\ 0 & 0 & 0 & 0 & 0 & 0 & 1 \end{pmatrix} \begin{matrix} \text{Goal A} \\ \text{Shot A} \\ \text{Zone B} \\ \text{Neutral} \\ \text{Zone A} \\ \text{Shot B} \\ \text{Goal B} \end{matrix}$$

A key Markov-chain result is that the probability of transition from state *i* to state *j* in *t* seconds is obtained by taking the relevant element when the matrix *P* is raised to the tth power. We let P_R denote the probability matrix for the regular-strength case and P_E, the probability matrix for the empty-net case. We seek to find the time *t** for which the probability of scoring for Team B is equal in both cases. If we start in the *neutral state* (row 4), then the relevant question is to determine *t** such that the (4,7) element (which corresponds to a transition from the *neutral zone* to *Goal B*) is the same for the regular-strength and empty net cases. In the article, Fig. 29.4 shows the (4,7) element for both scenarios as a function of time. The point at which they intersect is *t**.

Even strength

In the even-strength or regular-strength case both teams have five skaters and a goalkeeper. The number of transition probabilities to be determined in this case can be reduced due to symmetry. The symmetry is based on the assumption that the two teams are evenly matched. It follows that $d1 = b5$, $d2 = b4$, $d3 = b3$, $d4 = b2$, $d5 = b1$. Explaining each number in turn, $d1$ is the shooting percentage. According to National Hockey League (NHL) statistics, 9.82% of shots on goal end up in the net. All shots that are saved by the goalie travel the $d5$ link; therefore, $d5 = 1 - d5 = 1 - d1 = 0.9018$. The parameter $d2$ is the conditional probability of a shot in the next second given that we are in the offensive zone. Given the average number of shots in a game (28.9), the total number of seconds played (3,600), and the amount of time spent in the attacking zone (37.1%) (more on this later), this probability turns out to be $d2 = 28.9/(3{,}600*0.371) = 0.02163$.

After timing several games, it turns out that the average amount of time the puck spends in the *neutral zone* during a visit is 3.81 seconds; the probability that the puck will leave the *neutral zone* in any given second is therefore 1/3.81 = 0.262. The probability the puck will stay in the *neutral zone* is n = 1 − 0.262 = 0.738. If the puck does leave the zone, then it can leave with equal probability to *Zone A* or *Zone B*; therefore, the appropriate transition probabilities are $d3 = b3 = 0.262/2 = 0.131$. The timing data show that on average the puck stays in the offensive zone for 10.97 seconds. This yields $d4 = 1/10.97 = 0.091$. Since the occurrence of a shot is $b4 = 0.02163$, we can use subtraction to get $nb = 1 - b4 - d4 = 1 - 0.02163 - 0.091 = 0.8872$. The timing data also allow us to estimate that the puck spends 25.8% of the time in the neutral zone (3.81/(3.81 + 10.97)) and 37.1% in each defensive zone. This percentage was applied earlier to estimate the probability that a shot is taken.

These probabilities are included in our Markov chain in Fig. 29.2. We simulated 100 games using these probabilities and found that 2.75 goals/game were scored by each side on average

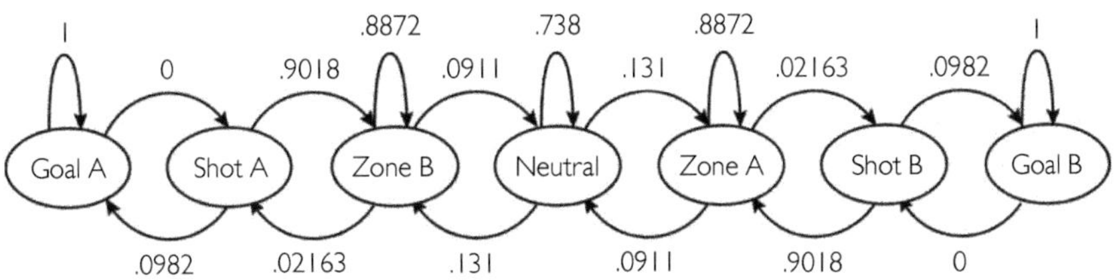

Figure 29.2 Graphical representation of the even-strength Markov model. Numbers on arcs from a zone indicate probability of a transition out of the zone in the direction shown during the next second of play.

Figure 29.3 Graphical representation of the empty-net Markov model. Numbers on arcs from a zone indicate probability of a transition out of the zone in the direction shown during the next second of play.

$$P = \begin{pmatrix} 1 & 0 & 0 & 0 & 0 & 0 & 0 \\ d1 & 0 & b1 & 0 & 0 & 0 & 0 \\ 0 & d2 & na & b2 & 0 & 0 & 0 \\ 0 & 0 & d3 & n & b3 & 0 & 0 \\ 0 & 0 & 0 & d4 & nb & b4 & 0 \\ 0 & 0 & 0 & 0 & d5 & 0 & b5 \\ 0 & 0 & 0 & 0 & 0 & 0 & 1 \end{pmatrix} \begin{array}{l} \text{Goal A} \\ \text{Shot A} \\ \text{Zone B} \\ \text{Neutral} \\ \text{Zone A} \\ \text{Shot B} \\ \text{Goal B} \end{array}$$

and an average of 28.9 shots were generated by each team. These figures are in good agreement with observed NHL statistics.

Empty net

In the empty-net case, in which the trailing team has substituted an additional offensive player in place of their goal-tender, the game takes on a different shape. More time is taken in the attacking zones, and teams are more successful at getting shots off. Previous researchers collected data that showed that Team B's rate of goal scoring increases by a factor of 2.67. This clearly shows that pulling the goalie does work; in fact, it's almost three times better than the status quo. The problem is that the same researcher found that the opposing team (which is now shooting at an empty net) also sees an increase in the rate of goal scoring by a factor of 7.83, a three times better chance of scoring than the team that has pulled the goalie. No wonder the fans' hearts race when the goalie goes to the bench! If we were to play an entire game with the empty net, the score would on average be 2.75*7.83 = 21.5 to 2.75*2.67 = 7.3 or 21 to 7!

There is not much data to guide the estimation of parameters in the empty-net case. It's easy to see that $d1 = 1$ and $b1 = 0$ since all of Team As shots at the unguarded net will score. I have assumed that Team B's shooting percentage is the same $b5 = 0.0982$. I have also assumed that Team A's ability to release a shot in the attacking zone is unchanged from regular strength ($d2 = 0.02163$) and that both teams were equally effective at clearing the zone in both the empty-net and regular-strength ($d4 = b2 = 0.0911$) cases. I assumed that $b4$, the conditional probability of Team B releasing a shot, is exactly multiplied by 2.67, the increased scoring rate of Team B. The other assumptions made include $d3 = b3$, the two teams have an equal chance of getting the puck into the other attacking zone. To find the common value, I adjusted $d3$ iteratively until the results of the entire-game empty-net Monte Carlo simulation yielded the 21 to 7 average score mentioned previously.

Note that the estimates in the empty-net case are somewhat speculative. For example, it is certainly plausible that Team B, with the extra attacker, has an advantage in the neutral zone so that $b3 > d3$. I return to that issue later in assessing how sensitive our results are to the various assumptions.

The asymmetric Markov chain in Fig. 29.3 corresponds to a hockey game when Team B has pulled its goalie. This is the *Empty-Net* case.

Using the model

It is possible using Markov-chain techniques (see sidebar) to compute the probability with t seconds remaining that the chain will end up in state *Goal A* (which sees Team B now

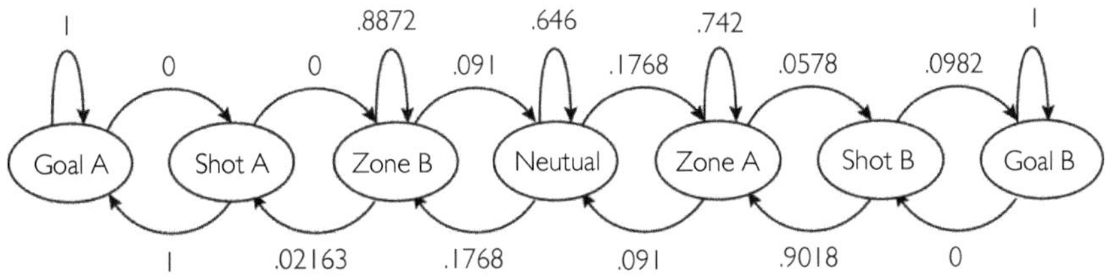

Figure 29.4 Probability that the next goal is a goal scored by Team B (the team behind by one goal) before the end of the game as a function of time remaining under two strategies.

trailing by two goals), *Goal B* (which means Team B has now tied the score), or with no further scoring. Since Team B is behind, it can compute the probability of scoring, given the current state/location of the puck, and determine if it is better off playing at regular strength or with an empty net. If the puck changes state, the coach's strategy can also change since he can recompute the probabilities. A graphical view of the results assuming we start in the neutral zone is shown in Fig. 29.4. At 399 seconds before the end of the game, the probability that the Team B wins is about 23% regardless of which strategy it chooses. This is the indifference point we seek. With less than 6:39 remaining, Team B is better off pulling the goalie. Prior to that point, it should play at even strength.

Of course there is much greater risk to the strategy of pulling one's goalie if the current situation is that Team B is in its own defensive zone. We can take this into account by considering different initial states for our Markov analysis.

If the puck is stuck in Team B's defensive zone, the coach should choose to hold off on retiring the goalie because the danger is too great that the other team will score. Once the clock drops below 5:09 remaining, he might as well pull his goalie since the *empty-net* scenario yields better probabilistic results. With the puck in B's offensive zone we can be more aggressive; the results suggest the optimal time to pull the goalie is 8:07.

SUMMARY

The results obtained by the Markov model are striking. They suggest pulling the goalie much earlier than other statistical analyses and much earlier than the general philosophy of most NHL coaches. There are several reasons to prefer this new result. First, having results that depend on the current location of the puck is appreciably more useful than the state-insensitive results of the Poisson model. Moreover, the Markov chain seems like a better model because its inherent structure is most similar to the flow of the game of hockey.

Still the result is surprising, and it is natural to question it. I have tried several modifications of the Markov model and a variety of assumptions about the empty-net probabilities. In every case, the answer comes out to be approximately six or seven minutes.

So next time your local NHL team is down a goal and the clock hits 6:39, draw a deep breath and shout out with all your lungs, "Pull the Goalie!" Neighboring fans are likely to look at you as if you are insane, but who knows; maybe your team will score the tying goal.

CHAPTER 30

The National Hockey League Entry Draft, 1969–1995

An application of a weighted pool-adjacent-violators algorithm

DON DAWSON AND LONNIE MAGEE

Abstract
We use NHL statistics for players selected in the National Hockey League Entry Draft to predict performance, which is measured by career games played, as a function of the player's position, year, and overall rank in the draft. This is done by inequality-constrained least squares using a weighted pool-adjacent-violators algorithm. The results are used to examine the value of having the *i*th versus the *j*th overall pick, to rank the drafting success of franchises, and to evaluate the performance of European players. The number of games actually played by European players drafted in the 1988–1992 period notably exceeded their expected number, as determined by the model.

Keywords: *isotonic regression model, multiplicative model, professional sports*

Introduction

As noted by Neale (1964), professional sports leagues are unique. Although cartel-like activity off the field is optimal from an economic point of view, it is anathema when it comes to activity on the field of play. Attendance and revenue tend to be maximized when game outcomes are uncertain. To facilitate this end, various leagues have instituted a draft system which, ideally, will allow for relatively random outcomes.

It is the aim of this article to analyze both the performance of groups of players and the relative success of the franchises in drafting players in the National Hockey League. Performance is measured by career regular season NHL games played. The expected games played of each drafted player is modeled as a function of his position on the ice, draft year, and his position in the draft. So as not to create confusion with two uses of the term "position," we will refer to the position in the draft as the "pick number."

With these estimates, we examine three issues:

1. The expected difference in games played between a player drafted *i*th overall and a player drafted *j*th overall for any *i*, *j* and position.
2. The drafting records of franchises, by comparing the actual career games played of players they drafted with their predicted games played.
3. The extent to which European players have been undervalued in the draft, again by comparing actual and expected career games played.

The year and pick number effects are estimated by least squares subject to a monotonic non-increasing pick number effect. These estimates are obtained by a weighted pool-adjacent-violators algorithm, taken from Barlow, Bartholomew, Bremner, and Brunk (1972).

The next section describes the data and the "Background" section gives some background on the NHL draft. The section "Using career games played as the performance measure" discusses this use. The model is presented in the section after that, followed by the estimator and algorithm. The sections that follow address the three issues listed here, and the final section concludes.

The data

The data were taken from Tredree and Bontje (1998). For all drafted players we entered the following variables, by year: draft position (referred to as "pick number" in this article); drafting franchise; location of team with whom the drafted player had last played when drafted (Canada, United States, or Europe); position (left wing, right wing, center, defenseman, or goaltender); and total regular season games played in the NHL as of the end of the 1997–1998 season. We also entered career scoring and penalty data for non-goaltenders and other career statistics for goalies, but we preferred to use career games as our performance measure as discussed in the section "Using career games . . . " These data are available in QuattroPro format from the authors upon request.

Team data have been consolidated in five cases to allow for team shifts which have occurred over the sample period (see Table 30.1). For example, the Kansas City Scouts, Colorado Rockies, and New Jersey Devils are considered to be one drafting unit.

We have the NHL career statistics as of the end of the 1997–1998 season for all players drafted from 1963–1998. Consequently, we have no NHL statistics for players drafted in 1998. Only a few players drafted in 1996 and 1997 had played in the NHL as of 1998, and they had played only a small number of games. Therefore we will only include players drafted up to and including 1995 in our analysis. Also, we only include players drafted in 1969 or after, for reasons given in the next section.

Table 30.1 ***Franchise drafting success.***

Rank	Franchise	Actual Games	Predicted Games	Percent Difference
1	Quebec/Colorado	22,365	18,264	22.5
2	Buffalo	39,323	34,055	15.5
3	Winnipeg	21,683	18,894	14.8
4	N.Y. Rangers	39,621	34,949	13.4
5	San Jose	3,141	2,783	12.9
6	Florida	1,212	1,081	12.2
7	Edmonton	17,475	16,420	6.4
8	Montreal	53,019	50,000	6.0
9	Ottawa	1,769	1,669	6.0
10	N.Y. Islanders	33,100	31,397	5.4
11	Boston	33,489	32,191	4.0
12	Atlanta/Calgary	30,719	30,395	1.1
13	Detroit	34,568	34,397	0.5
14	Chicago	33,033	32,890	0.4
15	Hartford	17,470	17,434	0.2
16	Toronto	34,408	34,456	–0.1
17	Anaheim	1,014	1,051	–3.5
18	Kansas City/Colorado/New Jersey	27,064	28,185	–4.0
19	Tampa Bay	1,662	1,757	–5.4
20	Washington	25,448	26,905	–5.4
21	Pittsburgh	27,323	29,505	–7.4
22	Los Angeles	24,689	26,847	–8.0
23	Vancouver	29,905	32,588	–8.2
24	Philadelphia	30,019	33,450	–10.3
25	St. Louis	24,739	28,617	–13.6
26	California/Oakland/Cleveland	9,931	12,986	–23.5
27	Minnesota/Dallas	25,939	35,633	–27.2

Background

The following information is drawn from Tredree and Bontje (1998) and Meagher (1998). The first Amateur Draft conducted by the National Hockey League was held in Montreal on June 5, 1963. Prior to this time there was no organized draft. Franchises, on their own initiative, signed young players on a "C-form." This effectively gave the franchise control of the player's hockey-related destiny for his life. Even after retirement, a player had to seek the franchise's permission to look for coaching or administrative positions with other NHL clubs. The Amateur Draft and sponsorship arrangements coexisted until 1969, when sponsorship was terminated. From 1969 on, the vast majority of players entering the NHL went through the Amateur Draft, which was renamed the Entry Draft in 1979. For this reason, we consider only draft data from 1969 and after.

In 1963, the six league members drafted 21 players. As the league expanded to 28 teams, so did the number of players drafted. In 1998, the last year in our dataset, 258 players were drafted. The focus of the draft has also expanded from Canadian players to ones playing anywhere in the world.

The rules of the drafting system have changed over time. The age of eligibility has varied from 17 to 20 but has been set at 18 since 1980. The exclusive right of the Montreal Canadiens to draft the best players from Quebec effectively expired in 1969, the last year they exercised it. The order of drafting varied in the early years but it was codified in 1969—the order of selection is based on the reverse order of standing in the league in the previous year. Since 1995, the first round has been an exception—a weighted lottery system has been introduced for the top 14 picks (Meagher 1998, p. 132). (Teams can, however, improve their drafting position by means of trades.) In 1969 it was decided that the drafting process would continue until every franchise declared that its participation was complete. Currently it consists of nine rounds plus up to one round of compensatory draft selections.

Using career games played as the performance measure

We use total NHL regular season games played as a performance measure for all three positions: forwards, defensemen, and goalies. In work not reported here (Dawson and Magee 2000), we also use total NHL regular season points as a performance measure for defensemen and forwards.

Career games played is not a perfect measure of a player's value, but we feel that it is the best statistical measure available from our data source. Other available measures such as points per game or points per season are misleading for players who have not played many games. These measures also do not reflect longevity. The length of a player's career indicates positive characteristics that may not be reflected in his scoring totals, or in his points per game or per season, such as the quality of his defensive play, physical durability, and other intangibles.

The use of total games played could be criticized on the grounds that, barring injury, most established NHL players at a given position play roughly the same number of games per year. The only exception is first-string versus second-string goalies. A star and a mediocre player could log a similar number of games in a season. However, we are comparing the success of all *drafted* players, not only players who become NHL regulars. The majority of drafted players do not play even a single game in the NHL. A drafted player who enjoys a lengthy or even a medium-length NHL career, even if he not considered a star, is indeed a very successful draft selection when compared to the group of all drafted players.

The model

We model the expected value of games played as a function of position, pick number, and draft year. The parameters are estimated separately for goaltenders, defensemen, and forwards. The following notation then applies separately to each of these three positions. To simplify notation we do not include a position subscript. Let p_{it} be the career NHL regular season games played as of 1997–1998 of a player who was drafted ith overall in year t. We seek a model of the conditional mean of p_{it} given i and t, $E(p_{it}|i, t)$.

We considered a range of nonlinear additive effect models. We modeled $(p_{it})^{\lambda}$ for various choices of λ, including data-driven ones, as an additive function of several powers of the pick number, and individual year effects. All specifications were found to be inadequate when tested using a variant of Ramsey's (1969) RESET test for specification error.

Given this failure to find an acceptable regression model, we turned to a multiplicative nonparametric model. The conditional mean is modeled as

$$E(p_{it}|i, t) = \pi_i \tau_t, \tag{30.1}$$

where π_i is the pick number effect and τ_t is the year effect. Since one expects that more promising players would be selected earlier in the draft, then as i increases, expected p_{it} (career games) should decrease. Therefore, we assume that π_i is nonincreasing in i. In other words, a later pick cannot be better, on average, than an earlier pick. This motivates an isotonic regression model with respect to π_i.

We model the year effect as multiplicative rather than additive since we expect the absolute size of the year effect to be greater for larger π_i. An additive isotonic regression model was developed by Bacchetti (1989).

Variation in τ_t over time can be expected for several reasons. Players drafted near the end of the 1969–1995 period will tend to have a lower p_{it} since their NHL careers as of the end of the 1997–1998 season will have been shorter than those of the older players. Players entering the NHL earlier in the 1969–1995 period were entering a league with fewer teams than those entering later. The number of games in the regular season changed over this period. Finally, through chance and other factors, some years had more good players in the draft than others.

A normalizing restriction is required. We restrict the average of the 27 τ_ts to equal one. π_i is then the expected value of p_{it} for a player drafted ith overall in an average year, when $\tau_t = 1$.

Once π_i and τ_t have been estimated for each position by $\hat{\pi}_i$ and $\hat{\tau}_i$, respectively, the three issues in the introduction can be addressed. The change in $\hat{\pi}_i$ over i tells us about the average performance of players across different pick numbers. The second and third issues can be explored by computing $\hat{p}_{it} = \hat{\pi}_i \hat{\tau}_t$, where $\hat{p}_{it}$ is the expected performance of a player drafted ith overall in year t. The residual, $p_{it} - \hat{p}_{it}$, tells us how that player actually performed compared to this prediction. These residuals can be added up within franchises to address the second issue. The third issue can be explored by summing these residuals over European players.

Estimation of pick number and year effects

π_i and τ_t for each position are estimated by least squares, minimizing

$$\sum_t \sum_i I_{it}(p_{it} - \pi_i \tau_t)^2 \tag{30.2}$$

subject to the restrictions $\pi_i \geq \pi_{i+1}$, $i = 1, \ldots, 286$, where 286 is the highest number of picks in any year in the sample. I_{it} is an indicator variable which equals one if a player was drafted ith overall in year t who plays at the position whose coefficients are being estimated, and equals zero otherwise.

We consider the estimation of $\hat{\pi}_i$ and $\hat{\tau}_t$ separately, which leads to a zigzag algorithm. Given $\hat{\pi}_i$, minimization of (30.2) with respect to τ_t yields:

$$\hat{\tau}_t = \sum_i I_{it} p_{it} \hat{\pi}_i \Big/ \sum_i I_{it} \hat{\pi}_i^2 \tag{30.3}$$

for each t.

To see how we obtain $\hat{\pi}_i$ given $\hat{\tau}_t$, replace τ_t with $\hat{\tau}_t$ and expand the square in (30.2):

$$\sum_i \sum_t I_{it} \left[p_{it}^2 - 2 p_{it} \pi_i \hat{\tau}_t + \pi_i^2 \hat{\tau}_t^2 \right].$$

Summing each term over t, dropping the first term, factoring out $\left(\sum_t I_{it} \hat{\tau}_t^2\right)$ and completing the square gives

$$\sum_i \left(\sum_t I_{it} \hat{\tau}_t^2 \right) \left[\left(\sum_t I_{it} \hat{\tau}_t p_{it} \Big/ \sum_t I_{it} \hat{\tau}_t^2 \right) - \pi_i \right]^2 = \sum_i w_i (b_i - \pi_i)^2, \tag{30.4}$$

where $w_i = \left(\sum_t I_{it} \hat{\tau}_t^2\right)$ and $b_i = \sum_t I_{it} \hat{\tau}_t p_{it} \big/ \sum_t I_{it} \hat{\tau}_t^2 = \sum_t \left(I_{it} \hat{\tau}_t^2\right) \left(p_{it}/\hat{\tau}_t\right) \big/ \sum_t I_{it} \hat{\tau}_t^2$. Minimizing the sum of squares with respect to π_i can be seen from (30.4) to correspond to choosing π_i to minimize the weighted sum of squared differences between b_i and π_i, where the weights are the w_i's. The last expression for b_i above shows that b_i itself is a weighted average of the year-adjusted performances, $p_{it}/\hat{\tau}_t$, of the players we wish to include in the sum, for whom $I_{it} = 1$. The weights in (30.4), w_i, are the sums of the squared year effects of the players we wish to include in the sum who were drafted in position i. Thus, the sum in (30.4) places more weight on draft positions in which many players were drafted in high-performance years.

We use a weighted pool-adjacent-violators (WPAV) algorithm (Barlow et al. 1972, p. 13), to obtain estimates of π_i from (30.4). This is an extension of the (unweighted) pool-adjacent-violators algorithm (PAV or PAVA) also discussed by Ayer et al. (1955), Hanson, Pledger, and Wright (1973) and Härdle (1990, pp. 218–223).

PAVA divides the original vector of pick-number performance indicators $\{b_1, b_2, \ldots, b_{286}\}$ into blocks, and replaces all members of each block by the average of the b_i's in that block. The blocks are chosen so that the members of the resulting vector, which are the $\hat{\pi}_i$'s, are nonincreasing in i. PAVA accomplishes this by starting with each b_i representing its own block. It then looks for violations of this monotonicity requirement, pools the violators into a single block, and replaces the original b_i's in that block by their average. WPAV generalizes PAVA by replacing the original b_i's instead by their *weighted* average, using the w_i's as weights.

Our WPAV algorithm moves from $i = 1$ to $i = 286$ in this manner, checking for violators on either side of any newly defined block along the way. For example, suppose the current block runs from $i = 45$ to $i = 51$, and the values of $\hat{\pi}_{45}, \ldots, \hat{\pi}_{51}$ are set equal to their weighted average of 63.2. If the adjacent block to the left, $i = 42$ to $i = 44$, say, has $\hat{\pi}_{42} = \hat{\pi}_{43} = \hat{\pi}_{44} = 61.8$, then monotonicity does not hold and this adjacent block is a violator. This is dealt with by pooling these two blocks into a single block running from $i = 42$ to $i = 51$, containing $\hat{\pi}_i$ values set equal to the weighted average of all members of this new larger block.

Our overall zigzag algorithm is then:

1. Set the year effects $\hat{\tau}_t = 1$ for all t.
2. Minimize (30.2) with respect to the $\hat{\pi}_i$s by minimizing (30.4) using the WPAV algorithm.
3. Minimize (30.2) with respect to the $\hat{\tau}_t$'s using (30.3).
4. Repeat Steps 2 and 3 until convergence.
5. Normalize the $\hat{\tau}_t$'s by dividing by their average, and multiply the $\hat{\pi}_i$'s by the same number.

Pick number and year effects

Figure 30.1 plots the year effects ($\hat{\tau}_t$'s) for each position, using games played as the performance measure p_{it}. Because fewer goaltenders are drafted than forwards or defensemen, there is more random year-to-year variation, causing the goaltenders' plot to be bumpier. There is a slight increase during the 1970s, a spike in 1979 for forwards and defensemen (1980 for goaltenders), a high period in the early 1980s, and finally a gradual decline from the mid-1980s onward. The spike in 1979 could be due to eligibility of some WHA talent in the entry draft, as the rival WHA league amalgamated with the NHL that year. The post-1984 decline is explained by some players still being in mid-career by the time the statistics end in 1998. The slightly higher level in the early 1980s compared to the 1970s could be explained by the increase in the number of teams from 17 to 21 in 1979. A similar pattern emerges when the plot uses points scored as the performance measure for forwards and defensemen, not shown here.

Figure 30.2 plots the pick number effects on games played for the three positions. The plots for forwards and defensemen are remarkably similar. The goaltenders' plot has longer flat sections, again partly reflecting the smaller sample size for goaltenders. A goaltender drafted

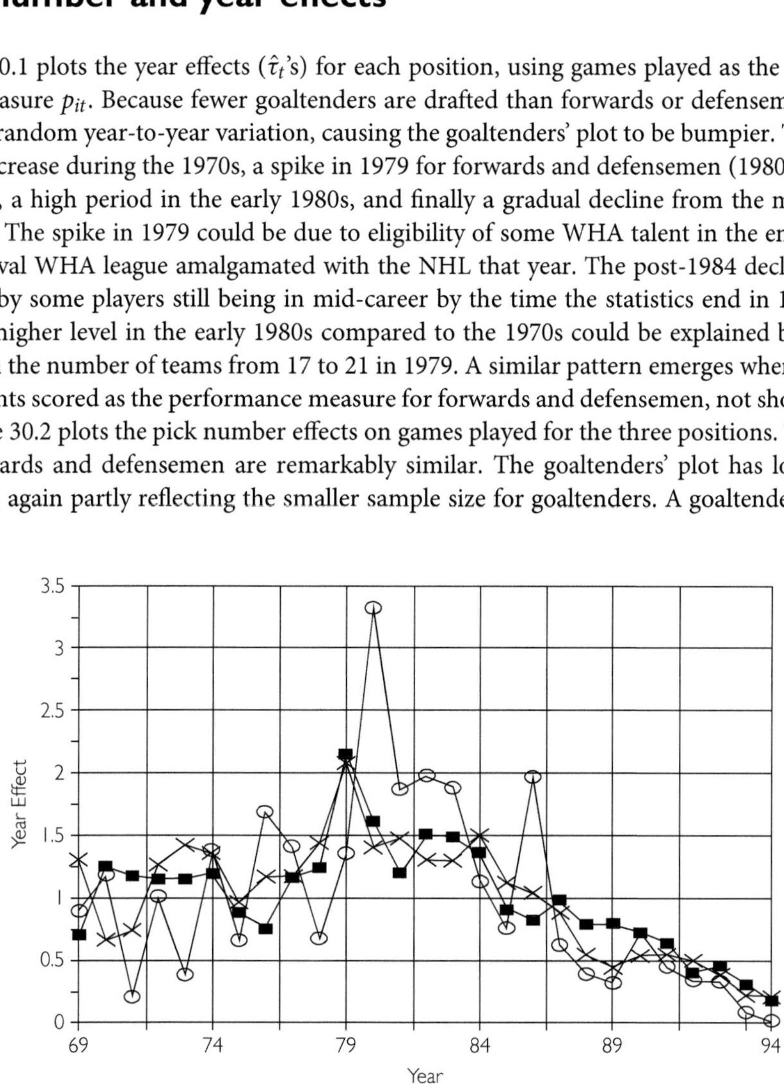

Figure 30.1 Year effects on games played. The year effects ($\hat{\tau}_t$s) are plotted against time, for forwards (solid squares), defensemen (Xs), and goalies (circles). The level of each plot has been normalized so that the average value is one.

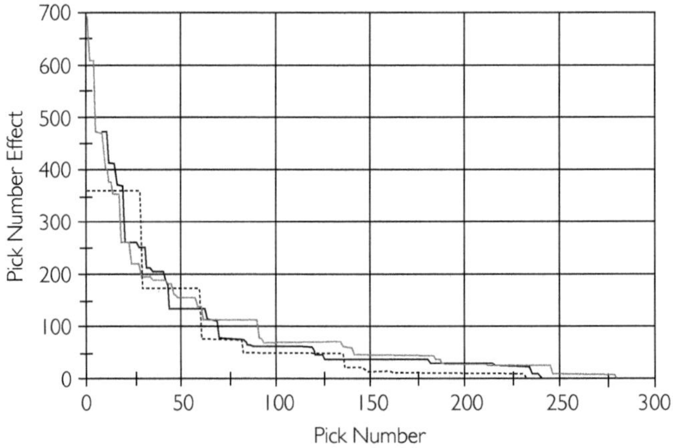

Figure 30.2 Pick number effects on games played. The pick number effects ($\hat{\pi}_i$s) are plotted against the pick number, i, for forwards (thick solid line), defensemen (thin solid line), and goalies (dashed line).

very high, say in the top 10, on average plays fewer games than do forwards and defensemen who were drafted that high. However, goaltenders drafted with a pick number in the early 20s play more games on average than the other players. Since regular NHL non-goalies tend to play more games per season than goalies, this finding implies that goalies drafted that high tend to have longer NHL careers than non-goalies. (We are looking at the expected number of career games of drafted players, many of whom do not play in the NHL at all. 57.2% of the players drafted during this 1969–1995 period had not played any games in the NHL as of the end of the 1997–1998 season. The maximum number of games in this dataset is 1,432 by Mike Gartner, a forward drafted number 4 in 1979.) Note that, even though about 10% of all drafted players are goaltenders, they comprise only 3.3% of the players drafted in the top 20 overall.

Franchise drafting performance

Table 30.1 lists the total number of predicted games played and actual games played of all players drafted during this period by franchise. Franchises are ranked by the percentage difference between actual and predicted games played.

The results for franchises that began operating in the 1990s must be interpreted with caution. They have drafted a small number of players, and those players have had at most only a short stint in the NHL by 1997–1998. It is possible that the better-than-average overall drafting performance of some the new franchises is due to their having had to rush their drafted players into the NHL sooner than have the more established franchises. Complete year-by-year franchise drafting records are available upon request.

Montreal has done for more drafting than any of the other franchises, as measured by predicted games. In terms of the differences between actual and predicted performance, the best drafting franchises have been Quebec/Colorado, Buffalo, Winnipeg (now Phoenix), and the New York Rangers. The only defunct franchise, California/Oakland/Cleveland, has a poor drafting record. Minnesota/Dallas, however, has the worst drafting record by a wide margin using this measure.

Average performance of players drafted from European teams

One might wish to divide players according to whether they were raised in Europe or in North America. However, we only know whether the team that the player belonged to when they were drafted was located in Europe or North America. In the vast majority of cases, this team location will correspond with the player's origin, so we will refer to a player drafted from a European team as European. In recent drafts, though, there is a small but growing number of European players drafted from Canadian junior teams.

We will not distinguish between Canadian and U.S. players. In work not reported here, the difference in performance of players drafted from Canadian versus U.S. teams was small compared to the differences between Europeans and North Americans described in the following.

In Figures 30.3, 30.4, and 30.5, we compare the total number of NHL games played by the European players to their expected number of games as defined by our model. Both are expressed as a percentage of total games played by all players drafted that year at that position. If the actual plot lies above the predicted, it indicates that the European players were undervalued in the sense that they did better than the model "expected" them to given when they were picked.

The European forwards and defensemen drafted in the 1988–1992 period have done much better than predicted. This is also a period in which the Europeans' predicted percentages of games played increased rapidly, possibly as franchises tried to catch up to the fact that these players were doing very well.

This persistent gap did not exist prior to 1988. It seems that, on average, franchises were not using any special knowledge to draft good European players "hidden" late in the draft before

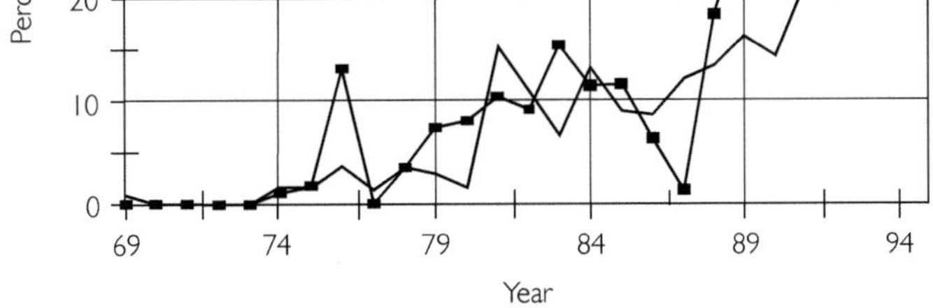

Figure 30.3 European forwards, games played. The actual (with solid squares) and predicted (no squares) games played by all European forwards drafted in a given year, expressed as a percentage of the total games played by all forwards drafted that year, are plotted against year.

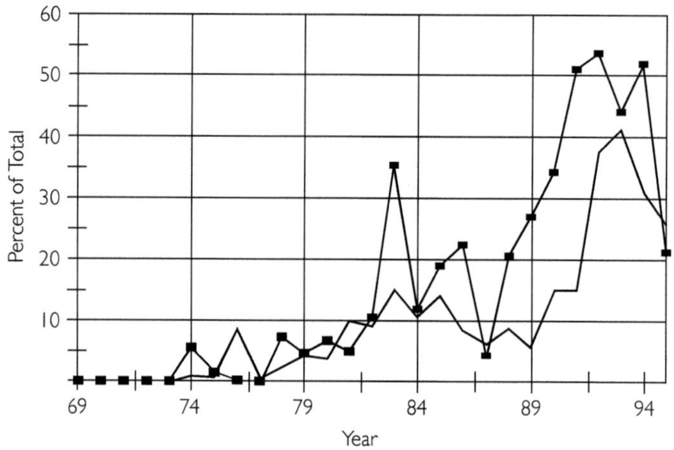

Figure 30.4 European defensemen, games played. Same as Figure 30.3, but for defensemen.

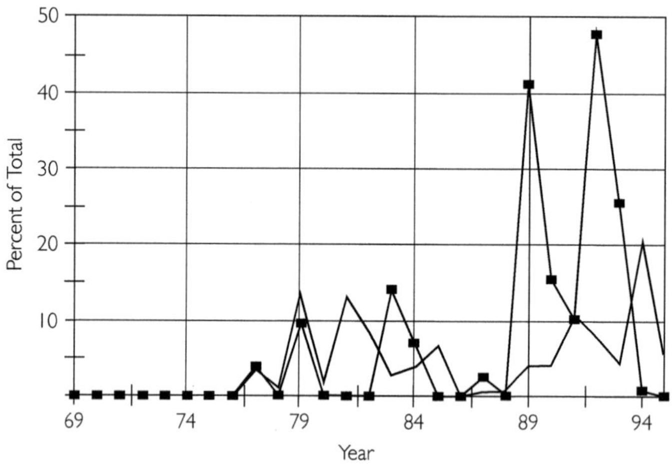

Figure 30.5 European goalies, games played. Same as Figure 30.3, but for goalies.

1988. Another possibility is that European players drafted pre-1988 tended to be older than the North American draftees, reducing their expected career games. The gap in the figures narrows after 1992, but since these more recently drafted players had not had much NHL experience as of the 1997–1998 season, this narrowing may or may not be meaningful.

This surge in Europeans' performances beginning in the 1988 draft year may be a result of a higher fraction of drafted players coming to North America from Europe than previously, due to political developments in Europe and escalating NHL salaries. As franchises realized that this probability had increased, they started choosing European players higher in the draft, causing the predicted plots in Figures 30.3 and 30.4 to increase in the early 1990s.

The 1988–1992 period also saw European goaltenders drafted who performed much better (i.e., played more games) than expected, although the year-to-year pattern, shown in Figure 30.5, is not as consistent as it is for non-goaltenders.

Conclusion

We have modeled the expected performance of hockey players selected in the NHL Entry Draft as the product of a year effect and a "pick number" effect, which is a nonincreasing function of their pick number, or rank in the draft. Performance is measured as the number of career regular season games played.

The estimated expected games played of forwards and defensemen are very similar functions of the pick number. This function is steeply declining at early picks and becomes flatter at later picks. This function is somewhat different for goalies. Buffalo, Quebec/Colorado, Winnipeg (now Phoenix) and the New York Rangers have drafted especially well, and Minnesota/Dallas has not. Europeans drafted during 1988–1992 performed much better than the model predicted.

The model is estimated by a least-squares algorithm that combines a weighted pool-adjacent-violators algorithm with multiplicative year effects. The algorithm is simple to program, converges quickly, and imposes a plausible monotonicity restriction on the pick number effect without further restricting the shape of the function.

The multiplicative year effects allow for the pick number function to be applied to various years of the draft, including those more recent years in which many of the successful players' careers were still in progress when the data were collected. In work not reported in this paper, we estimated the model separately for players drafted in 1969–1981 and 1982–1995. The shapes of the two estimated pick number functions were very similar, providing some support for the model. This model could be used to study drafts in other sports or leagues, and possibly in other contexts.

REFERENCES

Ayer, M., Brunk, H. D., Ewing, G. M., Reid, W. T., and Silverman, E. (1955), "An empirical distribution function for sampling with incomplete information," *The Annals of Mathematical Statistics*, 26, 641–647.

Bacchetti, P. (1989), "Additive isotonic models," *Journal of the American Statistical Association*, 84, 289–294.

Barlow, R. E., Bartholomew, D. J., Bremner, J. M., and Brunk, H. D. (1972), *Statistical Inference Under Order Restrictions*, London: Wiley.

Dawson, D. A., and Magee, L. (2000), "The National Hockey League Entry Draft, 1969–1995: An application of a weighted pool-adjacent-violators algorithm," Department of Economics Working Paper 2000–04, McMaster University.

Hanson, D. L., Pledger, G., and Wright, F. T. (1973), "On consistency in monotonic regression," *The Annals of Statistics*, 1, 401–421.

Härdle, W. (1990), *Applied Nonparametric Regression*, Cambridge: Cambridge University Press.

Meagher, G. (1998), "Inside the National Hockey League. Off-ice action: understanding the business side of hockey," in *Total Hockey*, eds. D. Diamond, J. Duplacey, R. Dinger, I. Kuperman, and G. Zweig, Kansas City, Andrews McMeel Publishing.

Neale, W. (1964), "The peculiar economics of professional sports," *Quarterly Journal of Economics*, 78, 1–14.

Ramsey, J. B. (1969), "Tests for specification errors in classical linear least squares analysis," *Journal of the Royal Statistical Society*, Series B, 31, 350–371.

Tredree, C., and Bontje, P. (1998), "The NHL Entry Draft," in *Total Hockey*, eds. D. Diamond, J. Duplacey, R. Dinger, I. Kuperman, G. Zweig, Kansas City: Andrews McMeel Publishing.

CHAPTER 31

Bowlers' hot hands

REID DORSEY-PALMATEER AND GARY SMITH

Abstract
Earlier analysis of basketball data debunked the common perception that players sometimes have "hot hands." That analysis, however, did not control for several confounding influences. Our analysis of professional bowling indicates that, for many bowlers, the probability of rolling a strike is not independent of previous outcomes and the number of strikes rolled varies more across games than can be explained by chance alone. For example, most bowlers have a higher strike proportion after *j* consecutive strikes than after *j* consecutive nonstrikes, and this difference becomes more pronounced as *j* increases from 1 to 4.

Keywords: *bowling, self-efficacy*

Bowlers' hot hands

The binomial model is often applied to athletic endeavors even though the model is surely an imperfect representation of athletic performances. Unlike a fairly flipped coin, athletes compete under varying game conditions and have aches, pains, and other human frailties. A baseball player's probability of getting a hit depends on the opposing pitcher. A basketball player's probability of making a basket depends on the location from which the shot is taken. Every athlete's performance is affected by physical health.

One interesting question is whether athletes' success probabilities are affected by physical or psychological factors of the sort described by Purvis Short, a professional basketball player: "You're in a world all your own. It's hard to describe. But the basket seems so wide. No matter what you do, you know the ball is going to go in." This quotation prefaces the second of three hot-hands articles based on research by Gilovich, Vallone, and Tversky (Gilovich, Vallone, and Tversky 1985; Tversky and Gilovich, 1989a, 1989b). These articles argue that the common belief that players sometimes get "hot" or "get in a zone" is erroneous, in that memorable performances happen no more often than would be predicted by the binomial model.

We argue that the basketball data used by Gilovich, Vallone, and Tversky have serious flaws and that statistical tests often have little power to detect hot hands. Our analysis of bowling data—a sport largely free of confounding influences—indicates that, for many bowlers, the probability of rolling a strike is not independent of previous outcomes and the number of strikes rolled varies more across games than can be explained by chance alone.

Good data

The primary data analysed by Gilovich, Vallone, and Tversky are the shooting accuracy of Philadelphia 76er basketball players during the 1980–1981 season. The probability of making a shot was usually somewhat *lower* after having made shots than after having missed shots, and the number of runs of hits and misses by individual players across all games and within individual games was slightly larger than would be expected if shots were independent (the opposite of the hot-hands phenomena), though these patterns were seldom statistically persuasive. For example, only one of nine runs tests had a p value less than 0.05, and this was for a player who had more runs than expected.

Unfortunately, their data do not identify how much time passed between shots. A player's two successive shots might be taken 30 seconds apart, 5 minutes apart, in different halves of a game, or even in different games. Another problem is that the shots a player chooses to take may be affected by his recent experience. A player who makes several shots may be more willing to take difficult shots than is a player who has been missing shots. In addition, the opposing team may guard a player differently when he is perceived to be hot or cold. Shot selection may also be affected by the score and the number of fouls accumulated by players on both teams.

The hot-hands question asks whether highly skilled and motivated athletes sometimes encounter hot and cold spells that are not due to confounding influences and cannot be easily explained as chance fluctuations about a constant success rate. Smith (2003) examined horseshoe pitching data from the 2000 and 2001 World Championships and finds variations in player performances within games and across games that indicate that success probabilities are not constant. Championship horseshoe data are much cleaner than basketball data in that every pitch is made from the same distance at regular, brief intervals with intense concentration and little or no strategy. Bowling is a sport with similar statistical characteristics. (Remarkably, Walter Ray Williams Jr., is one of the top athletes in both sports, having won six world horseshoe pitching championships and six professional bowling player of the year awards.)

Bowling data

Bowling lanes are slightly less than 63 feet long and between 41 and 42 inches wide, with gutters on either side to catch balls that fall off the lane. Ten target pins are arranged in a pyramid shape with 60 feet from the start of the lane to the pin at the pyramid tip closest to the bowler.

Each game has 10 frames. In tournament match play, two bowlers use two lanes to compete against each other. One bowler starts on the left lane and the other starts on the right. The player on the left bowls his first frame on the left lane. After that, each bowler (starting with the bowler on the right) bowls two frames, the first on the right lane and the second on the left lane, until the end, when the bowler who went first bowls his tenth frame on the right lane.

The ten pins are reset at the start of each frame. If a bowler knocks all ten pins down on the first throw, this is called a strike. If fewer than ten pins are knocked down, the fallen pins are cleared away and the bowler is given a second chance to hit the remaining pins. If the remaining pins are all knocked down on the second roll, this is called a spare; otherwise it is an open frame.

The bowler's base score in each frame is equal to the number of pins knocked down in that frame. If there is a spare, the score is increased by number of pins knocked down on the next roll; if there is a strike, the score is increased by number of pins knocked down on the next two rolls. If a spare or strike is rolled in the tenth frame, the bowler is allowed bonus shots (one for a spare and two for a strike) on that lane. A perfect game consists of 12 strikes (ten frames plus two bonus rolls), which gives a score of 300. Professional bowlers roll strikes around 60% of the time and average more than 200 points per game.

Professional Bowlers Association (PBA) tournaments start with 120 to 175 bowlers who bowl nine games in the first round. The top 64 from this round advance to the next round and again bowl nine games. The top 32 advance to the match-play rounds and are seeded based on their performance in the first 18 games. The Round of 32 consists of 4-of-7 match-play games between pairs of bowlers, with the winners advancing to the Round of 16, which is 3-of-5 match play. The Round of 8 is also 3-of-5. After the Round of 8, there is the Final Round, in which all games are single-elimination. In most tournaments, there is a wild-card match between the Round-of-8 loser with the best win-loss record in the tournament and the Round-of-8 winner with the worst win-loss record. The wild-card winner gets a place in the semifinals, and the winners of the two semifinal matches compete in the championship match.

In comparison to basketball data, bowling data have many valuable properties for testing violations of the binomial model. Every roll is made from the same distance at regular intervals with intense concentration and no strategic considerations. One possible confounding influence is the condition of the lanes—in particular, the way the lanes are oiled. Oil is usually applied to the first 35-to-45 feet of the lane with the last 15-to-25 feet being dry. The balls gradually wear down the oil and also cause the oil to move down the lane, which can affect a ball's trajectory. Years ago, the lanes were oiled only once a day and lane conditions could change substantially during the day. Left-handed bowlers were thought to have an advantage because the more numerous right-handed bowlers caused more serious changes in oil conditions on the right side of the lanes. Today, the lanes are stripped and oiled before each round in order to stabilize lane conditions. In PBA match-play rounds a maximum of seven games are rolled on a lane before it is re-oiled. One indicator of the improved stability of lane conditions is that only 1 of the top 10 money winners in 2002–2003 was left-handed.

Frame, Hughson, and Leach (2003) find evidence of hot hands in bowling using data from the Final Round of 1994–1998 PBA tournaments. During these years, the Final Round used a stepladder format in which the fifth-seeded and fourth-seeded bowlers played each other, with the winner moving on to play the third-seeded bowler. The winner of this match played the second-seeded bowler and the winner of this match played the first-seeded bowler. They find that the winners of each game won more than 50% of their next games, even though they were playing against higher seeds. This pattern may have something to do with the higher seeds sitting idly while the lower seeds bowl continuously.

Our data do not involve stepladder matches and, in addition, we look at each bowler's performance within games. Detailed frame-by-frame results for the 2002–2003 season are available from the PBA (2003) for all match-play games: the Round of 32, Round of 16, Round of 8, and Final Round.

The binomial model

We characterize each bowler's performance in each frame (or bonus shot) as either a strike or nonstrike, analogous to a coin flip with a success probability that is not necessarily 0.50. We use this notation:

p_i = *player i*'s probability of rolling a strike conditional on having just rolled *j* consecutive strikes.
q_{ij} = player *i*'s probability of rolling a strike conditional on having just rolled *j* consecutive nonstrikes.

The binomial model assumes, for each player *i*, that the trials are independent, so that

$$p_{ij} = q_{ij} = p_i \quad \text{for all} \quad j, \tag{31.1}$$

and that the success probability p_i is stationary. In the context of bowling, each player's chances of rolling a strike would (a) not depend on whether he has rolled strikes or nonstrikes in other frames; and (b) be the same in every frame. The success probability p_i can, of course, vary across bowlers.

Independence

One violation of the binomial model that might be interpreted as a hot hand occurs when an athlete's success probability depends on the outcomes of previous trials; for example, bowler *i*'s success probability might be higher after a success than after a failure ($p_{i1} > q_{i1}$) or higher after two consecutive successes than after two consecutive failures ($p_{i2} > q_{i2}$). Tversky and Gilovich (1989b) focused on independence as the key assumption disputed by those who believe in hot hands: "Many observers of basketball believe that the probability of hitting a shot is higher following a hit than following a miss, and this conviction is at the heart of the belief in the 'hot hand.'"

Stationarity

A very different violation of the binomial model that might be interpreted as a hot hand occurs when there is a sustained change in the success probability. For instance, bowler *i*'s success probability p_i might be 0.5 for 30 trials, then abruptly change to 0.7 for 10 trials, and then return to 0.5 for 60 trials. If such changes are unrelated to the outcomes of other trials, the binomial model's independence assumption is not violated. The hot hand described by Purvis Short seems to involve nonstationarity rather than nonindependence. If hot hands do occur, there may also be cold hands when an athlete's success probability temporarily falls.

Frame, Hughson, and Leach (2003) showed that the runs test used by Gilovich, Vallone, and Tversky has little power to detect nonstationarity in a regime-shifting model in which a player has a specified probability of switching back and forth between hot-shooting and cold-shooting regimes. For instance, a player might always have a 0.25 probability of switching between a hot regime with a 0.6 success probability and a cold regime with a 0.3 success probability.

Wardrop (1999) examined the power of several tests using a different model of nonstationarity. The athlete has a base success probability P_B (e.g., 0.5). At some random point in 100 Bernoulli trials, the athlete gets hot and the success probability rises to P_H for a specified number of trials (e.g., $P_H = 0.8$ for 10 trials). When the hot period ends, the athlete's success probability returns to P_B. The binomial model applies during each of the three subperiods in that the outcomes are independent and the success probability is constant within each subperiod. None of the tests Wardrop considered has much power to detect nonstationarity unless P_H is much larger than P_B.

Wardrop's model is an appealing representation of the phenomenon described by Purvis Short. We will investigate the power of several tests by using a generalized model in which bowlers have temporary hot and cold spells.

A simulation model

Our simulation model involves a single bowler who has temporary hot and cold spells. In our model, the bowler's probability of rolling a strike is usually P_B, but during every seven games there are randomly determined, nonoverlapping 10-frame hot and cold periods with respective success probabilities P_H and P_C. We use $P_B = 0.60$, which is approximately equal to the overall 0.59 strike proportion in our dataset, and two pairs of hot and cold probabilities: $P_H = 0.7$, $P_C = 0.5$, and $P_H = 0.8$, $P_C = 0.4$. For comparison, Walter Ray Williams Jr., the top bowler in 2002–2003, had a 0.66 strike proportion in his match-play games. For a PBA match-play bowler, a change in the strike probability from 0.6 to 0.4 or 0.8 seems enormous, but perhaps not impossible.

In each game the bowler has a 10% chance that the tenth frame will be open (neither a strike nor a spare), which is the proportion of the time this event occurred in our data.

We consider three tests which we illustrate with the following one-game simulation, where the result in each frame is characterized as either a strike, s, or nonstrike, n:

n n s s n s s s s s n.

The first test is based on the conditional strike frequencies: a tabulation of the number of strikes and nonstrikes immediately after a strike or nonstrike. In our example,

	current frame		
previous frame	strike	nonstrike	total
strike	5	2	7
nonstrike	2	1	3
total	7	3	10

Fisher's exact test uses the hypergeometric distribution to test the null hypothesis that the conditional strike probability does not depend on whether there was a strike or nonstrike in the previous frame; that is, Equation (31.1) with $j = 1$. Here with 10 previous frames, 7 of which contain strikes, the exact probability that 7 current frames chosen at random contain 5 or more strikes is 0.708.

The second test is a nonparametric test based on the number of runs, where a run is a sequence of successive strikes or non-strikes. (Gilovich, Vallone, and Tversky reported both a runs test and a serial correlation test, but Wardrop showed that in practice these two tests give virtually identical results.) For a fixed number of successes and failures, Stevens (1939) gave the formula for calculating the exact probability that the number of runs would be equal to or smaller than the value actually observed (indicating the presence of hot and cold streaks) for the null hypothesis that the given number of successes and failures are randomly arranged. In our example, there are seven strikes, four nonstrikes, and five runs. If these seven strikes and four nonstrikes were randomly arranged, the exact (one-sided) probability that there would be five or fewer runs is 0.333.

The third test is a nonparametric test based on the length of the longest run. For a fixed number of successes and failures, Bateman (1948) gave the formula for calculating the exact probability that the length of the longest run would be as long as actually observed for the null hypothesis that the given number of successes and failures are randomly arranged. In our example, the length of the longest run is equal to five strikes in a row. If the seven strikes and four nonstrikes were randomly arranged, the exact (one-sided) probability that there would be at least one run of length five or longer is 0.227.

The seven-game simulation model was replicated 10 million times for each of the three scenarios: $P_H = P_C = 0.6$; $P_H = 0.7$, $P_C = 0.5$; and $P_H = 0.8$, $P_C = 0.4$. The three statistical tests were applied to the data generated by each simulation and a tabulation was made of the frequencies with which the three statistical tests yielded p values less than or equal to 0.05. Results are shown in Table 31.1 for tests based on the first game, the first three games, the first five games, and all seven games in each simulation. The large number of simulations allows the reporting of the p value proportions to three decimal places. If the success probability is

Table 31.1 ***Simulation results. Proportion of p values less than 0.05 when the 0.6 success probability increases to P_H = 0.6 + D for 10 frames and falls to P_C = 0.6—D for 10 frames in every seven games.***

D	1 Game			3 Games			5 Games			7 Games		
	0.0	0.1	0.2	0.0	0.1	0.2	0.0	0.1	0.2	0.0	0.1	0.2
Fisher's exact test	0.009	0.009	0.009	0.019	0.019	0.019	0.025	0.025	0.026	0.028	0.028	0.030
Number of runs	0.025	0.025	0.025	0.034	0.035	0.037	0.038	0.040	0.045	0.040	0.042	0.051
Longest run	0.016	0.016	0.016	0.018	0.019	0.021	0.018	0.019	0.025	0.019	0.021	0.030

p values are based on these tests:
Fisher's exact test: compares the conditional strike and nonstrike frequencies after a strike or nonstrike in the preceding frame
Number of runs: exact probability that the number of runs would be at least as small as actually observed if the observed strikes and nonstrikes were arranged randomly.
Longest run: exact probability that the length of the longest run would be at least as long as actually observed if the observed strikes and nonstrikes were arranged randomly.

around 0.05, the simulation standard error for the sample success proportion is approximately equal to $(0.05(0.95)/10,000,000)^{0.5} = 0.00007$.

The case $P_H = P_C = 0.6$ corresponds to the null hypothesis that the success probability is constant. Because discrete data yield discontinuous p values, the chances of obtaining a p value less than 0.05 is less—sometimes substantially less—than 0.05. The values $P_H = 0.7$, $P_C = 0.5$ and $P_H = 0.8$, $P_C = 0.4$ represent substantial changes in a bowler's strike probability, yet the frequency with which the null hypothesis is rejected at the 5% level is not much higher than when the null hypothesis is true (and is usually less than 5%). These simulations make it especially pertinent to remember that the failure of a hot-hands statistical test to reject the null hypothesis does not prove that the binomial model is true.

Tests of many athletes

If the binomial model is not appropriate, it may be because the performances of a few unique athletes are dramatically inconsistent with the model, or it may be because many (if not most) athletes experience hot and cold spells. Hot-hands studies generally examine the performance of several athletes and our interpretation of the results must be tempered by the number of tests conducted. For example, if each test has a 0.05 probability of Type I error, it will not be surprising if 5 of 100 independent tests have p values less than 0.05. On the other hand, it may be noteworthy if 10 tests have p values less than 0.05. In order to gauge the results of many tests, we can calculate a "p value for the p values" by testing the null hypothesis that, on each test, there is a 0.05 probability of obtaining a p value less than or equal to 0.05:

$$H_0 : P[p \text{ value} \leq 0.05] = 0.05. \tag{31.2}$$

If there are n independent tests, each with 0.05 success probability, the binomial distribution gives the probability that x or more tests would have p values less than 0.05. For example, with $n = 100$ and $x = 10$, the p value for the p values is $P[x \geq 10] = 0.0282$.

Our simulation model demonstrates that the interpretation of a count of the number of p values below a prespecified level is complicated by the fact that discrete data yield discontinuous p values. Consider a game with seven strikes and four nonstrikes. If strike and nonstrike frames are independent, the probability of at least one run of six or more frames is 0.076 and the probability of at least one run of seven or more frames is 0.015: the probability of a p value less than 0.05 is 0.015.

Now suppose that we look at 100 independent games with seven strikes and four nonstrikes and find five games with at least one run of seven or more. At first glance, it seems unsurprising that 5 of 100 tests have p values less than 0.05. However, if the probability of a p value less than 0.05 is actually 0.015, the binomial distribution tells us that, in 100 independent tests each with a 0.015 success probability, there is only a 0.0177 probability of 5 or more successes. The p value for our p values is 0.0177; 5 of 100 games with at least one run of seven or more is noteworthy!

Matters are considerably more complicated than this because the number of strikes and nonstrikes and the probability of a p value less than 0.05 vary from game to game. Sometimes the probability of a p value less than 0.05 is 0.015; sometimes it is 0.025. Another complication is that the number of strikes and nonstrikes may be such that the number of runs cannot possibly be statistically persuasive because the probability of a p value less than 0.05 is 0. For example, if a bowler rolls all strikes, the statistical fact that such data will always yield a longest run of 12 means that a test based on the length of the longest run has a p value of 1.0. Now suppose we have 100 tests of which 20 cannot possibly have p values less than 0.05. If we find that 8 of 100 tests have p values less than 0.05, it is really 8 of the 80 tests that can have p values less than 0.05.

Thus, if the null hypothesis is true, we will tend to find that fewer than 5% of the test results have p values less than 0.05. A corollary is that, if the null hypothesis is false, we may find that not many more (or even fewer) than 5% of the tests have p values less than 0.05.

Analysis

Frame-by-frame data from the 2002–2003 PBA match-play rounds are used to investigate whether bowling data are consistent with the binomial model's assumptions of independence and stationarity. Because the success probability can vary across bowlers, we analyse data for individual bowlers and we will often use a p value for the p values to determine whether more bowlers get hot hands than would be predicted by the binomial model.

Independence

The binomial model assumes that a player's probability of rolling a strike is independent of whether he rolled strikes or nonstrikes previously. Table 31.2 summarizes data based on the frequencies with which bowlers rolled strikes immediately after having rolled 1–4 consecutive strikes or 1–4 consecutive non-strikes. All of these data are from within games; we do not consider whether the first roll of a game is influenced by the last roll of the previous game. For each category, we consider only bowlers who had at least 10 observations in both situations. For example, there were 43 different bowlers who had at least 10 occasions in which they bowled after rolling four consecutive strikes and also had at least 10 occasions in which they bowled after rolling four consecutive nonstrikes.

The first data row in Table 31.2 shows n_j, the number of bowlers who had least 10 opportunities to bowl immediately following j consecutive strikes and also had least 10 opportunities to bowl immediately following j consecutive nonstrikes. The second and third data rows are the unweighted averages across bowlers of the conditional strike proportions after j consecutive strikes and j consecutive nonstrikes:

$$\frac{\sum_{i=1}^{n_j} f_{ij}}{n_j} \quad \text{and} \quad \frac{\sum_{i=1}^{n_j} g_{ij}}{n_j},$$

where

f_{ij} = player i's strike proportion immediately following j consecutive strikes.
g_{ij} = player i's strike proportion immediately following j consecutive nonstrikes.

Because these are unweighted averages, they are not affected by the fact that good bowlers have more opportunities to bowl after strikes than after nonstrikes. These unweighted averages can help us assess whether the observed differences in conditional strike proportions are substantial. For example, a 0.612 strike probability (the 83rd percentile for unconditional strike

Table 31.2 ***Strike proportions immediately following consecutive strikes or nonstrikes, PBA Tour match-play rounds, 2002–2003.***

	Number of consecutive strikes/nonstrikes			
	1	2	3	4
Number of bowlers	134	111	81	43
Unweighted mean conditional strike proportion				
after strikes	0.571	0.582	0.593	0.612
after nonstrikes	0.560	0.546	0.533	0.492
Number of bowlers with higher conditional strike proportion				
after strikes	80	77	59	34
after nonstrikes	54	33	22	9
Wilcoxon signed rank test for paired difference in conditional strike proportions				
mean difference	0.011	0.036	0.060	0.120
p value	0.020	0.000	0.000	0.000
Binomial test for number of bowlers with Fisher exact $p < 0.05$				
number of bowlers	10	13	8	7
p value for the p values	0.135	0.004	0.050	0.005

proportions among PBA match-play bowlers) is much better than a 0.492 strike probability (the 17th percentile).

The fourth and fifth data rows show the number of bowlers who had a higher strike proportion after j consecutive strikes than after j consecutive nonstrikes. For example, 34 of the 43 bowlers who had at least 10 opportunities to bowl after four consecutive strikes and after four consecutive nonstrikes had a higher strike proportion after four consecutive strikes. The higher conditional strike proportions after strikes are evidently not due to a few bowlers getting very hot, but rather to many bowlers having hot and cold spells.

Table 31.2 shows the p values for two statistical tests described below of the null hypothesis that Equation (31.1) holds simultaneously for all bowlers; that is, each bowler's conditional success probability is the same after j consecutive strikes and after j consecutive nonstrikes. The conditional success probabilities can of course depend on j and vary across bowlers.

Matched-Pair Difference in Conditional Strike Proportions. The first statistical test is the Wilcoxon (1945) signed-rank test for paired differences. The paired difference for each bowler is the difference between his strike proportion after j strikes and after j nonstrikes: $d_{ij} = f_{ij} - g_{ij}$. For example, Paul Fleming had 123 opportunities to bowl after having bowled four consecutive strikes and he rolled a strike on 84 of these occasions ($f_{i4} = 0.683$); he also had 37 opportunities to bowl after having bowled four consecutive nonstrikes and he rolled a strike on 16 of these occasions ($g_{i4} = 0.432$). His paired difference is $d_{i4} = 0.683 - 0.432 = 0.251$. For $j = 4$, the mean difference for all 43 bowlers is

$$\frac{\sum_{i=1}^{n_4} d_{i4}}{n_4} = 0.120.$$

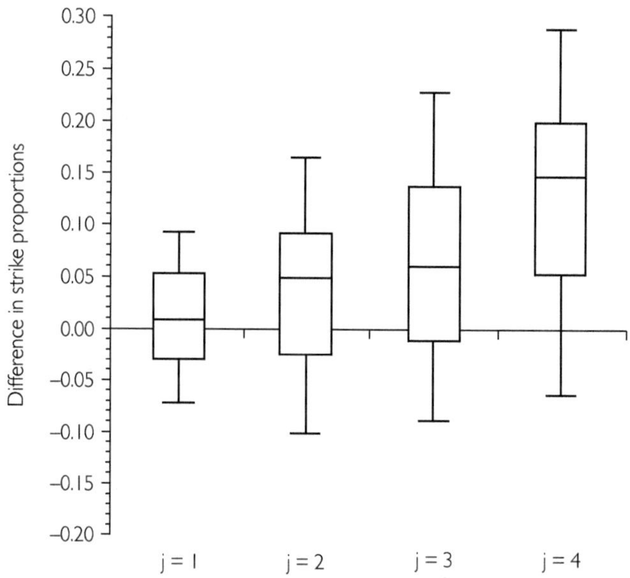

Figure 31.1 Distribution across bowlers of the difference in conditional strike proportions after j consecutive strikes and j consecutive nonstrikes (PBA Tour match-play rounds, 2002–2003).

The means of the paired differences increase as j increases from 1 to 4: 0.011, 0.036, 0.060, and 0.120, respectively.

Figure 31.1 shows four side-by-side boxplots for the values of d_{ij} across bowlers. For each value of j, the Wilcoxon test ranks the absolute values of the paired differences across bowlers, signs the ranks based on whether the difference is positive or negative, and sums these signed ranks R_{ij}. With more than 40 observations, the probability distribution of the Wilcoxon test statistic

$$W_j = \frac{\sum_{i=1}^{n_j} R_{ij}}{\sqrt{\sum_{i=1}^{n_j} R_{ij}^2}} \tag{31.3}$$

is well approximated by the normal distribution. As shown in Table 31.2, the p values are less than 0.05, especially for $j > 1$.

Number of Bowlers with Fisher Exact $p < 0.05$. The second test statistic is based on the application of Fisher's exact test to each bowler's conditional success frequencies after j consecutive strikes and after j consecutive nonstrikes. Unlike the Wilcoxon test, this test uses data on the number of observations for each bowler.

Using the example of Paul Fleming again, this bowler had 123 opportunities to bowl after having bowled four consecutive strikes and he rolled a strike on 84 of these occasions; he also had 37 opportunities to bowl after having bowled four consecutive nonstrikes and he rolled a strike on 16 of these occasions:

	font current frame		
previous 4 frames	strike	nonstrike	total
4 strikes	84	39	123
4 nonstrikes	16	21	37
total	100	60	160

Using the hypergeometric distribution, Fisher's exact test gives a p value of 0.0056.

For each value of j, we count the number of bowlers who had a Fisher exact p value less than 0.05 with a higher success proportion after strikes than after nonstrikes. For $j = 4$, for example, Table 31.2 shows that of the 43 bowlers who had at least 10 opportunities to bowl after four consecutive strikes and after four consecutive nonstrikes, 7 (16%) had differences in their strike frequencies in these situations that yielded a p value less than 0.05 with a higher strike proportion after four consecutive strikes.

The p value for the p values shown in Table 31.2 is for a test of the null hypothesis that each bowler has a 0.05 probability of having a p value less than 0.05 with a higher strike proportion after j strikes:

$$H_0 : P[p \text{ value} \leq 0.05 \quad \text{with} \quad f_{ij} > g_{ij}] = 0.05.$$

The binomial distribution gives the probability, were the null hypothesis true, that the number of bowlers who have a p value less than 0.05 with a higher strike proportion after j strikes would be as large or larger than actually observed. For example, for the $n = 43$ bowlers who had at least 10 opportunities to bowl after four consecutive strikes and after four consecutive nonstrikes, the binomial probability of having seven or more such cases is $P[x \geq 7] = 0.005$. This p value for the p values is quite low and even more impressive if we recall that the underlying p values are discontinuous.

Individual bowlers. We can also identify the individual bowlers whose strike proportions were most inconsistent with the independence assumption. Table 31.3 shows four such bowlers. As with the 2×2 table for Paul Fleming discussed earlier, the p values are based on the hypergeometric distribution for testing the null hypothesis that each bowler's conditional success probability is the same after j consecutive strikes and after j consecutive nonstrikes. To put these p values into perspective, we again need to consider the number of bowlers tested. For example, of the 111 bowlers who, on at least 10 occasions, bowled after two consecutive strikes and after two consecutive nonstrikes, the probability that two or more bowlers would have p values less than 0.0005 is 0.0015. For each of these four bowlers, the observed differences in strike proportions following consecutive strikes and nonstrikes are substantial.

Stationarity

One type of evidence of nonstationary success probabilities would be if a bowler's performance fluctuated from game to game within a round more than would be expected if bowling were a Bernoulli process with constant success probability.

Table 31.3 ***Four bowlers' strike proportions immediately following consecutive strikes or nonstrikes.***

	Number of consecutive strikes/nonstrikes			
	1	2	3	4
Total number of bowlers	134	111	81	43
Paul Fleming				
strike proportion after strikes	0.636	0.627	0.649	0.683
strike proportion after nonstrikes	0.559	0.463	0.423	0.432
p value	0.0106	0.0005	0.0007	0.0056
Bryon Smith				
strike proportion after strikes	0.636	0.665	0.695	0.684
strike proportion after nonstrikes	0.540	0.432	0.436	0.400
p value	0.0156	0.0002	0.0032	0.0197
Mike DeVaney				
strike proportion after strikes	0.651	0.637	0.667	0.632
strike proportion after nonstrikes	0.505	0.435	0.452	0.421
p value	0.0011	0.0015	0.0134	0.0830
Dave D'Entremont				
strike proportion after strikes	0.643	0.622	0.634	0.610
strike proportion after nonstrikes	0.530	0.463	0.512	0.529
p value	0.0027	0.0055	0.1025	0.3624

The *p* values are based on Fisher's exact test comparing each bowler's conditional strike and nonstrike frequencies after a strike or nonstrike in the preceding frame.

Table 31.4 shows two examples from the 2002–2003 PBA World Championship, the last tournament of the season. In his four Round of 16 games, Walter Ray Williams Jr. had an overall strike proportion of 0.739. If his chances of rolling a strike were the same in each game, the expected value of the number of strikes in each game would be 0.739 multiplied by the number of frames in that game. In practice, his actual strike proportions in these four games varied dramatically from 0.364 to 1.000. The chi-square statistic uses the squared differences between the observed values O_i, and the expected values E_i to gauge whether these differences are improbably large:

$$\chi^2 = \sum \frac{(O_i - E_i)^2}{E_i}. \tag{31.4}$$

The exact p value for this chi-square statistic can be computed using the multivariate hypergeometric distribution (Agresti and Wackerly 1977). For the first example in Table 31.4, the exact p value is 0.001. In his four Round of 8 games in this tournament, in contrast, Walter Ray's overall strike proportion was 0.66, with game strike proportions that varied from 0.50 to 0.75. In this case, the p value is an unpersuasive 0.631.

We computed p values for all bowlers in all rounds in this way and tabulated the number of p values that were less than 0.05. The null hypothesis is given by Equation (31.2)—for each test there is a 0.05 probability that the p value will be less than 0.05. If there are n independent tests and x tests have p values less than 0.05, the binomial distribution gives the p value for the

Table 31.4 ***Walter Ray Williams Jr's strikes by game in the round of 16 and round of 8 at the 2002–2003 PBA World Championship, with expected values in parentheses.***

Round	Game	Strike proportion	Strikes	Nonstrikes	Total	p value
16	1	0.364	4 (8.1)	7 (2.9)	11	
	2	0.917	11 (8.9)	1 (3.1)	12	
	3	0.636	7 (8.1)	4 (2.9)	11	
	4	1.000	12 (8.9)	0 (3.1)	12	
	Total	0.739	34	12	46	0.001
8	1	0.727	8 (7.3)	3 (3.7)	11	
	2	0.750	9 (7.9)	3 (4.1)	12	
	3	0.500	6 (7.9)	6 (4.1)	12	
	4	0.667	8 (7.9)	4 (4.1)	12	
	Total	0.660	31	16	47	0.631

The p value is the exact probability for the chi-square statistic based on the squared differences between the observed and expected values.

p values—the probability that x or more tests would have p values less than 0.05. Here, 70 of 1,001 tests had p values less than 0.05. If the null hypothesis were true, the probability that 70 or more of $n = 1{,}001$ tests would have p values less than 0.05 is $P[x \geq 70] = 0.0036$. Although 70 is only 7% of 1,001 tests, remember that we are using discrete data for individual tests that have little power to detect plausible violations of the binomial model.

An alternative way of looking at these data is to tabulate the number of tests with p values less than 0.50. If the null hypothesis were true, we would expect about half the tests to have p values less than 0.5 and half to have p values larger than 0.5. In practice, 555 of these 1,001 tests had p values less than 0.5. If the p values were equally likely to be above and below 0.5, the binomial distribution gives a 0.0003 probability that 555 or more p values would be below 0.5.

Perfect games. A perfect game of 12 consecutive strikes might be viewed as the ultimate hot hand. Yet, tests applied to such a game in isolation can provide no evidence of a hot hand. A game with all strikes always has undefined conditional proportions after nonstrikes, one run, and a longest run of 12. Instead, we compare the total number of perfect games with the number that might be expected if the binomial model applied.

There were 19 perfect games during match play on the 2002–2003 PBA tour. To determine whether this is more or fewer than might be expected, we did 100,000 simulations with each bowler rolling the number of match-play games that he actually rolled on the tour and having a constant strike probability equal to his actual strike proportion during his match-play games. In these simulations, there were an average of 11.7 perfect games and only 3.1% of the simulations had 19 or more perfect games.

Another way to look at the perfect games is to consider that there were a total of 42 occasions in which bowlers began games with 10 strikes in 10 frames, needing strikes in the two bonus rolls to complete a perfect game. Interestingly, the overall strike proportion for these 10-strike bowlers during the 2002–2003 season was 0.560, somewhat lower than the 0.590 strike proportion for all bowlers. If we assume that each of these 42 bowlers had a strike probability equal to his own season strike proportion, the probability that 19 or more of these 42 bowlers would roll the two strikes needed to complete a perfect game is 0.032.

Discussion

Gilovich, Vallone, and Tversky (1985) and Tversky and Gilovich (1989a, 1989b) argued that the performance of professional basketball players does not provide persuasive evidence of hot hands. Their studies have been widely cited, leading many to believe that the perception by fans and players that athletes sometimes get hot is an example of how people erroneously see patterns in random data.

One challenge in looking for statistical evidence of hot hands is that most athletic contests involve many confounding influences. Another hurdle is that statistical tests with relatively few observations generally have little power unless there are large violations of the binomial model's assumptions.

Bowling data are relatively clean in that, unlike basketball data, every roll in a game is from the same distance and made at regular, brief intervals. In our analysis of PBA match-play data, we find, unlike Gilovich, Vallone, and Tversky, evidence that individual success probabilities for many bowlers are neither independent of previous outcomes nor constant across games. In particular, many bowlers tend to do better after having done well and their performances tend to vary more across games than would be predicted by chance.

For example, most bowlers have a higher strike proportion after j consecutive strikes than after j consecutive nonstrikes, and this difference becomes more pronounced as j increases from 1 to 4. For $j = 4$, 34 of the 43 bowlers had a higher strike proportion after four consecutive strikes than after four consecutive nonstrikes and, looking at all 43 matched-pair differences in strike proportions, the average paired difference is 0.120 and the p value is less than 0.001. In addition, applying Fisher's exact test to the conditional strike frequencies for these 43 bowlers, 16% had individual p values less than 0.05.

The inherent nature of basketball games and many other athletic contests makes it very difficult to obtain data clean enough for conclusive tests of the hot hands theory. However, it seems unlikely that professional bowlers are the only athletes to experience hot hands.

[*Received March 2003. Revised November 2003.*]

REFERENCES

Agresti, A., and Wackerly, D. (1977), "Some exact conditional tests of independence for $r \times c$ cross-classification tables," *Psychometrika*, 42, 111–125.

Bateman, G. (1948), "On the power function of the longest run as a test for randomness in a sequence of alternatives," *Biometrika*, 35, 97–112.

Frame, D., Hughson, E., and Leach, J. C. (2003), "Runs, Regimes, and Rationality: The Hot Hand Strikes Back," working paper.

Gilovich, T., Vallone, R., and Tversky, A. (1985), "The hot hand in basketball: on the misperception of random sequences," *Cognitive Psychology*, 17, 295–314.

PBA Tour (n.d.), retrieved March 15, 2003, from http://www.pbatour.com.

Smith, G. (2003), "Horseshoe pitchers' hot hands," *Psychonomic Bulletin and Review*, 10, 753–758.

Stevens, W. L. (1939), "Distribution of groups in a sequence of alternatives," *Annals of Eugenics*, 9, 10–17.
Tversky, A., and Gilovich, T. (1989a), "The cold facts about the 'hot hand' in basketball," *Chance*, 2, 16–21.
—— (1989b), "The 'hot hand': statistical reality or cognitive illusion?" *Chance*, 2, 31–34.
Wardrop, R. L. (1999), "Statistical Tests for the Hot-Hand in Basketball in a Controlled Setting," working paper.
Wilcoxon, F. (1945), "Individual comparisons by ranking methods," *Biometrics*, 1, 80–83.

CHAPTER 32

Did Lennox Lewis beat Evander Holyfield?

Methods for analysing small sample interrater agreement problems

HERBERT K. H. LEE, DANIEL L. CORK AND DAVID J. ALGRANATI

Abstract

On March 13th, 1999, a highly anticipated prize-fight between heavyweight champions Evander Holyfield and Lennox Lewis was ruled a draw by the three official judges. Many observers of the fight felt that Lewis had clearly outperformed Holyfield; dissatisfaction with the result—particularly the pro-Holyfield score-card of judge Eugenia Williams—fuelled speculation that the fight had been fixed and prompted official investigations. In this paper, we examine whether the official judges scored the fight in a significantly different way from other professional observers of the fight. We do so by analysing the round-by-round scoring within the context of interrater agreement. The literature on interrater agreement typically considers a large number of samples rated by a small number of judges and relies on asymptotic results for tests. In our case, the sample size is too small to rely on asymptotics. Instead, we investigate several techniques that can be applied to small sample interrater agreement problems, including logistic regression, an exact test and some Bayesian approaches. We demonstrate these methods on both the March 1999 Holyfield–Lewis fight, as well as the September 1999 bout between welter-weights Oscar de la Hoya and Felix Trinidad.

Keywords: *Bayesian statistics, boxing, logistic regression*

Introduction

Boxing is a sport with a history that is rich in legendary champions and thrilling battles. Unfortunately for boxing, that history is also ridden with controversies and scandals. Although scandal sometimes erupts from actions by the referees or the fans, it seems that the most frequent source of controversy in boxing is the scoring of the bouts. Unless a boxing match ends in a knockout or other stoppage, the fight is decided on the basis of the score-cards of three official judges. The combination of many viewers' unfamiliarity with the boxing scoring system, the somewhat arcane nature of that scoring system and the passions that observers may feel for one fighter over another frequently leads to judges' decisions being hotly contested. In this paper, we take a statistical look at the score-cards of two fights with controversial outcomes. The underlying problem is one of interrater agreement, that of comparing the scores of different judges for each of the rounds of the fight. However, we face the complication of a limited amount of data, since boxing matches last at most 12 rounds. Traditional methods for interrater agreement which rely on asymptotic results clearly do not apply here. Thus we consider several methods which do apply to such small samples, including exact tests, logistic regression and Bayesian methods.

The first fight that we consider in this paper is the clash on March 13th, 1999, between American Evander Holyfield and British boxer Lennox Lewis at New York's Madison Square Garden for the first unified heavyweight championship in over 5 years. To many observers, Lewis appeared to outfight a sluggish Holyfield; however, when the judges' score-cards were read, the fight was ruled a draw. British judge Larry O'Connell scored the fight as even, whereas South African judge Stanley Christodoulu's score-card favoured Lewis. Most controversially, judge Eugenia Williams of New Jersey declared Holyfield the winner, seven rounds to five; among the rounds that Williams awarded Holyfield was the fifth round, in which Lewis appeared most dominant of a tentative, defensive Holyfield. A tremendous outcry over this seemingly wrong decision touched off investigations within boxing circles and in New York's state government, with charges of corruption and incompetence levelled at judge Williams in particular. Eventually a rematch between the fighters in November 1999 was won by Lewis by decision.

Several months after the first Lewis–Holyfield bout, boxing's best welter-weight (147 lb) fighters—Oscar de la Hoya and Felix 'Tito' Trinidad—put their undefeated records on the line on September 18th, 1999, in Las Vegas. The fight ended with a majority decision favouring Trinidad, a ruling that failed to disturb some observers because of de la Hoya's unusual tactic in the late rounds of the fight. Apparently believing himself to be well ahead of Trinidad on the judges' score-cards, de la Hoya largely kept away from Trinidad in the last three rounds, throwing relatively few punches and avoiding any possibility of a victory by Trinidad by a knockout. Still, the judges' decision was assailed as a 'bad decision' by some in the press who agreed with de la Hoya and his corner-men that he had won enough early rounds to secure a victory.

Such perceived bad decisions are commonly believed by boxing observers and insiders to be the result of underhandedness or foul play, usually by the promoter of the bout (see, for example, Newfield, 1995). Allegations of incompetence or foul play are easy to cast around, but they are difficult to prove. Boxing has no central panel or review board to investigate the decisions of world championship bouts. Moreover, the investigation of bad decisions—and, for that matter, statistical analyses of the sport of boxing in general—are hampered by the subjective, rather imprecise nature of boxing scoring. The intangible factors that lead to boxers winning rounds are difficult to quantify and easy to misconstrue. In this paper, we use data on 1999's two landmark 'bad decisions' to explore the variability in boxing scoring and to address the quality of the decisions.

The data that we consider here are the round-by-round score-cards for both bouts returned by the official judges, augmented by round-by-round score-cards reported by broadcast teams and other media outlets. The score-cards compiled by unofficial observers provide some frame of reference in assessing how extreme the official judges' scoring of particular rounds may be. Furthermore, much of the criticism of the official decisions came from the media, so comparing official score-cards with media score-cards allows us to assess the strength of these criticisms. Looking at both the official and the media score-cards, the problem that we face is that of comparing several raters (approximately 10 scorers) on a small number of observations (12 rounds of boxing).

Our analysis proceeds in the next section with further detail on the mechanics of boxing scoring. In the section 'Methods for analysing small sample interrater agreement', we outline methods for dealing that. The sections following that apply these methods to the Holyfield–Lewis and de la Hoya–Trinidad fights, whereas the final section offers concluding comments and directions for future research.

The boxing scoring system

The Marquess de Queensbury rules, which form the foundation of modern boxing, specify only the most elemental aspects of the sport—the contest is divided into rounds, the combatants are to wear gloves, and so forth. The exact rules governing the conduct of a particular boxing match depend in part on the official sanctioning body that authorizes the fight and on the boxing or athletics commission of the state or nation where the fight takes place.

Boxing's three major sanctioning bodies have standardized certain rules for world title bouts. Under these uniform rules, championship bouts are scheduled for 12 3-minute rounds. Three judges are appointed to score the bout on the '10-point must' system, which we discuss in more detail in the section on 'Scoring using the ten-point must system'. The judges are seated along the edge of the boxing ring; they do not sit together and cannot consult one another. After each round, the judges submit their score for that round to the referee (who does *not* score the bout), and the referee gives the scores to a clerk at the ringside to record and tally them.

If the fight goes the full 12 rounds without being stopped because of a knockout, disqualification or other reason, then the outcome of the fight is decided on the basis of the judges' score-cards. Each judge's round-by-round scores are summed together; if judge A has more points for boxer X than for boxer Y, then judge A has scored the bout for boxer X (the margin in points by which boxer X beats boxer Y on judge A's score-card does not matter). Alternatively, judge A's score-card can show victory for boxer Y (if boxer Y has more points) or an even contest (boxers X and Y have equal points).

On the basis of the individual judges' totalled scores, boxer X wins the fight if two of the ringside judges score the bout in his favour. A draw between the two boxers can result in one of two ways: either at least two of the judges score the bout as even (the same amount of points for boxer X and boxer Y) or one judge scores the bout for boxer X, a second judge scores it for boxer Y and the third scores the bout as even.

Scoring criteria

In determining which boxer wins a particular round, judges are instructed to consider four criteria. The first is the number of *clean punches landed* by each boxer, where clean punches are those thrown with a clenched fist that land above the waist on the stomach, chest or the front or side of the head. Typically, punches to the shoulders, elbows, arms or gloves of the opponent do not count as a clean punch landed (because the opponent has blocked the punch). The second criterion is *defence*, the avoidance of punches via blocking, bobbing and weaving (movement of the head and waist) and good footwork (Kaczmarek, 1996). Third, boxers are evaluated for their *effective aggressiveness*—whether the fighter forces the action of the bout by pursuing his opponent and initiating action, while landing punches. The final criterion is the concept of *ring generalship*. Professional boxing judge Tom Kaczmarek observed that

> 'ring generalship applies to the fighter who uses skills beyond his punching power to control the action in the ring'

(Kaczmarek, 1996); this includes the use of feints and footwork, as well as the ability to make the opponent fight a style that he would prefer not to fight. A judge's determination of how well a boxer has satisfied these criteria must be based only on what that judge observes and believes; for instance, although commercial parties compute breakdowns of punches thrown and landed and these statistics are frequently cited by on-air broadcasters, they are neither official fight statistics nor are they available to the ringside judges.

That there are four criteria for judging whether a boxer won a round is clear; how these criteria should be weighted in making the decision is not. A World Wide Web page on boxing scoring hosted by the cable network Home Box Office (HBO) and authored by former professional boxing judge Harold Lederman—who unofficially scores fights for the network's boxing telecasts—implies that 'each of these criteria is supposed to count for 25%'. But the page then contradicts the premise of equal weights on the criteria, noting that

'all good judges will tell you that clean punching counts a lot more than everything else in scoring a round of professional boxing'

(the Web page is accessible through the HBO boxing Web site at http://www.hbo.com/boxing). Kaczmarek (1996) considered effective aggressiveness and clean punches landed as one category and deemed this composite category the most important scoring criterion; ring generalship is the next most important criterion to Kaczmarek, followed by defence. In the absence of clearer standards, it appears that individual judges—like observers scoring boxing matches on their own—are free to weight the criteria as they choose. Consequently, different boxing styles and interpretations of the criteria introduce variability into judges' score-cards and the resolution of bouts.

Scoring using the 10-point must system

After considering the criteria described in the previous subsection, a judge must assign a score for a particular round in a boxing match using what is known as the 10-point must system. Under this system, the winner of the round receives 10 points, and the loser of the round receives 9 points or fewer. If the judge decides that each boxer is equally deserving—that there is no 'loser' of the round—the judge has the discretion to score the round as 'even', assigning 10 points to each fighter.

On the surface, the ability to assign the loser of a round 9 points or fewer gives individual judges wide latitude in their scoring; in practice, this is not the case. Kaczmarek (1996) suggested that a round won by a '"close" to a "definite" margin' should be scored 10–9. Rounds are almost always scored 10–9, unless a fighter is knocked down during the round, in which case the round would be scored 10–8. If a fighter were to be knocked down by his opponent two or three times, the corresponding score would be 10–7 and 10–6 respectively. Kaczmarek described 10–6 rounds as 'very rare'; typically, bouts are stopped if a fighter has been knocked down three times in one round. Judges vary in their propensity to score rounds even (10–10); additionally, if one fighter thoroughly dominates his opponent but does not score a knockdown, some judges will score the round 10–8.

Another factor which influences the scoring is that the referee can penalize fighters for infractions of the rules. These penalties are usually the deduction of 1 point off the offending fighter's score in the round in which the infraction occurred. Typically, point deductions only occur for repeated infractions; the referee usually warns the offending fighter first unless the infraction is particularly egregious.

An important feature of the boxing scoring system is that 'each round should be judged as an isolated event, regardless of what has transpired earlier' (Kaczmarek, 1996). Once a judge turns in a score for a particular round, the slip with the score is turned over to a ringside official via the referee. Because they are not keeping a running score-card in front of them, judges do not have the opportunity to change scores that they assigned in previous rounds or to base their scores on previously scored rounds. Indeed, judges may only become aware of the implications of their aggregate scores at the same time as the public does—when they are announced at the end of the fight. In the Holyfield–Lewis bout, judge O'Connell maintained after the fight that he was as surprised as anyone when his score-card turned out to show a draw.

Discrepancies in scores

Though they watch the same fight, weigh the same criteria and follow a common system of scoring, the score-cards that are turned in by official ringside judges can vary widely—and the official judges' opinion may be wildly at odds with the verdicts rendered by boxing commentators, journalists and the public. Why do these discrepancies exist?

Perhaps the greatest cause of discrepancy between the official scores of a bout and other observers' 'scores' is that many people are not scoring the fight round by round, on the 10-point must system. Instead, many people may watch a bout and form a general impression of the contest; they may rely on an intuition about which boxer won a round rather than considering all four criteria. We can readily conceive of scenarios under which a failure to commit round-by-round scores under the 10-point must system can clash with more general perceptions about a fight's winner. For example, boxer A wins five rounds by a 'definite' margin, and boxer B wins seven rounds by a 'close' margin. On the 10-point must system, boxer B wins the bout, 115–113, whereas an observer just making one overall summary impression is likely to favour boxer A.

Others—particularly viewers of televised boxing broadcasts—may be swayed by unofficial punch statistics, such as the summaries of 'jabs' and 'power punches' thrown and landed. These punch statistics are not considered by the judges in making decisions and are recorded from a vantage point that is different from (and perhaps inferior to) the ringside location of official judges. Moreover, they reflect—at best—only two of the four scoring criteria (clean punches landed and defence).

The most striking, and probably most costly, example of such scoring fallacies might be the tactical errors that many believe transpired in Oscar de la Hoya's corner during his fight with Felix Trinidad. By the last three rounds of the fight, de la Hoya and his corner-men believed themselves to be safely ahead on points on the judges' score-cards. After the fight was decided in Trinidad's favour, de la Hoya and his advisors frequently cited unofficial punch statistics as evidence that they had been wronged. Specifically, the statistics showed that de la Hoya landed 41% of his total punches (263 of 648) compared with Trinidad's 36% (166 of 462), but, by '[believing] that punch statistics are as irrefutable as second-grade arithmetic answers'—believing them to be a true and indisputable indicator of who won the fight—de la Hoya made a terrible 'miscalculation' (Fernandez, 1999).

Other discrepancies in boxing scoring may arise through the lack of standardization in applying the scoring criteria. As mentioned in the criteria subsection, the judges may assign different weights to the different scoring criteria. Another problem is that, whereas clean punches landed and defence are fairly objective, effective aggressiveness and ring generalship contain subjective elements (e.g. the point at which a boxer's aggression becomes 'effective' is a decision which each judge makes individually). Even the seemingly objective criteria of clean punches landed and defence could be legitimately viewed differently by judges because the judges are situated at different points at the ringside. When a fighter's back is to a judge, it becomes more difficult to discern whether punches are being landed or blocked.

As mentioned earlier, a common belief is that discrepancies in scores arise because the judges are either incompetent or corrupt. As a general rule, the more important a bout, the more senior and experienced the judges are who are assigned to that bout. This would seemingly decrease the chances that the officials in important bouts, such as those examined in this paper, are incompetent.

Methods for analysing small sample interrater agreement

In this paper, we consider round-by-round score-cards recorded by both official boxing judges and media observers. This is a problem of interrater agreement that differs substantially from more common problems in the field. Much of the extant literature on interrater agreement deals with agreement matrices—structures with one dimension per rater so that, for example, cell [1,1,2] of the matrix indicates the number of times that the first two raters awarded a score of 1 and the third rater awarded a score of 2. Statistics like κ (Cohen, 1960) are then applied to measure agreement.

Such agreement matrices can be very useful when the number of raters is small, and the number of observations is sufficiently large to invoke the asymptotic arguments that are typically used to evaluate the size of κ. But agreement matrices become less useful as the number of raters increases (making the matrix hopelessly sparse) or when the number of observations is small. In our case, the number of raters is roughly equal to the number of observations, which in turn is small; we must consider alternative methods.

The methods that we consider and apply in this paper are all designed to examine the legitimacy of individual score-cards. They vary in complexity in their assumptions; by performing the battery of procedures, we are—in a sense—exploring the robustness of the individual tests' findings. We begin with frequentist and Bayesian 'exact' tests that assume a particularly simple structure, yet still yield probabilities for individual score-cards. We then add more realism to the modelling with a logistic regression model that allows for individual judge effects and biases to be estimated; we fit both a classical logistic regression model as well as a Bayesian analogue. Finally, a direct Bayesian approach assumes underlying latent variables that correspond to the magnitude by which a fighter won a round; this approach allows us to estimate individual judges' biases and to account for rounds scored as even. These methods are described in more detail in the following sections.

Exact tests

A basic approach to modelling boxing score-card data begins by making two assumptions: that no even (tie) rounds are permissible and that the probability that a fighter wins a particular round depends only on the round, not on the judge. The first of these assumptions is not highly unrealistic—as discussed in 'Scoring using the ten-point must system', boxing scoring is effectively binary since most rounds in boxing are scored either 10–9 for fighter A or 10–9 for fighter B. Indeed, only six of the 252 round scores in the data sets that we use in this analysis are scored as even; all the others are scored 10–9 for one fighter over the other. The second assumption—that there are no individual judge effects—is less realistic, but we shall add this layer of complexity later.

Under these assumptions, we can model each judge's score on each round as conditionally independent Bernoulli variables. Let x_{ij} be 1 if judge j scores round i for fighter 1, and 0 if they score it for fighter 2. We can estimate p_i, the probability that a judge will score round i for fighter 1, by the proportion of the judges who score round i for fighter 1. Following the ideas of Fisher (1935) (such as in his tea-tasting example), we can compute the probability of observing each sequence of scores (score-card) and see how unlikely the particular score-card of a judge is. Since there are only 12 rounds in a championship boxing match, we can easily programme a computer to evaluate the probabilities of all 2^{12} possible score-cards and compute test statistics and p-values.

The key concern is computing the probability of observing a score-card as 'extreme' as that given by a particular judge. There are two obvious ways in which we could define extreme, and we shall explore both of them. First, a score-card could be extreme in that it has at least as many rounds scored for fighter 1 as the score-card in question (or symmetrically for fighter 2). Alternatively, a score-card could be considered extreme if it has at least as many rounds scored for fighter 1 and it has a probability of being observed that is no higher than that of the score-card in question.

There is a Bayesian analogue of this analysis. We model the rounds as independent and the judges as exchangeable. We treat the scores for round i as Bernoulli distributed with probability p_i and put a prior on the p_i. In the case of a uniform prior, we can obtain the posterior for p_i in closed form and again compute the probability of seeing a score-card as extreme as that given by a particular judge. For more complicated hierarchical models for p_i, we can use Markov chain Monte Carlo (MCMC) sampling either to obtain point estimates for p_i (followed by again computing all possible score-cards) or to estimate the probability of observing as extreme a score-card as that of a particular judge directly.

Logistic regression

We next consider methods that continue to assume that no even rounds are permissible but allow for individual judge effects to be estimated. Following the methods of Agresti (1990), we fit a logistic regression model to the scores with factor covariates for the judges and the rounds. Let p_{ij} be the probability that judge j scores round i for fighter 1. Let α_i be an indicator covariate for round i (omitting the first round for identifiability), and let β_j be an indicator covariate for judge j (also omitting the first judge). Then the model is

$$\log\left(\frac{p_{ij}}{1-p_{ij}}\right) = \mu + \alpha_i + \beta_j + \varepsilon_{ij} \qquad (32.1)$$

where μ is an overall level and ε is the error term. After fitting the model, we can use the fitted values for β_j to compare the judges. Judges with unusually large or small βs could be seen to be biased in favour of one of the fighters.

In evaluating the coefficients to attempt to determine significance or to select a model, we typically need to rely on asymptotic approximations. Clearly these are inappropriate for a sample as small as ours, so we cannot assume normality of the fitted coefficients or a χ^2-distribution for the deviance. However, McCullagh and Nelder (1989) pointed out that the χ^2-approximation for the difference in deviance (for comparing models) is usually still good even when the assumptions are not met. Our experience with these models shows no evidence to the contrary, so we rely on this approximation for a valid analysis of classical logistic regression models.

This model has the obvious Bayesian analogue, where a hierarchical prior is used for the parameters. We can then compare the posteriors for the βs to check for possible biases in the scores of the judges. The Bayesian approach has the additional advantage in that MCMC sampling provides a direct estimate of the posterior for parameters, so we do not need to rely on normality assumptions for them.

In theory, the generalized linear model framework allows us (in both the classical and the Bayesian cases) to add a level to the model to deal with ties in the scores. However, the relative scarcity of ties in this already small sample makes it impractical to incorporate the ties in the data into an analysis with these methods.

A direct Bayesian model

Our final approach follows that of Johnson (1996) and Johnson and Albert (1999). Consider a latent variable that measures the true difference in performance between the two fighters in a boxing match in a particular round—that indicates the 'true' winner of the round. Under this framework, an individual judge's score for that round reflects their perception of that underlying difference in performance; an assessment can be made about whether an individual judge is consistently biased in favour of one fighter or another.

For each round i, denote the difference in performance between fighter 1 and fighter 2 by y_i, and let $y_i \sim N(0, 1)$. The scale is arbitrary, so we fix the standard deviation at 1 for identifiability. If both fighters are truly even in performance in the round, then $y_i = 0$. Large positive y_i suggest dominance by fighter 1, whereas a negative y_i suggests that fighter 2 had the better performance.

Obviously, y_i is unobserved; instead, what we observe are individual judges' perceptions of the sign of y_i. For judge j, we model the perceived difference in performance on round i as $y_{ij} \sim N(y_i + b_j, \sigma_j^2)$, where b_j is a bias term for judge j. If $y_{ij} > 0$, judge j will score round i for fighter 1, i.e. $x_{ij} = I_{\{y_{ij}>0\}}$. Finally, we put priors on b_j and σ_j^2 to obtain the full model:

$$\begin{aligned} y_i &\sim N(0, 1), \\ y_{ij} &\sim N(y_i + b_j, \sigma_j^2), \\ b_j &\sim N(0, \tau^2), \\ \sigma_j^2 &\sim \Gamma^{-1}(a_1, a_2) \end{aligned}$$

where τ, a_1 and a_2 are hyperparameters chosen on the basis of historical data (or chosen to reflect prior beliefs).

The model is easily fitted with MCMC methods. We can then look at the posteriors for the various b_j and σ_j to compare the judges. Judges with b_j centred away from 0 can be deemed to be 'biased' in favour of one fighter. The variance parameter σ_j speaks to the consistency of an individual judge, in that a larger σ_j implies higher variability from the 'true' result, or at least from the result implied by the other judges.

Frequentist versions of this model also exist (Quinn, 1989; Henkelman et al., 1990; Uebersax and Grove, 1993). These tend to involve latent variables for class membership, where membership is determined by a score variable that is analogous to x_i above. The problem with the frequentist approach here is that more data are necessary to estimate all the parameters of the model than in the hierarchical model case. In fact, they generally require multiple observations for at least some of the combinations of covariates (here, judges and rounds). Since we have only one observation from each judge for each round, this form of model is not appropriate for this paper.

The terminology used to describe this model raises a subtle point that should be made clear. Bias is a somewhat unfortunate term, given that it could be interpreted pejoratively. We emphasize that we use the term bias in accordance with the usual statistical definition of the term. In noting that a particular judge's fitted bias favours one fighter, we emphasize that we mean that their score-card tends to favour one fighter relative to the true latent difference in performance y_i; it should *not* be interpreted as saying that this deviation from the latent y_i is necessarily the result of any corrupt or improper behaviour.

DEALING WITH TIES

This final, direct Bayesian approach is particularly advantageous because it provides a natural way of dealing with the few observed ties (even rounds) that occur in boxing scoring data. Specifically, a cut-off value t can be introduced, defining an interval around $y_{ij} = 0$ wherein a judge can decide that the two fighters were evenly matched and score the round as even. Judge j scores round i for fighter 1 if $y_{ij} > t$, scores the round as a tie if $-t \leq y_{ij} \leq t$ and scores the round for fighter 2 if $y_{ij} < -t$. Although we might want to use separate values of t for each judge, ties are relatively rare, and there are not sufficient data to estimate all these parameters. We elect to set a common t across all judges, and we put an exponential prior on t.

Analysis of the Holyfield–Lewis fight

Score-cards for the Holyfield–Lewis fight are shown in Table 32.1, in which an H indicates a round for Holyfield, an L a round for Lewis and E an even round (tie). The total score reported in the final column lists the score for Holyfield first and Lewis second. The first three rows are the score-cards of the official judges, and the last seven are other professional observers of the fight, including score-cards from unofficial scorer Harold Lederman (as displayed on the HBO broadcast of the fight) and the editor of *Boxing Monthly* Glyn Leach. Note that the first three of the unofficial judges returned exactly the same score-card, in favour of Lewis.

Exact tests

The exact test methodology outlined in 'Exact tests' does not allow for even rounds. Accordingly, we treat even rounds in Table 32.1 as rounds for Holyfield; since Lewis was perceived as the winner by the media and the public, a failure to score these rounds for Lewis could be reasonably interpreted as round victories for Holyfield.

As described in 'Exact tests', we calculate the probability of scoring round i for Holyfield; with the full set of p_i, we can then compute the probability of observing each of the 2^{12} possible score-cards. We find that the probability of observing official judge Williams's score-card (seven rounds for Holyfield and five for Lewis, in the exact order as on her score-card) is only 0.0019. The probability of observing judge O'Connell's even score-card is also 0.0019; treating

Table 32.1 ***Score-cards for the first Holyfield–Lewis fight.*** †

Scorer	Individual scores for the following rounds:												Total
	1	2	3	4	5	6	7	8	9	10	11	12	
Official judges													
Eugenia Williams (USA)	L	L	H	H	H	L	L	H	H	H	H	L	115–113
Larry O'Connell (UK)	L	L	H	L	L	H	E	H	H	E	H	L	115–115
Stanley Christodoulu (South Africa)	L	L	H	L	L	L	L	H	H	H	E	L	113–116
Media and unofficial score-cards													
HBO–Lederman	L	L	H	L	L	L	L	L	H	H	L	L	111–117
Sportsticker	L	L	H	L	L	L	L	L	H	H	L	L	111–117
Boxing Times	L	L	H	L	L	L	L	L	H	H	L	L	111–117
Sportsline	L	L	H	L	L	H	L	L	L	H	L	L	111–117
Associated Press	L	L	H	H	L	L	L	L	L	H	L	L	111–117
ESPN	L	L	L	L	L	L	L	L	L	H	L	L	109–119
Boxing Monthly–Leach	L	L	H	E	L	E	L	H	H	H	H	L	115–115

† Rounds in favour of Holyfield are denoted H, in favour of Lewis L and even rounds E. All non-even rounds were scored 10–9 in favour of the fighter indicated.

even rounds as rounds for Holyfield, O'Connell's score-card also features seven rounds for Holyfield and five rounds for Lewis, albeit permuted differently from that on Williams's card.

The fundamental question is how extreme these two official score-cards are, with respect to all 2^{12} possible score-cards. We find that the probability of observing a score-card with at least as many rounds scored for Holyfield as on the cards of Williams and O'Connell is 0.0256 (aggregating the individual probabilities of 1586 possible score-cards). Under an alternative definition of extreme, we find that the probability of observing a score-card with at least as many rounds for Holyfield as on the cards of Williams and O'Connell, but with a probability no greater than that for the Williams or O'Connell score-cards (0.0019), equals 0.0182 (based on 1585 score-cards, the only difference between the previous set being the sequence of rounds exactly equal to that on Leach's card). From this analysis, we see that both Williams and O'Connell have produced score-cards which are highly unlikely to be observed by chance (the p-value is less than 0.05).

We can also use the fitted probabilities to estimate the probability that Holyfield won at least six rounds—the probability that Holyfield tied or beat Lewis, assuming that round i is won by Holyfield with probability p_i and that no ties are permitted. Adding up the estimated probabilities of all possible score-cards with six or more rounds for Holyfield, we estimate this probability to be 0.1320. Accordingly, the likelihood that the bout should end in an outcome other than a victory to Lewis is small but non-trivial; it is not totally unreasonable for a judge to have scored the fight a draw or for Holyfield, even though the exact combinations of scores registered by Williams and O'Connell are quite improbable.

Putting a uniform prior distribution on p_i and estimating p_i by MCMC sampling—the Bayesian analogue of the classical exact test that has just been described—we find that the score-cards of Williams and O'Connell are not quite as extreme as the classical test suggests. The posterior probability of observing a score-card with at least as many rounds for Holyfield as Williams's is 0.072, and the probability of at least as many rounds on a card with no higher posterior probability than that of Williams is 0.067. These values are larger than the previous versions because the fitted values for p_i have been shrunk towards their prior mean of 0.5. Still, these probabilities are rather small, and we may legitimately question the objectivity of the scores of Williams and O'Connell. Several hierarchical models and prior specifications on p_i were tried, and the results were similar.

Logistic regression

The preceding exact tests are interesting but do not account for individual judge effects. To achieve a more realistic model, we start by fitting a logistic regression as in equation (32.1). We use a single parameter for rounds 1, 2 and 12, as all were scored unanimously for Lewis. We also use a single parameter for HBO, Sportsticker and *Boxing Times*, as all produced identical scorecards. Finally, we set the coefficients for round 1 and for the HBO–Sportsticker–*Boxing Times* score-cards to 0 for identifiability of the model (this coefficient was chosen because HBO is a common element to both this fight and the de la Hoya–Trinidad fight). An analysis-of-deviance table finds that the round effect is highly significant (p-value 5×10^{-12}) and that after taking the differences in rounds into effect the coefficients for judges are still highly important with a p-value of 1.5×10^{-6}. The coefficients themselves are not extremely large in magnitude; the effect for ESPN is the strongest negative at -18.5 (t-value 58.2), whereas the coefficients for Williams, O'Connell and Leach each register the largest positive value of 4.21 (t-value 1.86). The main result of this model is that the judges vary significantly in their propensity to score rounds for a particular fighter.

Bayesian logistic regression was fitted using MCMC sampling, with 10,000 draws from the posterior after 15,000 burn-in samples. Results are shown in Table 32.2 and are similar to those under the classical logistic regression model. Again, judge effects are relative to the HBO–Sportsticker–*Boxing Times* score-card, the effect of which is forced to equal 0.95% credible intervals for the judge effects show that Williams, O'Connell and Leach are more pro Holyfield than HBO is (the interval is strictly positive), and ESPN is more pro Lewis (strictly negative).

The logistic regression model is less useful than the exact test methodology in determining whether a particular judge's score-card is markedly different from their peers'—specifically, to assess whether judge Williams's score-card is fundamentally out of line compared with all other observers. To test for a single judge, we fitted the classical logistic regression model (equation 32.1) with parameters for the rounds and a single parameter for the judge in question. A model of this sort for examining the scores of Williams yields a coefficient on Williams with associated p-value 0.0127, indicating that judge Williams did seem to produce a score-card that was different from the rest of the sample.

Similarly, we can also fit a model to examine whether the official judges—collectively—are significantly different in their scoring from the unofficial judges. Again, we allow for the full set of round effects in the classical logistic regression model, but only a single judge effect corresponding to an indicator variable flagging whether the score was returned by an official judge or not. An analysis-of-deviance table in this case again shows that there is a significant difference, with the p-value being 1.3×10^{-4}.

Table 32.2 ***Bayesian logistic regression results for the judge effect coefficient β_i for the first Holyfield–Lewis fight, with coefficients relative to the HBO–Sportsticker–Boxing Times score-cards.***

Scorer	Posterior mean	Posterior standard deviation	Posterior median	Credible interval	
				2.5th percentile	97.5th percentile
Official judges					
Williams	3.885	1.396	3.864	1.228	6.663
O'Connell	3.882	1.399	3.859	1.217	6.688
Christodoulu	1.975	1.418	1.983	−0.823	4.732
Media and unofficial score-cards					
Sportsline	−0.344	1.568	−0.307	−3.500	2.643
Associated Press	−0.301	1.574	−0.276	−3.515	2.718
ESPN	−4.384	2.164	−4.238	−9.027	−0.567
Boxing Monthly–Leach	3.845	1.427	3.821	1.116	6.653

Direct Bayesian model

We now apply the direct Bayesian model of the earlier subsection, first to the data treating even rounds as wins for Holyfield, and then to the full data set including ties. Though the analysis that incorporates ties is more realistic, we start with the analysis where ties are treated as rounds for Holyfield to facilitate a direct comparison of the fitted biases from the direct Bayesian methods with the results of the exact test and logistic regression procedures applied earlier.

For a data set that is this small, the choice of prior obviously has a great influence on the fitted values of the parameters. However, our experience is that the prior mainly affects the scale of the parameters. The relative sizes of the parameters, as well as the conclusions drawn from the fitted model, are fairly robust to the choice of prior. For the results reported here, we chose a precision parameter of $\tau^2 = 0.1$ for the biases, and inverse gamma parameters $a_1 = 2.0$ and $a_2 = 0.04$ (prior mean 0.04) for the variance of the judges.

The key structural difference between this model and the preceding logistic regression models is the underlying latent variable measuring the fighters' true difference in performance. Hence, by examining the MCMC output for these y_i, we can estimate the underlying probability that Holyfield won at least six rounds. This estimate works out at 0.0856. So we find that it is possible, although somewhat unlikely, that Holyfield really won or tied the bout. This gives us further reason to look more closely at the possible biases of the judges.

Table 32.3 shows the posterior mean bias for each judge, the posterior standard deviation and the probability that their bias is larger than 0, i.e. that they are biased in favour of Holyfield. Table 32.3 confirms—and brings into stark relief—the patterns that emerged in the logistic regression analysis. First, the three official judges do appear to have a bias towards Holyfield relative to the unofficial scorers. Second, official judges Williams and O'Connell have

Table 32.3 ***Summary of fitted biases for judges of the Holyfield–Lewis fight.***

Scorer	Mean bias	Standard deviation	$P(b_j > 0)$
Official judges			
Williams	0.30	0.15	0.977
O'Connell	0.30	0.15	0.974
Christodoulu	0.09	0.14	0.758
Media and official score-cards			
HBO–Lederman	–0.14	0.15	0.165
Sportsticker	–0.14	0.15	0.172
Boxing Times	–0.14	0.14	0.163
Sportsline	–0.15	0.16	0.176
Associated Press	–0.14	0.16	0.183
ESPN	–0.54	0.20	0.003
Boxing Monthly–Leach	0.30	0.14	0.985

particularly pronounced estimated biases; the probability that each of their biases exceeded 0 (favoured Holyfield) is over 0.97. Finally, strong estimated biases are not unique to the official judges—ESPN stands out with the largest estimated bias in magnitude, emphatically favouring Lewis.

When we incorporate the additional parameter t into the direct Bayesian model and account for ties, we obtain the results summarized in Table 32.4. As in the non-tie analysis, the choice

Table 32.4 ***Summary of fitted biases for judges when accounting for ties.***

Scores	Mean bias	Standard deviation	$P(b_j > 0)$
Official judges			
Williams	0.32	0.15	0.983
O'Connell	0.13	0.19	0.762
Christodoulu	0.06	0.14	0.669
Media and official score-cards			
HBO–Lederman	–0.16	0.15	0.138
Sportsticker	–0.14	0.15	0.174
Boxing Times	–0.15	0.15	0.174
Sportsline	–0.15	0.17	0.186
Associated Press	–0.15	0.17	0.184
ESPN	–0.51	0.19	0.001
Boxing Monthly–Leach	0.22	0.13	0.953

of hyperparameters has an appreciable effect on the scale of the fitted parameters but has little influence on the conclusions drawn from the model.

We estimate the probability that Holyfield won at least six rounds at 0.0667. To put the judge biases in Table 32.4 in more context, the mean value for the threshold for scoring a tie was 0.0357. The fitted bias for Williams now stands out clearly over the other 'pro-Holyfield' judges, and her score-card is the most extreme in favour of Holyfield. The judges who scored rounds as ties (O'Connell, Christodoulu and Leach) are seen to have less extreme biases when we take the ties into account. ESPN is seen to be unusually biased in favour of Lewis, to a larger magnitude than the bias of Williams.

In particular, a key result of accounting for ties is that O'Connell's score-card now looks much more reasonable. He had scored round 7 as a tie, whereas everyone else in the sample scored it for Lewis. Counting his round 7 as a pro-Holyfield round makes his score-card much less probable by chance than counting it more accurately as a tie. Correctly assessing the ties lets us see that Williams has the viewpoint that is most biased in favour of Holyfield. But, as ESPN's card suggests, it was possible for an expert observer to view the proceedings of March 13th, 1999, even more strongly in Lewis's favour than Williams judged it in favour of Holyfield.

Analysis of the de la Hoya–Trinidad fight

We now replicate the analysis on a collection of score-cards from the Oscar de la Hoya–Felix Trinidad fight. Table 32.5 shows the score-cards of the three official judges along with eight unofficial score-cards. HBO broadcasted the fight, just as they had broadcasted the first Holyfield–Lewis bout, so Harold Lederman again provided the network's unofficial score-card; the Associated Press and Sportsticker are also common sources between the two data sets. Two of the *Los Angeles Times*'s sports columnists—Steve Springer and Tim Kawakami—both produced coverage of the fight that included round scores. One distinctive feature of the de la Hoya–Trinidad data set is that it includes the score of another current professional boxing judge with world title fight experience, Daniel van de Wiele of Belgium. Although van de Wiele was not an official judge for the de la Hoya–Trinidad fight, he scored the bout as he viewed it and shared his score-card with one of the authors via personal correspondence.

A D shows a round scored for de la Hoya, a T for Trinidad and an E an even round (a tied round). The total scores are for de la Hoya first and Trinidad second. Note that the official judges (and the unofficial professional who published a score-card) tend to favour Trinidad whereas the media tend to favour de la Hoya. The total scores are closer than in the Holyfield–Lewis fight. There is more dissension on individual rounds, and round 3 is the only round on which all 11 score-cards agree. It seems that this fight was a closer fight, and it may not be unreasonable to have scored it either way. We may hope that our methods will come to a similar conclusion.

There is only one tie in the entire data set, round 1 for official judge Logist. This tie is very important in practical terms; if he had scored this round for de la Hoya, his overall score would have been a tie score, and the bout would have been ruled a draw by majority decision. In applying our methods that preclude the use of even rounds, we consider Logist's even round as a round in favour of Trinidad (since it can be argued that this round score is the one that won the fight for Trinidad); of course, in our final Bayesian analysis, we shall handle the even round directly.

Table 32.5 *Score-cards for the de la Hoya–Trinidad fight.*†

Scores	Individual scores for the following rounds:												Total
	1	2	3	4	5	6	7	8	9	10	11	12	
Official judges													
Glen Hamada (USA)	D	T	D	T	D	D	D	T	D	T	T	T	114–114
Bob Logist (Belgium)	E	T	D	T	D	D	T	D	T	T	T	D	114–115
Jerry Roth (USA)	T	T	D	T	D	D	D	D	T	T	T	T	113–115
Media and unofficial score-cards													
HBO–Lederman	D	D	D	T	D	T	D	D	T	T	T	T	114–114
Associated Press	D	T	D	D	D	D	T	D	D	T	T	T	115–113
USA Today	D	D	D	T	D	D	T	D	D	T	T	T	115–113
Las Vegas Review–Journal	D	T	D	T	D	D	D	D	D	D	T	T	116–112
Los Angeles Times–Springer	D	D	D	T	D	D	T	D	D	T	T	T	115–113
Los Angeles Times–Kawakami	D	T	D	T	D	D	D	D	D	T	T	T	115–113
Sportsticker	T	D	D	T	D	D	D	D	D	T	D	T	116–112
van de Wiele	D	T	D	T	T	D	D	D	T	T	T	T	113–115

† Rounds in favour of Trinidad are denoted T, in favour of de la Hoya are denoted D, and even rounds E. The first score listed under the total is for de la Hoya, the second for Trinidad. All non-even rounds were scored 10–9 in favour of the fighter indicated.

Using the exact test methodology to investigate the score-cards of Logist and Roth shows that the probability of observing at least as many rounds for Trinidad (or tied) is 0.21, whereas performing the simple Bayesian analysis to generate round probabilities to use in the exact test yields a probability of a draw or victory to Trinidad of 0.26. Hence, the pro-Trinidad score-cards of two of the official judges are not extreme—the probability of observing score-cards even more partial to Trinidad is appreciable.

Fitting a logistic regression using both round and judge as covariates finds that, whereas the effect of round is significant, the effect of judge is not, and none of the coefficients for the individual judge effects has a significant t-value. Moreover, fitting the model for comparing a single judge (say, Roth) with the others also leads to an insignificant judge effect, as we might expect. Thus, logistic regression analysis suggests that there is not enough evidence to show that any judge was significantly out of line from the others.

Comparing the official judges with the rest of the sample also leads to an insignificant judge effect, although the p-value of the χ^2-test is borderline insignificant at 0.06. On a further review of the data, the most pro-Trinidad result of the non-official score-cards was that of van de Wiele, who has served as a professional judge in other fights. Regrouping van de Wiele with the official judges does lead to a significant model for a judge effect (p-value 0.0117). Thus, there is some evidence that people who are professional judges scored this fight differently from people who are professional media fight observers—an intriguing result, but one whose origins the data alone cannot unravel.

A Bayesian logistic regression also shows no large differences between the judges. Table 32.6 shows that all the 95% credible intervals overlap 0 (and each other). Again, the coefficients

Table 32.6 *Bayesian logistic regression results (judge effect β_i) for the de la Hoya–Trinidad fight.*

Scorer	Mean	Standard deviation	Median	Credible interval	
				2.5th percentile	97.5th percentile
Official judges					
Hamada	0.311	1.507	0.303	−2.624	3.243
Logist	0.860	1.509	0.855	−2.039	3.846
Roth	1.152	1.524	1.151	−1.874	4.168
Media and unofficial score-cards					
Associated Press	−1.002	1.502	−0.967	−3.987	1.870
USA Today	−0.885	1.526	−0.874	−3.921	2.028
Las Vegas Review–Journal	−2.014	1.560	−1.989	−5.140	1.002
Los Angeles Times–Springer	−0.854	1.502	−0.847	−3.814	2.104
Los Angeles Times–Kawakami	−0.885	1.538	−0.894	−3.870	2.127
Sportsticker	−1.954	1.514	−1.910	−4.988	0.914
van de Wiele	1.207	1.505	1.196	−1.712	4.131

are relative to the HBO–Lederman score-card, which has been given a coefficient of 0 for identifiability.

Next, we fit the direct Bayesian model, which is configured such that a positive underlying difference in performance—and a positive bias—suggests a victory to Trinidad, whereas negative differences and biases favour de la Hoya. The estimated biases from this model are shown in Table 32.7. We find that the fitted biases are all small in magnitude and have large standard

Table 32.7 *Summary of fitted biases for judges of the de la Hoya–Trinidad fight.*

Scorer	Mean bias	Standard deviation	$P(b_j > 0)$
Official judges			
Hamada	0.06	0.14	0.674
Logist	0.10	0.16	0.745
Roth	0.13	0.13	0.845
Media and unofficial score-cards			
HBO–Lederman	−0.07	0.14	0.314
Associated Press	0.06	0.15	0.652
USA Today	−0.04	0.13	0.364
Las Vegas Review–Journal	−0.16	0.14	0.125
Los Angeles Times–Springer	−0.04	0.13	0.382
Los Angeles Times–Kawakami	−0.04	0.13	0.369
Sportsticker	−0.15	0.15	0.159
van de Wiele	0.14	0.14	0.853

Table 32.8 *Summary of fitted biases for judges of the de la Hoya–Trinidad fight when accounting for ties.*

Scorer	Mean bias	Standard deviation	$P(b_j > 0)$
Official judges			
Hamada	0.06	0.15	0.675
Logist	0.06	0.15	0.659
Roth	0.14	0.13	0.851
Media and unofficial score-cards			
HBO–Lederman	−0.08	0.15	0.303
Associated Press	0.06	0.16	0.649
USA Today	−0.05	0.14	0.350
Las Vegas Review–Journal	−0.17	0.15	0.121
Los Angeles Times–Springer	−0.05	0.13	0.352
Los Angeles Times–Kawakami	−0.05	0.13	0.358
Sportsticker	−0.16	0.16	0.156
van de Wiele	0.15	0.14	0.853

deviations; again, no single judge stands out as unduly partial to one fighter. The single largest estimated bias in the collection of scorers is in the *Las Vegas Review–Journal* score-card, for which the probability of a bias towards de la Hoya is 0.875. But, again, the size of this probability and of the fitted bias does not suggest any form of impropriety. There is enough variability in the data to make all the observed score-cards reasonable. The estimated probability that Trinidad won at least six rounds is 0.379, which is not far from $\frac{1}{2}$.

Finally, we use the Bayesian model to handle Logist's even round directly; the results are summarized in Table 32.8. Not surprisingly, the only judge whose results are significantly altered by accounting for ties is Logist; his fitted bias drops towards 0, and his probability of bias for Trinidad moves towards $\frac{1}{2}$. All the score-cards are still reasonably consistent with a close fight and with each other. The estimated probability that Trinidad won at least six rounds is 0.377, which is again not far from $\frac{1}{2}$.

Conclusions

In summary, our analysis and battery of tests examines the variability of scoring in two of 1999's most high profile—and controversial—fights. The data suggest that two of the official judges in the first Holyfield–Lewis fight—O'Connell and, particularly, Williams—scored the fight in a significantly different manner from other professional observers. The exact tests suggest that Williams's card is extreme relative to the traditional cut-off at $\alpha = 0.05$, and her pro-Holyfield card stands out as extreme in the other analyses that we have performed. It is worth noting, however, that one of the unofficial judges—the reporter for ESPN—had an estimated pro-Lewis bias that was larger in magnitude than Williams's pro-Holyfield bias. The analysis of score-cards from the de la Hoya–Trinidad fight suggests a picture that is entirely consistent with random variation in scoring a very close fight. The p-values for Roth and Logist in the exact test, 0.21, are considerably larger than 0.05. The other methods used reveal that none of the judges have a clear bias, with credible intervals for the judge effects comfortably containing 0. Despite the small sample size, we could employ several different methods that all reached the same conclusions. This demonstrates both the validity of the methods and the practicality of analysing data sets of this size.

The quality of the conclusions does depend on the quality of the data, and one might wish to quibble with the choice of the score-cards that were used in the analysis. However, few people publish round-by-round scores, so we have used all the score-cards that we could find from qualified scorers (people professionally involved in boxing, either as a referee, judge or reporter). The small number in this paper reflects the fact that few people score fights round by round, something which might change if people understood the scoring system more clearly. Admittedly, valid criticisms can be levelled at media score-cards of boxing matches—for instance, fight broadcasters must necessarily divide their concentration between the action in the ring and the mechanics of providing interesting live commentary, and reporters may converse with (and be swayed by) peers during the fight, unlike the official judges. Still, professional boxing reporters have experience in watching and critiquing live boxing matches, and utilizing their round-by-round judgments—even though their training differs in most cases from that of the official judges—is a reasonable first step in examining variability in scoring. And, as the media were responsible for most of the criticism of the outcomes, it makes sense to compare the official judges with the reporters.

Let us be clear: in concluding that Williams's scoring was significantly different, we do not make any assertions of corruption; nor do we challenge the competence of any of the judges (official or unofficial) whose score-cards were considered here. Indeed, by the simple act of committing to paper (or, at least, Web page) round-by-round scores rather than aggregate impressions, the unofficial judges mentioned here contribute to an informed discussion of outcomes of fights and, for that, should be commended.

Rather, our primary finding—and the message that we hope is borne out by this analysis—is a simple truth that may seem trivial to statisticians but is often lost on many boxing observers: there is variation in the scoring of boxing rounds. It is possible for different people, viewing a fight from different angles and weighting (already vague) scoring criteria in slightly different ways, to watch the same round of boxing and to arrive at different conclusions; such is the nature of close rounds in boxing, and not necessarily the result of corruption. During the fight with Felix Trinidad, Oscar de la Hoya and his corner-men demonstrated the folly of not recognizing that scores can vary; de la Hoya's tactic of avoiding Trinidad in the late rounds seemed predicated on the assumption that he had won the first eight rounds with a probability at or near 1. But this was far from so; only one of these early rounds, round 3, was awarded to de la Hoya by all the scorers in our data set.

As fans of the sport, we are not so naïve as to believe that bribery and other impropriety are completely foreign to boxing and that fights have not been fixed; boxing's rich history is populated by both angels and scoundrels. Our analysis does clearly suggest that O'Connell's and Williams's score-cards in the first fight between Holyfield and Lewis were extreme observations; exactly why that is the case is not something which statistics can explain. But, in boxing, there is far too often a rush to equate disagreement over a decision with the contention that one or more judges was 'crooked', 'corrupt', 'incompetent' or 'on the take'. What is frequently forgotten by such critics is that a judge's score-card is effectively the sum over 12 binary variables where, frequently, there is no clear and unequivocal winner in a particular round. In a close fight—such as between de la Hoya and Trinidad—there is enough variability that it is entirely reasonable to have a total which favours either fighter.

In light thereof, this research speaks to the importance of continuing education on the boxing scoring system. Many criticisms of supposedly bad decisions would probably be averted if the spectator actually scored the bout by using the 10-point must system, committing scores to paper as the fight proceeds (and gaining an appreciation for how difficult it can be to judge close rounds). Our finding that professional judges scored the de la Hoya–Trinidad fight differently from the set of unofficial judges suggests that broadcasters and fight reporters—like official judges—may benefit from seminars in boxing scoring, to make sure that all parties are clear on the criteria that should be applied in scoring a round of boxing. Fundamentally, work is still necessary to achieve a consensus on those scoring criteria—whether the existing four criteria are meaningful in practice, and how they should be weighted.

Boxing celebrates and cherishes absolutes—decisive knockouts and champions who successfully defend their titles against any challenge. Yet it can be argued that boxing is most thrilling when two skilled, equally matched fighters are pitted against each other, where both fighters perform to the peak of their abilities, and when all participants—audience and fighters alike—hang on the words of the ring announcer to learn who won the fight. Ironically, such moments are also times when the sport is highly vulnerable to criticism, as a hard-fought battle can quickly be dubbed a bad decision. Scoring discrepancies will always occur by human nature—people will always see things in slightly different ways—but continuing to leap from disagreement about the outcome of a fight to accusations of another bad decision—without beginning to recognize the inherent variability in boxing scoring—would be the worst decision of all.

ACKNOWLEDGEMENTS

The authors would like to thank the Joint Editor, Associate Editor and referees for their helpful comments. This work was partially supported by National Science Foundation grants DMS 9819950 and DMS 9873275.

REFERENCES

Agresti, A. (1990) *Categorical Data Analysis*. New York: Wiley.

Cohen, J. (1960) A coefficient of agreement for nominal scales. *Educ. Psychol. Measmnt*, **20**, 37–46.

Fernandez, B. (1999) Trinidad's win matter of opinion. *Philadelphia Daily News*, Sept. 20th.

Fisher, R. A. (1935) *The Design of Experiments*. Edinburgh: Oliver and Boyd.

Henkelman, R. M., Kay, I. and Bronskill, M. J. (1990) Receiver operator characteristic (ROC) analysis without truth. *Med. Decsn Mak.*, **10**, 24–29.

Johnson, V. E. (1996) On Bayesian analysis of multirater ordinal data. *J. Am. Statist. Ass.*, **91**, 42–51.

Johnson, V. E. and Albert, J. H. (1999) *Ordinal Data Modeling*. New York: Springer.

Kaczmarek, T. (1996) *You be the Boxing Judge: Judging Professional Boxing for the TV Boxing Fan*. Pittsburgh: Dorrance.

McCullagh, P. and Nelder, J. A. (1989) *Generalized Linear Models*, 2nd edn. London: Chapman and Hall.

Newfield, J. (1995) *Only in America: The Life and Crimes of Don King*. New York: Morrow.

Quinn, M. F. (1989) Relation of observer agreement to accuracy according to a two-receiver signal detection model of diagnosis. *Med. Decsn Mak.*, **9**, 196–206.

Uebersax, J. and Grove, W. M. (1993) A latent trait finite mixture model for the analysis of rating agreement. *Biometrics*, **49**, 823–835.

PART VIII

Miscellaneous Topics

Introduction

TIM SWARTZ

This section of the anthology consists of noteworthy papers involving miscellaneous topics. One's first reaction may be "what is a miscellaneous sports topic?" A miscellaneous sports topic can be viewed as one with a focus beyond the domain of a specific sport. Consequently, the five outstanding statistical contributions considered in this section are far-reaching in the sense that their general applicability may be useful across various sporting activities. The papers deal broadly with data analysis issues (e.g., modeling, rating, and the detection of sporting phenomena) and with sociological issues (substance abuse and sports participation). The common thread among all of the papers is that they develop or use statistical methodology to address real issues that arise in sport.

The nature of sporting competitions between two teams requires a winner and a loser, although sometimes matches end in a draw. Match outcomes depend on the scoring mechanism of the sport, where all scoring is essentially discrete. At the one end of the spectrum, some sports have scores that are "very discrete" such as soccer where draws with no scoring are common. At the other end of the spectrum, discrete scoring involves a large number of points (e.g., basketball, cricket) where continuous approximations have been successfully used to model the scoring process. The first paper by Karlis and Ntzoufras (2003) is concerned with the former type of scoring where teams score few points and discrete modeling is necessary. Whereas independent Poisson distributions have historically been used to model the scoring for each of the two teams in these situations, Karlis and Ntzoufras (2003) recognize that the assumed independence between the scores of the two teams may be unrealistic. Their work develops bivariate Poisson models with appropriate correlation structures for modeling sports scores. The authors' models are applied and assessed to datasets from soccer and water polo, although one can envision the utility of the approach for other sports. Their models are appealing due to the simplicity of marginal distributions, the interpretability of parameters, the incorporation of meaningful covariates, the incorporation of an overdispersion parameter, and improved model fit. Bayesian extensions may be beneficial in future work in more sophisticated model building and in the implementation of prior beliefs based on past data.

In a match between two teams, sports enthusiasts and gamblers are most interested in predicting which team will win. A straightforward approach involves looking at the league standings and predicting that the team with the best record will win. This simplistic assessment ignores the current strength of each team that can provide an important determinant in prediction. The second paper in this section by Knorr-Held (2000) addresses the related problem of dynamically rating teams. Intuitively, when rating teams, more weight should be given to recent performances and the key contribution by Knorr-Held is the determination of a data-driven weighting rating procedure. The paper develops cumulative link models designed for ordinal response data (e.g., wins/draws/losses) where the parameters α_{ti} characterize the strength of team i at time t. The paper is ambitious in the sense that modeling is carried out in a semiparametric framework that permits freedom for the parameters α_{ti}. Various estimation schemes are investigated including one-step ahead prediction in an extended Kalman filter setup and a fully Bayesian approach using Markov chain methods. Several limitations are highlighted as well as possible extensions that may attract the attention of future researchers. The utility of the proposed methodologies is presented in both soccer and basketball.

There was a time (and it is still perhaps true today) when the discussion of "streaky" sporting behavior was a standard component of the sports lexicon. Descriptions such as players being "hot," "on fire," or conversely, "in a slump" were routine. However, many analyses across various sports have largely dispelled the notion of streakiness. The third paper by Albert and Williamson (2001) develops some novel techniques for the detection of streakiness with examples from baseball and basketball. Their introduction provides a nice summary of some of the key issues in studying streakiness. They then introduce some fascinating inferential procedures that permit flexibility in the specification of streaky models. The approach is simulation-based where both data and parameters are simulated. The methods rely on the need to propose parameters and statistics that characterize streakiness and the need to define appropriate priors. In a nutshell, they assess simulated parameters that yield data with streakiness comparable to the observed data. The general approach shares some common ground and is possibly a precursor to some of the approximate Bayesian computation and ABC methods that have gained popularity in recent years.

The fourth paper in this section by Berry and Chastain (2004) is a shorter, less technical contribution on inferences based on urine samples collected from athletes. While the safety of athletes and leveling of the playing field is of great concern, the accuracy of accusations involving doping, specifically the use of testosterone is also an important issue. The particular challenge to detection of testosterone doping is that testosterone is a naturally occurring hormone present in the human body. This paper provides an engaging analysis of the case of Mary Decker Slaney, an American world-class distance runner, who allegedly engaged in testosterone doping. The impact of the decisions made against Slaney highlight the importance of specificity versus sensitivity considerations, and for this reason, the paper may serve as illuminating reading for an introductory statistics course. The paper argues that conclusions concerning guilt should not be made on specificity alone. Rather, the paper demonstrates how inferences ought to be obtained, and should be based on specificity, sensitivity, and the use of Bayes' rule.

The fifth and final paper by Farrell and Shields (2002) investigates economic and demographic factors affecting sporting participation in England. Much has been written about the benefits of exercise (which can be provided through sport), and it follows that understanding the barriers and pathways to participation is important for general well-being. The paper makes use of the comprehensive 1997 health survey for England. They apply a broad interpretation of sport and their definition of participation requires a seemingly low standard leading to overall participation rates of 44.5%. After providing some descriptive summary statistics, a random-effects probit model is proposed. The model includes not only observable individual-specific covariates but also unobservable household-specific preferences. Among the more interesting and provocative general conclusions (i.e., across all sports) are: (1) ethnic minorities have lower participation rates than whites, (2) the presence of children in a household increases participation rates for men but not so for women, (3) drinking is positively correlated with participation in sport, and (4) participation rates increase with increasing income. Perhaps most importantly, household norms have a large impact on individual participation. From a policy and investment perspective, the authors state "this points to the importance of education in changing individuals' preferences towards sports."

REFERENCES

Albert, J. and Williamson, P. (2001). Using Model/Data Simulations to Detect Streakiness. *The American Statistician*, 55(1), 41–50.

Berry, D.A. and Chastain, L.A. (2004). Inferences about Testosterone Abuse among Athletes. *Chance*, 17(2), 5–8.

Farrell, L. and Shields, M.A. (2002) Investigating the Economic and Demographic Determinants of Sporting Participation in England. *Journal of the Royal Statistical Society: Series A (Statistics in Society)*, 165(2), 335–348.

Karlis, D. and Ntzoufras, I. (2003). Analysis of Sports Data by using Bivariate Poisson Models. *Journal of the Royal Statistical Society: Series D (The Statistician)*, 52(3), 381–393.

Knorr-Held, L. (2000). Dynamic Rating of Sports Teams. *Journal of the Royal Statistical Society: Series D (The Statistician)*, 49(2), 261–276.

CHAPTER 33

Analysis of sports data by using bivariate Poisson models

DIMITRIS KARLIS AND LOANNIS NTZOUFRAS

Abstract

Models based on the bivariate Poisson distribution are used for modelling sports data. Independent Poisson distributions are usually adopted to model the number of goals of two competing teams. We replace the independence assumption by considering a bivariate Poisson model and its extensions. The models proposed allow for correlation between the two scores, which is a plausible assumption in sports with two opposing teams competing against each other. The effect of introducing even slight correlation is discussed. Using just a bivariate Poisson distribution can improve model fit and prediction of the number of draws in football games. The model is extended by considering an inflation factor for diagonal terms in the bivariate joint distribution. This inflation improves in precision the estimation of draws and, at the same time, allows for overdispersed, relative to the simple Poisson distribution, marginal distributions. The properties of the models proposed as well as interpretation and estimation procedures are provided. An illustration of the models is presented by using data sets from football and water-polo.

Keywords: *bivariate Poisson regression, difference of Poisson variates, inflated distributions, soccer*

Introduction

The Poisson distribution has been widely accepted as a simple modelling approach for the distribution of the number of goals in sports involving two competing teams. Although several researchers (see, for example, Lee 1997 and Karlis and Ntzoufras 2000 and the references therein) have shown the existence of a (relatively low) correlation between the number of goals scored by the two opponents, this has been ignored in most modelling approaches since it demands more sophisticated techniques. Maher (1982) discussed this issue, and Dixon and Coles (1997) extended the independent Poisson model by introducing indirectly a type of dependence. In team sports, such as football and water-polo, it is reasonable to assume that the two outcome variables are correlated since the two teams interact during the game. Moreover, in some sports games, the two opponents try to score sequentially, and thus the speed of the game of one team leads to more opportunities for both teams to score. A typical example is basketball: the correlations for the National Basketball Association and the Euroleague scores for the 2000–2001 season are 0.41 and 0.38 respectively.

An alternative to the independent Poisson model can be constructed assuming that the two outcome variables follow a bivariate Poisson distribution (see Kocherlakota and Kocherlakota 1992 and references therein). The marginal distributions are simple Poisson distributions, whereas the random variables are now dependent. Maher (1982) mentioned the bivariate Poisson distribution but its use has been largely ignored, mainly because of the computational burden for fitting such a model.

The remainder of the paper proceeds as follows. Firstly, in the next section, we present briefly the bivariate Poisson distribution and discuss its applicability in modelling sports data, especially for football games. The bivariate Poisson distribution allows for improving the model fit of the number of draws, a problem reported by some researchers (for example see Maher 1982 and Lee 1997). An interesting feature of the bivariate Poisson model is the fact that the distribution of the difference of the two variates is the same as the distribution of the difference of two independent Poisson variates. However, the parameters have an entirely different interpretation. Moreover, an incorrect use of the independent Poisson case leads to significant differences. The effect of such a misspecification is illustrated by using a simple example. Maximum likelihood estimation of the parameters is made through an EM algorithm. In the section 'Inflated bivariate Poisson distributions', extensions through inflated models are proposed. Since a draw is represented by diagonal terms in a bivariate distribution, adding an inflation term on the diagonal allows for more precise modelling of the number of draws. In 'Application in sports', the models proposed are illustrated by using examples from football and water-polo. Finally, concluding remarks can be found the last section.

The bivariate Poisson distribution and its implementation in sports modelling

The bivariate Poisson distribution

Consider random variables X_κ, $\kappa = 1, 2, 3$, which follow independent Poisson distributions with parameters $\lambda_\kappa > 0$. Then the random variables $X = X_1 + X_3$ and $Y = X_2 + X_3$ follow jointly a bivariate Poisson distribution $BP(\lambda_1, \lambda_2, \lambda_3)$, with joint probability function

$$P_{X,Y}(x, y) = P(X = x, Y = y)$$
$$= \exp\{-(\lambda_1 + \lambda_2 + \lambda_3)\}\frac{\lambda_1^x}{x!}\frac{\lambda_2^y}{y!}\sum_{k=0}^{\min(x,y)}\binom{x}{k}\binom{y}{k}k!\left(\frac{\lambda_3}{\lambda_1\lambda_2}\right)^k. \quad (33.1)$$

This bivariate distribution allows for dependence between the two random variables. Marginally each random variable follows a Poisson distribution with $E(X) = \lambda_1+\lambda_3$ and $E(Y) = \lambda_2+\lambda_3$. Moreover, $\mathrm{cov}(X, Y) = \lambda_3$, and hence λ_3 is a measure of dependence between the two random variables. If $\lambda_3 = 0$ then the two variables are independent and the bivariate Poisson distribution reduces to the product of two independent Poisson distributions (referred to as the double-Poisson distribution). For a comprehensive treatment of the bivariate Poisson distribution and its multivariate extensions see Kocherlakota and Kocherlakota (1992) and Johnson et al. (1997).

It is plausible to adopt this distribution for modelling dependence in team sports. A natural interpretation of the parameters of a bivariate Poisson model is that λ_1 and λ_2 reflect the 'net' scoring ability of each team whereas λ_3 reflects game conditions (e.g. the speed of the game, the weather or stadium conditions).

The probability of the difference

Let us now define the difference $Z = X - Y$ of the goals scored by two opposing teams. Since $P(Z = z) = P(X - Y = z) = P(X_1 + X_3 - X_2 - X_3 = z) = P(X_1 - X_2 = z)$, the probability function of Z is independent of λ_3 and is the same as that derived from two independent Poisson variates. So Z follows the Poisson difference distribution with parameters λ_1 and λ_2, denoted as PD(λ_1, λ_2), given by

$$P_Z(z) = P(Z = z) = \exp\{-(\lambda_1 + \lambda_2)\} \left(\frac{\lambda_1}{\lambda_2}\right)^{z/2} I_z\{2\sqrt{(\lambda_1 \lambda_2)}\}, \tag{33.2}$$

$z = \ldots, -3, -2, -1, 0, 1, 2, 3, \ldots$, where $I_r(x)$ denotes the modified Bessel function (see Abramowitz and Stegun 1974, page 375) defined by

$$I_r(x) = \left(\frac{x}{2}\right)^r \sum_{k=0}^{\infty} \frac{(x^2/4)^k}{k!\Gamma(r + k + 1)}. \tag{33.3}$$

A special case of two independent Poisson distributions for the case of equal means was derived by Irwin (1937) whereas the general case was considered by Skellam (1946). Interesting references on the Poisson difference distribution can be found in Johnson et al. (1992, page 191). Keller (1994) calculated the probability of winning a game assuming independent Poisson distributions for both X and Y.

Although distribution (33.2) implies that the winning probability ($Z > 0$) does not depend on parameter λ_3, treating the number of goals independently for each team leads to an overestimation of model parameters. It should be kept in mind that, since the parameters λ_1 and λ_2 are estimated from the marginal distributions, the covariance parameter λ_3 is confounded. In the following section we examine the effect of such a misspecification.

The effect of model misspecification

Let us consider that the true underlying model is the bivariate Poisson model but we use instead the double-Poisson model. Then we assume that the difference $Z = X - Y \sim$ PD($\lambda_1 + \lambda_3$, $\lambda_2 + \lambda_3$) instead of the correct $Z \sim$ PD(λ_1, λ_2). This misspecification has quite a large effect even if the covariance λ_3 is as low as 0.10, which is about the observed covariance in football.

Fig. 33.1 depicts the relative change in the probability of a draw between the two competing teams, when independent Poisson distributions are considered compared with the bivariate Poisson model with small covariance parameter λ_3. Namely the independent Poisson model

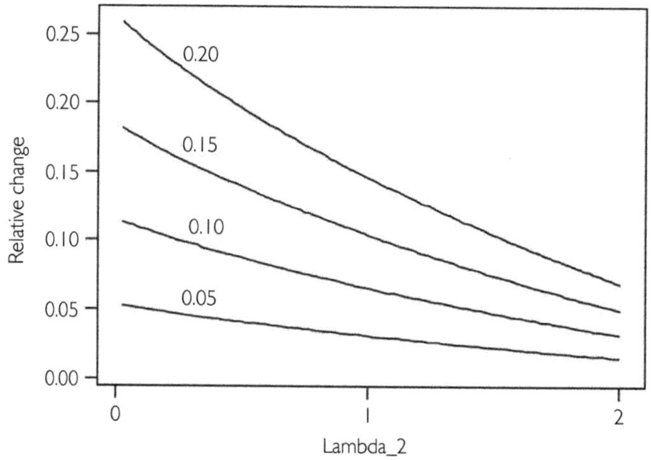

Figure 33.1 Relative change in the probability of a draw when the two competing teams have marginal means equal to $\lambda_1 = 1$ and λ_2 ranging from 0.1 to 2: the different lines correspond to different levels of the covariance λ_3.

assumes that $X \sim \text{Poisson}(1)$ whereas $Y \sim \text{Poisson}(\lambda_2)$. The competing model assumes that $(X, Y) \sim \text{BP}(1-\lambda_3, \lambda_2-\lambda_3, \lambda_3)$. Note that for both models the marginal means are the same, but the bivariate Poisson model assumes also the existence of the covariance λ_3. We let the value of λ_2 vary from 0.1 to 2, for different values of the covariance term $\lambda_3 = 0.05, 0.10, 0.15, 0.20$. These values of the covariance are rather small but are close to those that are observed in real football data.

Fig. 33.1 shows that, for the typically observed range of counts in football data, the probability of a draw under a bivariate Poisson model is larger than the corresponding probability under the double-Poisson model even if λ_3 is quite small. For example, if we consider a bivariate Poisson model with $\lambda_3 = 0.05$ and $\lambda_1 = \lambda_2 = 1$ then we expect almost 3.3% more draws than under the corresponding independent Poisson model, whereas if λ_3 increases to 0.20 we expect 14% more draws. It is also clear that the larger the λ_3 the larger the relative change is. This may explain the empirical fact that the observed number of draws is usually larger than that predicted under an independent Poisson model.

Estimation

In this section we focus on the estimation of the parameters of the bivariate Poisson distribution. For the simple, but unrealistic for sports data, model without covariates, standard estimation procedures have been proposed (see, for example, Kocherlakota and Kocherlakota, 1992). Here we consider more realistic models that include covariates. Bivariate Poisson regression models have been described recently in Kocherlakota and Kocherlakota (2001) and Ho and Singer (2001). The former presents a Newton–Raphson approach for maximizing the likelihood whereas the latter describes a generalized least squares method.

Let us consider the general case of a bivariate Poisson regression. For the ith observation the model takes the form

$$\begin{aligned} (X_i, Y_i) &\sim BP(\lambda_{1i}, \lambda_{2i}, \lambda_{3i}), \\ \log(\lambda_{\kappa i}) &= \mathbf{w}_{\kappa i}\beta_\kappa, \quad \kappa = 1, 2, 3, \end{aligned} \tag{33.4}$$

$i = 1, \ldots, n$ denotes the observation number, $\mathbf{w}_{\kappa i}$ denotes a vector of explanatory variables for the ith observation used to model $\lambda_{\kappa i}$ and β_κ denotes the corresponding vector of regression coefficients. It is clear that each parameter of the bivariate Poisson distribution may be influenced by different characteristics and variables. For this reason, the explanatory variables that are used to model each parameter $\lambda_{\kappa i}$ may not be the same. Parameter estimation for such a model is not straightforward. Hence, we make use of the EM algorithm to obtain maximum likelihood estimates.

To construct the EM algorithm for the bivariate Poisson regression model, we make use of the trivariate reduction derivation of the bivariate Poisson distribution. For this reason, for each observation i, we further introduce the latent variables X_{1i}, X_{2i}, and X_{3i} for which we assume a Poisson distribution with parameters λ_{1i}, λ_{2i} and λ_{3i}, respectively. Moreover, we assume that $X_i = X_{1i} + X_{3i}$ and $Y_i = X_{2i} + X_{3i}$.

The EM algorithm proceeds by estimating the unobserved data via their conditional expectations at the E-step and then it maximizes the complete-data likelihood at the M-step. Hence, at the E-step, we obtain the posterior expectation of X_{1i}, X_{2i}, and X_{3i} given the data and the current parameter values and then, at the M-step, we maximize the complete-data likelihood by fitting three Poisson regression models. The aim is now to estimate the regression coefficients β_κ for $\kappa = 1, 2, 3$. The full algorithm is available from the authors on request. This EM algorithm is very flexible and many variations of the bivariate Poisson model can be fitted with slight modifications.

Inflated bivariate Poisson distributions

The bivariate Poisson model introduces correlation between the variables, but the marginal distributions are still Poisson. As an improvement, relative to the simple bivariate Poisson model, we may consider mixtures of bivariate Poisson distributions, either finite or infinite. Such mixtures may have a variety of forms, depending on the varying parameters and the mixing distribution. However, such models incorporate very complicated structures and, therefore, are not very useful for practical application in sports modelling.

A type of inflated model was used by Dixon and Coles (1997) for modelling football games. They initially assumed two independent Poisson distributions and then they corrected the expected values for the outcomes (0–0, 1–0, 0–1, 1–1) by an additional parameter. We propose alternative models which, at the same time, incorporate correlation between the variables and overdispersed (relative to Poisson) marginal distributions while they further improve the fit on the counts of draws.

If the score 0–0 is underestimated by the model then we may inflate the probability at the (0,0) cell by adding a parameter. In such a case, a model similar to that proposed by Li et al. (1999) is specified. We propose a more general model formulation which inflates the probabilities of draws. A draw between two teams is represented by the outcomes on the diagonal of the probability table. To correct for the excess of draws we may add an inflation component on the diagonal of the probability function. This model is an extension of the simple zero-inflated model that allows only for an excess in (0,0) draws. We consider, for generality, that the starting model is the bivariate Poisson model.

Under this approach a diagonal inflated model is specified by

$$P_D(x,y) = \begin{cases} (1-p)\mathrm{BP}(x,y|\lambda_1,\lambda_2,\lambda_3), & x \neq y, \\ (1-p)\mathrm{BP}(x,y|\lambda_1,\lambda_2,\lambda_3) + pD(x,\theta), & x = y, \end{cases} \tag{33.5}$$

where $D(x,\theta)$ is a discrete distribution with parameter vector θ. Such models can be fitted by using the EM algorithm.

Useful choices for $D(x,\theta)$ are the Poisson, the geometric or simple discrete distributions such as the Bernoulli distribution. The geometric distribution might be of great interest since it has its mode at zero and decays quickly. As a discrete distribution we consider $P(X=j)=\theta_j$ for $j = 0,1,2,\ldots,J$, where $\Sigma_{j=0}^{J}\theta_j = 1$; $J \leq 3$ is usually sufficient for football data whereas $J = 0$ corresponds to zero-inflated models. Although univariate zero-inflated Poisson regression models have been developed and examined in detail (see, for example, Lambert 1992 and Böhning et al. 1999), multivariate extensions, similar to the models proposed in this paper, are relatively rare with the exception of Li et al. (1999), Gan (2000), and Wahlin (2001).

There are two important properties of such models. Firstly, the marginal distributions of a diagonal inflated model are not Poisson distributions but mixtures of distributions with one Poisson component. Secondly, if $\lambda_3 = 0$ (corresponding to the double-Poisson distribution) the resulting inflated distribution introduces a degree of dependence between the two variables under consideration. For this reason, diagonal inflation may correct both the overdispersion and the correlation problems that are encountered in modelling football games.

Application in sports

Modelling football games

Many references in this field assume that the number of goals scored by each team follows a Poisson distribution (Maher 1982, Lee 1997, and Rue and Salvesen 2000, among others). Such models take the general form

$$X_i \sim \mathrm{Poisson}(\lambda_{1i}),$$
$$Y_i \sim \mathrm{Poisson}(\lambda_{2i}),$$
$$\log(\lambda_{1i}) = m + \mathrm{home} + \mathrm{att}_{h_i} + \mathrm{def}_{g_i} + \mathrm{home.att}_{h_i} + \mathrm{home.def}_{g_i} + \mathrm{att.def}_{h_i g_i} + \mathrm{home.att.def}_{h_i g_i}$$
$$\log(\lambda_{2i}) = \mu + \mathrm{att}_{g_i} + \mathrm{def}_{h_i} + \mathrm{att.def}_{g_i h_i}$$

for $i = 1,2,\ldots,n$, where n is the number of games or observations, i is a game (observation) indicator, h_i and g_i indicate the home and the away team in game i, X_i and Y_i are the goals scored by the home (h_i) and the away (g_i) team in game i, λ_{1i} and λ_{2i} are the corresponding expected number of goals, μ is a constant parameter, home is the home effect parameter, and finally, att_k and def_k encapsulate the offensive (or attacking) and defensive performances of team k. Karlis and Ntzoufras (2000) examined such models in a general log-linear setting allowing also for model selection.

Although we have implemented our proposed models in various data sets, here we focus on the Italian serie A data for the 1991–1992 season and give some brief details for the Champions League data for the 2000–2001 season. The classical likelihood ratio test (LRT) and its asymptotic $\chi^2 p$-value as well as Bayes information criterion (BIC) and Akaike information criterion

(AIC) were used in the selection and fitting of models. We adopted a simpler structure for the parameters involved in the linear predictors of λ_1 and λ_2. Hence, for each game $i (i = 1, \ldots, n)$,

$$\begin{aligned}(X_i, Y_i) &\sim \mathrm{BP}(\lambda_{1i}, \lambda_{2i}, \lambda_{3i}),\\ \log(\lambda_{1i}) &= \mu + \mathrm{home} + \mathrm{att}_{h_i} + \mathrm{def}_{g_i},\\ \log(\lambda_{2i}) &= \mu + \mathrm{att}_{g_i} + \mathrm{def}_{h_i}.\end{aligned} \tag{33.6}$$

To achieve identifiability of the above model parameters, we may use any standard set of con straints. Here we propose to use either sum-to-zero or corner constraints, depending on the interpretation that we prefer. For this example, we chose sum-to-zero constraints for ease of interpretation. Therefore, the overall constant parameter specifies λ_1 and λ_2 when two teams of the same strength play on a neutral field. Offensive and defensive parameters are expressed as departures from a team of average offensive or defensive ability.

For the covariance parameters λ_{3i} we considered various versions of the linear predictor which can be summarized by

$$\log(\lambda_{3i}) = \beta^{\mathrm{con}} + \gamma_1 \beta^{\mathrm{home}}_{h_i} + \gamma_2 \beta^{\mathrm{away}}_{g_i},$$

where β^{con} is a constant parameter and $\beta^{\mathrm{home}}_{g_i}$ and $\beta^{\mathrm{away}}_{h_i}$ are parameters that depend on the home and away team, respectively. Parameters γ_1 and γ_2 are dummy binary indicators taking values 0 or 1, depending on the model that we consider. Hence when $\gamma_1 = \gamma_2 = 0$ we consider constant covariance, when $(\gamma_1, \gamma_2) = (1{,}0)$ we assume that the covariance depends on the home team only and so on.

The parameter λ_3 can be interpreted as a random effect which acts additively on the marginal mean and reflects game conditions. An alternative structure on the design matrix can be easily implemented if additional information is available, or if we assume that attacking abilities are different in home and away games, or if the home effect varies from team to team.

Diagonal inflated models are suitable for championships with an excess of draws which cannot be captured by double-Poisson models, or even bivariate Poisson models. Here we illustrate diagonal inflated models in the Italian serie A data for the 1991–1992 season. The scoring system in that season gave 2 points for a win and 1 point for a draw. This system often gives an excess of draws. We considered various models including the double-Poisson, the bivariate Poisson and the diagonal inflated models using several diagonal distributions; Table 33.1. The best-fitted model is the bivariate Poisson model with an extra parameter for the 1–1 score which was otherwise considerably underestimated: Table 33.2. The model selected is supported by the AIC, BIC and LRT for testing the hypothesis $H_0 : p = 0$, where p is the inflation proportion (p-value less than 0.01). Details for the model selection procedure are given in Table 33.1. Note that the zero-inflated and the geometric diagonally inflated models did not improve the likelihood since 0–0 scores were not underestimated. Moreover, the improvement that is offered by a Poisson diagonal component was statistically significant when added to the simple Poisson model but not when added to the bivariate Poisson model.

The goodness of fit was assessed by comparing our proposed model with the full or saturated model which fits the data exactly. According to the LRT our proposed model fits our data sufficiently well (p-value 0.85). Moreover, the AIC and BIC measures for the full model are 2204.0 and 4928.7 respectively. Both these criteria indicate the selection of our model against the alternative full or saturated model.

Table 33.1 ***Details of the fitted models for the Italian serie A 1991–1992 data.***

Model distribution	Additional model details	Log-likelihood	Number of parameters	*p*-value	AIC	BIC
1, double Poisson		−771.5	36		1614.9	1774.2
Covariates on λ_3						
2, bivariate Poisson	Constant ($\gamma_1 = \gamma_2 = 0$)	−764.9	37	$0.00^{\dagger}$	1603.9	1767.5
3, bivariate Poisson	Home team effect ($\gamma_1 = 1, \gamma_2 = 0$)	−758.9	55	$0.84^{\ddagger}$	1627.8	1871.1
4, bivariate Poisson	Away team effect ($\gamma_1 = 0, \gamma_2 = 1$)	−755.6	55	$0.41^{\ddagger}$	1621.2	1864.5
5, bivariate Poisson	Home and away team effects ($\gamma_1 = \gamma_2 = 1$)	−745.9	72	$0.33^{\ddagger}$	1635.7	1954.3
6, zero-inflated bivariate Poisson	Constant	−764.9	38	$1.00^{\$}$	1605.9	1773.9
Diagonal distribution						
7, diagonal inflated bivariate Poisson	Geometric	−764.9	39	$1.00^{\$}$	1607.9	1780.3
8, diagonal inflated bivariate Poisson§§	Discrete (1)	−756.6	39	$0.00^{\$}$	1591.1	1763.7
9, diagonal inflated bivariate Poisson	Discrete (2)	−756.6	40	1.00*	1593.1	1770.1
10, diagonal inflated bivariate Poisson	Discrete (3)	−756.4	41	0.54**	1594.8	1776.2
11, diagonal inflated bivariate Poisson	Poisson	−763.5	39	$0.25^{\$}$	1605.1	1777.5
12, diagonal inflated Poisson	Poisson	−767.0	38	$0.01^{\$}$	1610.0	1778.1

† $H_0 : \lambda_3 = 0$.
‡ $H_0 : \lambda_3 =$ constant.
\$ $H_0 : p = 0$.
\$§ Best-fitted model.
* $H_0 : \theta_2 = 0$.
** $H_0 : \theta_3 = 0$.

Table 33.3 provides the parameter estimates of a simple Poisson model and the selected diagonal inflated bivariate Poisson model. The expected number of goals from this model for game i with home team h_i and away team g_i are

$$\begin{aligned} E(X_i) &= \lambda_{1i}, \\ E(Y_i) &= \lambda_{2i}, \end{aligned} \tag{33.7}$$

where λ_{1i} and λ_{2i} are given by expression (33.6). For the inflated bivariate Poisson model, a Bernoulli distribution was used as inflation with parameter θ_1. The expected number of goals by using this model can be calculated as

$$\begin{aligned} E(X_i) &= (1-p)(\lambda_{1i} + \lambda_{3i}) + p\theta_1, \\ E(Y_i) &= (1-p)(\lambda_{2i} + \lambda_{3i}) + p\theta_1, \end{aligned} \tag{33.8}$$

Table 33.2 ***Estimated draws for every model.***

Model distribution	Additional model details	Estimates for the following scores				
		0–0	1–1	2–2	3–3	4–4
Observed data		38	58	10	4	1
1, double Poisson		38	33	9	1	0
Covariates on λ_3						
2, bivariate Poisson	Constant	49	35	11	2	0
3, bivariate Poisson	Home team effect	51	34	11	3	0
4, bivariate Poisson	Away team effect	49	34	11	2	0
5, bivariate Poisson	Home and away team effects	47	32	10	2	0
6, zero-inflated bivariate Poisson		49	35	11	2	0
Diagonal distribution						
7, diagonal inflated bivariate Poisson	Geometric	49	35	11	2	0
8, diagonal inflated bivariate Poisson†	Discrete (1)	43	58	9	2	0
9, diagonal inflated bivariate Poisson	Discrete (2)	43	58	9	2	0
10, diagonal inflated bivariate Poisson	Discrete (3)	43	58	9	3	0
11, diagonal inflated bivariate Poisson	Poisson	50	38	13	3	1
12, diagonal inflated Poisson	Poisson	45	40	14	3	1

† Best-fitted model.

where λ_{1i} and λ_{2i} are calculated by using expression (33.6) and p is the estimated mixing proportion.

Fitted counts for draws are shown in Table 33.2. Note that the simple Poisson model fits the number of 0–0 draws very well but considerably underestimates the number of 1–1 scores. In contrast, the bivariate Poisson model overestimates 0–0 scores since it reduces the estimated ability of the teams, as described in 'The bivariate Poisson distribution and its implementation in sports modelling'. Finally, the model selected seems to be a trade off between counts for 0–0 and 1–1 scores. It is closer to the observed data since it exactly estimates the number of 1–1 draws and slightly overestimates the number of 0–0 draws.

It is worth mentioning that the scoring system has been changed to encourage teams not to be satisfied by draws; a win is now worth 3 points and a draw just 1 point. This has led to a reduction in the number of draws in recent championships.

Note that the LRT for testing mixture models with different numbers of components is known to be inappropriate (see, for example, Lindsay, 1995). So the choice of model between such mixtures can be based on the AIC.

A variety of models was fitted in the Champions League data for the 2000–2001 season. The best-fitted model was the bivariate Poisson model with constant λ_3 supported by the LRT which rejects the hypothesis $H_0 : \lambda_3 = 0$ (p-value 0.042) and the AIC. The zero- and diagonal inflated models did not improve the model fit. This is mainly because these models are useful only when the model selected underestimates the number of draws.

Table 33.3 ***Estimated parameters for the Poisson and bivariate Poisson models for 1991–1992 Italian serie A data.***[†]

Team	Results for model 1, double Poisson		Results for model 2, bivariate Poisson	
	Attack	Defence	Attack	Defence
1, Milan	0.68	−0.50	0.84	−1.18
2, Juventus	0.18	−0.50	0.22	−0.70
3, Torino	0.11	−0.60	0.18	−0.86
4, Napoli	0.43	0.12	0.51	0.19
5, Roma	0.00	−0.16	0.02	−0.17
6, Sampdoria	0.02	−0.16	0.10	−0.16
7, Parma	−0.15	−0.27	−0.14	−0.34
8, Inter	−0.29	−0.28	−0.37	−0.29
9, Foggia	0.49	0.50	0.57	0.63
10, Lazio	0.16	0.10	0.28	0.21
11, Atalanta	−0.18	−0.11	−0.21	−0.11
12, Fiorentina	0.18	0.13	0.29	0.28
13, Genoa	−0.04	0.25	−0.09	0.40
14, Cagliari	−0.21	−0.08	−0.21	−0.01
15, Verona	−0.40	0.43	−0.51	0.57
16, Bari	−0.33	0.24	−0.50	0.33
17, Cremonese	−0.29	0.28	−0.36	0.45
18, Ascoli	−0.34	0.61	−0.64	0.75
Other parameters				
Intercept μ	−0.18		−0.57	
Home team effect	0.36		0.50	
λ_3	0.00		0.23	
Mixing proportion	0.00		0.09	
θ_1			1.00	

[†] Expected number of goals can be calculated by using equations (7) and (6) for model 1 and equations (8) and (6) for model 2.

Modelling water-polo outcomes

In this section we present an implementation of the bivariate Poisson models on water-polo games. This sport was selected because of the relatively small scores (so that it is plausible to use discrete distributions) and large correlations between the scores of the competing teams. Useful information can be found at www.usawaterpolo.com and www.hickoksports.com. The main aim of the game is to score goals. Usual scores are around 8 goals for each team, with strong correlation between the scores of the competing teams.

Here we implemented the bivariate Poisson models to the data of the European national team cup held at Florence in September 1999. Twelve national teams played a total of 50 games.

In our analysis we considered only full-time scores, ignoring extra time. The effect of this truncation is minimal with only two draws observed.

The model that we consider was similar to the corresponding model for football games without the parameter estimating the home effect:

$$(X_i, Y_i) \sim \text{BP}(\lambda_{1i}, \lambda_{2i}, \lambda_{3i}),$$

$$\log(\lambda_{1i}) = \mu + \text{att}_{o_{1i}} + \text{def}_{o_{2i}}, \tag{33.9}$$

$$\log(\lambda_{2i}) = \mu + \text{att}_{o_{2i}} + \text{def}_{o_{1i}}, \tag{33.10}$$

for $i = 1, \ldots, 50$; here o_{1i} and o_{2i} are indicators corresponding to the first and second mentioned team opponents in game i. For λ_3 we consider two cases; in the first λ_3 is constant whereas in the second λ_3 is given by

$$\log(\lambda_{3i}) = m + \text{team}_{o_{1i}} + \text{team}_{o_{2i}} \tag{33.11}$$

for each game between teams o_{1i} and o_{2i}, where team_k is the effect of team k on λ_3. For these data we did not consider the zero-inflated and diagonal inflated models since draws in water-polo are rare. As in the football implementation, we may use either sum-to-zero or corner constraints depending on the interpretation that we prefer. Here we considered corner constraints with Germany as the base-line team. The constant parameter specifies parameters λ_1 and λ_2 when Germany plays a team with identical offensive and defensive ability. Moreover, the offensive (att_k) and defensive (def_k) parameters express differences in each team's offensive or defensive abilities from those of Germany (the base-line team).

As a consequence three models were fitted (the double Poisson, bivariate Poisson with constant covariance and bivariate Poisson with covariance depending on opposing teams). All three criteria used (the LRT, AIC and BIC) indicated as best model the bivariate Poisson model with constant λ_3; Table 33.4. The AIC and BIC values for the full or saturated model (508.4 and 726.4 respectively) also indicated selection of our model against the full model. The covariance parameter λ_3 was found to be equal to 5.55, which indicated significant covariance between the scores of the opposing teams.

Table 33.4 ***Details of the fitted models for the water-polo 1999 European National Cup.***

Model	Covariates on λ_3	Log-likelihood	Number of parameters	p-value	AIC	BIC
1, double Poisson		−186.67	23		417.3	474.3
2, bivariate Poisson	Constant	−171.91	24	0.000†	391.8	451.3
2, bivariate Poisson (Hungary base-line; $\mu = 0$)	Constant	−172.31	23	0.376‡	390.62	447.60
3, bivariate Poisson	Equation (11)	−167.31	35	0.602§	404.6	491.3

† $H_0 : \lambda_3 = 0$.
‡ $H_0 : \mu_3 = 0$.
§ $H_0 : \lambda_3 = \text{constant}$.

The parameter estimating the defensive ability of Hungary was found to be a very large negative number. This implies that the fitted values of the goals conceded by Hungary tend to be constant for all opposing teams and equal to the covariance parameter λ_3 (instead of $\lambda_2 + \lambda_3$). Any value for the defensive ability of Hungary lower than −20.0 results in identical fitted values and likelihood. For this reason, and to avoid unidentifiability, we have set the defensive ability of Hungary to −20.0. Although this may imply that the model does not provide a good or sensible estimation of the Hungarian team's defensive ability this is not so. If we isolate the data concerning the defence of the Hungarian team (7,3,6,3,6,4,5,12) we observe that they are under-dispersed relative to the Poisson distribution (mean, 5.75; variance, 8.50). The variance test (see Karlis and Xekalaki 2000 for a critical review) does not reject the hypothesis of the Poisson distribution. Hence, the constant mean for the Hungarian team's defence seems plausible. For the rest of the teams, the observed data are overdispersed and therefore it is plausible to assume that the mean is not constant. Note that this numerical problem appears mainly because of the large number of parameters relative to the amount of data that is available in such competitions and to the unbalanced design of the data. For this reason, in full season leagues we shall not have such problems.

To avoid possible overparameterization we also considered a model with Hungary as a baseline team and the corresponding constant (which is a measure of the overall performance of

Table 33.5 ***Estimated parameters for the double-Poisson and bivariate Poisson models for the 1999 water-polo European National Cup Data.***†

Team	Final ranking	Results for model 1, double Poisson		Results for model 2, bivariate Poisson ($\mu = 0$)	
		Attack	Defence	Attack	Defence
1, Germany	8	−0.80	0.23	−2.03	2.28
2, Greece	4	−0.48	0.06	−0.79	1.81
3, Hungary	1	0.00	0.00	0.00	0.00
4, Italy	3	−0.45	−0.03	−1.68	0.12
5, Croatia	2	−0.15	0.18	−0.76	0.81
6, Netherlands	12	−0.56	0.50	−2.31	2.59
7, Romania	9	−0.74	0.16	−2.65	1.71
8, Russia	5	−0.34	0.39	−1.50	1.76
9, Slovakia	10	−0.54	0.29	−2.63	1.95
10, Slovenia	11	−0.63	0.34	−2.70	2.22
11, Spain	6	−0.28	0.35	−0.88	2.08
12, Yugoslavia	7	−0.54	0.17	−1.37	1.89
Other parameters					
Intercept		2.18		0.00	
λ_3		0.00		5.50	

† Base-line level Hungary. The expected numbers of goals are $\lambda_{1i} + \lambda_{3i}$ and $\lambda_{2i} + \lambda_{3i}$; λ_{1i} and λ_{2i} are given by equations (33.9) and (33.10) for both models. For model 1 $\lambda_{3i} = 0.0$.

Table 33.6 ***Estimated parameters for the double-Poisson and bivariate Poisson models for the 1999 water-polo European National Cup Data.***

Game	Phase	Opponents		Observed score	Expected goals for the following models:			
		Team 1	Team 2		Double Poisson		Bivariate Poisson†	
					Team 1	Team 2	Team 1	Team 2
1	Group A	Italy	Hungary	7–7	5.7	8.6	5.7	6.7
2	Group A	Slovenia	Hungary	3–11	4.7	12.4	5.6	14.8
3	Group A	Hungary	Croatia	7–6	10.6	7.7	7.8	6.0
4	Group A	Greece	Hungary	3–8	5.5	9.3	6.0	11.7
5	Group A	Hungary	Slovakia	13–6	11.8	5.2	12.6	5.6
6	Quarter-final	Hungary	Germany	15–4	11.1	4.0	15.4	5.7
7	Semi-final	Hungary	Italy	7–5	8.6	5.7	6.7	5.7
8	Final	Croatia	Hungary	12–15	7.7	10.6	6.0	7.8

† Hungary base-line; $\mu = 0$.

Hungary) constrained to be 0. This may be interpreted in the following way: if Hungary plays with a team which has the same attacking and defensive ability then the expected number of goals scored will be 1 goal added to the covariance parameter λ_3, i.e. $E(X) = E(Y) = 1 + \lambda_3$. The advantage of such a model is that it does not assume a constant expected number of goals conceded by Hungary. Comparing the two models with a χ^2-test ($H_0 : \mu = 0$; p-value 0.38), the BIC or AIC we conclude that the model with Hungary as the base-line team and its corresponding constant equal to 0 should be selected. For details of the model parameters of the two finally selected models see Table 33.5; fitted values for the games of Hungary are given in Table 33.6.

Discussion

In the present paper the bivariate Poisson distribution and its extensions were used to model sports data. The bivariate Poisson distribution allows for correlation between the scores of the competing teams, which is plausible for certain team sports. Diagonal inflated models are also proposed to improve the modelling aspects further. The models proposed provide a better fit of the football data since they can handle both correlation and overdispersion. Furthermore, they improve the fit on the diagonal of the observed table of results, which reflects ties between the two opponents. According to the models proposed we can extend the double-Poisson model by either considering a bivariate Poisson model or by inflating the diagonal elements of the joint probability function. Both models incorporate correlation, but the latter introduces overdispersion as well.

Maximum likelihood estimation for bivariate Poisson regression models was described in detail. The EM algorithm that is proposed in this paper can be easily extended to incorporate more complicated models. Such models were described in Kocherlakota and Kocherlakota (2001). Additionally, we have extended the zero-inflated multivariate Poisson models of Li et al. (1999) by defining more general inflation models, with potential implementation in manufacturing or marketing. The EM algorithm that was proposed for such models can be quite helpful for real data applications of such models.

ACKNOWLEDGEMENTS

The authors thank the Joint Editors and the referees for their helpful comments on previous versions of the paper.

REFERENCES

Abramowitz, M. and Stegun, I. A. (1974) *Handbook of Mathematical Functions*. New York: Dover Publications.

Böhning, D., Dietz, E., Schlattmann, P., Mendonça, L. and Kirchner, U. (1999) The zero-inflated Poisson model and the decayed, missing and filled teeth index in dental epidemiology. *J. R. Statist. Soc.* A, **162**, 195–209; correction, **163** (2000), 121–122.

Dixon, M. J. and Coles, S. G. (1997) Modelling association football scores and inefficiencies in the football betting market. *Appl. Statist.*, **46**, 265–280.

Gan, N. (2000) General zero-inflated models and their applications. *PhD Thesis*. North Carolina State University, Raleigh.

Ho, L. L. and Singer, J. M. (2001) Generalized least squares methods for bivariate Poisson regression. *Communs Statist. Theory Meth.*, **30**, 263–277.

Irwin, J. O. (1937) The frequency distribution of the difference between two independent variates following the same Poisson distribution. *J. R. Statist. Soc.* A, **100**, 415–416.

Johnson, N., Kotz, S. and Balakrishnan, N. (1997) *Discrete Multivariate Distributions*. New York: Wiley.

Johnson, N., Kotz, S. and Kemp, A. W. (1992) *Univariate Discrete Distributions*. New York: Wiley.

Karlis, D. and Ntzoufras, I. (2000) On modelling soccer data. *Student*, **3**, 229–244.

Karlis, D. and Xekalaki, E. (2000) A simulation comparison of several procedures for testing the Poisson assumption. *Statistician*, **49**, 355–382.

Keller, J. (1994) A characterization of the Poisson distribution and the probability of winning a game. *Am. Statistn*, **48**, 294–299.

Kocherlakota, S. and Kocherlakota, K. (1992) *Bivariate Discrete Distributions*. New York: Dekker.

Kocherlakota, S. and Kocherlakota, K. (2001) Regression in the bivariate Poisson distribution. *Communs Statist. Theory Meth.*, **30**, 815–827.

Lambert, D. (1992) Zero-inflated Poisson regression, with an application to defects in manufacturing. *Technometrics*, **34**, 1–14.

Lee, A. J. (1997) Modeling scores in the Premier League: is Manchester United really the best? *Chance*, **10**, 15–19.

Li, C. S., Lu, J. C., Park, J., Kim, K. and Peterson, J. (1999) Multivariate zero-inflated Poisson models and their applications. *Technometrics*, **41**, 29–38.

Lindsay, B. G. (1995) Mixture models: theory, geometry and applications. *Regl Conf. Ser. Probab. Statist.*, **5**.
Maher, M. J. (1982) Modelling association football scores. *Statist. Neerland.*, **36**, 109–118.
Rue, H. and Salvesen, Ø. (2000) Prediction and retrospective analysis of soccer matches in a league. *Statistician*, **49**, 399–418.
Skellam, J. G. (1946) The frequency distribution of the difference between two Poisson variates belonging to different populations. *J. R. Statist. Soc.* A, **109**, 296.
Wahlin, J. F. (2001) Bivariate ZIP models. *Biometr. J.*, **43**, 147–160.

CHAPTER 34

Dynamic rating of sports teams

LEONHARD KNORR-HELD

Abstract

We consider the problem of dynamically rating sports teams on the basis of categorical outcomes of paired comparisons such as win, draw and loss in football. Our modelling framework is the cumulative link model for ordered responses, where latent parameters represent the strength of each team. A dynamic extension of this model is proposed with close connections to nonparametric smoothing methods. As a consequence, recent results have more influence in estimating current abilities than results in the past. We highlight the importance of using a specific constrained random walk prior for time-changing abilities which guarantees an equal treatment of all teams. Estimation is done with an extended Kalman filter and smoother algorithm. An additional hyperparameter which determines the temporal dynamic of the latent team abilities is chosen on the basis of the optimal one-step-ahead predictive power. Alternative estimation methods are also considered. We apply our method to the results from the German football league *Bundesliga* 1996–1997 and to the results from the American National Basketball Association 1996–1997.

Keywords: *constrained random walk prior, cumulative link model, dynamic model, invariance of estimators, ordinal response, paired comparisons, rating*

Introduction

When sports teams compete in pairs, they collect scores or goals within a game. Typically, the winning team, the team with more scores at the end of the game, is rewarded with a certain number of points whereas the losing team remains empty handed. Often there is also the possibility of a draw, where both teams have the same score and therefore are rewarded with the same number of points. For example, in football a winning team receives 3 points and a draw is rewarded with 1 point for each team. In general, with or without a draw, results from those paired comparisons are essentially given in ordered categories and those categories determine the standing in the league.

Many approaches to rating sports teams use the difference in scores as the response variable within standard linear model methodology. Other approaches, especially in football, use log-linear Poisson models for the number of goals of both teams. However, it is clear that a rating system should reward a team for winning per se and not for running up the score (Harville, 1977). This has led many researchers to propose robust versions for rating sports teams on the basis of the difference in scores. For example, Harville (1977, 1980) proposed to truncate the difference at some predefined cutpoint, whereas Bassett (1997) used L_1-norm regression. Taking Harville's argument to the limit, we use only result categories as the basis to rate teams within a regression model for ordinal categorical data. Thus the main goal here is rating and not prediction, where the score of each team and many other factors should be considered as possible predictors such as the percentage of ball possession of each team or the number of spectators. Nevertheless, we use the predictive power of our formulation for model fitting.

Let y_{ij} be the result from a paired comparison of team i and team j, where team i is the home team and team j is the visiting team. For example, results from football matches are given in three categories, say

$$y_{ij} = \begin{cases} 1 & \text{if the home team } i \text{ wins,} \\ 2 & \text{for a draw and} \\ 3 & \text{if the visiting team } j \text{ wins.} \end{cases}$$

Cumulative link models are a popular framework to analyse paired comparisons. These models assign a latent ability α_i to each team i, representing the strength of the team. The model is constructed in such a way that the difference in ability of the competing teams α_i—α_j affects the probabilities of the results y_{ij} through a response function F; see for example Agresti (1992).

An important special case is the cumulative logistic model, which boils down to the model of Bradley and Terry (1952) for two possible outcome categories (win or loss). An extension to more than two categories is discussed in Tutz (1986). We give a short review of the classical non-dynamic cumulative link model in the next section.

When paired comparisons are observed over time, often the teams' performances vary over time because of injuries of important players, changes of the coach or for other reasons. Fahrmeir and Tutz (1994a) introduced dynamic models for longitudinal paired comparison data where abilities are allowed to vary smoothly over time. These models can be seen as Bayesian dynamic models with specific smoothing priors and have close connections with nonparametric smoothing methods, since no functional form is specified for the temporal development of the now time-changing abilities. An extended version of the Kalman filter algorithm for categorical data is used to estimate unknown parameters. Similar dynamic versions of the Bradley–Terry model were proposed independently in Glickman (1993), who used Markov chain Monte Carlo (MCMC) methods for a full Bayesian analysis.

In 'Dynamic cumulative link model' we introduce a dynamic model which is based on the approaches above but has certain amendments. One crucial point is that our model treats all teams symmetrically, which is guaranteed by a specific singular multivariate Gaussian distribution as a smoothing prior for the temporal development of the abilities of the teams. Additional threshold parameters, which represent a possibly existing home advantage, are assumed as time constant and team independent. Estimation and prediction are outlined in section after that. Posterior mode estimators of the abilities are calculated with the extended Kalman filter and smoother (EKFS) algorithm of Fahrmeir (1992). A hyperparameter, determining the temporal smoothness of the abilities of each team, can be chosen by maximizing one-step-ahead prediction criteria, natural by-products of the EKFS algorithm. Such an approach has an optimality feature, which is understandable, at least in principle, to the public. Alternative ways to estimate the smoothing parameter are an EM type of algorithm and a fully Bayesian analysis by MCMC methods, which are also considered. The corresponding software is available from the author on request.

We apply our method to data from the German football league in 1996–1997 in the section that follows. A comparison with estimates by MCMC simulation suggests that inference by the EKFS gives reliable estimates. Furthermore, we analyse a larger data set from the American National Basketball Association, season 1996–1997. Here we have found interesting and pronounced temporal trends in the estimated strengths of several teams. The final section concludes with possible modifications and generalizations of our model and with some other comments.

Cumulative link model

Let n denote the number of teams in the league, and let R denote the number of categories in the ordinal response scale. A cumulative link model for a comparison y_{ij} of a home team i with a visiting team j is defined by

$$\Pr(y_{ij} \leq r) = F(\theta_r + \alpha_i - \alpha_j), \quad r = 1, \ldots, R-1, \tag{34.1}$$

where F is a distribution function, $\theta_1 < \ldots < \theta_{R-1}$ are so-called threshold parameters and α_i, is the latent ability of team i. For notational convenience we furthermore introduce $\theta_0 = -\infty$ and $\theta_R = \infty$ so that the probability of observing a result $y_{ij} = r$ can be written as

$$\Pr(y_{ij} = r) = F(\theta_r + \alpha_i - \alpha_j) - F(\theta_{r-1} + \alpha_i - \alpha_j), \quad r = 1, \ldots, R.$$

The threshold parameters can represent a home advantage, which is an important factor for nearly all kinds of sports. For illustration consider a match where both teams have the same ability $\alpha_i = \alpha_j$. The probabilities $\Pr(y_{ij} = r) = F(\theta_r) - F(\theta_{r-1})$ depend now only on the thresholds θ_r and θ_{r-1}. For example, in the case of three categories, the larger θ_1 is, the larger is the probability $\Pr(y_{ij} = 1) = F(\theta_1)$ that the home team wins. Note that the home advantage is assumed to be the same for all teams.

The estimation of $\theta = (\theta_1, \ldots, \theta_{R-1})'$ and $\alpha = (\alpha_1, \ldots, \alpha_n)'$ is done by maximizing the likelihood, the product of individual contributions $\Pr(y_{ij} = r)$ of all matches. Computation can be done conveniently with standard software. Note that the model is unidentifiable because only differences of abilities enter the likelihood. Adding a constant to α_i, $i = 1, \ldots, n$, will not change the likelihood. It is therefore necessary to impose an additional constraint so that the level of abilities is specified. Usual constraints are $\Sigma_{i=1}^{n} \alpha_i = 0$ or $\alpha_n = 0$, say. Invariance of the maximum likelihood (ML) estimator (e.g. Cox and Hinkley, 1974) ensures that estimated abilities are equivalent, whatever constraint was used. For example, the ML estimator $\hat{\alpha}_1, \ldots, \hat{\alpha}_{n-1}$

under the constraint $\alpha_n = 0$ can be used to calculate the ML estimator $\tilde{\alpha}_1, \ldots, \tilde{\alpha}_n$ under the constraint $\Sigma_{i=1}^{n} \alpha_i = 0$ by

$$\begin{aligned} \tilde{\alpha}_i &= \hat{\alpha}_i - \frac{1}{n} \sum_{j=1}^{n-1} \hat{\alpha}_j, \quad i = 1, \ldots, n-1, \\ \tilde{\alpha}_n &= -\frac{1}{n} \sum_{j=1}^{n-1} \hat{\alpha}_j. \end{aligned} \tag{34.2}$$

In the 'Dynamic cumulative link model' section we shall show that this invariance no longer holds for dynamic cumulative link models.

The estimated abilities α_i, $i = 1, \ldots, n$, are the basis for rating teams. This approach has certain advantages compared with the standard rating based on adding points in football or counting the number of wins in basketball. In particular, because abilities are estimated simultaneously, the approach automatically adjusts for the strength of the corresponding opponents and for the home advantage. Furthermore, future games can be predicted.

The ML estimator might not exist for some constellations of data, owing to the discrete nature of the data. For example, a team which wins or loses all its matches will have a positive or negative infinite estimated ability respectively. It is therefore advisable to check before the analysis that all teams did not win or lose all their matches. Singularities will also arise if teams can be partitioned into two subsets in which none of the teams in one subset competes against any other team in the other subset. Comparisons within a league, however, are usually scheduled in a way that every team competes against every other team in the league, so this type of singularity will not arise.

Dynamic cumulative link model

The basic model

Suppose now that paired comparisons y_{tij} are observed over time t. We consider time $t = 1, \ldots, T$ as discrete valued and equally spaced, such as days, weeks or months. Our starting-point is to allow in model (34.1) for time-dependent abilities α_{ti}, $t = 1, \ldots, T, i = 1, \ldots, n$:

$$\Pr(y_{tij} \leq r) = F(\theta_r + \alpha_{ti} - \alpha_{tj}), \quad r = 1, \ldots, R-1. \tag{34.3}$$

This specification allows us to rate sports teams dynamically by estimating time-dependent abilities. We assume Gaussian first-order random walks

$$\alpha_{ti} \sim N(\alpha_{t-1,i}, \sigma^2) \tag{34.4}$$

as smoothing priors for the abilities of each team i. This is a common assumption for the dynamic modelling of paired comparisons; see Glickman (1993, 1999) and Fahrmeir and Tutz (1994a). The corresponding prior distributions neither impose stationarity nor assume a specific parametric form; in fact model (34.4) is related to semiparametric and nonparametric smoothing methods as reviewed by Fahrmeir and Knorr-Held (2000). This paper also indicates how to generalize the prior to observations which are not made on a regular time grid.

Model (34.4) implies that recent matches have more weight for estimating current abilities than results way back in time, which is a natural assumption. The parameter σ^2 determines the weights and hence the loss of memory rate of the random walks. For the limiting case $\sigma^2 = 0$ the model reduces to the classical non-dynamic model (34.1).

Ensuring exchangeability by a constrained random walk prior

The crucial point is how to impose a smoothing prior on the abilities $\alpha_t = (\alpha_{t1}, \ldots, \alpha_{tn})'$ without destroying an exchangeable treatment of the teams. This problem occurs, since, as in the time-independent case, we must impose an additional restriction on $\alpha_t, t = 1, \ldots, T$, to assure identifiability. It is, however, not as straightforward as in the time-independent case, because the posterior mode estimator in dynamic models is *not* invariant with respect to the identifiability constraint. We therefore propose a construction based on a specific multivariate singular Gaussian distribution for α_t, which ensures that, marginally, all components of α_t follow a first-order random walk (34.4), but where the sum $\sum_{i=1}^{n} \alpha_{ti}$ is 0 for each time point t. Harvey (1989, pp. 41–44), has described a related approach, where seasonal dummy variables sum to 0.

More formally, we assume that α_t follows a constrained multivariate Gaussian random walk

$$\alpha_t = \alpha_{t-1} + u_t, \quad u_t \sim N(0, Q), \quad t = 1, \ldots, T, \tag{34.5}$$

with independent disturbances $u_t, t = 1, \ldots, T$, and initial value α_0 fulfilling $\mathbf{1}'\alpha_0 = 0$. Here $\mathbf{1}$ denotes the vector $(1, 1, \ldots, 1)'$. We now specify a specific singular dispersion matrix Q of rank n—1, which ensures that $\mathbf{1}'u_t = 0$ and hence $\mathbf{1}'\alpha_t = 0$ for $t = 1, \ldots, T$. In general there are many such matrices Q but—apart from a proportionality constant—there is only one which treats components of u_t as exchangeable. A detailed discussion can be found in Knorr-Held (1997). For the case where all components of u_t have the same variance σ^2, Q is given by

$$Q = \sigma^2 \left(I - \frac{1}{n}\mathbf{1}\mathbf{1}' \right).$$

Here I denotes the identity matrix. The exchangeable treatment is easily seen, because all non-diagonal entries, and hence all covariances between components of u_t, are equal. Furthermore, all diagonal entries are equal so, marginally and ignoring a multiplicative factor $(n—1)/n$ for the variance, every component of α_t follows a regular univariate random walk (34.4). Note that our model implies that components of u_t are a priori negatively correlated.

It is easily seen that the sum of components of u_t is 0, because $\mathbf{1}'u_t$ has variance

$$\mathbf{1}'Q\mathbf{1} = \sigma^2\mathbf{1}' \left(I - \frac{1}{n}\mathbf{1}\mathbf{1}' \right) \mathbf{1} = 0$$

and is therefore equal to $E(\mathbf{1}'u_t) = 0$ with probability 1. For the more general case with individual variances σ_i^2 for each team, Q is given by $Q = M\Sigma M$ with

$$M = I - \frac{1}{n}\mathbf{1}\mathbf{1}'$$

and $\Sigma = \text{diag}(\sigma_1^2, \sigma_2^2, \ldots, \sigma_n^2)$.

It is computationally convenient to consider a linear transformation of u_t, say Lu_t, where the first n—1 components are Gaussian with *regular* dispersion matrix and the last component is 0 with probability 1. For example,

$$L = \begin{pmatrix} I & -1 \\ \mathbf{1}' & 1 \end{pmatrix}$$

in such a transformation matrix. The first n—1 components of Lu_t now have dispersion $P = \sigma^2(I + \mathbf{1}\mathbf{1}')$. We can now perform inference for these n—1 a priori positively correlated components. The transformation $M(Lu_t)$ with M as defined above, which corresponds to equation (34.2), retransforms Lu_t to u_t. The whole approach can also be used in the general case where each component has its own variance σ_i^2, $i = 1, \ldots, n$; here $P = \text{diag}(\sigma_1^2, \sigma_2^2, \ldots, \sigma_{n-1}^2) + \sigma_n^2\mathbf{1}\mathbf{1}'$.

Fahrmeir and Tutz (1994a) assumed *independent* Gaussian first-order random walk priors for all except one team, say

$$\alpha_{ti} = \alpha_{t-1,i} + u_{ti} \quad u_{ti} \sim N(0, \sigma_i^2), \quad i = 1, \ldots, n-1, \quad t = 1, \ldots, T, \tag{34.6}$$

and put the ability of the last team to 0 for each time point t. They also used a similar strategy for other smoothing priors such as the local linear trend model. After estimation, they recentred the estimates to mean 0 for every time point t by using the transformation matrix M from above. However, this approach does not treat teams symmetrically because of the prior independence of the random walks: a change of the reference team will lead to different ratings. This can be best seen for the case where all random walks have the same variance σ^2: the recentred increments u_{ti}, $i = 1, \ldots, n$—1, now have variance $\sigma^2\{1—2/n + (n—1)/n^2\}$ whereas increments of the reference team have variance $\sigma^2(n—1)/n^2$, which is smaller for $n > 2$. The estimated abilities of the reference team will show less temporal variation than all the others. For example, for $n = 18$ as in the German football league example in the relevant subsection, the standard deviation of the recentred u_{ti} is 0.970σ, $i = 1, \ldots, n$—1, whereas the standard deviation of u_{tn} is 0.229σ. This difference increases as n increases. In the more general case with team-specific variances σ_i^2, the variance σ_n^2 of team n does not even occur in the model specification (34.6) and therefore cannot be estimated from the data.

Model (34.6) can easily be modified to achieve a symmetric treatment of the teams. The replacement of the independent random walk priors for the n—1 components with the correlated multivariate random walk with covariance matrix P will give a model which is equivalent to the constrained random walk model (34.5), as outlined above.

Recently, Glickman and Stern (1998) have used a related, but slightly different, approach for fixing the overall level of time-dependent abilities. They proposed that not α_t but the mean of $\alpha_t|\alpha_{t-1}, \sigma^2$ is centred at 0:

$$\alpha_t|\alpha_{t-1}, \sigma^2 \sim N(M\alpha_{t-1}, \sigma^2 I) \tag{34.7}$$

with M as defined above. Thus the sum of components of α_t, $\mathbf{1}'\alpha_t$, has expectation 0 and variance $n\sigma^2$ whereas in our model $\mathbf{1}'\alpha_t$ has both expectation and variance equal to 0.

There is an important difference between the Glickman–Stern and our approach. Both are equally valid to predict future outcomes of games, as only differences in team abilities enter the likelihood, implied by equation (34.3). Also, for a given time t, teams can be ranked or rated on the basis of the Glickman–Stern estimates as well as on our estimates. However, the Glickman–Stern model is not readily usable for judging the temporal development of a given team. For example, suppose that team i has no match scheduled at time t. We would then expect the team's ability to be the same as at time $t-1$, no matter what the abilities of the other teams are at time $t-1$, and this is exactly what our formulation implies. However, in the Glickman–Stern model, the expected ability of this team is

$$\alpha_{ti} = \alpha_{t-1,i} - \frac{1}{n}\sum_{j=1}^{n} \alpha_{t-1,j}.$$

Hence, the expected ability might rise or drop, although the team has not performed in any games at all. Strictly speaking, estimated abilities in the Glickman–Stern model can only be interpreted for a given time t, but not for a given team i. Our model has the advantage that here estimated abilities are valid quantities to assess the performance of a specific team over time. Note that the primary focus in Glickman and Stern (1998) is prediction, and, for this, model (34.7) is equally well suited.

Estimation and prediction

The extended Kalman filter and smoother

For the moment, consider σ^2 as fixed. We use the EKFS algorithm of Fahrmeir (1992) and Fahrmeir and Tutz (1994a) to estimate time-dependent abilities. A detailed description can be found in Fahrmeir and Tutz (1994b), Chapter 8. Threshold parameters θ are estimated by ML given the current (smoothed) estimates of α. Both steps (EKFS estimation of α for fixed θ and ML estimation of θ for fixed α) are alternated until convergence, which takes only a few seconds on a standard workstation or a Pentium processor in both of our applications.

The EKFS algorithm starts with an initial value α_0, a preseason estimate of the abilities, and then subsequently estimates α_t, $t = 1, \ldots, T$, on the basis of all games played until time t. This is called the filtering step and gives filtered estimates $\hat{\alpha}_t$. In a second step, the smoothing step, filtered estimates $\hat{\alpha}_t$, $t = T-1, \ldots, 0$, are smoothed on the basis of all games played until time T. The smoothed estimate of α_0 is used as a new initial value in the next iteration. The algorithm requires an initial value for the prior dispersion Q_0, say, of α_0. In our applications we used $Q_0 = M$, which is weakly informative but avoids numerical problems with more diffuse priors. The final estimates of α_0 are virtually identical whatever starting value for α_0 was used.

This algorithm can be derived as an approximate posterior mode estimator; see for example Fahrmeir and Tutz (1994b). Alternatively, we could use *iterative* EKFS (Fahrmeir and Wagenpfeil, 1997), which gives the exact posterior mode. The computation time increases, however, because an additional level of iteration is required. Furthermore, differences between estimates by non-iterative and iterative EKFS are typically small. See the references above for more details on the properties of posterior mode estimators.

The estimates of α_T and θ can be used to predict future matches. For example, suppose that team j is scheduled to visit team i at the next round ($t = T+1$) to play a match. The probabilities of the outcomes $\Pr(y_{T+1,ij} = r)$, $r = 1, \ldots, R$, can be estimated by model (34.3):

$$\Pr(y_{T+1,ij} = r) = F(\hat{\theta}_r + \hat{\alpha}_{Ti} - \hat{\alpha}_{Tj}) - F(\hat{\theta}_{r-1} + \hat{\alpha}_{Ti} - \hat{\alpha}_{Tj}), \tag{34.8}$$

because the first-order random walk assumption for α_t gives $\hat{\alpha}_T$ as the predicted ability at time $T + 1$. Note that the filtered and smoothed estimates of α_T coincide. More general one-step-ahead predictions are used later to assess the quality of the prediction and to estimate the smoothing parameter σ^2.

Estimating the smoothing parameter by maximizing the predictive power

In the following, we propose to estimate the smoothing parameter σ^2 on the basis of one-step-ahead prediction. Alternative ways are outlined afterwards.

Above it was sketched how filtered estimate $\hat{\alpha}_T$ can be used to predict future games. Filtered estimates $\hat{\alpha}_t$ are also available for $t = 0, \ldots, T—1$ from the EKFS, so we can perform a retrospective one-step-ahead prediction to assess the model fit. This is done by subsequently predicting outcomes at time $t + 1$ based on filtered estimates $\hat{\alpha}_t$ and comparing the predicted probabilities with the actual observed result by some criterion.

Let N be the total number of paired comparisons over the whole time period. We suppress the dependence on time t and opponents i and j and denote the predicted probabilities $\Pr(y_k = s)$ of outcomes $s = 1, \ldots, R$ by $\hat{p}_k(s), k = 1, \ldots, N$. These predictions are calculated on the basis of filtered estimates $\hat{\alpha}_t$, similar to equation (34.8), given only game information before period $t + 1$.

A comparison of the actual observed result $y_k = r$ with the predicted probabilities can be done by various criteria. We have implemented the following four concordancy measures:

$$C_1 = \sum_{k=1}^{N} \mathbf{1}\left[\arg\max_{s=1,\ldots,R}\{\hat{p}_k(s)\} = r\right];$$

$$C_2 = \frac{1}{N}\sum_{k=1}^{N} \log\{\hat{p}_k(r)\};$$

$$C_3 = -\frac{1}{N}\sum_{k=1}^{N}\left[\{1 - \hat{p}_k(r)\}^2 + \sum_{s\neq r}\hat{p}_k(s)^2\right];$$

$$C_4 = \frac{1}{N}\sum_{k=1}^{N}\hat{p}_k(r).$$

Criterion C_1 is the number of correctly predicted games, where those outcomes are predicted, which have the highest predictive probability. Criterion C_2 is a log-likelihood criterion and is used also by Glickman (1999) in a similar approach to select hyperparameters. Measure C_3 is based on quadratic loss whereas C_4 is equivalent to the corresponding measure based on absolute loss. Similar criteria are used in discriminant analysis and nonparametric regression for estimating smoothing parameters by cross-validation; see for example Fahrmeir and Tutz (1994b, page 174). We perform a separate estimation of α and θ, as outlined in the subsection 'The extended Kalman filter and smoother', for each of a large number of values of σ^2 between 0 and 1. The smoothing parameter σ^2 will then be chosen on the basis of maximal predictive concordancy with respect to the corresponding criterion.

When T is small, often some or all of the criteria are maximized for the limiting case $\sigma^2 = 0$. For larger T, there will often be evidence for time-changing abilities and (at least some of) the criteria are maximized for truly positive values of σ^2. Note that the '0–1' criterion C_1 has the slightly unattractive feature that it is not continuous as a function of σ^2 so that optimal values of the smoothing parameter σ^2 are typically within a certain interval. All the other criteria are continuous functions of σ^2.

Alternative estimation methods

Alternatively, an EM type of algorithm can be implemented to estimate σ^2; see for example Harvey (1989) or Fahrmeir and Tutz (1994b). A disadvantage of this method is rather slow convergence. We have, nevertheless, implemented an EM type of algorithm for σ^2 and report the corresponding estimates in our applications for comparison with the estimates based on predictive concordancy.

For a fully Bayesian analysis, MCMC methods can be used to make simultaneous inference about all the unknown parameters. Such methods require the specification of a prior distribution for the threshold parameters θ and the variance σ^2. Whereas, for the former, improper priors (uniform on the whole real line) can be chosen, for the latter, proper priors must be used, to avoid problems with improper posteriors. Computationally convenient are inverse gamma priors, say $\sigma^2 \sim \mathrm{IG}(a, b)$ with fixed values of a and b. Typical 'weakly informative' choices are $a = 1$ and b small, say 0.001, 0.01 or 0.1 (see for example Besag et al. 1995).

An advantage of MCMC sampling is that the uncertainty about the estimated parameters θ and σ^2 is incorporated in the estimation of α and that samples from predictive distributions can be generated. However, standard MCMC methods do not give *filtered* estimates, an issue that will be further discussed in the concluding section. We have also implemented an analysis of dynamic paired comparison models by MCMC simulation to assess the accuracy of our algorithm. The updating of components of θ was done by Gaussian Metropolis proposals (see for example Smith and Roberts 1993), whereas the updating of abilities was done by (multivariate) conditional prior proposals (Knorr-Held, 1999) for each vector α_t. For inference by MCMC methods, the initial value α_0 can be omitted in the formulation of the model.

Applications

In the following we illustrate our method with two applications. We use the dynamic cumulative link model (34.3) with the logistic response function $F(x) = 1/\{1 + \exp(-x)\}$ together with the constrained random walk prior (34.5). We have also tried the extreme minimal value distribution function $F(x) = 1 - \exp\{-\exp(x)\}$, which, however, did not fit the data as well as the logistic model in terms of predictability. For simplicity of presentation we do not display (approximate) pointwise credible intervals, which are available from the EKFS.

The German football league 1996–1997

In the 1996–1997 season of the German football league, the *Bundesliga*, $n = 18$ teams competed for the German championship. The teams met each other twice within the season giving each team a home advantage once. In total, $N = 306$ matches were performed between August 16th, 1996, and May 31st, 1997. We have categorized the timescale in $T = 42$ calendar weeks which gives roughly one match per team and per time point. Note that there was a winter break between December 8th, 1996, and February 13th, 1997, where no matches took place. As noted in the first section, possible outcomes are given in $R = 3$ categories: win ($y = 1$), draw ($y = 2$) and loss ($y = 3$) of the home team. On the basis of these results, points are assigned to the teams (3 for a win; 1 for a draw) which determine the standings in the league table. Table 34.1 gives the final standing for the 1996–1997 season.

We have estimated the smoothing parameter σ^2 based on all four prediction criteria. Criterion C_4, the absolute loss criterion, and criterion C_1 agree very closely in fitting the optimal model: C_4 is maximized for $\sigma^2 = 0.0171$ with a value of 0.4351, compared with 0.4329 for $\sigma^2 = 0$. Criterion C_1 is maximized around $\sigma^2 = 0.02$ with 160 correctly predicted games (157 for $\sigma^2 = 0$). The EM-type estimate is slightly lower with $\hat{\sigma}^2 = 0.0114$. Convergence was very slow. This indicates that there is not much information on temporal variation of abilities in the data. In fact, the other two criteria C_2 and C_3 are maximized for the limiting case $\sigma^2 = 0$. The reason might be that they both give relatively more weight to small values of $\hat{p}(k)$ than do C_1 and C_4. Small estimated probabilities are more likely for large values of σ^2 where estimated abilities have more temporal variation and cause estimated probabilities to be more extreme.

Figs 34.1 and 34.2 show filtered and smoothed estimated abilities of all 18 teams for $\sigma^2 = 0.0171$. The smoothed estimates show rather different patterns for the various teams and demonstrate the advantages of our nonparametric dynamic model. Note also that within the winter break ($t = 18, \ldots, 26$) filtered estimates are horizontal lines because of the prior model.

Table 34.1 ***Final ranking for the 1996–1997 season.***

Rank	Team	Total points	Points at home	Points away
1	Bayern München	71	43	28
2	Bayer Leverkusen	69	46	23
3	Borussia Dortmund	63	40	23
4	VfB Stuttgart	61	38	23
5	VfL Bochum	53	38	15
6	Karlsruher SC	49	29	20
7	1860 München	49	30	19
8	Werder Bremen	48	35	13
9	MSV Duisburg	45	23	22
10	1. FC Köln	44	28	16
11	Borussia Mönchengladbach	43	32	11
12	Schalke 04	43	22	21
13	Hamburger SV	41	28	13
14	Arminia Bielefeld	40	23	17
15	Hansa Rostock	40	22	18
16	Fortuna Düsseldorf	33	20	13
17	SC Freiburg	29	22	7
18	St Pauli	27	19	8

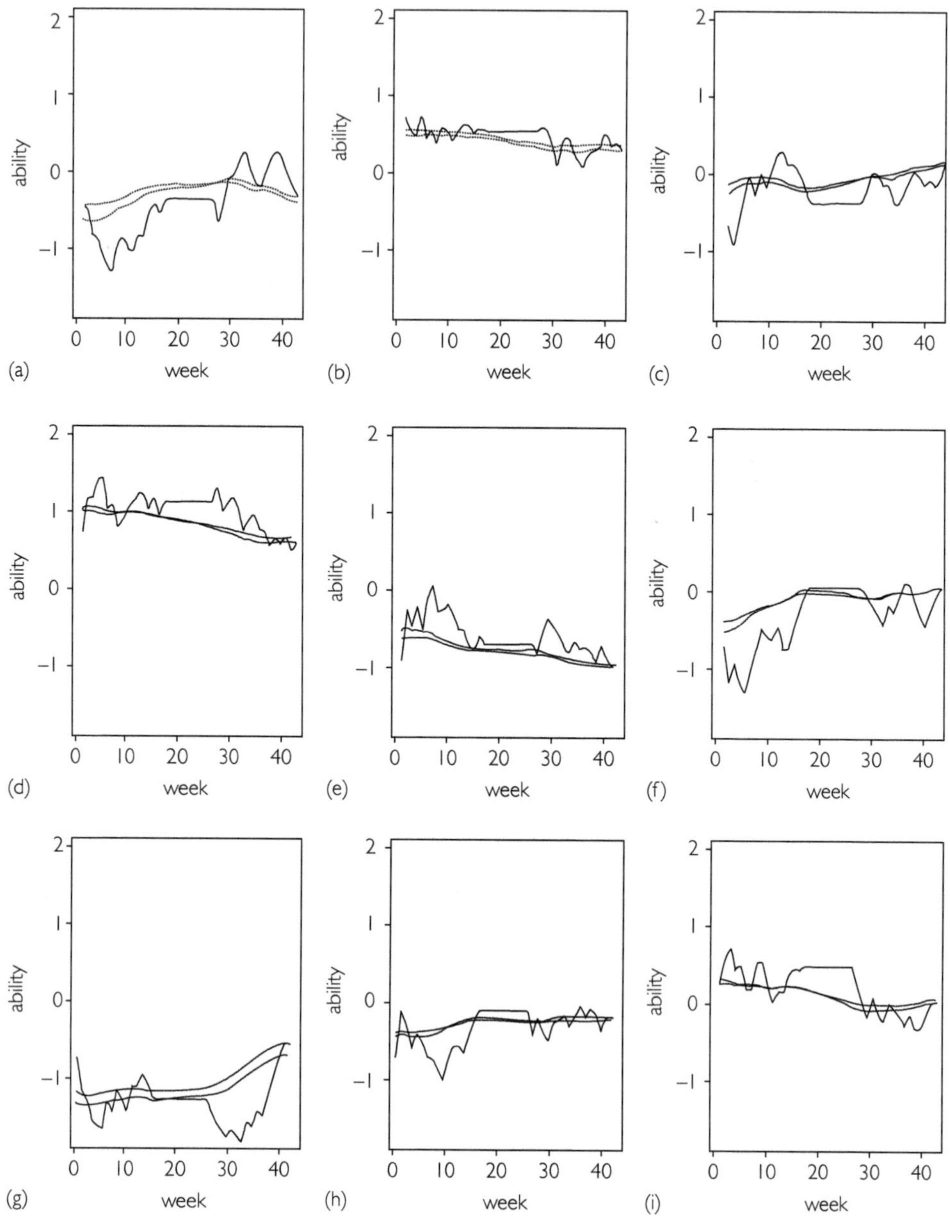

Figure 34.1 Football data (——, filtered time-changing abilities; ...., smoothed time-changing abilities;——, MCMC estimates): (a) Arminia Bielefeld; (b) VfL Bochum; (c) Werder Bremen; (d) Borussia Dortmund; (e) Fortuna Düsseldorf; (f) MSV Duisburg; (g) SC Freiburg; (h) Hamburger SV; (i) Karlsruher SC.

Later champions Bayern München had quite a time constant performance with smoothed abilities between 1.1 and 1.2. Other teams, however, had a substantial time-dependent performance. For example, Borussia Mönchengladbach had a rather poor performance before the winter break whereas its estimated ability at the end of the season was even slightly above average with a value of 0.06. Interestingly, this change of performance coincided with the dismissal

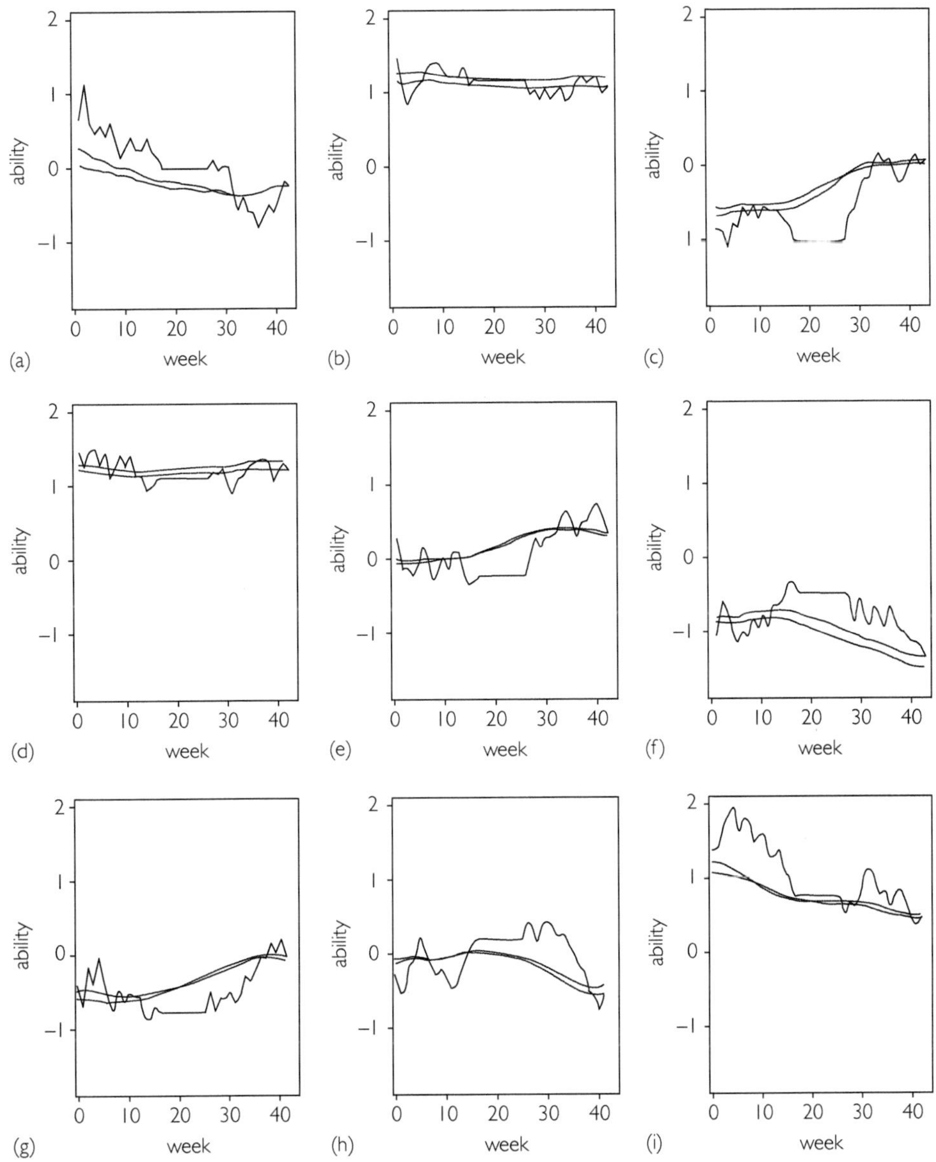

Figure 34.2 Football data (——, filtered time-changing abilities; ...., smoothed time-changing abilities;——, MCMC estimates): (a) 1. FC Köln; (b) Bayer Leverkusen; (c) Borussia Mönchengladbach; (d) Bayern München; (e) 1860 München; (f) St Pauli; (g) Hansa Rostock; (h) Schalke 04; (i) VfB Stuttgart.

of the head coach shortly before Christmas. SC Freiburg showed a similar dynamic but, in contrast with Borussia Mönchengladbach, the better performance towards the end of the season did not pay off: the team was relegated from the first to the second division.

The estimates of the threshold parameters are $\hat{\theta} = (0.050,\ 1.33)$ and reflect the strong home advantage that was already apparent in the raw data: 51.1% of all matches were won by the home team and only 26.1% by the visiting team.

Table 34.2 ***Parameter estimates (posterior means) by MCMC sampling for various hyperprior specifications.***

a	b	$\hat{\sigma}^2$	$\hat{\theta}_1$	$\hat{\theta}_2$
1	0.01	0.0005	0.068	1.24
1	0.1	0.0049	0.072	1.26
1	0.4	0.0183	0.076	1.30
1	1	0.0424	0.080	1.35

For comparison, we have analysed this data set also by MCMC simulation. Table 34.2 gives posterior mean estimates of σ^2 and θ for $a = 1.0$ and various values of b. The estimates have a strong sensitivity, especially $\hat{\sigma}^2$ with respect to the prior for σ^2. Consequently, the estimated abilities differ very much, with the degree of smoothness determined by $\hat{\sigma}^2$. Unfortunately, there are no clear guidelines about how to choose the prior for σ^2. For $a = 1$ and $b = 0.4$, which comes closest to the C_4-estimate $\hat{\sigma}^2 = 0.0171$, we have calculated posterior mean estimates of the ability parameters, and these are also displayed in Figs 34.1 and 34.2. From these pictures it can be seen that the MCMC estimates are quite similar to the smoothed estimates from the EKFS. Hence, the Kalman filter algorithm gives quite reliable results here. The small differences, which seem to depend on the absolute value of $\hat{\alpha}_{it}$, may be caused by the approximateness of the EKFS algorithm, the slightly different model formulation for the MCMC algorithm with all the parameters stochastic and without α_0, or by skewed posterior distributions where the posterior means and modes do not coincide. In a second MCMC analysis, we fixed $\sigma^2 = 0.017$ and the small differences between means and modes decreased slightly.

1996–1997 season of the National Basketball Association

In the American National Basketball Association $n = 29$ teams performed paired comparisons in the 1996–1997 season. We analyse $N = 1189$ games excluding results from the play-off matches. These games took place between November 1st, 1996, and April 20th, 1997, which gives a total of $T = 171$ calendar days. Note that in basketball there is not the possibility of a draw because of the overtime rule. Games can only end with $R = 2$ categories.

Our model fit criteria show behaviour similar to that for the football data: C_1 and C_4 again agree very closely. Criterion C_1 is optimal around $\sigma^2 = 0.005$ with 842 correctly predicted games (837 for $\sigma^2 = 0$). C_4 has an optimal value of 0.6366 for $\sigma^2 = 0.00595$, compared with 0.6334 for $\sigma^2 = 0$. The EM-type estimate is also quite close with $\hat{\sigma}^2 = 0.00425$. The log-likelihood criterion C_2 and the quadratic loss function criterion C_3, however, again prefer the non-dynamic model $\hat{\sigma}^2 = 0$.

The following results are based on the C_4-optimal value $\sigma^2 = 0.00595$. The threshold parameter was estimated by $\hat{\theta} = 0.37$, reflecting a substantial home advantage. Filtered and smoothed estimated abilities of selected teams are shown in Fig. 34.3. Interestingly the Chicago Bulls, the later champions, showed a steadily declining performance. They might have not played with their full force towards the end of the season, having already qualified for the play-off matches. Other teams such as the Houston Rockets, the Phoenix Suns or Utah Jazz have a very remarkable dynamic which would have been overlooked by a parametric model, where, for example, abilities are assumed to develop linearly or quadratically in time.

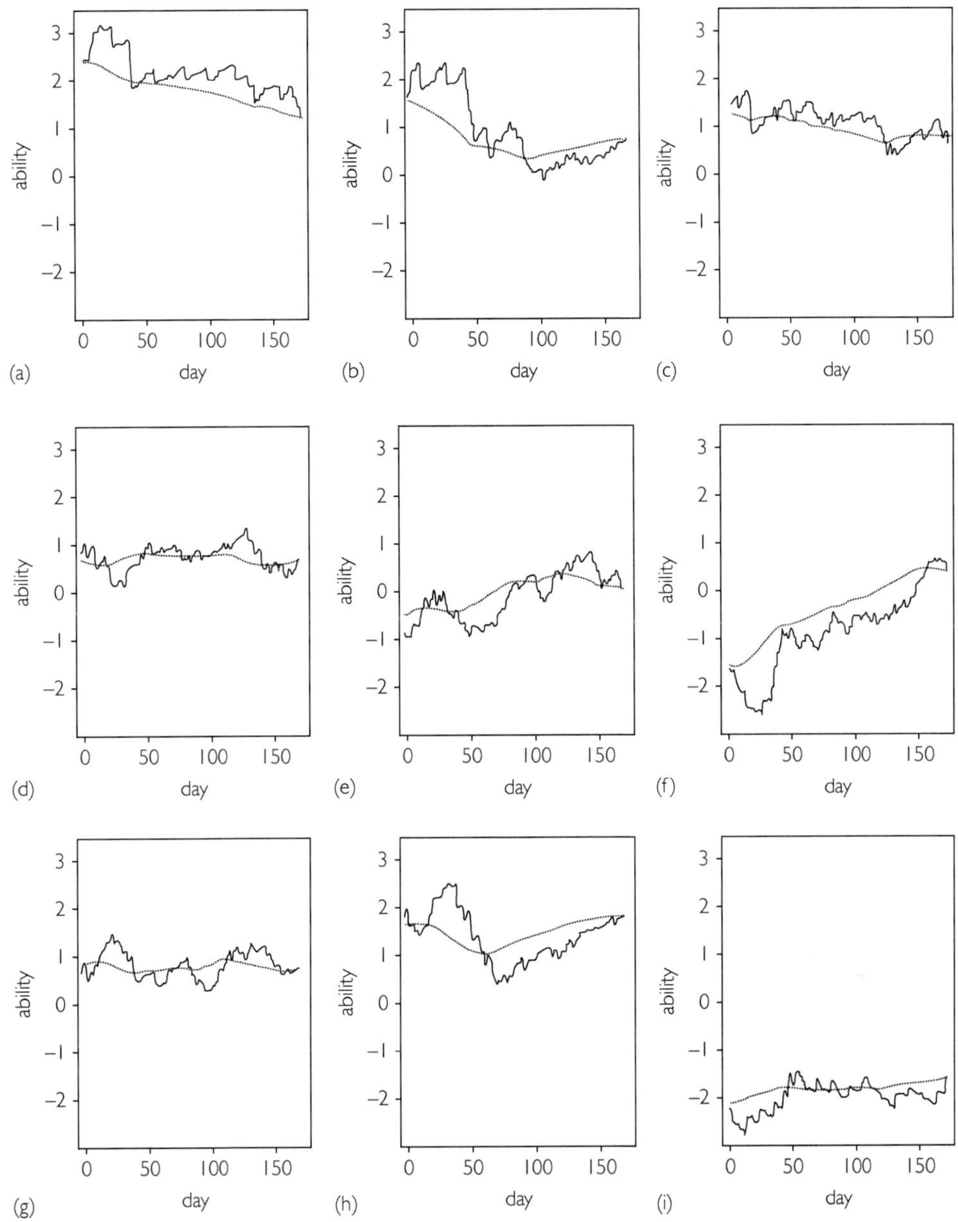

Figure 34.3 Basketball data (——, filtered time-changing abilities; ...., smoothed time-changing abilities): (a) Chicago Bulls; (b) Houston Rockets; (c) Miami Heat; (d) New York Knicks; (e) Orlando Magic; (f) Phoenix Suns; (g) Seattle Supersonics; (h) Utah Jazz; (i) Vancouver Grizzlies.

Concluding remarks

This paper has discussed the application of dynamic cumulative link models for rating sports teams. Our prior model ensures a symmetric treatment of all teams, assuming a multivariate singular Gaussian random walk prior with exchangeable components. Similar priors can also be used if the response variable is the difference in scores between the home and the visiting team, which is the more traditional approach to rating and prediction, because estimation can be done within the standard linear model; see for example Harville (1977, 1980) or Harville and Smith (1994). A dynamic approach for modelling the difference in scores within the state space model is proposed in Sallas and Harville (1988). There are also areas outside dynamic models for paired comparisons where constrained random walk priors are potentially useful. For example, the dynamic modelling of categorical covariate effects with constrained random walk priors is used in Knorr-Held and Besag (1998) for space–time modelling of disease risk data.

The estimation of the smoothing parameter turned out to be difficult. The reason seems to be that the type of categorical data does not provide much information about the temporal variation in teams' abilities. Among the four different measures, both the 0–1 and the absolute loss function criterion showed good performance whereas the other two choices have been disappointing. As an alternative, an EM type of algorithm can be used, which, however, has rather poor convergence properties. Fully Bayesian estimation by MCMC sampling was very sensitive with respect to hyperprior specifications.

A referee suggested considering (independent) stationary autoregressive processes as a prior model for the abilities of each team $i = 1, \ldots, n$:

$$\alpha_{ti} \sim N\{\mu_i + \phi(\alpha_{t-1,i} - \mu_i), \sigma^2\},$$

$$\alpha_{0i} \sim N\left(\mu_i, \frac{\sigma^2}{1-\phi^2}\right).$$

The parameter μ_i could be interpreted as the long-term average ability of team i. Such a model has the advantage that, a priori, the abilities of the teams will not in the long run tend towards ∞ or $-\infty$. Estimation can be done by MCMC sampling. Also, identifiability can easily be achieved by fixing the mean ability of a particular team, say the nth, at 0, i.e. $\mu_n = 0$, and it does not matter here which team's mean performance is fixed. However, such a model is less parsimonious and therefore problems of sensitivity with respect to now two hyperparameters σ^2 and φ are likely to increase. For data over a rather short time period, like in both of our applications, it seems that our approach will be sufficiently flexible and the long-run behaviour properties of the prior model are not as crucial from an applied point of view.

There are several possible generalizations of our model. As already noted in the 'Cumulative link model', each team can have assigned a team-specific smoothing parameter σ_i^2. This might be appropriate if there are substantially more data than in both of our applications, where information about the temporal variation, summarized in σ^2, seemed to be rather small. Furthermore, it will be very difficult, for moderate to large sizes of n, to find the optimal values of $\sigma_1^2, \ldots, \sigma_n^2$ on the basis of one-step-ahead prediction. Similar comments apply to more enhanced smoothing priors such as the second-order random walk, where independent first differences $\alpha_t — \alpha_{t-1}$ are replaced by second differences $\alpha_t — 2\alpha_{t-1} + \alpha_{t-2}$, or the local linear trend model (Harvey, 1989). On the basis of our experience with these priors, we believe that they have limited use as long as the rating of sports teams within one season is considered as in our two applications. They might be useful if considerably more data are observed over a longer period.

Throughout this paper we have used constant threshold parameters. This assumption can be relaxed, allowing for team-specific or time-dependent threshold parameters. Harville and Smith (1994) analysed college basketball results with various models and found team-to-team differences in the home advantage to be relatively small. Knorr-Held (1997) analysed the *Bundesliga* (1995–1996 season) by MCMC methods with additional team-specific random effects for the thresholds, but there was not much evidence for home advantage heterogeneity here either. Fahrmeir and Tutz (1994a) allowed for time-dependent threshold parameters in an analysis of the *Bundesliga* but found threshold estimates to be stable even over a period of 23 years. We have therefore used the simple model with constant threshold parameters. If necessary, it would be no problem to extend our approach to more general models.

For the application considered, we prefer the EKFS for statistical inference rather than MCMC methods. Although MCMC sampling is a more elaborate approach, it seems to be much easier to obtain *filtered* estimates by the EKFS. These filtered estimates can be used to assess the predictive power and to choose the smoothing parameter. Furthermore, in a full Bayesian analysis, there often seems to be strong sensitivity with respect to hyperprior specifications. However, recently several interesting proposals have been made to obtain filtered estimates by *dynamic* MCMC methods, e.g. Gordon et al. (1993), Berzuini et al. (1997) and Pitt and Shephard (1999). Applications of these approaches to dynamic ordered paired comparison systems are, because of the high dimensionality of the model, beyond the scope of this paper.

ACKNOWLEDGEMENTS

The author expresses thanks to Ludwig Fahrmeir, the Associate Editor and two referees for several suggestions which have improved the paper.

REFERENCES

Agresti, A. (1992) Analysis of ordinal paired comparison data. *Appl. Statist.*, **41**, 287–297.

Bassett, Jr, G. W. (1997) Robust sports ratings based on least absolute errors. *Am. Statistn*, **51**, 99–105.

Berzuini, C., Best, N. G., Gilks, W. R. and Larizza, C. (1997) Dynamic conditional independence models and Markov chain Monte Carlo methods. *J. Am. Statist. Ass.*, **92**, 1403–1412.

Besag, J. E., Green, P. J., Higdon, D. and Mengersen, K. (1995) Bayesian computation and stochastic systems (with discussion). *Statist. Sci.*, **10**, 3–66.

Bradley, R. A. and Terry, M. E. (1952) Rank analysis of incomplete block designs: I, The method of pair comparisons. *Biometrika*, **39**, 324–345.

Cox, D. R. and Hinkley, D. V. (1974) *Theoretical Statistics*. London: Chapman and Hall.

Fahrmeir, L. (1992) Posterior mode estimation by extended Kalman filtering for multivariate dynamic generalized linear models. *J. Am. Statist. Ass.*, **87**, 501–509.

Fahrmeir, L. and Knorr-Held, L. (2000) Dynamic and semiparametric models. In *Smoothing and Regression: Approaches, Computation and Application* (ed. M. Schimek), Ch. 18. New York: Wiley. To be published.

Fahrmeir, L. and Tutz, G. (1994a) Dynamic stochastic models for time-dependent ordered paired comparison systems. *J. Am. Statist. Ass.*, **89**, 1438–1449.

—— (1994b) *Multivariate Statistical Modelling Based on Generalized Linear Models*. New York: Springer.

Fahrmeir, L. and Wagenpfeil, S. (1997) Penalized likelihood estimation and iterative Kalman smoothing for non-Gaussian dynamic regression models. *Comput. Statist. Data Anal.*, **24**, 295–320.

Glickman, M. E. (1993) Paired comparison models with time-varying parameters. *PhD Dissertation*. Department of Statistics, Harvard University, Cambridge.

—— (1999) Parameter estimation in large dynamic paired comparison experiments. *Appl. Statist.*, **48**, 377–394.

Glickman, M. E. and Stem, H. S. (1998) A state-space model for National Football League scores. *J. Am. Statist. Ass.*, **93**, 25–35.

Gordon, N. J., Salmond, D. J. and Smith, A. F. M. (1993) A novel approach to non-linear and non-Gaussian Bayesian state estimation. *IEE Proc.* F, **140**, 107–133.

Harvey, A. C. (1989) *Forecasting, Structural Time Series Models and the Kalman Filter*. Cambridge: Cambridge University Press.

Harville, D. A. (1977) The use of linear-model methodology to rate high school or college football teams. *J. Am. Statist. Ass.*, **72**, 278–289.

—— (1980) Predictions for National Football League games via linear-model methodology. *J. Am. Statist. Ass.*, **75**, 516–524.

Harville, D. A. and Smith, M. H. (1994) The home-court advantage: how large is it, and does it vary from team to team? *Am. Statistn*, **48**, 22–28.

Knorr-Held, L. (1997) *Hierarchical Modelling of Discrete Longitudinal Data; Applications of Markov Chain Monte Carlo*. Munich: Utz.

—— (1999) Conditional prior proposals in dynamic models. *Scand. J. Statist.*, **26**, 129–144.

Knorr-Held, L. and Besag, J. (1998) Modelling risk from a disease in time and space. *Statist. Med.*, **17**, 2045–2060.

Pitt, M. K. and Shephard, N. (1999) Filtering via simulation: auxiliary particle filters. *J. Am. Statist. Ass.*, **94**, 590–599.

Sallas, W. M. and Harville, D. A. (1988) Noninformative priors and restricted maximum likelihood estimation in the Kalman filter. In *Bayesian Analysis of Times Series and Dynamic Models* (ed. J. C. Spall), pp. 477–508. New York: Dekker.

Smith, A. F. M. and Roberts, G. O. (1993) Bayesian computation via the Gibbs sampler and related Markov chain Monte Carlo methods. *J. R. Statist. Soc.* B, **55**, 3–23.

Tutz, G. (1986) Bradley–Terry–Luce models with an ordered response. *J. Math. Psychol.*, **30**, 306–316.

CHAPTER 35

Inferences about testosterone abuse among athletes: How does one teach anti-doping officials about evidence-based decision making?

DONALD A. BERRY AND LEEANN CHASTAIN

Drug testing of athletes has two purposes: to prevent artificial performance enhancement (known as doping) and to discourage the use of potentially harmful substances. Significant advancements in laboratory tests for doping have developed during the 30-year history of such organized efforts, and competitors at the Olympic Games of 2002 benefited from such scientific advancements. However, developing a good test is only part of an effective and fair process.

As is well known in medicine, the outcome of a test never plays a solo role in diagnosis. It represents only one factor to be considered. Physicians arrive at a diagnosis by weighing the likelihood that a test result correctly identifies a disorder or disease in the context of other relevant information. Such information includes the patient's symptoms, if any, the patient's characteristics, such as age and gender, prior history of screening and disease, etc. The characteristics of the test procedure must also be taken into account, including its sensitivity, specificity, and inherent variability. In the best of worlds, physicians start from a set of differential diagnoses based on what is known at that point in time. They may plan further tests, depending on the results of the first test, and they update the likelihood of each of the possible diagnoses as new information becomes available. Simply put, a medical diagnosis is a statistical inference.

Similarities between the processes of diagnosing a complex medical disorder and determining whether or not an athlete's test result is indicative of doping are especially strong when the alleged infraction is the use of a substance that is a normal physiologic component, such as testosterone. In interpreting diagnostic tests, however, physicians have an inferential advantage: if a patient's sample shows an elevated marker that might indicate a serious disease, the truth becomes known when the patient turns out to have or not to have the disease. This contributes to databases of information from both diseased and healthy individuals. Whether or not an athlete has abused a substance will rarely become known by other objective evidence. There is no "gold standard" when testing for substance abuse.

Mary Decker Slaney at the 1996 Summer Olympics. (AP Photo/Doug Mills)

The appropriate tool for decision making under uncertainty is Bayes' rule. Its contribution to evidence-based medicine has been described by many authors, and is thoroughly supported in two articles by Steven N. Goodman, published in the *Annals of Internal Medicine* in 1999. The application of Bayes' rule is well understood in medicine, and is appropriate for anti-doping science, as well. Failure to recognize relevant statistical issues when making inferences from laboratory tests leads to flawed conclusions in the case of an accusation of testosterone use. These issues will be discussed in the context of the case of the USA Track and Field (USATF) and runner Mary Slaney.

Mary Decker Slaney was a world-class runner for more than 20 years. She was the 1983 world champion at 1,500 and 3,000 meters, and held the fastest times for American women in distances from 800 to 10,000 meters. At the time of her induction into the USATF Hall of Fame in 2003—nearly 20 years after she captured these records—Slaney still held the North and Central American women's record for out-door track at 1 mile, 1,500 meters, and at 3,000 meters. She also continued to hold the 1-mile record for this world region on an indoor track. In the mid-1980s, Slaney underwent treatment for exercise-induced asthma, in addition to multiple surgeries to repair bones in her left foot and remove degenerated tissue from her Achilles tendon. Rehabilitation and a return to elite competition were slow, but in 1996, at the age of 37, Slaney was making a comeback in distance running. Her efforts were halted by a suspension due to alleged doping consistent with testosterone use. In this article we raise questions about the appropriateness of such suspensions.

Performance enhancement and the anti-doping movement

Combating doping among athletes is a worldwide endeavor, involving the establishment in 1999 of a World Anti-Doping Agency (WADA) to work with the Olympic committees and sports federations. More than 25 accredited laboratories are involved in the analysis of over 90,000 athletes' urine samples annually, at a cost of millions of dollars.

Athletes may be tested for all forms of doping, and face sanctions ranging from several months of suspension from competition to permanent ineligibility. The finding of a foreign, banned substance or its metabolites in a bodily sample is legal evidence of doping, and is a relatively straightforward inference. A rather different circumstance exists, however, when the substance in question is not foreign, but is normally present in the athlete's body. Such is the case with testosterone.

Scientists have attempted to detect artificial increases in testosterone concentrations through the establishment of a "normal urinary range" for the ratio of testosterone glucuronide to epitestosterone glucuronide (known as the T/E ratio). Human physiologic processes do not convert testosterone to epitestosterone, its 17 alpha-epimer, and doping scientists have found that the urinary concentrations of the two analytes are usually about the same. Urinary testosterone concentrations are known to fluctuate significantly in males and females. And little is known about epitestosterone because it has not been studied clinically as an indicator of health or disease. Despite these limitations, following the 1980 Olympic Games, doping scientists set the threshold for positive testosterone doping at a T/E ratio greater than 6:1. (Descriptions of the complicated methods used to measure testosterone and epitestosterone concentrations in urine have been published elsewhere; some references are listed at the end of this article.)

Biological issues

Many issues associated with urinary excretion of testosterone and epitestosterone are poorly understood, especially in the female. Research is limited, and some authors have urged caution when assessing the female T/E profile. Variations in steroid profiling (resulting in altered T/E ratio) are associated with ethnicity; age; sex; circadian rhythm; training and competition; diet; nutritional supplementation; environmental factors; alcohol ingestion; enzyme deficiencies; decreased epitestosterone excretion; menstruation; pregnancy; birth control pills; other hormonal therapy; consumption of meat from animals supplemented with anabolic steroids; polycystic ovary syndrome (a common endocrine disorder); and other pathologic medical conditions.

Christiane Ayotte, Professor and Director of the Montreal Doping Control Laboratory, member of WADA's Health, Medical and Research Committee, and expert witness for the USATF at Mary Slaney's hearing, reported in her research studies that some women exhibit physiologic T/E ratios greater than 6. Professor Ayotte's studies were published in the International Amateur Athletic Federation's (IAAF) *New Studies in Athletics*, 1997. Another expert witness for the USATF at Mary Slaney's hearing was Don H. Catlin, Director of the Olympic Analytical Laboratory at UCLA. In an article published in *Clinical Chemistry* in 1997, Catlin reported that the T/E ratio above which could not be attributed to normal physiologic variations was a mean ratio of 15. Mary Slaney's T/E ratios, which resulted in her suspension, were 9.5 and 11.6.

The case of Mary Decker Slaney

Mary Slaney submitted a routine urine sample at the 1996 U.S. Olympic Trials in Atlanta. Urinalysis at the accredited laboratory in Montreal, Canada, indicated that her T/E ratio exceeded the upper limit of 6.

Sports officials regarded this to be proof of testosterone doping. However, in September 1997 a Doping Hearing Board of the USATF exonerated Slaney. (The Hearing Board was presented with the arguments given in this article by one of the authors [DAB].) In April 1999 the IAAF (known officially as the International Association of Athletics Federations as of 2001), exerting its stated right to overrule a member federation, ruled that Slaney 'was guilty of a doping offense . . . and that the USATF's Doping Hearing Board decision . . . was erroneous.' Slaney filed a suit against the IAAF and the U.S. Olympic Committee (USOC) in U.S. District Court, seeking to reverse this ruling and to obtain an injunction against the use of the T/E ratio on women athletes until a better understanding of the causes of altered ratios could be reached. The decision from the court, issued on March 27, 2001, was that it did not have subject matter jurisdiction over the state-law claims against the USOC. The court ruled that it was obligated to recognize the valid arbitration in which Slaney had participated with the IAAF, and that although arbitrators were not bound by rules of evidence, the issue decided in the arbitration could not be relitigated.

Anti-doping officials behave as though a T/E ratio greater than 6 documents illegal steroid use to the exclusion of other possible explanations. This implication was made explicit by the IAAF in the April 1999 ruling declaring Mary Slaney's wins and achievements from June 17, 1996, through June 16, 1998, null and void. Such a conclusion based solely on this indirect laboratory measurement is inconsistent with common clinical or forensic practice.

False positives

Any laboratory test can give false positive results, whether because of inherent variability in the testing procedure, inherent biologic variability, or inappropriate handling of source material. If 1% of nonusers are found to have a urinary T/E ratio above the established threshold of 6:1 (a figure consistent with at least some of the testimony presented in the Slaney case), then an average of one of 100 athletes' tests will result in a positive reading even if none have used an illegal substance. If 90,000 urine samples are tested (the annual rate referenced previously) then 900 will test positive, even if none of the athletes used illegal substances. Using a more stringent cut-point to identify only 0.1% of users will still incorrectly identify 90 nonusers.

A standard statistical practice is to make adjustments in cut-points to account for the fact that many tests were carried out. To ensure that the probability of one or more false positives in a program of 100 tests is less than 1% a particular individual's test result would have to be in the upper 99.99th percentile of nonusers tested. That is, no more than 1 in 10,000 nonusers tested should have a T/E ratio above the cut-point. If 90,000 urine samples are tested, then the appropriate cut-point is the 99.99999th percentile. The implications are profound. And there are possibly insurmountable problems in applying such a standard. One is statistical: such an extreme percentile is not even approximately known. Mounting a study to identify a value of T/E for which only one in 10 million nonusers tests higher would be an enormous undertaking.

The trade-off between finding an innocent person guilty and protecting the integrity of sport is far from obvious. It is a societal issue beyond the scope of this article. But it seems unreasonable to impugn 1% or even a tenth of a percent of nonusers who happen to be tested.

Sensitivity and specificity of laboratory tests

The diagnostic accuracy of any laboratory test is defined as the ability to discriminate between two types of individuals—in this case users and nonusers. *Specificity* and *sensitivity* characterize diagnostic tests. In the anti-doping analogy regarding the T/E ratio, specificity is the proportion of nonusers with $T/E < 6$ (as discussed above) and sensitivity is the proportion of users with $T/E \geq 6$. Estimating these proportions requires collecting and tabulating data from the two reference samples, users and nonusers, as well as data from the appropriate subpopulation for that athlete. ("Users" will, of course, vary in dose, duration, and substance used. Bayes' rule allows for learning about the degree of use on the basis of test results, but we do not consider this more general setting here.)

Sensitivity must be addressed in the context of the effect of a particular banned substance upon the T/E ratio. In an article published in the *Journal of Mass Spectrometry* in 2000, David J. Borts and Larry D. Bowers, studying the use of advanced laboratory methods to measure the T/E ratio, stated that the "measurement of a prohibited substance [in the urine matrix] requires identification of the compound either by full scan or by consistency of ion abundance ratios between a reference material and the unknown" (p. 56). Anti-doping laboratories, however, fail to address sensitivity or to recognize its relevance. Specificity, generally addressed by keeping records of tests performed on athletes, introduces a clear bias of unknown magnitude because some athletes in their databases may be users but have "normal" T/E. For example, Asian males have baseline urine concentrations of testosterone and T/E that can be so low that testosterone doping may not elevate their T/E to that of some nonusers among Caucasian athletes.

Bayes' rule and the probability of guilt

Bayes' rule is a necessary inferential tool for relating experimental evidence to conclusions, such as whether someone has a disease or has used a particular substance. Applying Bayes' rule requires determining the test's sensitivity and specificity. It also requires a pre-test (or prior) probability that the athlete has used a banned substance.

Consider a population of 1,000 athletes with the following characteristics: 100 are users (prior probability = 10%), and of these, 50 would test positive (sensitivity = 50%). Of the 900 nonusers, nine would test positive (specificity = 99%). If an athlete tests positive, then the probability she is a user is 50/59 = 84.7% (positive predictive value, or PPV).

Where "prior" indicates the prior probability of being a user, Bayes' rule says

$$\text{PPV} = \frac{\text{sens} \times \text{prior}}{\text{sens} \times \text{prior} + (1 - \text{spec}) \times (1 - \text{prior})}$$

In the above example with 1,000 athletes:

$$\frac{50}{59} = \frac{(50/100) \times (100/1000)}{(50/100) \times (100/1000) + (9/900) \times (900/1000)}$$

If the sensitivity is 1 (in which case all users have positive tests) and the other factors are unchanged, then the PPV is

$$\frac{1 \times 0.1}{1 \times 0.1 + 0.01 \times 0.9} = \frac{100}{109} = 91.7\%$$

Bayes' rule relates "inverse probabilities": the probability of substance abuse given the test result, with the probability of the test result given substance abuse. But these two probabilities are not equal. Nor is the probability of substance abuse given the test result (posterior probability) the same as the probability of testing negative given that the athlete is a nonuser (specificity). In the example, the latter is 99% and the former is 84.7%.

An essential factor in Bayes' rule is the prior probability of disease, or guilt of substance abuse, depending on the application. In the above example, changing the prior probability from 10% to 90% means changing the posterior probability from 84.7% to 450/451 = 99.8%.

Figure 35.1 shows the relationship between posterior probability, prior probability, and sensitivity, for a fixed specificity of 99%. Such a figure allows for using Bayes' rule in reverse. For example, a hearing board might not want to rule against an athlete unless her probability of guilt is at least 95%. The prior probability must be at least 20%, even for perfect sensitivity. And if the standard is 99%, then the prior probability must be at least 50%.

Prevalence of disease is relatively easy to estimate, depending on the patient population. Prevalence of substance abuse is not. There is an inevitable subjective aspect of assigning a prior probability. A hearing board made up of an athlete's peers is especially appropriate for making such assignments. For example, assuming they know nothing about the athlete beyond the information presented at the hearing, they might regard testosterone substance abuse to be rare and so the positive predictive value would be at most moderately large. On the other hand, if they know testosterone substance abuse to be widespread, then the probability of abuse would be larger, based on a positive test.

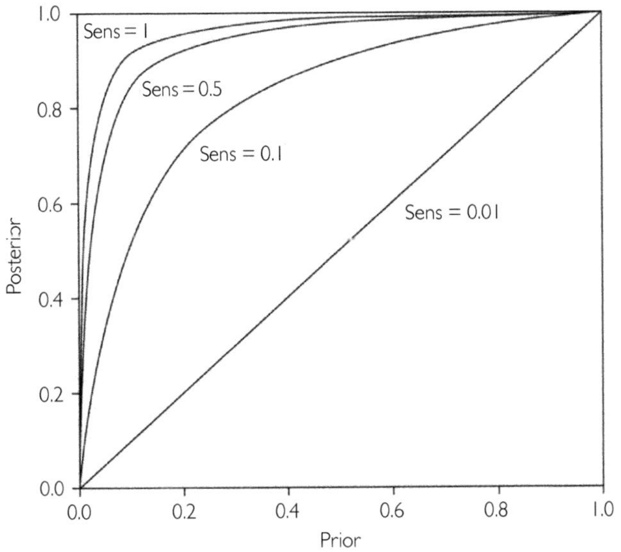

Figure 35.1 Relationship between posterior probability and prior probability for four different values of sensitivity. All curves assume specificity is 99%.

Conclusion

Conclusions about the likelihood of testosterone doping require consideration of three components: specificity and sensitivity of the testing procedure, and the prior probability of use. As regards the T/E ratio, anti-doping officials consider only specificity. The result is a flawed process of inference. Bayes' rule shows that it is impossible to draw conclusions about guilt on the basis of specificity alone. Policy-makers in the athletic federations should follow the lead of medical scientists who use sensitivity, specificity, and Bayes' rule in interpreting diagnostic evidence.

Judging whether Mary Slaney was guilty as alleged is not the point of this article. Our focus has been the flawed inferential process used in such cases. Wonderfully modern technological advances in the laboratory should not be marred by erroneous statistical interpretations. In the short term, the bar for urinalysis of testosterone doping must be set much higher (T/E much higher than 6). The threshold should be determined in light of the number of athletes to be tested, pegged to the number of false positives that are acceptable to society, and based on the limitations of current laboratory methods. If the sizes of current databases are too small to allow for good estimates of proportions above large thresholds, then the laboratory test should be suspended until such information becomes available.

ADDITIONAL READING

American Medical Association. 1987. "Scientific issues in drug testing." *Report J of the Council on Scientific Affairs, JAMA*; 257(22):3110–3114.

Berry, D.A. 1991. "Inferences using DNA profiling in forensic identification and paternity cases." (with discussion) *Stat Science*; 6:175–205.

Berry, D.A. 1994. "DNA, statistics and the Simpson case." *Chance*; 7(4):9–12.

Berry, D.A., Hochberg, Y. 1999. "Bayesian perspectives on multiple comparisons." *J Stat Planning Inference*; 82:215–227.

Black, D.L. 2001. "Doping control testing policies and procedures: a critique." In Wilson, W., Derse, E., eds. *Doping in Elite Sport: The Politics of Drugs in the Olympic Movement*. Champaign, Ill: Human Kinetics Pub; 29–42.

Kammerer, R.C. 2001. "What is doping and how is it detected?" In Wilson, W., Derse E., eds. *Doping in Elite Sport: The Politics of Drugs in the Olympic Movement*. Champaign, Ill.: Human Kinetics Pub; 3–28.

Mary Decker Slaney v. The International Amateur Athletic Federation and the United States Olympic Committee. 244 F.3d 580 (U.S. 7th Cir. 2001).

Sackett, D.L. 1997. *Evidence-Based Medicine: How to Practice and Teach EBM*. New York: Churchill Livingstone.

Strauss, R.H., Liggett, M.T., Lanese, R.R. 1985. "Anabolic steroid use and perceived effects in ten weight-trained women athletes." *JAMA*; 253(19):2871–2873.

van de Kerkhof, D.H., de Boer, D., Thijssen, J.H.H., Maes, R.A.A. 2000. "Evaluation of testosterone/epitestosterone ratio influential factors as determined in doping analysis." *J Anal Toxicol*; 24:102–115.

Zweig, M.H., Campbell, G. 1993. "Receiver-operating characteristic (ROC) plots: a fundamental evaluation tool in clinical medicine." *Clin Chem*; 39(4):561–577.

CHAPTER 36

Investigating the economic and demographic determinants of sporting participation in England

LISA FARRELL AND MICHAEL A. SHIELDS

Abstract

We provide a detailed statistical investigation into the economic and demographic factors that determine sporting participation in England. Using data from the 1997 health survey of England we fit random-effects probit models that take into account unobservable household preferences for sporting activities, as well as the economic and demographic characteristics of respondents. Our main results from the multivariate analysis are that sporting participation is positively related to household income, the educated participate in sports to a greater extent than the uneducated, there is no evidence of regional differentials in sporting participation and household preferences play an important role in the decision to participate in sports.

Keywords: *Health; Household preferences; Random effects; Social inclusion; Sporting participation*

Introduction

Increasing sporting participation in the UK has recently been emphasized by the Government as an important policy tool in meeting the population health targets set in the 'Health of the nation', and promoting social inclusion. The recent consultative White Paper 'A sporting future for all' (Department of Culture, Media and Sports, 2000) sets out a clear vision for sport in the UK based on broad participation and greater access to facilities across all income groups and life stages. Improving access to sporting facilities and activities is expected to increase peoples' emotional well-being and general sense of belonging (Steptoe and Butler, 1996).

Recent policy papers have also placed great emphasis on the role of sport in community regeneration (Department of Culture, Media and Sports, 2000).

However, there has been only limited analysis of the economic and demographic factors that affect sporting participation. A recent report by MINTEL (2000) provides a descriptive analysis of sporting participation in the UK. Sporting participation was found to fall with age and family commitments (mainly due to the presence of children). Many households with limited leisure time changed their leisure activities from sports-based activities to more home-orientated activities (such as home entertainment and 'do-it-yourself'). In the case of females who withdraw from the labour force following childbirth, the report noted an increased importance of sporting participation in terms of maintaining social contact outside the household even though the intensity of participation falls.

Although information about the life-cycle patterns of sporting participation provided by such descriptive reports can tell us much about how the demand for sports facilities will change as the demographic and social composition of the population changes, the primary objective of this paper is to provide a comprehensive statistical investigation into sporting participation, which can quantify the relative influence of economic and demographic factors on the demand for sporting activities in England. Specifically, we are interested in establishing the relative effect of characteristics such as age, ethnicity, education and health on the probability of participating in sporting activities. Given the recent attention given to sports as a means of promoting social inclusion, we also address the questions of whether or not low income acts as a substantial barrier to playing sports (as suggested in Argyle 1987), and whether there are regional differentials in sporting participation in England. These relationships are examined for a set of the most popular sports undertaken by the English population. Moreover, given that family influences have been found to play an important role in determining sporting participation (see, for example, Wold and Anderssen 1992), we examine the importance of household preferences in determining sporting participation by using a statistical framework (i.e., the random-effects probit model) that allows for within-household clustering of sporting participation. We use data from the 1997 health survey for England (HSE), which is unique in providing information about participation for over 80 types of sports.

The paper proceeds as follows. The next section presents our data and highlights the incidence of sporting participation in our sample. Our statistical framework for modelling sporting participation and our hypotheses are described in 'Statistical framework'. The results of this analysis are presented in 'Empirical results', and the final section concludes with a discussion of the results.

Data and sample characteristics

Our data source is the HSE, commissioned by the Department of Health and carried out by the Joint Health Surveys Unit of Social and Community Planning Research and the Department of Epidemiology and Public Health at University College London. Beginning in 1992, the HSE is an annual survey and is designed to monitor trends in the nation's health. We use data from the 1997 sweep of the survey, which was the first to include questions about household income. This is important because it allows us to investigate the economic as well as the demographic factors that impact on sporting participation. For our purposes, information on participation in over 80 sporting activities is collected, together with a host of individual and demographic characteristics. The HSE is a household survey that covers the adult population aged 16 years or over living in England, and data are collected by a combination of face-to-face interviews, self-completion questionnaires and medical examinations. Using the *Postcode Address File* as a sampling frame, the HSE is generally considered to be representative of England and additional details of the sampling procedures can be found in Prescott-Clarke and Primatesta (1998). The focus of this paper is the 3007 men and 3460 women, living in 3811 households, who were aged 16–65 years at the time of the survey. This sample consists of 1550 observations on single-adult households and 4917 observations on individuals in households for which we observe two or more adults. We have excluded those respondents who were still in full-time education, since access to sporting facilities for students is likely to be different from access for the general population. For example, universities generally heavily subsidize sporting activities, but students will also be concentrated in the lowest income groups.

To exclude sporting episodes of very short duration, we define an individual as being a sporting participant if, over the 4 weeks before the interview, they report having played in at least one episode of sport which lasted a minimum of 15 minutes. Overall, summing over the 80 types of sporting activity coded in the survey, 44.5% of the sample reported participating in some form of sport over the last 4 weeks, with males (48.1%) reporting a significantly higher participation rate than females (41.5%). In the following analysis we focus on the 10 most popular types of sporting activities in England. These are swimming (16.2%), cycling (11.8%), going to the gym (11.2%), general exercise (10.9%), dancing (9.9%), aerobics (8.2%), running (6.7%), football or rugby (6.3%), racket sports (5.6%) and golf (2.9%). General exercise is defined vaguely in the survey as 'press-ups, sit-ups etc.'. Football and rugby are not distinguished between in the survey, and rackets sports consist of badminton, tennis and squash.

The data reveal considerable differences in the types of sports that men and women play. Fig. 36.1 shows the percentage of men and women playing each of the 10 sports. A statistically significant higher proportion of males than females participate in cycling, football or rugby, gym visits, general exercise, golf, racket sports and running, whereas the converse is true for aerobics, dancing and swimming. Table 36.1 presents the distribution of sports by

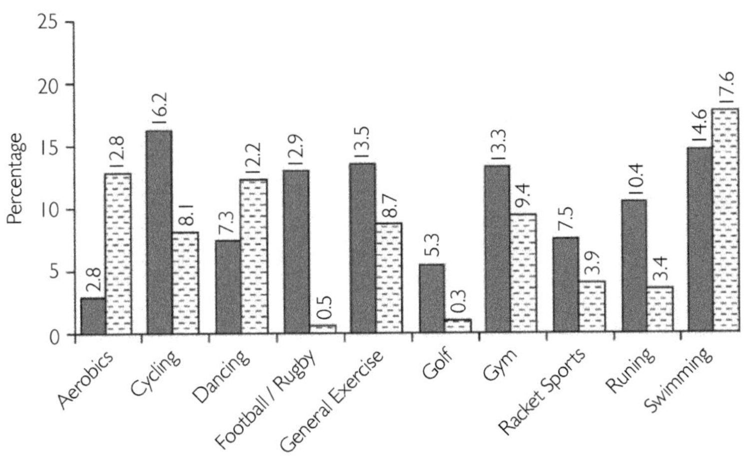

Figure 36.1 Sporting participation by gender: ■, males; <spotty box> females

Table 36.1 ***Sporting participation by selected demographic characteristics.*** †

Sport	Distributions by the following age groups:			Distributions by ethnicity		Distributions by qualifications		Distributions by household income	
	< 25 years	25–44 years	≥ 45 years	White	Non-white	None	Degree	<£10,400	>£52,000
All sports	62.4	52.6‡	30.1‡	45.2	35.6‡	24.2	63.8‡	30.9	60.7‡
	(1.9)	(0.9)	(0.9)	(0.6)	(2.3)	(1.0)	(1.6)	(1.3)	(1.3)
Aerobics	10.9	9.9	5.3‡	8.3	6.2	3.4	10.3‡	6.3	11.2‡
	(1.2)	(0.5)	(0.4)	(0.4)	(1.2)	(0.4)	(1.0)	(0.7)	(1.5)
Cycling	15.6	14.6	7.4‡	12.2	6.7‡	6.7	20.7‡	9.2	16.9‡
	(1.4)	(0.6)	(0.5)	(0.4)	(1.2)	(0.6)	(1.3)	(1.0)	(1.8)
Dancing	19.3	9.6‡	7.9‡	10.1	8.1	6.8	10.7‡	8.8	9.6
	(1.5)	(0.5)	(0.5)	(0.4)	(1.3)	(0.6)	(1.0)	(0.8)	(1.4)
Football or rugby	17.4	7.9‡	1.4‡	6.3	5.5	3.0	8.1‡	4.1	9.4‡
	(0.5)	(0.5)	(0.2)	(0.3)	(1.1)	(0.4)	(0.9)	(0.5)	(1.4)
General exercise	19.3	13.5‡	5.6‡	10.7	14.4‡	4.4	16.4‡	7.5	17.8‡
	(1.5)	(0.6)	(0.5)	(0.4)	(1.7)	(0.5)	(1.2)	(0.7)	(1.8)
Golf	1.3	2.6‡	3.8‡	3.1	1.0‡	1.1	3.7‡	0.7	7.3‡
	(0.4)	(0.3)	(0.4)	(0.2)	(0.5)	(0.3)	(0.6)	(0.2)	(1.2)
Gym	18.8	13.8‡	6.1‡	11.2	11.0	5.2	16.1‡	5.8	18.0‡
	(1.5)	(0.6)	(0.5)	(0.4)	(1.5)	(0.5)	(1.2)	(0.6)	(1.8)
Racket sports	7.5	6.8	3.6‡	5.6	5.3	1.4	12.6‡	2.0	11.2‡
	(1.0)	(0.5)	(0.4)	(0.3)	(1.1)	(0.3)	(1.1)	(0.4)	(1.5)
Running	13.5	8.3‡	2.8‡	6.7	5.7	2.0	12.4‡	3.6	11.0‡
	(1.3)	(0.5)	(0.3)	(0.3)	(1.1)	(0.4)	(1.1)	(0.5)	(1.5)
Swimming	17.1	20.7‡	10.6‡	16.9	6.5‡	7.8	23.5‡	10.0	24.9‡
	(1.4)	(0.7)	(0.6)	(0.5)	(1.2)	(0.7)	(1.4)	(0.8)	(2.1)
N	680	3177	2610	6049	418	1694	955	1320	438

† Standard errors are given in parentheses.

‡ Statistically significant difference at the 5% level between age groups relative to those aged less than 25 years, whites relative to non-whites, those with a degree level qualification relative to those with no qualifications and low household income (i.e. less than £10,400 per annum) relative to high household income (i.e. greater than £52,000).

some key demographic indicators. As we would expect, age plays an important role in determining sporting participation. Less than a third (30.1%) of individuals aged 45 years or over participate in sports compared with just over 60% of those aged under 25 years (62.4%). This differential is maintained for all the main sports but is particularly pronounced for football or rugby, gym visits and general exercise. The sole exception to this rule is golf, which appears to increase in popularity with age. In comparison with whites, England's non-white ethnic minorities appear to be under-represented in sports. Only 35.6% of non-whites participate in sports, which compares with 45.2% of whites. Focusing on specific sports, this ethnic minority differential is statistically significant for cycling (12.2% compared with 6.7%), golf (3.1% compared with 1%) and swimming (16.9% compared with 6.5%). The exception is general exercise for which greater proportions of non-whites than whites report participation (10.7% compared with 14.4%).

Education is clearly an important determinant of sporting participation. If we compare the two extremes of the educational spectrum, having no qualifications and possessing a degree (or higher degree), we find a significantly greater participation rate for the latter (63.8%) relative to the former (24.2%) group. This level of disparity is maintained across all 10 sports.

Of particular interest, in terms of social inclusion, is the relationship between household income and sporting participation. Household income is recorded in the survey in grouped format (31 bands). Consequently, in the following analysis we have controlled for the effect of household income on sporting participation by defining six dichotomous variables, representing as closely as possible increments of £10,000 (with an upper band of greater than £52,000 per annum). Moreover, for 10.7% of the sample, household income data are missing. In the statistical models presented in 'Empirical results', we have additionally included a separate control variable for the missing observations.

It is immediately apparent from Table 36.1 that income plays a key role in determining access to sporting activities. For illustration, comparing respondents in the lowest household income group (i.e. below £10,400 per annum) with those with household incomes greater than £52,000, we see a participation differential of around 100% (60.7% compared with 30.9%). The positive influence of income on sporting participation appears to hold for all 10 sports but is particularly pronounced for golf, racket sports, running and swimming. This corresponds to the general perception that golf and racket sports (but perhaps not running and swimming) are played primarily by the more affluent in society.

Statistical framework

The model that we specify needs to account for three characteristics of the data. Firstly, our dependent variables are binary in nature, i.e. taking the value of 1 if an individual participates in sports (or a particular type of sport) and 0 otherwise. Secondly, we expect that individual attributes and preferences, family circumstances and health will be important in explaining variations in individuals' probabilities of participating in sport. Thirdly, we wish to explore the importance of unobservable household preferences in determining sporting participation across household members. We use the word 'preferences' generically as meaning unobservable household-specific characteristics which affect sporting participation within the household.

We therefore adopt the following random-effects probit model:

$$S^*_{ih} = x'_{ih}\beta + \nu_{ih} \quad i = 1, 2, \ldots, n, \quad h = 1, 2, \ldots, H, \tag{36.1}$$

$$\nu_{ih} = \alpha_h + \mu_{ih} \tag{36.2}$$

and

$$S_{ih} = \begin{cases} 1 & \text{if } S^*_{ih} > 0, \\ 0 & \text{otherwise} \end{cases}$$

where S^* is an individual's unobservable propensity to participate in sports; S is observed sporting participation; x is a vector of exogenous characteristics which are expected to influence S^*; β is the associated vector of coefficients; α_h is the household-specific unobservable effect which captures differences in preferences towards sports across households; μ_{ih} is a random-error term. We assume a random-effects specification, where $\mu_{ih} \sim \mathrm{IN}(0, \sigma^2_u)$, and to marginalize the likelihood it is assumed that, conditionally on the x_{ih}s, α_hs are $\mathrm{IN}(0, \sigma^2_\alpha)$ and are independent of the μ_{ih}s and the x_{ih}s. This implies that the correlation between the error terms of individuals in a household is a constant given by

$$\rho = \mathrm{corr}(\nu_{ij}, \nu_{ik}) = \frac{\sigma^2_\alpha}{\sigma^2_\alpha + \sigma^2_\mu} \quad j \neq k. \tag{36.3}$$

A fuller discussion of the random-effects probit model and the associated likelihood function can be found in Arulampalam (1999). The likelihood is computed by using 20-point Gauss–Hermite quadrature (see Butler and Moffitt, 1982) and all our calculations were done using LIMDEP version 7 (Greene, 1998).

We fit several models of sporting participation. Firstly, we fit a general model of sporting participation, i.e. whether or not an individual participates in any sporting activities or not. Secondly, we fit separate models for the seven most popular sporting activities undertaken in England: swimming, cycling, gym exercise, running, aerobics, football or rugby and racket sports. Although we have data on all the sports given in Table 36.1, we exclude dancing and general exercise from the statistical analysis because of concerns over their exact meanings. Dancing may be dominated by activities undertaken in night-clubs (rather than in ballrooms), and general exercise is vaguely defined in the survey. We have also omitted golf owing to the small number of positive outcomes in the sample (i.e. less than 3% of the sample).

We include in the vector x characteristics which are typically expected to play an important role in determining an individual's sporting participation. These capture individuals' opportunities for participating in sports as well as characteristics which attempt to capture variations in individuals' attitudes and preferences towards health. Therefore, we control for age, gender, ethnicity, marital status, dependant children and infants (i.e. aged under 2 years) and education. Consistent with the descriptive analysis that is presented in the data section, we expect that participation in sports will decline with age and that the sporting choices of men and women will differ. To allow for different age profiles for men and women, in the models we interact gender with age. The preferences of Britain's ethnic minorities towards sport generally, or specific types of sports, perhaps for cultural or religious reasons, may lead to differential levels of participation relative to whites. The effect of ethnicity on sporting participation is an empirical question that has not been previous addressed, the answer to which will be important for both ethnic differences in health and issues of social inclusion.

As found in the MINTEL (2000) report, we expect that sporting participation will be influenced by family commitments. However, the effect of children on adult sporting participation is not clear. Children may limit the amount of time that is available for participating in sporting activities that are typically adult only in nature (e.g. aerobics and gym visits) but lead to higher participation in sports such as swimming, cycling and football that are often child-orientated activities. Moreover, there may be a difference in the effect of children and infants in determining sporting participation, which may differ by men and women. Consequently, we have added both children and infant–gender interactions into our statistical models. Greater education might lead to a greater awareness of the health benefits of sports, and the associated longer duration of schooling might have led to greater sporting participation in youth (thus developing lifelong preferences for sport). Thus, we would expect a positive relationship between education and sports, and we attempt to capture this effect through variables indicating highest qualification level achieved. In developing parsimonious models of sporting participation we have also examined the interaction between qualification level and gender. However, unlike the case of age, number of children and infants, we have found no evidence of any gender differences in the relationship between qualifications and sporting participation.

We also control for region of residence (i.e. eight National Health Service regions, as coded in the survey) and the degree of urbanization (i.e. rural, suburban or city). We might expect that access to certain types of sporting facilities would be less represented in rural areas (e.g. gyms and swimming-baths) than in the suburbs or city. As with the case of ethnicity, investigating whether or not there are regional differences in sporting participation will be important for the formulation of policy initiatives aimed at increasing sporting participation across the population. Because of the cyclical timing of many sports we also include controls for season.

Most importantly, we control for household income, which enables us to investigate whether low income is a barrier to sporting participation. We also include controls for employment status. Having controlled for the effect of income on participation, our employment status variables may capture differences in the amount of leisure time that is available for sports. For example, we might expect that the unemployed and labour market non-participants would have more time for sports than those currently employed or self-employed.

Importantly, our data have the special advantage of allowing us to control comprehensively for health status. In this respect, we control for subjective (i.e. 'do you think you are in good, fair or bad health?') and more objective ('do you have a long-term illness?') health status. It is expected that participation in sports overall will decline with bad health and having a long-term illness. However, experiencing bad health may lead to an increased participation in certain sports on medical advice (e.g. swimming). Finally, we attempt to capture individual preferences by including controls for drinking and smoking status. Smokers may have a lower preference for 'health' than non-smokers. The effect of currently being a drinker is less clear. If one of the benefits of participating in sport is the associated social environment, we might expect to find a positive relationship between drinking and participation.

Empirical results

The results from our models are generally consistent with our prior beliefs about sporting participation and our earlier descriptive analysis. The parameter estimates are presented in Tables 36.2 and 36.3. Owing to the difficulties in directly evaluating the quantitative effect of

Table 36.2 ***Determinants of sporting participation by sporting type: random-effects probit estimates.***[†]

Effect	Estimates for all sports			Estimates for aerobics			Estimates for cycling			Estimates for football or rugby		
	B	t-statistic	ME	β	t-statistic	ME	β	t-statistic	ME	β	t-statistic	ME
Age * male	–0.038	–12.74	–0.012	–0.012	–2.08	–0.001	–0.020	–4.52	–0.002	–0.053	–9.75	–0.002
Age * female	–0.024	–7.86	–0.007	–0.011	–2.68	–0.001	–0.018	–3.39	–0.002	–0.028	–2.48	–0.001
Male	0.694	4.48	0.219	–1.030	–3.69	–0.091	0.652	2.71	0.076	2.540	6.88	0.090
Married or cohabiting	–0.181	–2.68	–0.057	–0.165	–1.63	–0.015	–0.029	–0.28	–0.003	–0.060	–0.56	–0.002
Separated–divorced–widowed	0.116	1.26	0.037	0.080	0.58	0.007	0.171	1.19	0.020	0.260	1.45	0.009
Children * male	0.083	2.58	0.026	0.061	0.79	0.005	0.157	3.42	0.018	0.070	1.73	0.002
Children * female	–0.004	–0.12	–0.001	0.019	0.46	0.002	0.135	2.56	0.016	0.060	0.52	0.002
Infants * male	–0.262	–2.51	–0.083	–0.744	–2.01	–0.066	–0.057	–0.40	–0.007	–0.215	–1.58	–0.008
Infants * female	–0.323	–3.44	–0.102	–0.166	–1.28	–0.015	–0.880	–3.86	–0.102	–0.493	–0.90	–0.017
Non-white	–0.192	–1.97	–0.061	–0.141	–0.93	–0.013	–0.559	–3.24	–0.065	–0.174	–1.05	–0.062
Degree	0.677	8.61	0.214	0.518	4.02	0.046	0.550	4.70	0.064	0.143	1.08	0.005
Higher vocational	0.316	4.13	0.100	0.494	3.89	0.044	0.169	1.39	0.020	0.087	0.67	0.003
A-levels	0.455	5.67	0.144	0.685	5.41	0.061	0.101	0.81	0.012	0.264	1.97	0.009
O-levels	0.357	5.85	0.113	0.460	4.40	0.041	0.147	1.50	0.017	0.053	0.44	0.002
Other qualification	0.248	2.21	0.078	0.588	3.40	0.052	–0.046	–0.21	–0.005	–0.190	–0.46	–0.007
Self-employed	0.037	0.49	0.012	0.099	0.83	0.009	–0.032	–0.30	–0.004	0.050	0.47	0.002
Employed part time	0.095	1.36	0.030	–0.083	–0.82	–0.007	0.179	1.68	0.021	–0.070	–0.35	–0.002
Unemployed	0.263	2.61	0.083	0.125	0.75	0.011	0.410	2.90	0.048	0.147	0.97	0.005
Permanently sick	–0.481	–3.55	–0.152	–0.286	–1.10	–0.025	–0.388	–1.79	–0.045	–0.247	–0.76	–0.009
Non-participant	0.008	0.11	0.002	0.031	0.31	–0.007	–0.064	–0.53	–0.007	0.080	0.32	0.003
Drinker	0.457	5.50	0.144	0.206	1.68	0.018	0.076	0.57	0.009	0.100	0.55	0.004
Smoker	–0.311	–6.46	–0.098	–0.190	–2.54	–0.017	–0.217	–2.77	–0.025	–0.201	–2.47	–0.007

Fair health	−0.268	−4.43	−0.084	−0.237	−2.22	−0.021	−0.260	−2.50	−0.030	−0.195	−1.67	−0.007
Bad health	−0.404	−2.82	−0.128	−0.207	−0.93	−0.018	−0.230	−0.96	−0.027	−0.333	−0.92	−0.014
Long-term illness	−0.086	−1.85	−0.027	−0.148	−1.95	−0.013	−0.102	−1.39	−0.012	−0.135	−2.21	−0.007
>£52,000	0.543	4.80	0.171	0.336	2.12	0.030	0.002	0.10	0.000	0.198	1.09	0.007
£41,601–£52,000	0.500	4.05	0.145	0.181	1.10	0.016	−0.119	−0.67	−0.014	−0.037	−0.20	−0.001
£31,201–£41,600	0.440	4.45	0.139	0.225	1.61	0.020	0.066	0.44	0.008	0.182	1.10	0.006
[3pt] £20,801–£31,200	0.257	3.13	0.081	0.227	1.91	0.020	−0.071	−0.55	−0.008	0.115	0.77	0.004
£10,401–£20,800	0.139	1.89	0.044	0.040	0.35	0.004	−0.067	−0.55	−0.008	0.009	0.07	0.000
Income missing	0.110	1.22	0.035	−0.024	−0.17	−0.002	−0.050	−0.34	−0.006	−0.186	−1.10	−0.007
Constant	0.306	1.67		−1.261	−4.55		−1.119	−3.73		−2.204	−4.89	
ρ	0.357	11.72		0.344	4.96		0.536	13.33		0.187	1.80	
Log-likelihood	−3781			−1588			−2052			−1060		

[†] ME is the marginal effect calculated at the means of the explanatory variables. We also control for eight regions, degree of urbanization (i.e. city, urban or rural) and seasonal effects in each of the models. The t-statistic column refers to the estimate divided by its estimated standard error. The omitted categories are female, single, no children, no infants, white, no qualifications, employed full time, non-drinker, non-smoker, in good health, no long-term illness and household income less than £10,401 per annum.

Table 36.3 ***Determinants of sporting participation by sporting type: random-effects probit estimates.***†

Effect	Estimates for gym			Estimates for racket sports			Estimates for running			Estimates for swimming		
	B	t-statistic	ME	β	t-statistic	ME	β	t-statistic	ME	β	t-statistic	ME
Age * male	−0.026	−6.38	−0.003	−0.022	−3.99	−0.001	−0.037	−6.78	−0.002	−0.015	−3.51	−0.002
Age * female	−0.018	−3.77	−0.002	−0.004	−0.55	−0.000	−0.022	−3.17	−0.001	−0.024	−5.53	−0.004
Male	0.550	2.65	0.067	1.162	3.63	0.064	1.220	4.36	0.076	−0.625	−3.03	−0.095
Married or cohabiting	−0.306	−3.26	−0.037	−0.257	−2.05	−0.014	−0.213	−1.96	−0.013	0.065	0.69	0.010
Separated–divorced–widowed	0.020	0.15	0.002	0.149	0.83	0.008	−0.060	−0.35	−0.038	0.227	1.81	0.035
Children * male	−0.025	−0.57	−0.003	−0.013	−0.23	−0.001	0.058	1.23	0.004	0.133	2.83	0.020
Children * female	−0.091	−0.20	−0.001	0.019	0.26	0.001	0.073	1.20	0.005	0.075	1.77	0.011
Infants * male	−0.313	−2.09	−0.038	−0.023	−0.13	−0.001	−0.036	−0.25	−0.002	−0.106	−0.67	−0.016
Infants * female	−0.402	−2.66	−0.049	−0.842	−2.27	−0.046	−0.967	−2.73	−0.060	−0.103	−0.76	−0.016
Non-white	0.084	0.62	0.010	0.093	0.47	0.005	−0.106	−0.67	−0.007	−0.720	−4.27	−0.109
Degree	0.242	2.17	0.030	0.937	5.50	0.052	0.647	4.47	0.040	0.518	4.70	0.079
Higher vocational	0.063	0.53	0.008	0.642	3.59	0.035	0.387	2.56	0.024	0.445	4.06	0.068
A-levels	0.239	2.09	0.029	0.523	2.88	0.029	0.420	2.76	0.026	0.412	3.68	0.063
O-levels	0.244	2.62	0.030	0.524	3.20	0.029	0.328	2.44	0.020	0.317	3.49	0.048
Other qualification	−0.108	−0.54	−0.013	0.267	0.90	0.015	0.389	1.43	0.024	0.071	0.39	0.011
Self-employed	−0.136	−1.29	−0.017	0.077	0.61	0.004	0.005	0.05	0.000	−0.001	−0.01	−0.000
Employed part time	−0.031	−0.30	−0.004	0.278	2.07	0.015	−0.065	−0.42	−0.011	0.012	0.13	0.002
Unemployed	−0.020	−0.13	−0.002	−0.330	−1.39	−0.018	0.144	0.89	0.009	0.125	0.87	0.019
Permanently sick	−0.652	−2.65	−0.079	−0.509	−1.22	−0.028	−0.764	−1.81	−0.048	−0.431	−2.05	−0.065
Non-participant	−0.076	−0.67	−0.009	0.016	0.09	0.001	−0.065	−0.42	−0.004	0.039	0.38	0.006
Drinker	0.301	2.35	0.037	0.558	2.53	0.031	0.293	1.79	0.018	0.463	3.66	0.070
Smoker	−0.394	−5.39	−0.048	−0.388	−3.77	−0.021	−0.477	−4.84	−0.030	−0.310	−4.49	−0.047

Fair health	−0.202	−2.10	−0.025	−0.167	−1.16	−0.009	−0.354	−2.60	−0.022	−0.229	−2.43	−0.035
Bad health	−0.080	−0.37	−0.010	−0.197	−0.48	−0.011	−0.364	−0.94	−0.023	−0.174	−0.87	−0.026
Long-term illness	−0.14	−2.04	−0.017	−0.232	−2.35	−0.013	−0.166	−2.03	−0.010	−0.009	−0.15	−0.000
>£52,000	0.735	4.37	0.090	0.600	2.65	0.033	0.159	0.82	0.010	0.479	2.95	0.073
£41,601–£52,000	0.557	3.37	0.068	0.450	2.03	0.025	0.019	0.10	0.001	0.319	1.99	0.048
£31,201–£41,600	0.470	3.24	0.057	0.539	2.63	0.030	0.269	1.63	0.017	0.294	2.00	0.045
£20,801–£31,200	0.408	3.16	0.050	0.244	1.26	0.013	0.093	0.63	0.006	0.270	2.18	0.041
£10,401–£20,800	0.246	2.04	0.030	0.293	1.59	0.016	−0.079	−0.56	−0.005	0.153	1.34	0.023
Income missing	0.298	2.07	0.036	0.382	1.83	0.021	−0.050	−0.28	−0.005	0.177	1.25	0.027
Constant	−1.446	−4.82		−3.295	−6.67		−1.989	−5.47		−1.045	−3.94	
ρ	0.443	10.07		0.490	7.49		0.389	5.51		0.550	16.01	
Log-likelihood	−2015			−1196			−1316			−2547		

† ME is the marginal effect calculated at the means of the explanatory variables. We also control for eight regions, degree of urbanization (i.e. city, urban or rural) and seasonal effects in each of the models. The *t*-statistic column refers to the estimate divided by its estimated standard error. The omitted categories are female, single, no children, no infants, white, no qualifications, employed full time, non-drinker, non-smoker, in good health, no long-term illness, and household income less than £10,401 per annum.

the parameters of the probit model, we have also calculated marginal effects ME, $\partial F(\bar{X}, \beta)/\partial x_i$, evaluated at the means of the explanatory variables, $\bar{X}$.

Demographic factors

The gender coefficient shows us that men are significantly more likely to participate in any sporting activities (being male increases the probability of participation by 0.219 relative to females) and in all sports separately studied, except aerobics (ME = –0.091) and swimming (ME = –0.095) where the converse is true. This suggests that men and women therefore have very different preferences over the type of sporting activities that they enjoy.

In all cases the probability of sporting participation is predicted to decrease with age. We have also tested for a non-linear relationship between sports and age by including a quadratic and cubic specification. The polynomial terms were found to be statistically insignificant for all the sports examined. The estimated age profiles are highlighted in Figs 36.2, 36.3 and 36.4 for all sports (both men and women), and for the four most popular sports played by men and women separately. It is predicted that the decline in sporting participation with age is sharper for males than for females, with an additional 1-year increase in age reducing the probability of participation by –0.012 and –0.007 respectively. Interestingly, males have higher

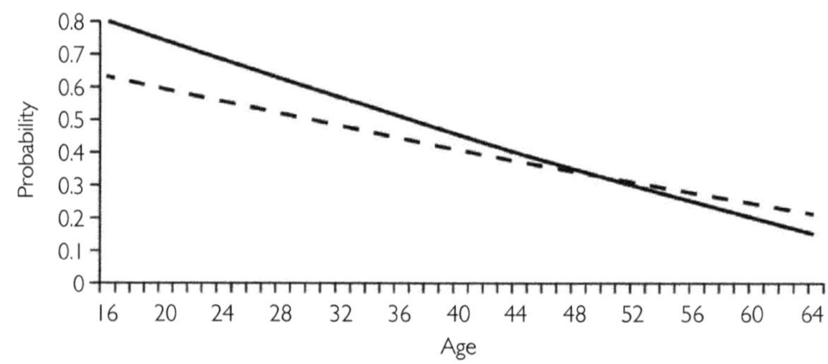

Figure 36.2 Predicted sporting participation by age and gender:——, males (all sports);——, females (all sports).

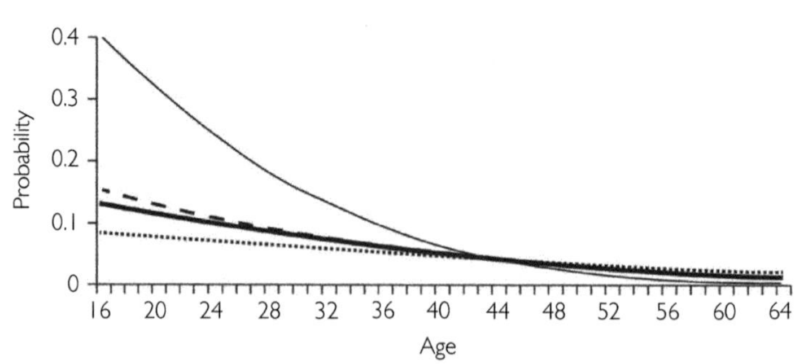

Figure 36.3 Predicted sporting participation by age for males (four most popular sports):——, cycling;——, football or rugby;——, gym;——, swimming.

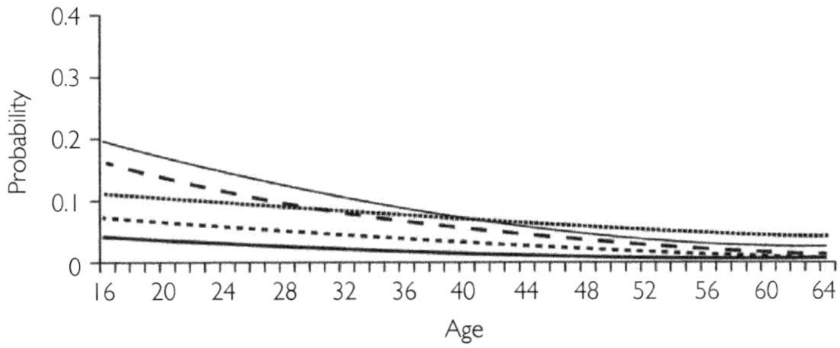

Figure 36.4 Predicted sporting participation by age for females (four most popular sports):——, aerobics;——, cycling;——, gym;—-, swimming.

sporting participation than females until age 50 years, whereas participation by females appears to dominate that by males in the later years of working age. Married people participate significantly less in sports overall relative to single people (ME = –0.057). However, at the level of the individual sports, being married only significantly affects participation in going to the gym, playing racket sports and running. There are generally no significant differences between participation for the widowed–divorced–separated group relative to single people.

The observation that married people play less sport is consistent with the MINTEL (2000) study, which states that household commitments mean that less time is available for participation in sport. If this were the explanation then we would expect to see that the introduction of children into the household has a further significant negative impact on the probability of playing sport. Importantly, we find key differences in the effect of children and infants, by gender. For males we find a significantly positive effect of number of children on overall sporting participation (each child increases the probability of participation by 0.026); this is true also for cycling (ME = 0.018), football or rugby (ME = 0.002) and swimming (ME = 0.011). In contrast, we find that children have no effect on the probability of overall participation for women, but have a positive effect for cycling (ME = 0.016) and swimming (ME = 0.021). The results with regard to cycling and swimming appear intuitive. Unlike the other sports examined, parents and children will often participate in these activities together. This result is very different from the story of 'time poor' households with children participating less in sport, which was revealed in the MINTEL (2000) study.

In contrast, having a dependant infant reduces the probability of playing sports for both men and women. The quantitative effect is much greater than for children, with an infant reducing the probability of overall sporting participation by –0.083 for men and –0.102 for women relative to respondents with no infants. This effect is also apparent for aerobics and gym exercise for males, and cycling, gym exercise, racket sports and running for females.

The descriptive statistics in the data section suggested an under-representation of England's ethnic minorities in many of the most popular sports. After controlling for other explanatory factors, this finding is confirmed for overall participation (ethnic minorities members have a –0.061 lower probability of participating than whites). This result is also found for cycling (ME = –0.065) and swimming (ME = –0.109). In contrast, certain sports seem to be preferred by non-whites such as going to the gym and racket sports (although these positive coefficients are not statistically significant). However, given the small numbers of ethnic minorities in our sample, we should be cautious when generalizing these results.

Education is positively correlated with overall participation in sport. The higher your level of qualifications the more likely you are to participate in at least one sport. For example, a degree level qualification increases the probability of participation in sports by 0.214 relative to individuals with no qualifications. This could be due to a better understanding of the importance of sport, or due to habits developed as students, where access to facilities is good and relatively inexpensive. This result is generally confirmed across all the types of sporting activities examined. In general, education is highly correlated with income and we might suspect that the significance of the education variables is due to a correlation with household income. We have therefore fitted the models without the inclusion of the education variables and found that the positive effect of household income remains. There is therefore no evidence to suggest that education is acting as a proxy for our economic variable of interest.

Importantly, for the focus of policy initiatives, indicators of the individuals' place of residence in terms of rural, suburban and urban appear to have little effect on participation, suggesting that there is access to sporting opportunities across all types of communities. An analysis of the data, controlling only for age and gender, suggests that residents of southern England have a significantly higher sporting participation rate compared with those living in the north or Midlands. The percentage of respondents participating in sports in the last 4 weeks was, in order of the highest region health authority first, 49.0% for South London, 48.2% for Anglia and Oxford, 46.6% for South England, 43.3% for North London, 42.5% for North Yorkshire and Humberside, 42.4% for the North West, 42.1% for Trent and 42.0% for the West Midlands. However, once we control for economic and other demographic characteristics in our statistical models, in particular household income, we find that geographical location within England has no significant effect on participation. We have tested the robustness of this finding by refitting the sporting participation models replacing the eight broad regional categories with a more disaggregate set of control variables for region. This was achieved by dividing the 100 local health authority areas observed in the data into 16 then 30 areas, based on geographical location and sample size permitting. Our finding of no geographical differences in the probability of participating in sports within our multivariate framework could not be rejected from this further analysis (these additional results are available from the authors on request).

The individual preference variables give a very interesting picture about the types of people who participate in sporting activities. In all cases drinking is positively correlated with participation in sport (being a drinker increases the probability of participation in any sport by 0.144 over non-drinkers). The exceptions are cycling and football where we see a positive but insignificant effect at the 90% level of confidence. This may in part be reflecting the social attribute of participation in sport and be capturing individuals who like to engage in activities which take place outside the home. Smoking, however, always has a significantly negative effect on participation in sport, reducing the probability of participation in any sports by −0.098 over non-smokers. This is reflective of life style choices. People who smoke might have a lower discount rate (i.e. a greater preference for the current than the future) and consequently could be less likely to care about staying fit and healthy through participation in sport.

Our data also allow us to control for the individuals' self-reported health status; the results show that people who report that they are in fair or bad health are significantly less likely to engage in sporting activities, relative to those who report good health (the reductions in the probability of participation in any sport are −0.084 for those reporting fair health and −0.128 for those in bad health, relative to the omitted category—good health). Of course we cannot identify whether the poor health is a result of a lack of exercise in the past, but it is negatively correlated with current participation in sport. We also have a more objective health status variable that accounts for having a long-term illness. We find that long-term illness is negatively correlated with all types of sporting activity but does not reach statistical significance for cycling and swimming.

Economic factors

How income impacts on participation in sport is of particular interest. We find clear evidence that income has a positive effect on the probability of participation in sport. Compared with low income households (i.e. below £10,400 per annum), the probability of participation increases by 0.044, 0.081, 0.139, 0.145 and 0.171 as we move up the five income groups. Therefore, having a household income between £20,801 and £31,200, and greater than £52,000 per year is equivalent respectively to a reduction in age of participation of about 7 and 14 years for men and 11 and 24 years for females. Separating by sporting type, the effect of income is significantly positive for gym visits, racket sports and swimming, and marginally significant for aerobics. We find no consistent income effect for cycling, football or rugby or running. Overall, these results provide some evidence that household income may act as a barrier to sporting participation and that, if sport is to be used as a policy instrument to enhance social inclusion, then better access programmes may be required. We have investigated the robustness of this finding by also fitting the models using a measure of equivalized rather than actual household income. Our equivalent income measure was total household income divided by the McClement score (which is based on the number, age and relationships of adults and children in the household, and is already coded in the 1997 HSE). As with observed household income, using this measure we found a positive and significant effect on sporting participation.

Conditionally on household income, the controls for the individuals' employment status are an attempt to capture the effect of the amount of time that the individual has to participate in sport. The estimates show that the unemployed do significantly more exercise than the employed (the marginal increase in the probability of participation being 0.083, relative to full-time employees). Being unemployed, however, is only significant for cycling when we look at the individual sports modelled separately here. Interestingly, however, we do not find such a result for non-participants in the labour market, who are mostly mothers at home. As expected, individuals who are not in the labour market owing to permanent sickness participate less in sports relative to full-time employees, whereas we find no evidence of differences in sporting probabilities between the self-employed and the reference group.

Household preferences

Finally, we report the estimated intrahousehold correlation. Since our data are based on households it is possible to exploit this panel element of the survey. The random-effects framework allows us to establish how much of the variation in the data can be explained by unobservable intrahousehold correlations. Do some households enjoy participation in sports whereas other households have 'preferences' for different types of leisure activities? We find that the intrahousehold correlation is substantial in all eight models (e.g. for participation in any sport $\rho = 0.357$ and $t = 11.72$). It might be that households fall into two groups: those who actively participate in sport and those who choose to abstain. This result is especially interesting as it could imply preference matching, or contagion of sporting activity. Sporty individuals may choose to live in the same households or sport could be infectious within the household. This again draws attention to the social nature of involvement in sport. Given that our sample includes a number of single-person households we have examined the robustness of our intrahousehold correlation findings by fitting the models including only those households for which we have data on at least two members. The estimated ρs are all quantitatively similar in magnitude and significance to those presented in Tables 36.2 and 36.3.

The high degree of intrahousehold correlation in sporting participation has important implications for policies aimed at increasing access. Our estimates suggest that increasing relative income (or reducing relative prices) may lead to increased sporting participation by improving the access to sporting facilities for poorer households. However, the success of such a policy would depend on the preferences of low income households towards sports. If these households have preferences to abstain from sports, then increasing access may have more limited success than expected. This points to the importance of education in changing individuals' preferences towards sports and developing the sporting preferences of the young in schools.

Conclusion

In this paper we have provided a detailed statistical analysis of the economic and demographic factors that determine sporting participation in England. To enable this we have used a unique data source, the HSE, which contains detailed information on a wide range of sporting activities. In particular, our aim has been to establish the role that factors such as family commitments, ethnicity and household income play in determining access to sport and to contribute to the debate regarding sport as a means of enhancing social inclusion. In addition, we have adopted a statistical framework that allows for the importance of unobservable household attitudes towards participation in sport.

Sporting participation is found to decline significantly with age for each of the seven most popular sports played in England. Males participate in sports to a greater extent than females; however, the types of sporting activities undertaken by men and women differ considerably. Males have significantly higher participation rates in cycling, football or rugby, gym exercise, rackets sports and running, whereas the opposite is true for aerobics and swimming. We find evidence that children increase participation in cycling and swimming, which are sports that can be undertaken with children. In contrast, dependant infants tend to reduce participation, but this effect is stronger for men than women.

Sporting participation is found to be positively related to education, with the exception of cycling and football or rugby. Drinking is positively related to sporting participation, whereas the opposite is true for smoking. The quantitative effects of these life style indicators are large. As expected, sporting participation is significantly and substantially reduced by poor health. Conditionally on other explanatory variables, we have found no evidence of any regional differentials in sporting participation in England. This result is robust to varying degrees of regional aggregation.

With respect to income and employment status, we find that low income households participate to a lesser extent in sports than do high income households. For example, having a household income greater than £52,000 per annum increases the probability of participating in sport by 0.171 relative to those with household incomes less than £10,400 per annum.

After controlling for income, we find that the unemployed participate in sport to a greater extent than those who are currently employed or labour market non-participants. This effect, however, is only significant for cycling when the sports are modelled separately.

Finally, using within-household variations in sporting participation in our data, we find robust evidence that unobservable household 'preferences' play an important role in determining sporting participation. This could be due to 'matching' of individual preferences within households, or the contagion of preferences for sporting behaviour.

In terms of policy prescriptions, our analysis provides evidence that income plays an important role in determining sporting participation in England. This provides some support for current Government policy aimed at the need to make sporting facilities more broadly available across all income groups. Increased participation in sport could be one method to help to promote the social inclusion, and to improve the health, of the less well off in society.

ACKNOWLEDGEMENTS

We would like to thank Stephen Creigh-Tyte, Ziggy Macdonald, Steve Pudney, Steve Thompson, Stephen Wheatley Price and two referees for valuable comments. The data were accessed through the Data Archive at the University of Essex.

REFERENCES

Argyle, M. (1987) *The Psychology of Happiness*. London: Methuen.

Arulampalam, W. (1999) A note on estimated effects in random effect probit models. *Oxf Bull. Econ. Statist.*, **61**, 597–602.

Butler, J. and Moffitt, R. (1982) A computationally efficient quadrature procedure for the one factor multinomial probit model. *Econometrica*, **50**, 761–764.

Department of Culture, Media and Sports (2000) National Lottery White Paper. *Command 3709*. London: Stationery Office.

Greene, W. (1998) *Limdep: User Manual Version 7.0*. New York: Econometric Software.

MINTEL (2000) *Family Leisure Trends*. London: MINTEL.

Prescott-Clarke, P. and Primatesta, P. (1998) *Health Survey for England: the Health of Young People '95–97'*. London: Stationery Office.

Steptoe, A. and Butler, N. (1996) Sports participation and emotional wellbeing in adolescents. *Lancet*, **347**, 1789–1792.

Wold, B. and Anderssen, N. (1992) Health promotion aspects of family and peer influences on sport participation. *Int. J. Sport Psychol.*, **23**, 343–359.

CHAPTER 37

Using model/data simulations to detect streakiness

JIM ALBERT AND PATRICIA WILLIAMSON

Abstract

A simulation-based approach is proposed for approximating a Bayesian analysis. Parameters and data are simulated from a Bayesian model and inference about a parameter is performed by exploring the set of simulated parameter values conditional on a set of values of a simulated statistic. The approach is used to learn about parameters of a streaky model on the basis of a statistic used to measure streakiness. The method is illustrated to detect streakiness in baseball hitting data and basketball shooting data.

Keywords: *Bayesian inference; Coin-tossing; Hot-hand; Markov switching model; Overdispersion, Quasi-binomial.*

Introduction

There has been much recent interest in the detection of streakiness or the "hot hand" in the performance of athletes in baseball, basketball, and other sports. Gilovich, Vallone, and Tversky (1985) and Tversky and Gilovich (1989) discussed the existence of the hot hand in basketball and concluded that any observed streakiness in data is simply one's misperception of the patterns inherent in random sequences. Larkey, Smith, and Kadane (1989), in their analysis of basketball shooting data from a number of professional players, took a different view. They concluded that there is some statistical evidence for the hot hand. Albright (1993) performed a number of statistical tests to detect streakiness in hitting data for a large group of professional players. Although some players appeared to exhibit streaky behavior for particular seasons, Albright found little support for the hot hand theory in baseball. In their comments on Albright's article, Stern and Morris (1993) criticized some of the tests that were conducted by Albright, and Albert (1993) presented a hidden Markov switching model that could be used to model streakiness. Wardrop (1999) also discussed the hot-hand debate. Specifically, Wardrop criticized particular tests used by Gilovich et al. (1985) and Tversky and Gilovich (1989) to judge if a player has the hot hand. See Stern (1997) for a recent discussion of the statistical detection of streakiness.

In the basketball and baseball settings, one observes a sequence of binary observations $y_1, \ldots, y_n$, where $y_i = 1$ if a successful (baseball) hit or (basketball) shot is observed, and $y_i = 0$ otherwise. The basic *coin-tossing model* assumes that the y_i are independent Bernoulli trials with a constant probability of success p. The *hot hand* or *streaky hypothesis* represents a deviation from the coin-tossing model. In this alternative hypothesis, there is either *nonstationarity* where the probability of success does not stay constant over the trials, or *autocorrelation* where the probability of success on a given trial depends on the player's success in recent trials. We refer to streakiness as a presence of nonstationarity and/or autocorrelation in the sequence.

In the usual approach to searching for the hot hand, a value of a statistic T is observed, such as a long run of successes, which seems to support the hot hand theory. Then one computes a p value of this statistic assuming the coin-tossing model. If this p value is small, then one rejects the coin-tossing model.

However, this approach generally is not effective in detecting the hot hand theory. One problem is that it is difficult to gauge the smallness of the p value. One will typically find significant p values if a large number of tests are made, or if the test is performed on a sample sequence with many observations. A more serious criticism of this approach is that there is little investigation into the ability of the test to detect deviations from the coin tossing model. Often many tests are performed without understanding the power of the tests to detect the type of streaky patterns that are of interest. By doing a power study, Wardrop (1999) demonstrated that particular "streaky statistics" used by Gilovich et al. (1985) are almost useless in detecting deviations from coin-tossing. One reason why power calculations are not made is that it can be hard to construct alternative models that represent a belief in a hot hand theory. Barry and Hartigan (1993) and Albert (1993) represented two of the relatively short list of articles that used particular hot hand models to analyze sequences of win/loss or hit/not-hit data.

This article proposes a simple Bayesian method, based on simulation, for detecting streakiness in sports data. Suppose that one can construct a streaky model that can possibly generate the sequence of successes and failures. We assume that this model is parameterized by a vector of parameters θ, and initial beliefs about θ are represented by a prior distribution. Suppose that a streaky statistic $T = t^{obs}$ is observed—what have we learned about the parameter θ? The section "Simulating models and data" describes a simple scheme proposed by Albert (1997) in a teaching context, which simulates from the joint distribution of the parameter and the data. By exploring the set of simulated parameter values conditional on a set of values of the simulated statistic T, we perform inferences on the streakiness parameter θ.

"Fitting an overdispersion model" introduces the use of the model/data simulation scheme for learning about possible nonstationarity in home run data for Mark McGwire. An overdispersion model is used to model nonstationarity, and the approximate posterior distribution on the overdispersion parameter based on the proposed simulation method is compared with the exact posterior distribution. The next section, "How streaky was Javy Lopez in 1998?," illustrates the fitting of a Markov switching model to detect streakiness in day-to-day baseball hitting data for a particular player. A single parameter λ is introduced to measure the distance from coin-tossing toward a streaky model, and we investigate the use of six "streaky statistics" in learning about this parameter. We measure the usefulness of a statistic in inference by means of the correlation of the simulated pairs of parameter and statistic values. "Detecting streakiness in free-throw shooting data" considers the issue of streakiness for basketball shooting data for a particular college player. A quasi-binomial distribution is used to model shooting proportions in groups of five shots—this distribution can model variability in these proportions that is greater or smaller than what is predicted from a binomial distribution. In a plot of dispersion parameters across time, the shooter appears to be more consistent over time, and a model/data simulation algorithm is used to assess the significance of the observed pattern in the data.

Simulating models and data

Suppose that one observes data $y_1, \ldots, y_n$ from a sampling model $f(y|\theta)$ and beliefs about the parameter θ are described by means of a proper prior distribution $g(\theta)$. (In the following, y will denote the vector of observations and θ is a vector of parameters.) The object is to learn about θ after the data $y = y^{obs}$ are observed, and, by Bayes's rule, this learning is accomplished by computation of the posterior distribution

$$g(\theta|y^{obs}) \propto \prod_{i=1}^{n} f(y_i^{obs}|\theta)g(\theta).$$

Now suppose that this posterior density is awkward to compute directly due to the complicated form of the likelihood. However, there exists a statistic $t(y)$ that we believe summarizes most, if not all, of the information contained in the data y. (If $t(y)$ is sufficient, then one can show that the posterior distribution will depend on y only through the value of the sufficient statistic, and so t will summarize *all* of the information about θ.) In this case, the posterior density can be approximated by the conditional density

$$g(\theta|t^{obs}) \propto f(t^{obs}|\theta)g(\theta),$$

where $f(t|\theta)$ is the sampling density of the statistic t.

In our simulation approach, it will be difficult to obtain a large sample of simulated values of θ conditional on a specific value of t, since t takes on many discrete values. However, one can approximate the likelihood

$$f(t^{obs}|\theta) \approx \Pr(t^{obs} - \epsilon < t < t^{obs} + \epsilon)\frac{1}{2\epsilon}$$

for a small value of ϵ, and it follows that the conditional density of interest is approximated by

$$g(\theta|t^{obs}) \approx g(\theta|t^{obs} - \epsilon < t < t^{obs} + \epsilon).$$

To summarize, if the statistic t is nearly sufficient for θ, then we can approximate the posterior density of interest by the density of θ conditional on the statistic t falling in a small interval about the observed value t^{obs}.

This approximation method is straightforward to apply in practice. In the following algorithm, we use the prior and sampling density to simulate sets of the parameter θ and the data y. From each simulated dataset y, the statistic t is computed. Then inference about θ is performed by focusing on the simulated values of θ where the simulated t values fall in an interval about the observed value t^{obs}.

The basic algorithm

1. **Simulate models and data.** Repeat the following three steps to obtain N model/data simulated pairs $\{(\theta_j^S, t_j^S)\}$.
   (a) Simulate a value of the parameter θ from the prior density $g(\theta)$—call this simulated value θ^S.
   (b) Simulate data $y_1^S, \ldots, y_n^S$ independently from the sampling density $f(y|\theta^S)$.
   (c) Compute the value of the statistic $t^S = t(y_1^S, \ldots, y_n^S)$ based on the simulated data.
2. **Perform the inference.** If $t = t^{obs}$ is observed, one learns about the parameter θ by focusing on the set of simulated pairs for which t^S falls in the interval $(t^{obs} - \epsilon, t^{obs} + \epsilon)$, for a suitably chosen value of the interval half-width ϵ.

How should ϵ be chosen? One wishes to choose ϵ small so that one has a good approximation to the posterior density which conditions on the value $t = t^{obs}$. On the other hand, one needs to collect enough simulation draws $\{\theta^S\}$ so that one gets accurate summaries of the posterior distribution. In the examples, ϵ was chosen so that the interval $(t^{obs} — \epsilon, t^{obs} + \epsilon)$ represented a small fraction of the simulated values of t^S, and the algorithm was iterated until at least 500 simulated values of t in this range were collected.

Fitting an overdispersion model

We first illustrate our method for a problem where the exact posterior density is straightforward to obtain. Mark McGwire has been the most prolific home run hitter in recent baseball history. In each of the five seasons 1995–1999, McGwire's observed rate in hitting home runs has remained in the 9–10% range, where the home run rate is defined as the number of home runs divided by the number of plate appearances. It is interesting to see if McGwire's pattern of home run hitting in this five-year period can be represented by a simple binomial model with a constant probability of success.

For each of the 698 games that McGwire played in this five-year period, we observe the number of plate appearances, m_i, and the number of home runs, x_i. If McGwire was indeed a streaky home run hitter, one might expect the home run counts to exhibit some clustering. In this streaky scenario, McGwire might hit a large number of home runs during some weeks, and have other weeks where he hits few or no home runs. To detect this type of clustering pattern, we group the data using a window of five games, which corresponds approximately to a week of games. Let y_i and n_i denote, respectively, the number of home runs and the number of plate appearances in the ith group of five games, where $i = 1, \ldots, 140$. Figure 37.1 graphs the number of home runs y_i for McGwire as a function of the group number i.

A basic model for this data assumes that the counts $\{y_i\}$ are independently distributed from binomial distributions with respective sample sizes $\{n_i\}$ and a constant probability p. The maximum likelihood estimate for this data is $\hat{p} = \sum y_i / \sum n_i = .0976$. To assess the suitability of a binomial fit, home run data were simulated for 1,000 five-year periods using McGwire's actual plate appearances and the estimated binomial probability. Table 37.1 displays a frequency table of the observed counts $\{y_i\}$ and the mean simulated counts based on the binomial model. The third row of the table gives the Pearson residuals $\{(\text{observed} - \text{expected})/\sqrt{\text{expected}}\}$. We see some differences between the two sets of counts. Specifically, the numbers of observed periods

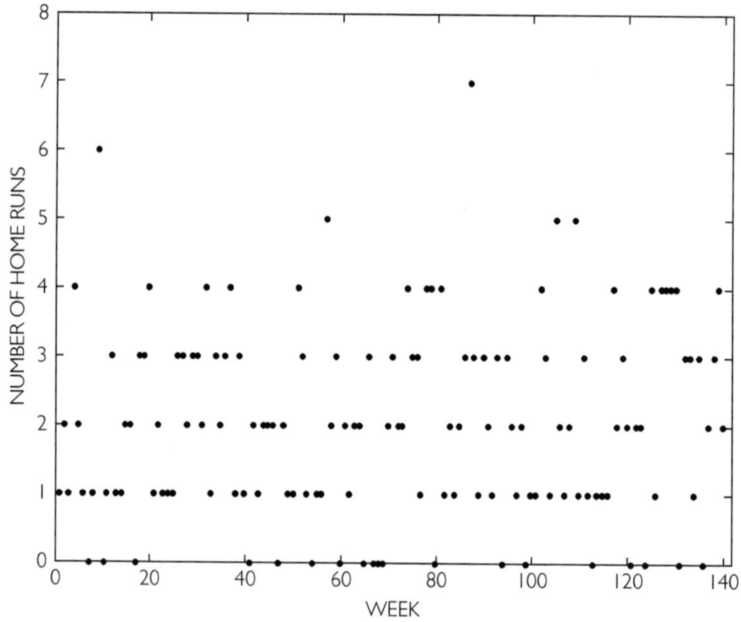

Figure 37.1 Graph of number of home runs hit by Mark McGwire in sequence of five-game periods during the seasons 1995–1999.

Table 37.1 Observed five-game home run counts, expected counts under a binomial model, and Pearson residuals for McGwire data.

	Number of home runs						
	0	1	2	3	4	5	6 +
Observed	19.0	38.0	33.0	28.0	17.0	3.0	2.0
Expected	16.9	37.5	39.5	26.6	12.9	4.8	1.9
Residual	0.51	0.08	−1.03	0.27	1.14	−0.82	0.07

with 0 and 4 home runs are somewhat higher, and the number of periods with 2 and 5 home runs are somewhat lower than what are predicted from the binomial model.

This brief analysis suggests that McGwire's home run counts may be more dispersed than predicted from a binomial model, and a binomial/beta distribution will be used to model the possible overdispersion. Suppose that the counts y_i are independently distributed from binomial distributions with parameters n_i and p_i. We suppose that the probabilities $p_1, \ldots, p_{140}$ are a random sample from a beta distribution with mean η and precision parameter K.

$$g(p_i) = \frac{1}{B(K\eta, K(1-\eta))} p_i^{K\eta-1}(1-p_i)^{K(1-\eta)-1},$$

where $B(a, b)$ is the beta function. Uncertainty about the Beta hyperparameters (η, K) is modeled by means of a prior density $g(\eta, K)$.

In this binomial/beta distribution, the hyperparameter η represents an average value of the probabilities $\{p_i\}$ and overdispersion is measured by the precision parameter K. As K approaches infinity, the model approaches a binomial model with common mean p. Here we place a uniform prior on the mean parameter η and assign K the proper, but vague, prior $(1 + K)^{-2}$.

A straightforward calculation shows that the posterior distribution of the hyperparameters (η, K) is given by

$$g(\eta, K|\text{data}) \propto \prod_{i=1}^{N} \frac{B(K\eta + y_i, K(1-\eta) + n_i - y_i)}{B(K\eta, K(1-\eta))} g(\eta, K),$$

where N is the number of periods. One makes inferences on the overdispersion parameter K by means of its marginal posterior

$$g(K|\text{data}) = \int_0^1 g(\eta, K|\text{data})d\eta.$$

To learn about overdispersion using the model/data simulation scheme, we first think of statistics that can be used to estimate the parameters η and K. Obvious choices for these statistics are the mean $\bar{y}$ and the standard deviation s of the y_i. The statistic s, in particular, measures the "larger than binomial" variation that we see in the observed counts. Then we simulate one set of models and data as follows:

1. **(Simulate models.)** Simulate values of the binomial probabilities $\{p_i\}$ from the prior model, by simulating values of (η, K) from the prior $g(\eta, K)$, and then simulating $\{p_i\}$ independently from a beta(η, K) distribution. Call the simulated probabilities $p_1^S, \ldots, p_N^S$.
2. **(Simulate data.)** Simulate data $y_1^S, \ldots, y_N^S$ independently from binomial distributions with respective sample sizes $n_1, \ldots, n_N$ and probabilities $p_1^S, \ldots, p_N^S$.
3. **(Compute statistics.)** Compute the mean $\bar{y}^S$ and the standard deviation s^S of the simulated data $\{y_i^S\}$.

We repeat Steps 1–3 of the algorithm a large number of times, obtaining the model/data simulated sequences $(\eta^S, K^S, \bar{y}^S, s^S)$.

The mean and standard deviation values from the sample are $\bar{y}_{\text{obs}} = 2.029$ and $s_{\text{obs}} = 1.414$. We make inferences about the parameters (η, K) by restricting attention to the set of simulated statistic pairs $(\bar{y}^S, s^S)$, where $\bar{y}^S$ is in the interval $(\bar{y}_{\text{obs}} - .2, \bar{y}_{\text{obs}} + .2)$ and s^S is in the interval $(s_{\text{obs}}—0.05, s_{\text{obs}} + 0.05)$.

To assess the accuracy of this method, Figure 37.2 shows the histogram of the simulated log K values and the exact marginal posterior of log K is overdrawn on the histogram.

Looking at Figure 37.2, the histogram of simulated values matches up well with the exact posterior density of log K. The histogram is shifted to the left from the exact posterior. (The mean and standard deviation of the simulated draws are given by 5.17 and 1.10, respectively, which can be compared with the exact values of 5.69 and 1.09.) This behavior is due partly to the fact that the mean $\bar{y}$ and the standard deviation s are not sufficient for the parameters in this setting. Also, the fact that s is in the set $(s_{\text{obs}}—0.05, s_{\text{obs}} + 0.05)$ is not as informative

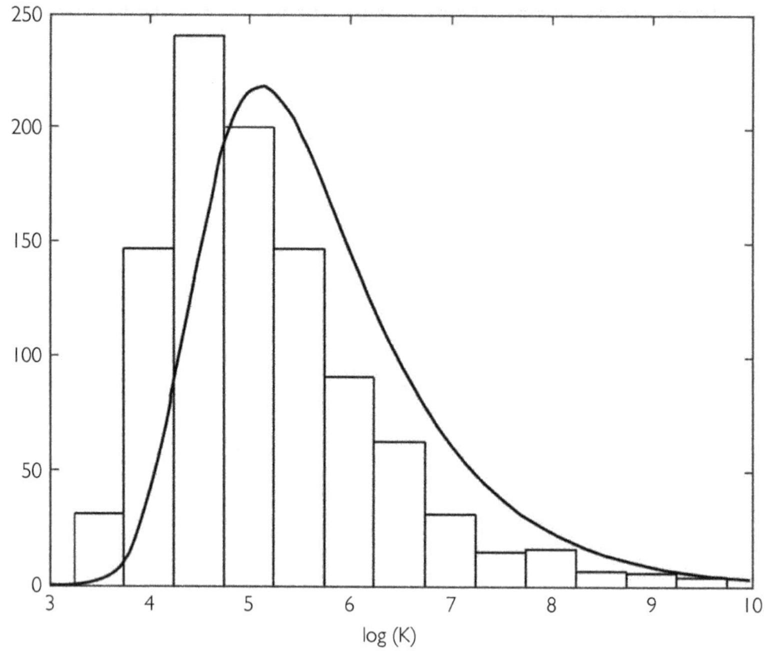

Figure 37.2 Posterior histogram of log K conditional on s in (1.364, 1.464) and $\bar{y}$ in (1.829, 2.229). The posterior density of log K is drawn on top of the histogram.

as knowing the exact value of s in the posterior. However, inferences based on the simulated sample will provide a good approximation to inferences based on the exact marginal posterior density.

Since the estimate of the beta hyperparameter K is large, there is little evidence that McGwire's recent pattern of home run hitting differs from a coin-tossing model. Suppose that one is interested in estimating McGwire's home run probability during the ith period p_i. In this binomial/beta model, this probability can be estimated by

$$\hat{p}_i = \frac{\hat{K}}{n_i + \hat{K}}\hat{\eta} + \frac{n_i}{n_i + \hat{K}}\frac{y_i}{n_i},$$

where $\hat{\eta}$ is an estimate at McGwire's home run probability in recent years and $\hat{K}$ is an estimate at the beta precision parameter. Here $n_i \approx 20$ and $\hat{K} \approx \exp(5.69) = 296$, which indicates that the estimate at the ith period home run probability is approximately equal to the global estimate $\hat{\eta}$.

How streaky was Javy Lopez in 1998?

For a second example, consider the hitting data for Javy Lopez for the 1998 major league baseball season. We observe the number of hits and the number of at-bats for each of the 132 games he played during the season. Figure 37.3 graphs the number of hits and at-bats as a function of the game number.

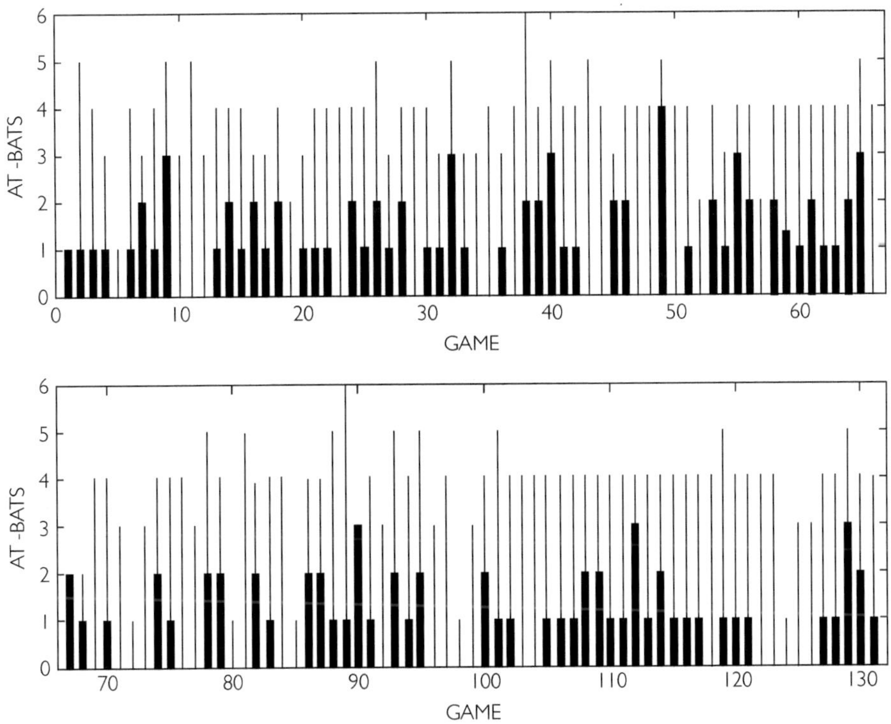

Figure 37.3 Graph of sequence of game at-bats (thin line) and hits (thick line) of Javy Lopez for the 1998 season.

Lopez was generally considered by the Atlanta Braves to be a streaky hitter during this season. In an Internet article published on April 17, 1998, Bill Zack commented that "If there's a more streaky hitter in baseball than Javier Lopez, the Braves don"t want to meet him." The reason for this belief in streakiness was based on his poor and good performance in relatively short periods. At the time the article was published, Lopez was in a 1-for-17 streak. It was said in the article that Lopez can hit 0.400 one week and 0.200 the next week.

We will address three basic questions. First, what is a reasonable way of measuring the degree of streakiness or non-homogeneity in Lopez's hitting data? Second, what is a realistic "streaky model" that can represent the type of streakiness that we might expect for this baseball hitter? Last, given that we have found a suitable streaky statistic, what is the evidence provided by our statistic in support of the streaky model?

Construction of a streaky statistic

From the baseball article, it appears that the belief in streakiness is supported by the extreme performance of Lopez during short sequences of games. To look at this performance graphically, one can plot the moving batting average of Lopez against the game number. Figure 37.4 displays a moving average plot of Lopez's hitting using a window width of 10 games. A plotted point (a, b) represents the batting average b of Lopez in a neighborhood of ten games about the

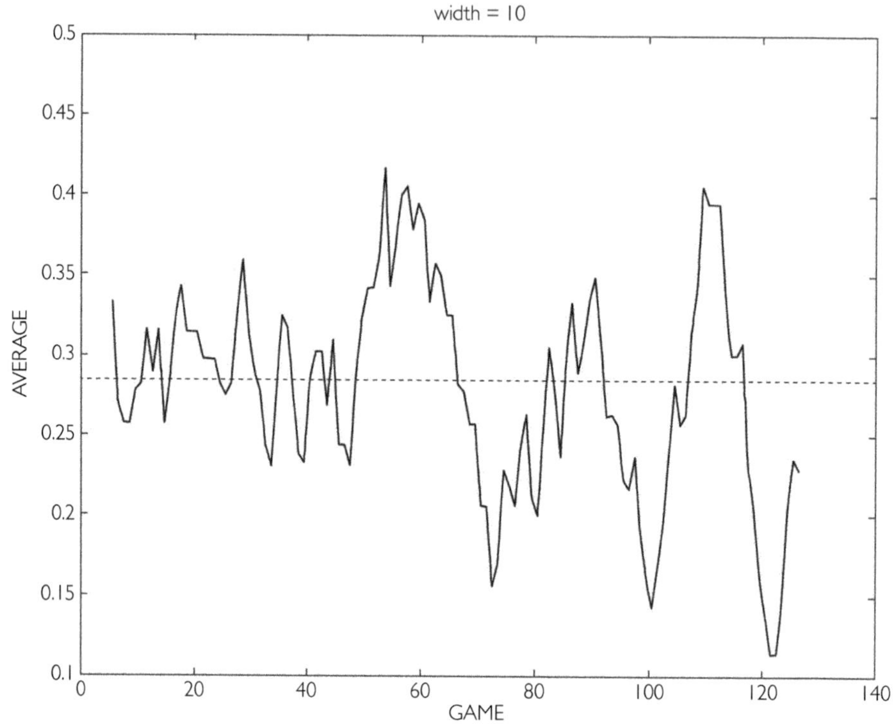

Figure 37.4 Graph of moving batting average of Javy Lopez using a window width of 10 games.

game number a. His batting average for the entire 1998 season is graphed using a dashed horizontal line. This graph shows interesting patterns. Lopez's hitting is pretty consistent for the first 40 games, he is hot for an interval about game 60, and then he oscillates between periods of poor hitting and good hitting at the end of the season.

One can quantitatively measure streakiness by using "interesting" statistics from this moving average plot. Specifically, streakiness is indicated by the following statistics.

- T_1 = difference between the largest and smallest moving average. Streakiness is indicated by a large value of T_1.
- T_2 = sum of the absolute differences between the moving averages and the season batting average over all games. Large values of T_2 indicate streakiness with a large number of cold and hot spells.

Suppose that each game is classified as "hot" (1) or "cold" (0) if the corresponding moving average (corresponding to a window width w) is larger or smaller than the hitter's season average. Then the data are reduced to the following binary sequence and one can measure streakiness by interesting patterns in this sequence.

```
1000001110111111111000111100001
1000111010001111111111111111110
0000000000000001001111111000000
00000000011111111110000000000
```

We will consider the use of two statistics based on runs of hot or runs of cold in the binary sequence:

- T_3 = number of runs (either runs of 0's or runs of 1's) in the sequence; and
- T_4 = length of the longest run (either a run of 0's or a run of 1's) in the sequence.

One would expect streaky hitters to display a relatively small number of runs and occasionally have long runs of good or poor games.

A third set of statistics is based on the observed autocorrelation structure in the sequence of hitting data. If a hitter is streaky, then one would expect his hitting performance on a given day to depend on his hitting in the previous k games. We observe y hits in n at-bats in a particular game with hitting probability p. Suppose that the batting average of the player in the previous k games is given by x_k. The hitting probability of the present game can be modeled as a function of x_k using the logistic function

$$\log \frac{p}{1-p} = \beta_0 + \beta_1 x_k.$$

We can use the slope estimate

$$T_5^{(k)} = \hat{\beta}_1$$

as an estimate of the dependence of the current day's hitting on the hitting on the previous k days. If this slope estimate is positive, then it provides some evidence that Lopez's hitting probability is higher when he has been successful in recent games.

The final statistic focuses on the homogeneity in the hitting probability of the player across games. Suppose that the games are subdivided by time into N groups, each of size b. If the true hitting probability does change over a season, one would expect significant variation in the observed group batting averages $\hat{p}_1, \ldots, \hat{p}_N$. So one measure of the non-homogeneity of the sequence would be

$$T_6^{(b)} = \text{standard deviation of subgroup batting averages}\{\hat{p}_j\}$$

Table 37.2 lists values of these six streaky statistics for Javy Lopez's data.

Note that some of these statistics, by themselves, give some evidence about streakiness. The regression coefficients of the logistic regression on the previous five and ten games ($T_5^{(5)}, T_5^{(10)}$) are negative in sign, indicating a negative association between Lopez's hitting and his hitting in recent games.

A streaky model

Here we focus on the use of one particular model, the Markov switching model (Albert 1993), for describing streakiness, since it seems to be a reasonable representation of the streakiness that is discussed in the Internet article. Suppose that we represent Lopez's hitting as a sequence of independent Bernoulli trials with hitting probabilities that are game dependent. Suppose that Lopez has two possible hitting states for each game. He is either a hot hitter with hitting probability p_H, or a cold hitter with hitting probability p_C, where $p_H > p_C$.

Table 37.2 ***Values of six streaky statistics for Javy Lopez's data.***

Name	Value
Range of moving averages	$T_1 = 0.302$
Sum of abs. diff. of moving averages	$T_2 = 6.27$
Number of runs	$T_3 = 22$
Length of longest run	$T_4 = 18$
Logistic regression slope	$T_5^{(5)} = -.861$
Logistic regression slope	$T_5^{(10)} = -1.217$
Standard deviation of subgroup bavg	$T_6^{(10)} = .0472$
Standard deviation of subgroup bavg	$T_6^{(5)} = .1005$

To complete the model, we assume that Lopez switches between the hot and cold hitting states for different games according to a Markov chain. We suppose that, given he is in a hot or cold state in a game, the probability that he remains in the same state (either hot or cold) for the next game is a, and so the probability that he switches from cold to hot (or hot to cold) in the next game is $1—a$. By choosing the staying probability a large, we are inducing positive autocorrelation in the sequence of hitting probabilities across games, and this will induce streakiness in the observed hitting sequence.

Prior distribution on parameters of the streaky model

The Markov switching model is parameterized by three numbers: the hot and cold probabilities, p_C and p_H, and the staying probability a. The basic coin-tossing model (independent trials and constant probability of success) is a special case of the Markov switching model where $p_H = p_C$. In this situation, we let the staying probability a be equal to 0.5—this indicates that there is no propensity to stay or leave from the same hot or cold state. At the other extreme, "significant" streakiness implies that there will be a large difference d between the hot and cold probabilities and there is a large positive probability of staying in a given state. If we regard p as an average probability, we can parameterize this streaky model by $p_H = p + d/2$, $p_C = p—d/2$, and the staying probability will have a large positive value, say $a = a^*$.

To define a continuum of models between coin tossing and significant streakiness, we define a compromise model M_λ which is defined by parameters that are weighted averages of the parameters in the coin tossing and streaky models:

$$M_\lambda = \{p_C = p - \lambda d/2, p_H = p + \lambda d/2, a = .5 + \lambda(a^* - .5)\}, 0 \leq \lambda \leq 1.$$

Note that M_0 corresponds to coin-tossing, M_1 corresponds to high streakiness, and the degree of streakiness is defined by the parameter λ. Values of λ close to one imply significant streakiness, and values of λ near zero are similar to independent coin tossing. We will call λ the streaky parameter, since it defines the distance away from coin-tossing towards the streaky alternative hypothesis.

In this example, we first define what we feel is significant streakiness for Lopez. We believe that Lopez is indeed streaky if his hot and cold probabilities are 100 points higher and lower, respectively, from his overall batting average 0.284 (i.e., $d = 0.1$). In addition, we view Lopez as very streaky if his staying probability is equal to $a^* = 0.8$. To complete the model, we place a uniform prior on the compromise parameter λ, reflecting little prior information about where Lopez falls on the streakiness continuum. This prior could easily be adjusted to model alternative beliefs about streakiness. For example, a strong belief in coin-tossing behavior would be represented by a prior that placed a large probability about the value $\lambda = 0$.

Simulating models and data

Suppose that we are interested in doing inference about the streakiness parameter λ based on a particular streaky statistic T. We perform the model/data simulation as follows:

1. We simulate a value of λ, λ^S, from the uniform prior. This value of λ defines values of the hot and cold probabilities and the staying probability. Call these simulated values p_C^S, p_H^S, and a^S.
2. Based on the simulated values p_C^S, p_H^S, a^S, we simulate the states (hot and cold) for the hitter for all of his games. This is done by flipping a coin to decide the state of the first game and then simulating the Markov chain to get the states of the remaining games. These states determine the set of hitting probabilities for all games.
3. We simulate binomial data based on the set of hitting probabilities.
4. We compute the value of the streaky statistic T from the simulated binomial data.

We performed this model/data simulation for 10,000 iterations. In Step 4 of the algorithm, we computed values of each of our proposed streaky statistics $T_1, \ldots, T_6$. When we plotted values of the simulated model/data pairs (λ^S, T_i^S) for each i, we found that the parameter and statistic were approximately linearly related. So we can summarize the relationship between the streaky parameter and the streaky statistic by means of the correlation of the model/data pairs. Looking at the correlations in Table 37.3, we see that there is a wide range of values. Values of the correlation that are close to zero correspond to statistics that are not informative about the value of the streaky statistic λ. In contrast, large absolute values of the correlation correspond to statistics that are useful in learning about the degree of streakiness in the Markov switching model.

To help interpret the size of the correlations in Table 37.3, consider the simple binomial/beta situation, where data y is binomial(n, p), and p has a beta distribution proportional to $p^{K\eta-1}(1-p)^{K(1-\eta)-1}$, as used in "Fitting an overdispersion model." Here the sample proportion $\hat{p} = y/n$ is used to learn about the probability p. A straightforward calculation gives that the correlation between $\hat{p}$ and p is

$$\mathrm{corr}(\hat{p}, p) = 1/\sqrt{1 + K/n}.$$

Suppose that a uniform prior is chosen for p where $K = 2$. In this case, the correlation between $\hat{p}$ and p increases rapidly as a function of n—the correlation is equal to 0.85, 0.91, and 0.95 for

Table 37.3 ***Correlation between simulated values of the streaky parameter λ and the statistic T for 8 choices of streaky statistics.***

Name	Correlation
T_1	0.30
T_2	0.33
T_3	−0.14
T_4	0.07
$T_5^{(5)}$	0.22
$T_5^{(10)}$	0.16
$T_6^{(10)}$	0.28
$T_6^{(5)}$	0.34

$n = 5, 10, 20$, respectively. So the correlations between statistics and parameter in Table 37.3 appear relatively small to those in the binomial/beta model.

We first look at the statistic T_3, the number of runs. The correlation of T_3 and λ is a small negative number, indicating that this statistic is not especially helpful in detecting streakiness.

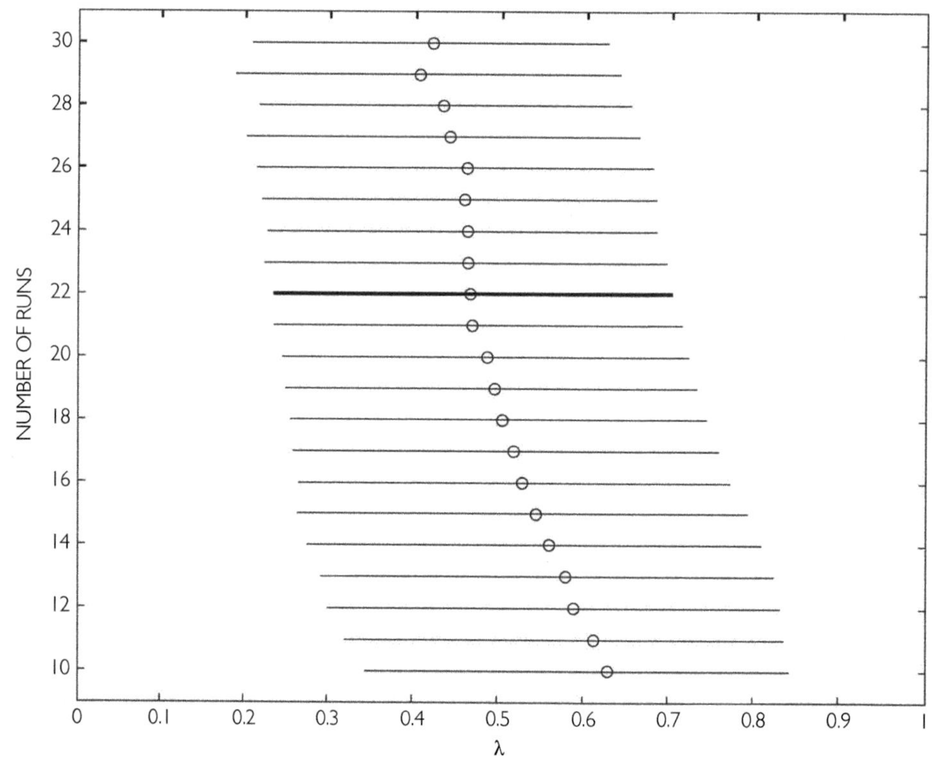

Figure 37.5 Posterior median and quartiles of streakiness parameter λ given value of number of runs statistic T_3.

This point is amplified by Figure 37.5 which displays the posterior median and quartiles of λ for a range of values of T_3. Note that the location of the middle 50% of the posterior distribution barely changes as the number of runs in the data increases. For Lopez's data, the number of runs is 22, and the posterior 50% probability interval of (0.23, 0.70) is shown by a bold line. This interval is similar in location to the 50% prior probability interval of (0.25, 0.75), which says that we have learned little about streakiness from this particular statistic.

Figure 37.6 displays posterior probability intervals for a more useful statistic, $T_6^{(5)}$, the standard deviation of subgroup batting averages. In contrast to Figure 37.5, the posterior median of λ increases as a function of the standard deviation, which shows that the subgroup standard deviation is informative for detecting streakiness. In our example, Lopez's value of $T_6^{(5)} = .1005$. The location of the 50% posterior probability, indicated on the figure by a bold line, is (0.21, 0.67), which can be compared to the prior probability interval of (0.25, 0.75).

Was Lopez a streaky hitter in 1998? Using both of the statistics T_3 and $T_6^{(5)}$, the posterior distribution for the streaky parameter λ was essentially the same as the prior distribution, indicating that the data have provided little evidence in support of or against streakiness. In other words, none of the streaky statistics proposed here appear to provide any support to the belief that Lopez was a streaky hitter during this season.

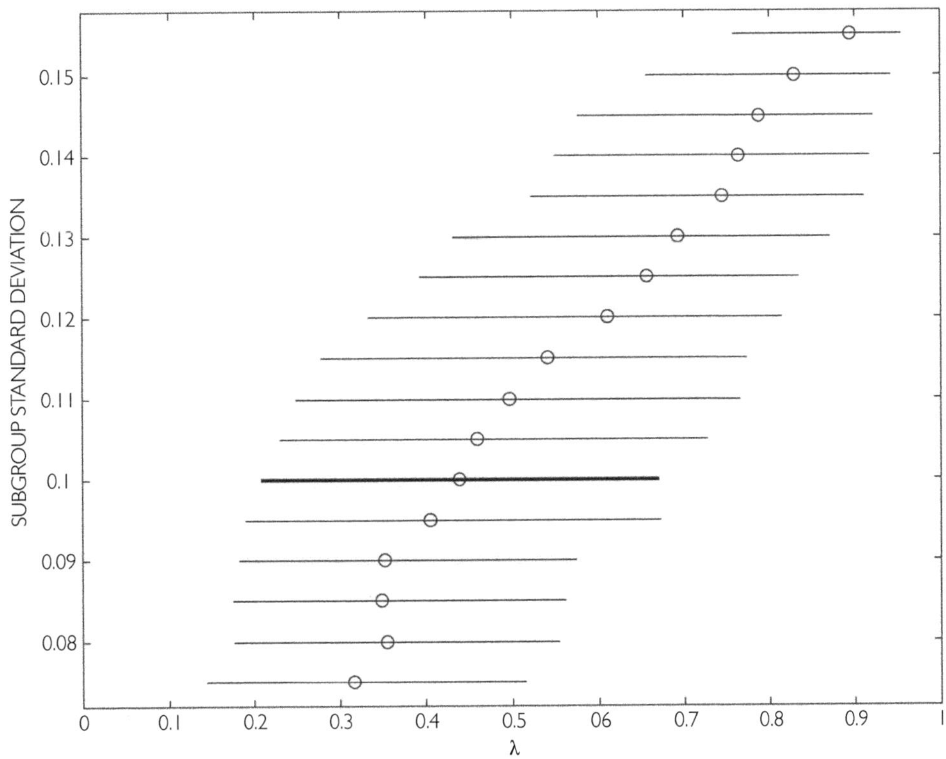

Figure 37.6 Posterior median and quartiles of streakiness parameter λ given value of standard deviation statistic $T_6^{(5)}$.

Detecting streakiness in free-throw shooting data

Wardrop (1999) described a controlled experiment to detect streakiness in basketball shooting. A college player took a sequence of shots from beyond the three-point line for 20 practices during the summer. On each of the 20 practices, the player attempted 100 shots and the results of the shots were recorded. The data can be represented as a matrix of 20 rows and 100 columns of 0's and 1's, where the rows correspond to practices, and a data value of 1 indicates a made shot and 0 indicates a miss.

These data resemble the baseball hitting data discussed in the earlier section and one could fit a Markov switching model to represent streakiness. We will instead fit a dispersion model to these data to illustrate the generality of the algorithm of 'Simulating models and data'.

If the player is indeed streaky, then one would expect the data to be somewhat clustered. That is, the player should have a larger number of cold and hot spells in shooting than what would be predicted from a binomial model. As an initial investigation, we group the 100 shots for each practice into 20 groups of size 5. For each group, the proportion of made shots is recorded. The data are summarized by the proportions $\{\hat{y}_{ij}\}$, where i indicates the practice number and j denotes the group within practice ($1 \leq i \leq 20, 1 \leq j \leq 20$). Figure 37.7 displays boxplots of the proportions $\{\hat{y}_{ij}\}$ grouped by practice number. Looking at the graph, we see that

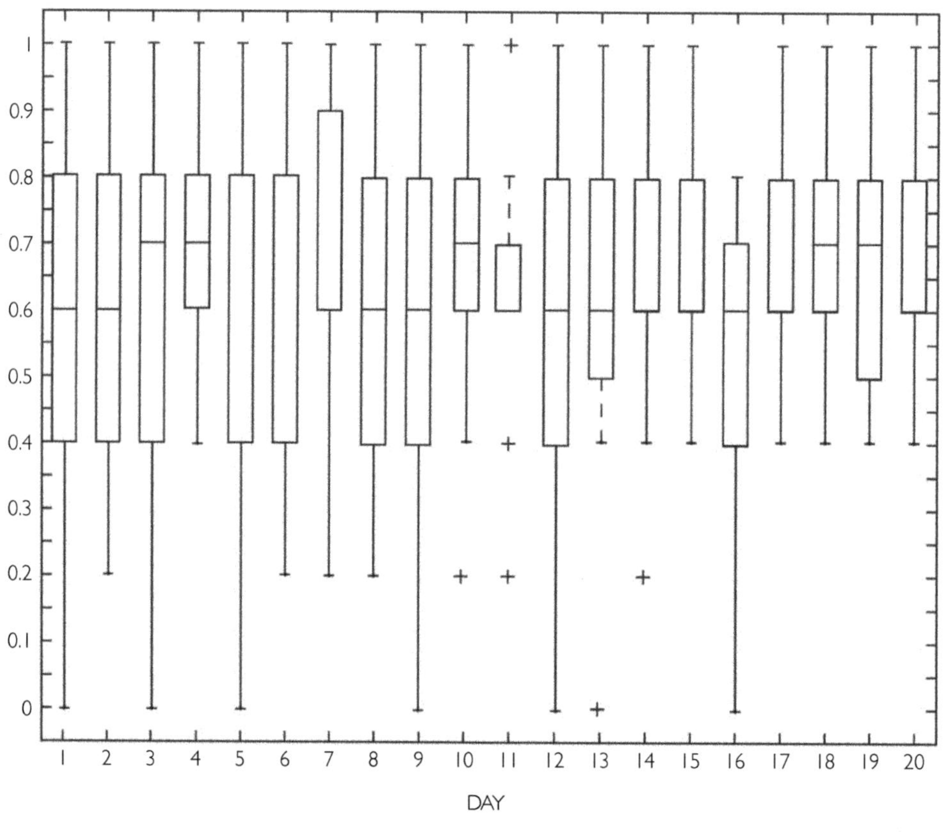

Figure 37.7 Boxplots of proportions of made shots grouped by practice number.

the spread of the boxplots tends to decrease over time, indicating that the player is becoming more consistent in her shooting performance.

Consider a set of group proportions for a given practice, say $\hat{y}_{i1}, \ldots, \hat{y}_{i20}$. If the shots were independent Bernoulli trials with constant probability p, then the group proportions would be independent with common mean p and standard deviation $\sqrt{p(1-p)/5}$. This binomial standard deviation can be estimated by $\sqrt{\hat{p}(1-\hat{p})/5}$, where $\hat{p}$ is the proportion of made shots for this practice.

We can assess this binomial assumption by comparing the actual standard deviation of the proportions $\{\hat{y}_{ij}\}$ with the estimated binomial standard deviation by means of a ratio

$$\frac{\text{standard deviation}\{\hat{y}_{ij}\}}{\sqrt{\hat{p}(1-\hat{p})/5}}.$$

Suppose we express the actual standard deviation as

$$\text{standard deviation}\{\hat{y}_{ij}\} = \sqrt{\hat{p}(1-\hat{p})/n_i^o},$$

where n_i^o is the "observed binomial sample size." Then the above ratio of standard deviations is equivalent to the ratio of sample sizes

$$\frac{\text{standard deviation}\{\hat{y}_{ij}\}}{\sqrt{\hat{p}(1-\hat{p})/5}} = \sqrt{\frac{5}{n_i^o}}.$$

For a particular practice, the statistic n_i^o measures how close the data resemble a binomial model. A value of n_i^o smaller than 5 indicates data that show more dispersion than predicted under the binomial model (called overdispersion), and a value of n_i^o larger than 5 indicates underdispersion.

Figure 37.8 plots the values of the logarithms of the sample size statistics, $\log n_i^o$, against the practice number i. The horizontal line at the value log 5 is shown for reference. After some initial instability for practices 1–4, note that the sample sizes generally increase as a function of the practice number. This indicates that the data are generally overdispersed for early practices and underdispersed for later practices. A least-squares regression on the data for practices 5–20 gives the fit

$$\log n^o = 1.04 + .0447 \times (\text{practice number}).$$

One way to confirm this visual impression is to separately fit least-squares lines to the two datasets corresponding to practices $1, \ldots, c$, and practices $c + 1, \ldots, 20$, respectively, and then choose the value of c which minimizes the sum of squared residuals. Using this procedure, we confirm that $c = 4$ is a suitable breakpoint for these data.

The observed variation in spread that we see from Figure 37.8 motivates the consideration of a family of distributions that can accommodate more or less spread than a binomial. A convenient choice is the quasi-binomial density. [A general class of quasi-likelihood densities

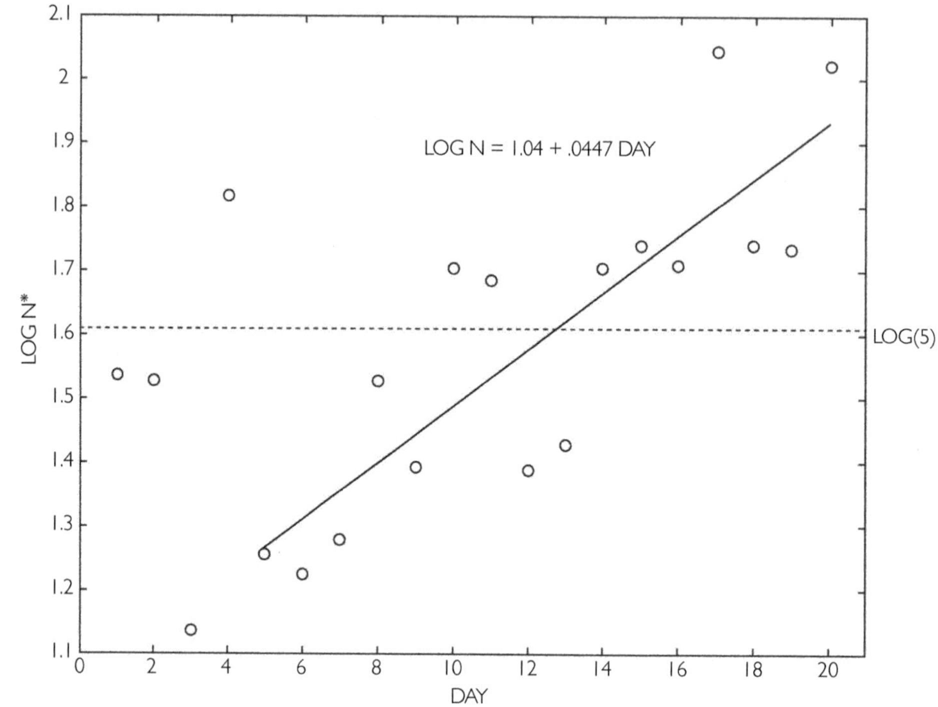

Figure 37.8 Plot of logarithms of sample sizes n_i^o as function of practice number. A least-squares line to the data for practices 5–20 is displayed.

was described by McCullagh and Nelder (1989, Ch. 9).] If y denotes the proportion of successes in n trials, then the quasi-binomial density with parameters n, p, n^* is given by

$$f(y|n,p,n^*)$$
$$= \frac{C\Gamma(n^*+1)}{\Gamma(n^*y+1)\Gamma(n^*(1-y)+1}p^{n^*y}(1-p)^{n^*(1-y)},$$
$$y = 0, 1/n, 2/n, \ldots, 1,$$

where C is a constant to ensure that the density sums to one. When $n^* = n$, the density reduces to the binomial density. A value of $n^* < n(n^* > n)$ gives a density with greater (smaller) variability than the binomial.

We can model a pattern in the spread of the proportions by means of a regression model on the quasi-binomial sample sizes. Let y_{ij}, the proportion of shots made in the jth subgroup of the ith practice, have a quasi-binomial density with parameters $(5, p_i, n_i^*)$, where n_i^* measures the spread of the group proportions. Assume, for generality, that there is an initial period of practice time, $i \leq c$, where the dispersion parameters follow the regression relationship

$$\log n_i^* = \beta_0^{(1)} + \beta_1^{(1)}(i - I^{(1)}),$$

where i is the practice number ($1 \leq i \leq c$) and $I^{(1)}$ is the mean of the practice numbers. Likewise, after the practice time is over ($i > c$), the dispersion parameters follow the new model

$$\log n_i^* = \beta_0^{(2)} + \beta_1^{(2)}(i - I^{(2)}).$$

To complete the model, prior distributions need to be placed on the success probabilities $p_1, \ldots, p_N$, the regression parameters $\beta_0^{(j)}, \beta_1^{(j)}, j = 1, 2$, and the breakpoint c. We suppose that the probabilities of making a shot can vary between practices and we assume that the p_i are a random sample from a beta(26, 14) distribution. This prior assumes that the shooting probabilities fall, with probability 0.95, in the interval (0.52, 0.77). We assume $\beta_0^{(1)}, \beta_1^{(1)}, \beta_0^{(2)}, \beta_1^{(2)}$ independent with each assigned a normal distribution with mean 0 and standard deviation 0.1. This prior is relatively non-informative and allows for increasing and decreasing relationships between the dispersion parameter and the practice number. Finally, we assign a uniform prior for the break-point c on the times $\{0, 1, \ldots, 9\}$, reflecting little prior knowledge about the length of the practice time.

One iteration of the model/data simulation for this example proceeds as follows.

1. Simulate a breakpoint c^S from the uniform prior on $\{0, 1, \ldots, 9\}$.
2. Simulate regression parameters $\beta_0^{(1)S}, \beta_1^{(1)S}, \beta_0^{(2)S}, \beta_1^{(2)S}$, and success probabilities $\{p_i^S\}$ from the above prior distributions.
3. Compute dispersion parameters $\{n_i^{*S}\}$ using the regression relationships.
4. Simulate group proportions $\{y_{ij}^S\}$, where y_{ij}^S is simulated from the quasi-binomial$(5, p_i^S, n_i^{*S})$ distribution.
5. Compute the estimated sample sizes $\{n_i^{oS}\}$, where as described earlier, n_i^{oS} is found by computing the ratio of the standard deviation of the $\{y_{ij}^S\}$ to the estimated binomial standard deviation.
6. Regress the log n_i^{oS} on the practice numbers using the procedure described earlier. Two separate least squares lines are fit using all possible breakpoints, and the optimal breakpoint $\hat{c}^S$ is found which minimizes the sum of squared residuals. A regression slope estimate $b_1^{(2)S}$ is found for the later practices using this optimal regression procedure.

This algorithm was repeated 1,000 times, resulting in a simulated sample of model/data sequences $(c^S, \beta_1^{(2)S}, \hat{c}^S, b_1^{(2)S})$. For the observed data, we observe $\hat{c} = 4$ and $b_1^{(2)} = .0447$. To perform inference about the slope parameter $\beta_1^{(2)}$, we focus on the simulated values $\beta_1^{(2)S}$ conditional on $\hat{c}^S = 4$ and the estimated slope $b_1^{(2)S)}$ in a small interval centered about the observed value 0.0447.

Figure 37.9 displays histograms of the posterior distributions of the regression slope $\beta_1^{(2)}$ (top) and the true breakpoint c (bottom) that are based on the simulated values produced using the above algorithm. Note that the histogram of the true breakpoint c is relatively diffuse, indicating that we have learned little about the location of the true breakpoint in these data. However, looking at the location of the histogram, we see that the posterior density of $\beta_1^{(2)}$ is centered about 0.05 and most of the probability lies on positive values. (The probability that the true slope exceeds 0 is approximately 93%.) Thus, there appears to be good evidence that the quasi-binomial sample size increases as a function of the practice number. In other words,

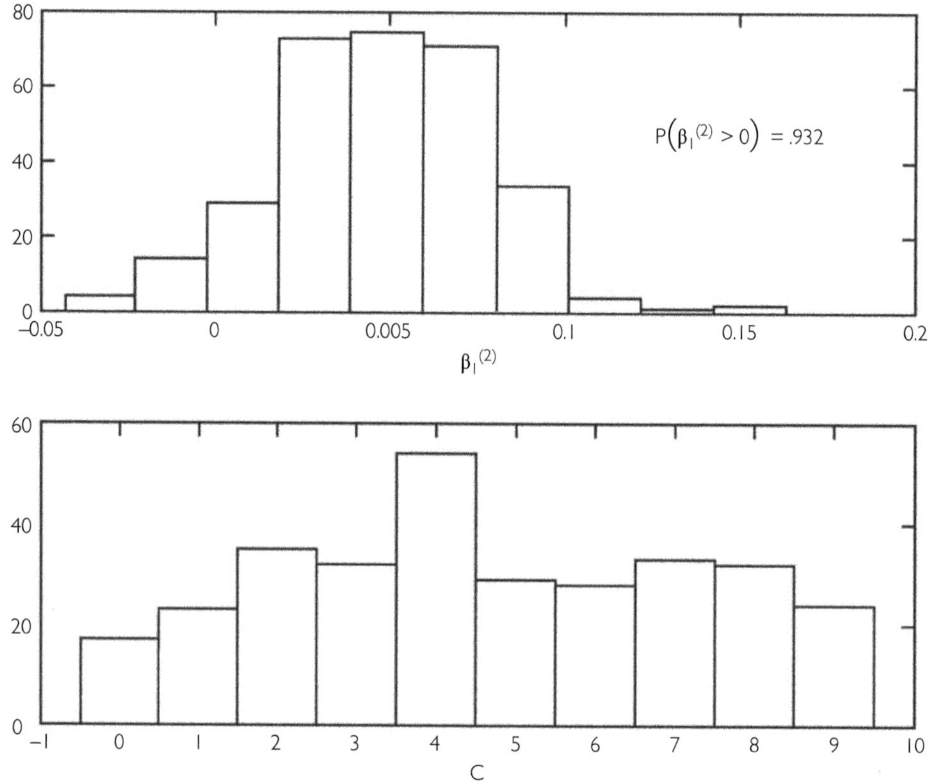

Figure 37.9 Histogram of simulated values of true breakpoint c and regression slope $\beta_1^{(2)}$ conditional on $\hat{c} = 4$ and $b_1^{\text{obs}} \in (.0447 - .01, .0447 + .01)$.

the true variation in the shooter's day performance from practices 5–20 appears to decrease over time, which means that the shooter performance is more consistent with repeated practice. Wardrop (1999), using methods that did not involve grouping data, also found significant evidence against coin-tossing in these data.

Closing remarks

The model/data simulation algorithm has several desirable features. First, when a complete Bayesian analysis is difficult to implement and there exists a statistic t which is believed to summarize most of the information about the parameter θ, this simulation algorithm provides a convenient way to obtain an approximate Bayesian analysis. Second, the examples illustrate that this algorithm allows much flexibility in the choice of alternative models, which is especially important in the detection of streakiness. Last, one obtains simulated ordered pairs of (parameter, statistic), and the relationship that one finds in these simulated pairs is instructive on how one learns about the parameter based on the statistic. We saw in "How streaky was Javy Lopez in 1998?" that the observed correlation coefficient between a streaky statistic and the streakiness parameter led to a better choice of a statistic to detect streakiness.

Although this methodology appears potentially useful in the detection of streakiness, the first two examples demonstrate that coin-tossing is a good model for data that appear streaky. There is little evidence for nonstationarity in McGwire's home run data and, although Javy Lopez might appear streaky in his pattern of hitting across games, the Markov switching model was unable to pick up any significant streakiness. The basketball data seem to deviate the most from coin-tossing. In this example, streakiness was manifested by a change in the variation of shooting behavior across time. This methodology is flexible to allow for innovative ways of detecting streakiness, such as illustrated in Section 5, and in the construction of plausible models to explain the observed streakiness.

[Received July 1999. Revised March 2000.]

REFERENCES

Albert, J. (1993) Comment on "A Statistical Analysis of Hitting Streaks in Baseball" by S. C. Albright, *Journal of the American Statistical Association*, 88, 1184–1188.

—— (1997), "Teaching Bayes's rule: a data-oriented approach," *Journal of the American Statistical Association*, 51, 247–253.

Albright, S. C. (1993) "A statistical analysis of hitting streaks in baseball," *Journal of the American Statistical Association*, 88, 1175–1183.

Barry, D., and Hartigan, J. A. (1993), "Choice models for predicting division winners in major league baseball," *Journal of the American Statistical Association*, 88, 766–774.

Gilovich, T., Vallone, R., and Tversky, A. (1985), "The hot hand in basketball: on the misperception of random sequences," *Cognitive Psychology*, 17, 295–314.

Larkey, P., Smith, R., and Kadane, J. (1989), "It's okay to believe in the 'hot hand,'" *Chance*, 2, 22–30.

McCullagh, P., and Nelder, J. A. (1989), *Generalized Linear Models*, London: Chapman and Hall.

Stern, H. S. (1997), "Judging who's hot and who's not," *Chance*, 10, 40–43.

Stern, H. S., and Morris, C. N. (1993), Comment on "A Statistical Analysis of Hitting Streaks in Baseball" by S. C. Albright, *Journal of the American Statistical Association*, 88, 1189–1194.

Tversky, A., and Gilovich, T. (1989), "The Cold Facts About the 'Hot Hand' in Basketball," *Chance*, 2, 16–21.

Wardrop, R. (1999), "Statistical Tests for the Hot Hand in Basketball in a Controlled Setting," Technical Report, Department of Statistics, University of Wisconsin-Madison.

AUTHOR INDEX

SUBJECT INDEX

Sports Statistics: Index Locators in *italic* refer to figures; those in **bold** to tables

A

B

E

F

G

H

I

J

K

L

M

N

O

P

Q

R

S